UNIVERSITY BOTANY-3

UNIVERSITY BOTANY-3

(Angiosperms, plant embryology and plant physiology)

Editor
Prof. S.M. Reddy
Department of Botany
Kakatiya University
WARANGAL

Authors

Dr. M. Madhusudana Rao
Reader, Dept. of Botany
Kakatiya Govt. Degree College
Hanamkonda, Warangal Dist.

Dr. M. Mahender Reddy
Reader, Dept. of Botany
Lal Bahadur College
Warangal Dist.

Dr. A. Srinivas Reddy
Lecturer Dept. of Botany
S.K.E. Degree College
Sirpur Khagaz Nagar, Adilabad Dist.

Dr. S. Janardhana Chary
Principal (Retd.)
S.R.A.S. Govt. College
Kothagudem, Khammam Dist.

NEW AGE

NEW AGE INTERNATIONAL (P) LIMITED, PUBLISHERS
New Delhi • Bangalore • Chennai • Cochin • Guwahati • Hyderabad
Jalandhar • Kolkata • Lucknow • Mumbai • Ranchi

Published by New Age International (P) Ltd., Publishers
First Edition: 2004
Reprint: 2025

GLOBAL OFFICES

- **New Delhi** **NEW AGE INTERNATIONAL (P) LIMITED, PUBLISHERS**
  7/30 A, Daryaganj, New Delhi-110002, (INDIA), **Tel.:** (011) 23253472, 23253771
  Visit us at www.newagepublishers.com • e-mail: contactus@newagepublishers.com
- **London** **NEW AGE INTERNATIONAL (UK) LTD.**
  27 Old Gloucester Street, London, WC1N 3AX, UK
  Visit us at www.newacademicscience.co.uk • e-mail: info@newacademicscience.co.uk

BRANCHES

- **Bangalore** 37/10, 8th Cross (Near Hanuman Temple), Azad Nagar, Chamarajpet, Bangalore- 560 018
  Tel.: (080) 26756823, **E-mail: bangalore@newagepublishers.com**
- **Chennai** 26, Damodaran Street, T. Nagar, Chennai-600 017, **Tel.:** (044) 24353401
  E-mail: chennai@newagepublishers.com
- **Cochin** CC-39/1016, Carrier Station Road, Ernakulam South, Cochin-682 016, **Tel.:** (0484) 4051304
  E-mail: cochin@newagepublishers.com
- **Guwahati** Hemsen Complex, Mohd. Shah Road, Paltan Bazar, Near Starline Hotel, Guwahati-781 008,
  Tel.: (0361) 2513881, **E-mail: guwahati@newagepublishers.com**
- **Hyderabad** 105, 1st Floor, Madhiray Kaveri Tower, 3-2-19, Azam Jahi Road, Near Kumar Theater, Nimboliadda Kachiguda, Hyderabad-500 027, **Tel.:** (040) 24652456, **E-mail: hyderabad@newagepublishers.com**
- **Kolkata** RDB Chambers (Formerly Lotus Cinema) 106A, 1st Floor, S N Banerjee Road, Kolkata-700 014
  Tel.: (033) 22273773, **E-mail: kolkata@newagepublishers.com**
- **Mumbai** 142C, Victor House, Ground Floor, N.M. Joshi Marg, Lower Parel, Mumbai-400 013
  Tel.: (022) 24927869, 24915415, **E-mail: mumbai@newagepublishers.com**
- **New Delhi** 22, Golden House, Daryaganj, New Delhi-110 002, **Tel.:** (011) 23262368, 23262370
  E-mail: sales@newagepublishers.com

C-25-02-14594

Printed in India at Inkfinity Prints (P) Ltd., Delhi.

NEW AGE INTERNATIONAL (P) LIMITED, PUBLISHERS
7/30 A, Daryaganj, New Delhi - 110002
Visit us at **www.newagepublishers.com**
(CIN: U74899DL1966PTC004618)

PREFACE

In the process of providing the text book for B.Sc. students studying Botany in different Universities, this book was written for students studying B.Sc. III year. A concept of common core syllabus for Universities of Andhra Pradesh was developed by Andhra Pradesh Council of Higher Education during 1993-94. In order to provide entire syllabus material as compact compendiam, this book is written include Angiosperm taxonomy and embryology and plant physiology.

A Sincere efforts were made to present subject in simple and precise manner and provide an update information to the student.

Part - A which includes taxonomy and embryology of angiosperms, provides history and nomenclature of angiosperms and different systems of classification. Recent trends in taxonomy and ethnobotany are also included. The different families included in the syllabus and described briefly and provided with distinguishing characters and economic importance. A note was also added on the mechanism of pollination. A comprehensive accomplish embryology of angiosperms and palynology is also included in Part-B. Plant physiology from the Part-C. At the end of each chapter a review questions on the pattern of University examination are also given. Each chapter is amply illustrated and provided diagrams specially drawn for this book.

We hope that the book will fulfill the aspirations of teachers and students. However there may be some short comings which may kindly bring to the notice of authors for improvement of this book which will be duely acknowledged.

Our sincere thanks to Mr. Siva Kumar for helpful suggestions. Our thanks to Mr Giridhar for neat diagrams. Our heartfelt thanks are due to M/s New Age International (P) Limited. Publisher, Hyderabad for its neat publication.

Editor & Authors

B.Sc. PAPER - III YEAR

(Taxonomy, Embryology, Physiology)

A. ANGIOSPERM TAXONOMY

1. Evolution of systems of classification
2. A detailed account of Bentham & Hookers' system of classification
3. Basic rules of nomenclature and International Code of Botanical nomenclature.
4. Taxonomy and economic importance of the following families.

1. Annonaceae	2. Malvaceae	3. Rutaceae
4. Fabaceae	5. Caesalphinaceae	6. Mimosaceae
7. Cucurbitaceae	8. Umbelliferae	9. Rubiaceae
10. Compositae	11. Sapotaceae	12. Asclepiadaceae
13. Convolvulaceae	14. Labiatae	15. Euphorbiaceae
16. Orchidaceae	17. Paimae	18. Graminae

5. Pollination mechanisms of Fabaceae, Asclepiadaceae, labiatae and Orchidaceae
6. A brief account of modern trends in taxonomy : Chemotaxonomy and Cytotaxonomy
7. Ethnobotany : An elementary account.

B. EMBRYOLOGY OF ANGIOSPERMS

1. Micro and Megasporogenesis. Monosperic, Bisperic and Tetrasporic types of female gametophytic development. Monocot and dicot embryo development.
2. Development of endosperm and Polyembryony
3. General account of polynology with special reference to the pollengrains of Hibiscus, Aceacia and grass.

C. PLANT PHYSIOLOGY

1. Enzyme action and enzyme kineties (outlines)
2. Cell organelles and their separation by differential centrifugation technique for in vitro studies (elementary knowledge)
3. Elementary stress physiology (concepts of water potential and physiology of stomatal transpiration should also be touched)
4. Membrane. Theory of absorption of mineral salt ion, criteria for their essentiality sand cultures, Hydroponics, and chemical fertilzers.
5. Photosynthesis : Related Photophysiology, ultra structure of chloroplast, Photophosphorylation. Diversity in carbon reduction pathways (C3,C4,CAM) Photorespiration and transport of metabolites.
6. Elementary account of bio-energetics and entropy. Glycolysis, Kreb's cycle, Oxidative Phosphorylation. Anaerobic respiration and Fermentation.
7. Nitrogen motabolism - Biological Nitrogen fixation, Nitrate reduction, Amino acids and Protein Synthesis.
8. Growth and Development - Patterns of growth, Phytohormones - Auxins, Gibberellins. Cytokinins and growth inhibitors (brief account), Phytochrome
9. Seed Dormancy, germination and storage
10. Fruit storage, Physiology (brief account).

CONTENTS

C. PLANT PHYSIOLOGY

PART - A

PLANT TAXONOMY

By

Dr. M. Madhusudana Rao
Reader, Dept. of Botany
Kakatiya Govt. Degree College
Hanamkonda, Warangal A.P.

Dr. M. Mahender Reddy
Reader in Dept. of Botany
L.B. College, Warangal A.P.

1	# INTRODUCTION TO TAXONOMY OF ANGIOSPERMS
PLANT TAXONOMY	

1.1. INTRODUCTION

The term Taxonomy (In Greek Taxis means arrangement and nomous is to name) for the first time is being introduced by the French Botanist A.P. de Condolle (1813) in his book 'The Theorie Elementary dila Botanique' (Theory of elementary Botany). According to him, taxonomy is concerned with 'The study of principles and practices of classification of plants based on their relatinships'.

Taxonomy, however, has been variously defined, according to a well known plant taxonomist, Cronquist (1968), "Taxonomy is a study aimed at producing a system of classification of organisms which best reflects the totality of their similarities and differences".

Jeffrey (1968) states that 'The Plant Taxonomy otherwise known as the systematic botany is concerned with identification, nomenclature and classification of plants'. Heywood (1967) opines that the term systematics is used to cover the scientific study of the diversity and differentiation of organisms and the relationships which exist between them. Taxonomy is that part of systematics which deals with the study of classification including its basis, principles, procedures and rules. Classification in a biological sense is the process of ordering plants into groups, which are arranged hierarchically.

The term systematics commonly considered as synonym with taxonomy, although biologists prefer to draw a line of demarcation between them. In the latter case, taxonomy is limited to a classificatory investigation and systematics on the other hand, gives a wider definition and includes many other aspects.

Heywood (1977) stated that one of the major roles of taxonomy is to produce a classification or a system of classification of organisms that best reflects the totality of their similarities and differences. Plason (1950) consideres plant taxonomy as a synthesis of systematic botany, classification, nomenclature and documentation.

The basis for classification of plants depend upon their morphological similarities (Alpha taxonomy). However, in recent times data from other related fields like plant anatomy, embryology, cytology, phytochemistry, physiology, genetics, ecology and phytogeography (Omega taxonomy), are taken into consideration for determining the inter-relationships among various groups.

The unit of taxonomic study is the 'Taxon' The term 'taxon' referes to any category of plants, i.e. species, genus, family or order, where the members show distinct genetic relationships. The species is the basic taxon and other ranks of higher rank in ascending series are genus, family, order, class, division and kingdom.

1.2 HISTORY OF TAXONOMY

Taxonomy had its earliest beginning in the writings of Theophrastus (370 - 285 BC) and ancient Greek Philosopher Aristotle (382 - 332 BC). Theophrastus, 'The Father of Botany' in his 'Historia Plantarum' roughly classified and described over 500 kinds of plants. His work contained a detailed description of plants mixed up with the Greek folkore and philosophy. It is of interest to know that the names given by him still find validity even in the present day taxonomy.

For many years man depended on plants for his many primary needs like food, cloth and living. This resulted in the exploration of different plants. In India, people of vedic period (800 BC- 200 BC) cultivated several crop plants like wheat, dates, vegetables, cotton and also number of medicinally important plants. Information regarding these plants was mentioned in the Atharwana Veda and Vrikshayruveda compiled by Parasara in Sanskrit. These works also contained data on morphology and keys for the identification of medicinal plants. This lead to the foundation of taxonomy where different families then also known as ganas were identified on the basis of their morphological peculiarities. For example, 'swastika ganam' as classified then is now referred as Brassicaceae. However, due to limited understanding of Sanskrit language, various works of Parasara could not get popularized.

During the middle ages, people used diverse types of plants to prepare tinctures, decoctions for medicinal purposes, to help these people, publication of reference books became necessary to identify the medicinally desired plants. Hence, books were compiled giving descriptive accounts of these plants. Such books were known as herbats and the authors of these books were known as herbalists. Herbalists contributed in a big way towards the improvement of descriptive taxonomy. A German herbalist, Otto Brunbels (1489-1534) was one among them and gave a vivid descriptions of medicinal and food plants with accurate sketches.

Brunbels and Jercome Bock (1498-1554) Leonart Fuchs (1501-1566), and Valerias Cordus (1515-1544) and others also contributed immensely for the development of plant taxonomy.

Taxonomy as a scientific study started in the 16th century with the works of Jean Bauhin and Gaspard Bauhin (1560 - 1624). Gaspard Bauhin first attempted to devise a binominal system of classification in his book 'Prodromes Theatri Botanici (1620) and also in Pinax Theatri Botanici (1963). He gave generic and specific epithets for the plants. This system was later adopted by Linnaeus as binomial nomenclature.

During the age of herbals (1470-1670) a doctrine called 'The doctrine of signature' was developed. This was based on the belief that, plants or plant parts resemble the portions of human body. They believed that they must have been created with the object of providing cures to diseases to those portions. Plants were named based on this assumption, which were supposed to have remedial properties.

Andrea Caesalpino (1519 - 1603) another herbalist gave a compendium containing descriptions about 1,500 plants in his book 'The De Plantis'. Caesalpino's descrptions were based on the organs of fructifications than the habit of plants. He made an intensive study of seeds, fruits and other plant parts.

An English naturalist John Ray (1627) classified nearly 18,000 species in 'Historia Plantarum'. His classification was mainly based on form and gross morphology of plant structures and later it was found to be superior over the artificial Linnaeus system.

Joseph Pitton de Tournefort (1656 - 1708), a French botanist was credited with the foundation of modern concepts of genera. Many of his generic names like *Betula, Castanea, Fagus, Quercus* and *Ulnus* are well known even today.

In India, a Portugeese botanist Garcia d'orta (1565) first published a comprehensive account of Indian medicinal plants. Van Rhede, the Dutch governor of Malabar described a large number of plants in his thirteen volumed series called 'The Hortus Malabaricus'. Roxburgh, Waltich and Weight were the poineers of taxonomy in India, who described more than 3,000 plants along with their sketches.

Hooker visited India between 1847 and 1851 explored the flora of Sikkim and Khasi hills. Gamble's flora of Madras presidency was written during the latter half of 19th century. Fyson of presidency college of Madras explored the flora of South Indian Hills and published two volumes containing illustrated account of the plants of area.

Riyhius, a German botanist studied the flora of Indonesia in the 19th century. Blatter amd Milerd (1929) published a profusely illustrated account of some beautiful Indian trees. Ran (1963) brought out an illustrated account of Himalayan flowering plants.

The Botanical Survey of India (BSI) with its headquarters in Howrah along with four regional centres at Pune, Gauhati, Coimbatore and Dehra Dun, is presently engaged in the exploration, classification of plants in India. The botanists of Forest Research Institute, (FRI) at Dehra Dun are also responsible for the exploration of forest flora of India.

It is of interest to know that during and after year 1960 was marked by the development of rapid and relatively simple techniques and ideas in plant taxonomy. Heywood (1967) referred this as the period of taxonomic revolution and explosion. It is characterized by the advances made in two areas, viz., chemotaxonomy and taximetrics or numerical taxonomy. In the taximetrics, usage of computers gave a fillip to plant taxonomy.

1.3 PRINCIPLES OF TAXONOMY

Taxonomy of plants is mainly aimed at the designing of a convenient method of classification. The earliest methods of classifications were mainly based on a set of easily observable characters (artificial classification). However, during 19th century, attempts have been made to classify plants on natural characters or on phylogenetic relations (Natural or phylogenetic classification). With this shift, there has been a rapid change in the principles of classification also.

Cronquist (1968) formulated certain basic principles of taxonomy. These include,

a) Taxa should be established only on the basis of multiple correlation of distinguishing characters. Further, no character of taxon should be considered singly for the establishment of its relationships with other taxa and always its should be considered along with other taxa.

b) Taxonomic importance of character be determined on the basis of its correlation with other characters.

1.4 FUNCTIONS OF TAXONOMY

Taxonomy has three basic functions. These are 1) Identification 2) Nomenclature and 3) Classification.

1.4.1 Identification :- Identification is a process of assigning of a plant to a specified group - ultimately to the species. Identification is generally done by comparing a representative specimen of a given taxon with that of another which has already been authoritaively determined. For this, morphological similarities are sufficient for the provisional determination and identification.

1.4.1.1. Nomenclature : Once, the taxon is identified, the next is to name it scientifically. The nomenclature is done in accordance with International Code of Botanical Nomenclature approved by the International Botanical Congresses from time to time.

1.4.1.2. Classification : The placing of newly identified and named new plants into a hierarchy of ranks or categories such as species, genera, family and so on is described as classification. However, criteria for classification followed by different taxonomists vary. There are basically three kinds of classifications followed by the taxonomists. These are 1) Artificial classification 2) Natural classification and 3) Phylogenetic classification.

Artificial classification is based upon one or a few easily comparable morphological characters. For example, the division of plants into trees, shrubs, herbs and climbers is an artificial classification while, natural classification gives emphasis to overall resemblances mostly in gross morphology.

Phylogenetic classification based on known or inferred phylogeny of taxon and are expressed in the form of a phylogenetic tree showing the supposed evolution of different groups.

Natural systems of classification are also described as horizontal classifications, as they are based on data available at any one time, on the other hand, phylogenetic classification rely on presumed ancestry, hence termed as vertical classification.

QUESTIONS

A. Answer the following in about 25 lines.

1. Define taxonomy, briefly outline the history of taxonomy.
2. Write an account on principles and functions of taxonomy.

B. Answer the following in about 10 lines.

1. Age of herbals
2. Development of taxonomy in India.
3. Botanical Survey of India.

C. Answer the following in 2 or 3 lines.

1. Taxon
2. Nomenclature
3. Classification
4. Phylogenetic classification
5. Artificial classification
6. Natural system of classification
7. Omega taxonomy
8. Theophrastus

2

PLANT TAXONOMY

SYSTEMS OF CLASSIFICATION

2.1 INTRODUCTION:

Todate, various classificatory system were proposed world over taking into consideration different factors. Jeffrey (1982) has identified four main categories of systems of classification. These are:

1. Classification based on habit
2. Artificial systems of classification
3. Natural systems of classification
4. Phylogenetic systems of classification

The first three systems were proposed earlier than the well known Darwin's theory of evolution. Hence, they are also known as pre-Darwinian systems. The Phylogenetic systems were proposed after the Darwin's theory of evolution in 1858. A brief account of these systems is given in this chapter.

2.2 CLASSIFICATIONS BASED ON HABIT

Taxonomy of plants based on habit was proposed by Philosophers, Herbalists, Botanists during 3rd to 18th century BC. They classified plants as trees, shrubs, small shrubs and herbs.

2.2.1 Theopharastus Classifications : Theopharastus (370 - 285 BC) was a pupil of Aristotle and Plato, classified plants as trees, shrubs, undershrubs and herbs and treated them as annuals, biannuals and perrenials depending upon the length of their life cycle. In the same way, some important taxonomic characters like racemose and cymose inflorescences, superior and inferior ovary, polypetalous and gamopetalous conditions were also described. About 480 different types of plants were classified in his publications 'Historia Plantarum' and 'The causes of plants'.

2.2.2 Dioscorides Classification : Pedanior Dioscorides (1st century AD) was a military physician and was interested in the medicinal properties of plants. His main contribution, 'The Materia Medica' was a vivid description of about 600 plants based largely on first hand observations. His contributions were considered to be a ready made reference to the botanists for the next 1,500 years.

2.2.3 Albertus Magnus Classification : Magnus (1193 - 1280) was in agreement with Theophrastus classification based on their habit. Magnus divided plants into dicotyledons (tunicate) and monocotyledons (corticate) based on the difference in their stem anatomy.

2.2.4 Andrea Caesalpino Classification : Caesalpino (1519 - 1603) divided plant kingdom into fifteen heterogenous groups. Caesalpino considered fruit, seed characters, position of the ovary superior/inferior. The nature of the stem (under ground bulb / aerial) and the presence of latex / absence of latex.

2.2.5 John Gerards Classification : Gerards (1597) in his two volumed work 'General History of Plants' described about grasses, corn, rushes, flags, bulbous or rooted plants in the first volume. In the second volume all sorts of herbs were included used in medicine as also meat and sweet smelling. Further, the plants groups were divided into trees, shrubs, bushes, fruit bearing plants, resins, gums, roses, mushrooms, mosses and cereals. Thus, Gerards classification was based on superficial resemblences and on the properties of plants in relation to man.

2.2.6 Gaspard Bauhin's Classification: Bauhin (1623) a Swiss botanists, published 'The Pinax' which described over 6,000 plants. He was the first to distinguish nomenclaturally between species and genera. The binomial nomenclature infact was founded by Bauhin.

2.2.7 John Ray's Classification : John Ray an English Botanist presented a classification of plants in his book 'The Historia Plantarum'. He was the first taxonomist to divide plants into Monocotyledons and Dicotyledons based on the presence of one or two cotyledons. Further, he divided plants into Arborae (woody plants) and Herbae (herbaceous plants). Following is the outline of John Ray's Classification.

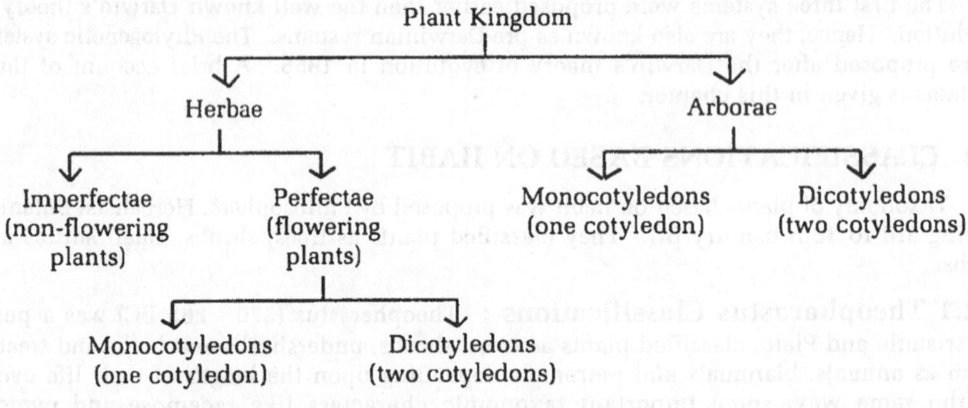

Lawrence (1951) opines that John Ray's classification was a system based on form and gross morphology of plant structure and in many respects was superior to the artificial Linnaeus System that came later.

2.2.8 Joseph Pitton de Tournefort's Classification : Tournefort (1656 - 1708) of France proposed a system of classification in his book 'Elements de Botanique'. He classified flowering plants into two large groups, namely, trees and herbs. Each of these was further divided into many sub-groups based on characters such as the presence or absence of petals, poly or gamopetalous and regular or irregular flowers. Tournefort also for the first time advanced the genus concept and many of his generic names are well known even today. He aptly described as "The Father of modern concept of genus".

Tournefort's system was however was found to be inferior to that of John Ray since no distinction between phenerogams and cryptogams and also between monocotyledons and dicotyledons, had been recognised

2.3 ARTIFICIAL SYSTEMS

Artificial systems of classifications were purely based on one or a few easily comparable morphological characters without giving any importance to affinities or interrelationships. The pioneers in the artificial systems were Carolus van Linnaeus (1707 - 1778), Pehr Kalm (1716-1779) and Hassel Quist (1722 - 1752). A detailed account of Linnaeus classification is given below.

2.3.1 Linnaeus system of classification : Carolus Van Linnaeus, a sweedish naturalist, was born on May 23, 1707 in Roshult, Sweeden. As a student, he published his first research paper on the sexuality of plants (1729). Later in 1730 he broughtout a list of plants classified following the system of Tournefort under the title 'Hortus Uplandica'. In the year (1737) Linnaeus published an account of 1,336 genera in his 'Genera Plantarum'. These plants were grouped into 24 classes identified on the basis of number and morphological arrangement of stamens.

Linnaeus very aptly called as the 'Father of Taxonomy' and also as the 'Father of modern Botany' for his epoch making work on the 'Binomial system of nomenclature' and also for his classification based on floral characters chiefly on the number of stamens (numerical relationships of the sex organs). His system thus based on the differences in the number of sex organs rather than on the similarities of sex organs. To commemorate his works in the history of Botany, a plant is named after Linneus, '*Linnaeaborealis.*

Following is the Linnaeus sexual system of classification appeared in the genera plantarum, which includes 24 divisions / classes based on the number of stamens, their adhesion and the length (Table 2.1). Another 401 genera were also classified by Linnaeus in the subsequent volumes of genera plantarum.

Table 2.1 Linnaeus system of classification

Sno.	Division / Class	No. of stamens / their adhesion	Eample
1.	Monandria	Flowers with one stamen	Eg. *Lemna, Scirpus*
2.	Diandria	Stamens two	- *Veronica, Salvia*
3.	Triandria	,, three	- *Iris, Sisyrinchium*
4.	Tetrandria	,, four	- *Mentha, Ulmus*
5.	Pentandria	,, five	- *Primula, Solanum*
6.	Hexandria	,, six	- *Rumex, Asparagus*
7.	Heptandria	,, seven	- *Aesculus*
8.	Octandria	,, eight	- *Fagopyrum*
9.	Enneandria	,, nine	- *Rananculus Rheum*
10.	Decandria	,, ten	- *Acer, Kalmia*

Sno.	Division / Class	No. of stamens / their adhesion	Eample
11.	Dodecandria	Stamens eleven to Nineteen	- *Euphorbia*
12.	Icorandria	Stamens twenty or more in episepalous condition	- *Rosa, Rubus*
13.	Polyandria	Stamens twenty or more, attached to receptacle	- *Tilia, Papaver, Nymphaea*
14.	Didynamia	Stamens didynamous (2+2)	- *Ocimum, Linnaea*
15.	Tetradynamia	Stamens tetradynamous (4+2)	- *Brassica*
16.	Monadelphia	Stamens in one bundle	- *Malva, Althea*
17.	Diadelphia	Stamens in two bundles	- *Pisum, Lathyrus*
18.	Polyadelphia	Stamens in many bundles	- *Citrus, Hypericum*
19.	Syngenesia	Stamens united with one another by anthers	- *Viola, Tagetus*
20.	Gynandria	Stamens adnate to the gynoecium	- *Orchids, Calotropis*
21.	Monoecia	Plants monoecious	
22.	Dioecia	Plants dioecious	- *Carex, Morus*
23.	Polygamia	Plants polygamous	- *Salix, Urtica*
24.	Cryptogamia	Flowers concealed	- *Acer, Ermpetrum*
			- Algae, Fungi Mosses and Ferns

2.3.2 Merits and demerits of Linnaeus system

Though the system of classification advanced by Linnaeus was an artificial and often quoted as classic example of an artificial system of classification. It gained the popularity due to its simplicity and easy way of identification of plants. It was followed by botanists world over for more than one hundred years after the death of Linneaus. Linnaeus was aware of demerits of his system and made an attempt to make it more natural but he could not complete this task in his life time.

When one considers the demerits of the system, Linnaeus arranged genera into 24 artificial classes (divisions) having remote relationships.

The cryptograms (plants with concealed flowers) in which he had little interest were all grouped in the twenty-four class.

Linnaeus listed 77 orders and in many of which he found natural affinity of certain genera with one another eg. genera of Orchidaceae and Palmae. On the other hand, many of them were organized without any definite relationships and affinities. This resulted in mixing up genera belonging to different families under one group. For example, genera belonging to Zingiberaceae and Anacardiaceae have been placed in Monandria just for one character i.e. both the families are characterized by the presence of only one stamen.

Monocotyledons and Dicotyledons were placed in the same systematic position. Finally grasses were grouped along with others without considering the advanced features of the former. Inspite of above drawbacks, the system of classification of Linneus is a great contribution to systematics if one considers the limited scientific know how of those times.

2.4. NATURAL SYSTEMS OF CLASSIFICATION BASED ON FORM - RELATIONSHIPS

2.4.1 Adanson System of Classification : Michel Adanson (1727 - 1806) a French botanist was first to formulate a natural system of classification. In his book 'Families des plants' (1763), Adanson considered different characters and also affinities among various groups while arranging and classifying plants. He was poineer in identifying orders and families. Adanson also known for his contributions in the field of numerical taxonomy and described as the 'Father of numerical taxonomy'.

2.4.2 DeJussieu's System of Classification : Bernard de Jussieus a contemporary of Linnaeus (1699 - 1717) advanced a natural system of classification in his publication entitled 'The Genera Plantarum Scandum ordines Naturales Disposita' (1789). In this book, plants were classified into 100 orders (now known as families) and grouped into 15 classes.

In the first class called Acotyledonae Jussieu included toad stools, Mosses, Ferns, liverworts (all cryptogams) and some flowering plants whose reproduction was not known at that time.

The rest of 14 classes contained all other flowering plants. The flowering plants were divided into Monocotyledonae and Dicotyledonae. Following is the outline of de Jussieu classification.

I. Acotyledonae

Cryptogams like Toadstools, Mosses, Ferns, Liverworts etc.,

II. Monoctyledonae

a) Stamina Hypogyna
b) Stamina Perigyna
c) Stamina Epigyna

III. Dicotyledonae

A) Apetalae (flowers without petals)
a) Stamina epigyna
b) Stamina perigyna
c) Stamina hypogyna

B) Monopetalae (Corolla fused)
a) Corolla hypogyna
b) Corolla perigyna
c) Corolla epigyna
i) Antheris connatis
ii) Antheris distinctis

C) Polypetalae
a) Stamina hypogyna
b) Stamina perigyna
c) Stamina epigyna

D) (class no. 15) : Diclines irregularis
(Dicotyledons with unisexual flowers like Euphorbiaceae, Cucurbitaceae and also Gymnosperms).

2.4.3 de Candolle's system of classification A French botanist AP de Condolle (1778 - 1884) devised another system of classification in his book 'Theorie Elementaire de la botanique' (1813). He coined the term 'Taxonomy' to designate the theory of plant communication. The de Candolle's classification was subsequently revised by Alphonose (1806 - 1893).

Division - I VASCULARES (Vascular plants with cotyledons)

Class - I : Exogenae (Dicotyledons)

Sub-class a) Dichlamydeae (calyx and corolla present)

Series i) Thalamiflorae (hypogynous flowers) (orders 1-54)
ii) Calyciflorae (peri or epigynous flowers) (orders 55-118)
iii) Carolliflorae (epigynous and sympetalous flowers) (orders 119-155)

b) Monochlamydeae (uniseriate perianth) (orders 156 - 207).

Class - II : Endogeneae (Monocotyledons)

Sub-class -I - Phanerogamae (flowers present)

Sub-class-II - Cryptogamae (flowers absent)

Division - II CELLULARES (plants without vascular bundles and cotyledons)

Class -I Foliaceae (leafy mosses)

Class -II Aphyllae (no Leaves) (Algae, Fungi and Mosses)

2.4.4 Robert Brown's Classification : A Scottish Naturalist Robert Brown (1773 - 1858) mainly followed the de Condollean system ignoring Linneaus. Even though he has not proposed any system of classification, however, established several families. Brown for the first time proved that the gymnosperms were clearly distinct from angiosperms in having naked flowers.

2.4.5 John Lindly's (1799 - 1855) system of classification: Lindley advanced a classification system in which a distinction was made between flowering plants (phanerogams) and non-flowering plants (cryptogams). He also considered number of physiological characters for his classification. Lindley published his work in 'The Vegetable Kingdom' (1846). Lindley classified the phanerogams into to the following five classes;

1. Rhizogenes
2. Endogenes
3. Dictyogenes
4. Gymnogenes
5. Exogenes

2.4.6 Stephen Endlisher's System of Classification : Endlisher (1805 - 1849) of Vienna described about 6,835 plants in his 'Genera Plantarum'. He separated cryptogams from phanerogams. Gymnosperms were placed near the dicotyledons. Achlamydous flowers were considered primitive. The dicotyledons were classified into Epipetalae, Gamopetalae and Dyalepetalae.

2.4.7 Bentham and Hooker's System of Classification : George Bentham (1800 - 1884) and Sir Joseph Hooker (1817 - 1911) two great British systematists adopted a natural system of classification in their 'Genera Plantarum'. In real terms, Bentham and Hooker system is a modification of deCandolle and deJussuieu classification. In this system, a total number of 97,205 genera were grouped under 202 orders (presently families) and cohorts (orders). Bentham and Hooker Classified Spermatophyta (seed plants) into Dicoyledons, Gymnosperms and Monocotyledons. Further, Dicoltyledons included three divisions 1) Polypetalae 2) Gamopetalae and 3) Monochlamydeae. The polypetalae was futher divided into three series, namely, Thalamiflorae, Disciflorae and Calyciflorae on the basis of the structure of thalamus and the position of floral parts. The gamopetalae (sympetalae) was classified into three series. These are, Inferae (Inferior ovary); Hetermerae (superior ovary) and Bicarpellatae (carpels two). The monochlamydeae divided into 8 series and monocotyledons into 7 series. (for details refer next chapter).

2.5 . CLASSIFICATORY SYSTEM BASED ON PHYLOGENY :

The evolutionary theory of Charles Darwin (1859) published in 'The Origin of Species' had tremendous influence on the systematic botanists world over. This lead to the advancement of new classification systems giving importance to evolutionary advancements from simple to more complex form. Genetic as well as ancestral relationships were also considered while assigning them to different taxonomic groups, such as orders and cohorts.

Spraig (1967) opined that the number of classificatory systems that were proposed considering phylogeny is based on two important concepts, accordingly systematic botanists can be grouped under two categories.

A. ENGLARIAN SCHOOL

(Those who believed Engler's views on phylogeny). Systematists who belonged to this school of thought are,

Eichler, Engler and Prantl, Rendle, Wetstein and Lawrence

B. RANALIAN SCHOOL

(Those who believed Ranalian origin of Angiosperms) Botanists who belonged to this school of thought are,

Bessey, Hallier, Hutchinson, Tippo, Cronquist, Takhtajan and Throne

Ranalian School believes that the monocotyledons are evolved from dicotyledons and on the other hand Englarian School considers dicots are evolved from monocots.

A few of the important phylogenetic classificatory systems are described here.

2.6 AUGUST WILHELM EICHLER'S SYSTEM OF CLASSIFICATION:

A German Botanist, Eichler (1839 - 1887) published his evolutionary system of classification in his book 'Bluthen Diagrame', Though his system mainly considered evolutionary tendencies of plants, however, emphasis was given to genetic relationships.

The Dicotyledons were considered as advanced over monocotyledons. The Dicotyledons were grouped into choripetalae and sympetalae.

2.7 ENGLER AND PRANTL SYSTEM OF CLASSIFICATION:

Heinrich Gustav Adolf Engler (1844 - 1930) and Karl Anton Prantl (1840 - 1893) the two German botanists proposed a system of classification based on that of Eichler. They published their account in 'Die Naturlichen Pflanzen familien' in 23 volumes covering the entire plant kingdom. Engler and Prantl provided exhaustive keys and descriptions of all families. For each family, summary concerning to the embryology, morphology, anatomy and geography of the genera is given.

Engler and Prantl divided the plant kingdom into 14 major divisions. These include,

1. Schizophyta
2. Myxomycetes
3. Flagellatae
4. Dinoflazellatae
5. Heterocostae
6. Bacillariophyta
7. Conjugatae
8. Chlorophyceae
9. Charophyta
10. Phaeophyceae
11. Rhodophyceae
12. Eumycetes
13. Archegoniotae
14. Embryophyta Siphonogama

The seed plants (Embryophyta Siphonogama) were classified in the following way,

Embryophyta Siphonogama
- Angiosperms
  - Dicotyledons
    - Archichlamydeae
    - Metachlamydeae
  - Monocotyledons
- Gymnosperms

The difference between Engler's system with of Bentham is the amalgamation of polypetalae and monochlamydeae. Another significant feature is, the placing of monocotyledons before dicots.

The Angiospermous families have been arranged in ascending order in accordance with the increasing complexity of the flower in general and the character of the flower envelope in particular.

According Engler and Prantl system, the most primitive type of flower has no perianth; then only one whorled in the highest type. Two whorls of perianth in the latter case, the two whorls may be indistinguishable or distinguishable into calyx and corolla (homiochlamydeous / heterochlamydeous). The gamopetalous condition was considered more advanced than the polypetalous condition. Besides this, evolution is traced from hypogyny to epigyny via forms like perigyny and also indefinite number of stamens and carpels to a definite number.

The other salient features of this system are :

1. Providing of abundant illustrations of the types described,
2. Exhaustive keys were provided to the orders and families, and
3. Exhaustive summeries on embryology, anatomy and morphology were provided for families.

Demerits of Engler and Prantl system

1. It gave precedence to monocotyledons which is contrary to our present day knowledge of systematics.
2. The Ranalian families like Ranunculaceae, Nymphaceae and Magnoliaceae have been universally accepted as the most primitive by the modern taxonomists, on the other

hand, Engler and Prantl considered these as advanced and found to be derived from Anmentiferae (Catkin inflorescence bearing families) and also Centrospermae like Chenopodiaceae, Amarantaceae and Caryophyllaceae.

3. Engler and Prantl opined that the Dicotyledons are derived from Gymnosperms. However, evidences from floral anatomy, embryology infact suggested that the most primitive flower is found in Ranales (eg. *Magnolia*) in which indefinite numbered free sepals, petals, stamens and carpels are arranged on a conical receptacle.
4. The system of Engler and Prantl has been subjected to heavy criticism since it developed a tendency to obscure phylogeny of the Angiosperms.

Merits of Engler and Prantl's classification system

1. Various groups of plants have been built up step by step and were arranged phylogenetically to form a generally progressional and morphological series.
2. It is a natural system, since it was proposed a subsequent to the acceptance of the theory of descent. It throws light on the phylogeny of various groups and traces the increasing complexity of flower particularly, the essential parts
3. Evidences from internal anatomy were duly considered for the first time.

2.9 CHARLES EDWIN BESSEY'S SYSTEM OF CLASSIFICATION

Charles Bessey, a great American taxonomist (1945-1915)proposed a phylogenetic system of classification which appeared in 'The Annals of the Missouri Botanical Garden' under the title 'The phylogenetic taxonomy of Flowering plants'. The Besseyan system was based on a series of dicta or statements or guiding principles to determine the degree of primitiveness or evolutionary advancements of different groups.

Bessey opined that, the angiosperms were originated from gymnospermous ancestors like the Bennetittales (strobiliferous group) which in turn arose from cycadeoiod forms. He further feels that, the Ranales are the most primitive angiosperms which gave rise to three separate phyla of flowering plants. One of these gave rise to the monocotyledons and other two to dicots. The two evolutionary lines of dicots are,

1. Strobiloideae 2. Cotyloideae

2.10 HUTCHINSON'S SYSTEM OF CLASSIFICATION

John Hutchinson (1884 - 1972) of Royal Botanic Gardens, Kew, England proposed another phylogenetic system of classification in his book 'The families of Flowering Plants'. The first volume of this work described dicotyledons, appeared in 1926 and the second volume dealing with the monocotyledons published in 1934. These publications were subsequently revised several times and the final treatment appeared in 1973.

Hutchinson's system was based on 24 principles or dicta. Some of these include,

A. General dicta :

1. Hutchinson believed that evolution is both upwards and downwards, while the upward evolution tend towards preservation, the downward evolution brings degeneration.
2. Evolution will not uniformly affect all organs of an individual while some organs are advancing, others may remain static or even undergo retrogression.
3. Evolution tends to be consistent, and when a particular progression or retrogression takes place, interestingly it continues in the same fashion till the end of the phylum.

B. Principles relating to general habit

4. Trees and shrubs are regarded more primitive than herbs.
5. Annuals and biennials are advanced than perennials.

C. Principles relating to the flowering plants :

6. Opposite and whorled phyllotaxy is more advanced than alternate phyllotaxy.

D Principles relating to flowers and fruits :

7. Dioecious plants are more advanced than monoecious, unisexual flower are derived from bisexual flowers.
8. Apetalous flowers are derived from petaloid ones.
9. Polypetaly is more primitive than sympetaly.
10. Aggregate fruits are advanced and simple fruits are primitive. Capsule preceeds the drupe or berry.

Hutchinson's system mainly based on the works of Bentham and Hooker and Bessey. However, it was primarily gave importance to the habit of plants, Hutchinson thus classified Dicotyledons into Lignosae (woody plants) and Herbaceae (herbaceous plants) and monocotyledons into three series i.e. Calyciferae, Corolliferae and Glumiflorae.

Lignosae : Lignosae were further grouped into 54 orders beginning with Magnoliales and are followed by Annonales, Laurales, Dilleniales and ending with Verbenales and Bignoniales.

Herbaceae : It consists of 28 orders with predominently herbaceous families derived from Ranales. It starts with Ranales and ends with Lamiales.

Monocotyledons : Monocotyledons with three distinct groups based on the nature of perianth, Calyciferae, Corolliferae and Glumiflorae. These were further divided into 29 orders and 104 families.

Hatchinson believes that woody taxa (Lignosae) are distinct and phylogenetically older than the herbaceous groups. The dicotyledons according to him, arose from the proangiosperms and were first to make their appearance on the earth's surface. Regarding the origin of monocotyledons, he states that the group was derived from an early dicot like Ranales.

The Hutchinson's system faced much criticism as it rested mainly on gross morphological data. The division of dicots into Lignosae and Herbaceae was based primarily on habit of the plants, due to this related families were put apart and unrelated families were closely arranged.

Hutchinson failed to supply any explanatory notes as to the delimitations of the orders and families. Further, he did not give any importance to recent conclusions on the phylogeny of gymnosperms and also no proper consideration to morphological, anatomical and cytological evidences.

In view of some drawbacks in the Hutchinsons system of classification, it did not receive much attention from the systematic botanists, Nevertheless, it has been used as a guiding system by others like Oswald Tippo, Takthajan, Cronquist and others. His classification of monocotyledons considered as the best by the taxonomists world over.

2.11 OTHER CONTEMPORARY PHYLOGENETIC SYSTEMS

2.11.1 Rendle's system: Alfred Rendle (1865 - 1936) of British Museum of Natural History, London published an excellent two volume work, 'The classification of flowering plants'. His classification essentially ran on Engler and Prantl's line of classification with some minor differences. This system however, has the added advantage of having dicots with three categories. These are monochlamydeae (with 13 orders and 31 families), dialypetalae (17 orders and 95 families) and sympetalae (11 orders and 47 families).

Sympetalae is divided into two groups, these are 1) Penta cyclicae (Flowers with five whorls) 2) Tetracyclicae (flowers with four whorls). The tetracyclical group further divided into superae and inferae where the ovary is superior or inferior respectively.

Rendle believed in the polyphyletic origin of Angiosperms. He laid much emphasis on wind pollination versus insect pollination as phyletic criteria. Rendle firmly believed that the woody habit is more primitive than the herbaceous habit. He does not consider ranales as the most primitive order among dicotyledons.

2.11.2 Takhtajan's System : Armen Takhtajan of Botanical Institute of Academy of Sciences at Leningard, has presented a system of classification based on phylogenetic principles (1954). It was revised again several times and published in both Russian and English. In 1980 a new version appeared in 'The Botanical Review' (46 : 225-359) under the title 'The classification of plants (Magnoliophyta)'.

Takhtajan opined that the origin of Angiosperms in monophyletic rather than polyphyletic and accordingly angiosperms arose from some very ancient group of gymnosperms.

Takhtajan named angiosperms as Magnoliophyta and divided them into two classes.

1. Magnoliopsida (Dicotyledons)
2. Liliopsida (Monocotyledons)

Magnoliopsida is further divided into seven sub-classes. These are 1) Magnolidae 2) Ranunculidae 3) Hamamelidae 4) Caryophyllidae 5) Dillenidae 6) Rosidae and 7) Asteridae.

Liliopsida (Liliatae) consists of 3 sub-classes. These are 1) Alismidae 2) Lilidae 3) Arecidae.

Takthajan while classifying plants gave importance to anatomical, cytological, chemical, palynological and ultrastructural characters. The main objection to his classification is regarding the origin of monocotyledons from Nymphaeles.

2.11.3 Benson's System of Classification : Benson (1957) advanced a new system of classification of flowering plants. He mainly followed Bentham and Hooker with some minor changes. This made his classification more natural and phylogenetic. He divided Angiosperms into two sub-classes namely, 1) Dicotyledonae 2) Monocotyledonae, which were further divided in different groups, orders and families.

2.11.4 Gunderson's System : Gunderson (1950-57) in his 'Families of Dicotyledons' proposed a phylogenetic scheme of classification. This system taken into account, the basic chromosome number of a taxon along with anatomical, cytological, morphological attributes. The main drawback of Gunderson's system is the complete ignorance of polypetalous and gamopetalous conditions of the flower. He also did not deal with monocotyledons. He divided dicotyledons into ten large groups according to the scale of evolution and consequent primitiveness. These ten groups are;

1) Magnoliflorae	2) Cistiflorae	3) Thea group	4) Rosiflorae
5) Ulmus group	6) Malva group	7) Geranium group	8) Diatheflorae
9) Jasminiflorae and	10) Rubiflorae.		

2.11.5 Cronquist's System : Arthur Cronquist, Professor of Botany, California University and Curator of New York Botanic Garden proposed a classification in his publication, 'The Evolution and Classification of flowering plants' (1968). The latest version of Cronquist System appeared in another volume called 'An integrated system of classification of flowering plants' (1981).

Cronquist discussed a wide range of characters important to phylogenetic classification. He provided synoptic keys and charts showing relationships of orders and sub-classes. Cronquist considered the Pteridosperms (seed ferns) as possible ancestors of Angiosperms.

A comprehensive form of Cronquist classification is as follows.

Division : Magnoliophyta (Angiosperms)

↓ ↓

Class : Magnoliatae	Class: Liliatae
↓	↓
Sub-classes	Sub-classes
1. Magnoliadae	1. Alismatidae
2. Hamamelidae	2. Commelinidae
3. Caryophyllidae	3. Arecidae
4. Dilleniidae	4. Lillidae
5. Rosidae	
6. Asteridae	

The Magnoliopsida consists of 64 orders 318 families and with about 1,65,000 species. The Liliopsida has 19 orders and 65 families with about 50,000 species.

The class Magnoliatae under Magnoliopsida considered as basal complex and from it the rest of Angiosperms have evolved.\

2.11.6 Dahlgren's System : Rolf Dahlgen, a botanist from Denmark published his 'A system of classification of Angiosperms to be used to demonstrate the distribution of characters' in the Botanische Noiser Journal (1975). Subsequently revised version of this appeared in 1980 and 1983 in the Nordiac Journal of Botany. Dahlgren presented a method of illustrating an Angiosperm system as an imaginary phylogenetic shrub.

Dahlgren extensively used various morphological and chemical characters to delimit the taxa, these included apocarpy, syncarpy and monocarpellate condition. Further, choripetaly, sympetaly, mono-aperaturate pollen, sclered idioblasts, types of microsporogenesis, bi-and uni-tegmic ovules. Tenucellate, pseudocrassinucellate and crassinucellate ovules. Presence of raphides, silica bodies, ellagic acid, ellagitanims, benzylisoquionoline alkaloids, tropane alkaloids, polyacytylenes and various classes of flavonoids.

Dahlgren believed that the Angiosperms have monophyletic evolution. The Magnoliflorae have been regarded as most primitive dicotyledons because of their woody habit, presence of oils and benzylisquinoline alkaloids.

Asper Dahlgren's classification, the Angiosperms comprise a class Magnoliopsida. It has been further divided into sub-class Magnoliidae and Lillidae (Dicotyledons and Monocotyledons respectively). Magnoliliidae are further divided into 24 super orders. 80 orders and 346 families and the Lillidae into 7 super orders, 26 orders and 92 families.

2.11.7 Thorne's System : Robert Thorne, a well known American taxonomist proposed a system of classification in his publication titled as 'A synopsis of a putatively phylogenetic system of classification of flowering plants' in 1968. A revised version system was appeared in 'The Journal of Phytochemistry and Phylogeny' and also in 'The Nordiac Journal of Botany' in 1983.

Thorne laid emphasis on phytochemical details in his classification. He also considered host - parasite relationships, pollen and seed morphology, comparative anatomy, embryology, plant-geography, paleobotany and cytology.

Thorne believed that angiosperms have monophyletic origin. He divided them into sub-classes i.e. Dicotyledoneae (Annonidae) and Monocotyledoneae (Lillidae). The former includes 19 super orders and the later 9 orders.

Thorne expressed confidence that his system of classification is superior over classificatory system of his contemporaries like Takthajan and Cronquist because of importance given to closely related character than to variations in different groups.

2.11.8 Mitra's System : J.N. Mitra, an Indian Botanist, suggested a new classification of plants. He believed that the evolution of angiosperms took place in both upward and downward lines. Mitra did not claim that his classification as purely phylogenetic. Further, he considered the characters of unisexuality and achlamydous flowers are advanced being derived from bisexual dichlamydous ancestors. In this respect Mitra's system ran on the lines of Bessey, Hallier and Hutchinson Systems and opposed to those of Engler, Rendle and Cronquist.

2.12 BENTHAM AND HOOKER SYSTEM OF CLASSIFICATION

2.12.1 Introduction: George Bentham (1800-1884) and Joseph Dalton Hooker (1817-1911), the two English Botanists, associated with the Royal Botanic Garden, Kew, England presented their natural system of classification in their book 'The Genera Plantarum'. The first volume of this book appeared in Latin and was published in July, 1862 and the second volume of the Genera Plantarum and the rest was written by Hooker.

The Bentham and Hooker system shows many similarities with the classificatory systems of Bessey and Hutchinson.

Bentham and Hooker took great pains to define and to delimit various taxa. They examined each and every species in the Kew botanic gardens. Furher, Bentham and Hooker provided first hand descriptions of the families and genera of seed plants then known. Hence, this system was considered to be the best taxonomic work ever produced in the United Kingdom.

The Bentham and Hooker's system is now in wide use in number of Commonwealth countries including Britain. In India also this system is followed. Numbers of regional floras are prepared like, Flora of Bombay and Flora of Madras presidency following Betham and Hooker's system.

2.12.2 Bentham and Hooker's Scheme of Classification:

Bentham and Hooker described over 97,205 species of flowering plants spread over 202 orders or families, which inturn were grouped into cohorts (now organised orders). They divided the plants into two sub-kingdoms (Table 2.2).

i) Cryptogamia

ii) Phanergamia

Cryptogamia reproduce mainly by spores, which are unicellular. Cryptogamia are further divided into Thallophyta, Bryophyta and Pteridophyta.

The phanerogams are seed producing plants. These were further sub-divided into three groups.

Class: { Group -I - Docotyledons (Exogenes)
Group-II - Gymnospermae
Group-III - Monocotyledons (Endogenes)

Group - I : Dicotyledons, (Exogenes)

1. Herbs, shrubs or trees of different habits.
2. Floral parts free or united, usually five or multiples of this number.
3. Ovules enclosed in an ovary.
4. Embryo with two cotyledons
5. Radical forms the primary root.

Group - II: Gymonospermae:

1. Shrubs or trees.
2. Leaves needle like or scaly
3. Flowers in cones, spikes or clusters and unisexual.
4. Perianth absent.
5. Ovules naked and not enclosed in ovary.

Group -III: Monocotyledons (endogenes)

1. Herbs, more rarely shrubs or arborus as in palms and bamboos.
2. Leaves large, parallel-veined.
3. Perianth generaly trimerous free or united.
4. Embryo with one cotyledon.
5. Radical short-lived and is replaced by adventitious roots.

The Dicotyledons (Group-I) were further divided into 3 **divisions (Sub-classes).**

Division - I : Polypetalae
II : Gamopetalae
III : Monochlaymydeae.

Division (Sub-Class) -I : **Polypetalae:** Flowers bear perianth (Calyx and Corolla) in two whorls. Petals free, stamens are not united with petals.

Polypetalae sub-divided into three series.

Series- I : Thalamiflorae

Series-II: Disciflorae

Series-III: Calyciflorae.

Series - I: Thalamiflorae:

1. Flowers hypogynous
2. Perianth lobes in several whorls.
3. Stamens and carpels are usually indefinite and inserted on the receptacle.
4. Carpels most frequently free.

Thalamiflorae includes Cohorts (orders) and 34 orders (families).

Cohort -I **Ranales** :- Stamens indefinite,3- a uni-seriate, carpels apocarpous (free) or inserted on torus. Embryo minute, seeds endospermic and with abundant fleshy endosperm. It includes 8 families. (orders)

Orders (families) :

1. Ranunculaceae
2. Dilleniaceae
3. Calycanthaceae
4. Magnoliaceae
5. Annonaceae
6. Menispermaceae
7. Berberideae
8. Nymphaceae

Cohort-II: **Parietales:** Flowers bear definite number of perianth lobes, stamens are definite or numerous, carpels definite in number and fused (syncarpous). Ovary unilocular or divided into locules by false septa with parietal placentation, seeds non-endospermic or fleshy. It includes 9 families (orders)

Orders (families):

9. Sarraceniaceae
10. Papavaraceae
11. Cruciferae
12. Capparideae (Capparidaceae)
13. Resedaceae.
14. Cistineae
15. Violaceae
16. Comellaceae
17. Bixineae

Cohort - III: **Polygalinae:** Stamens definite, ovary usually perfectly or imperfectly bilocular. Fruit very often compressed laterally. Endosperm abundant and fleshy. It includes following 4 orders.

Orders (families):

18. Pittosporeae
19. Tremandreae
20. Polygalaceae
21. Vochysiaceae.

Cohort -IV: **Caryophyllineae:** Stamens definite or rarely indefinite in number, ovary is unilocular, placentation free-central, or rarely parietal. Embryo curved or straight. Seeds endospermic. It includes 4 orders.

Orders (families):

22. Frankenieceae
23. Caryophyllaceae
24. Portulaceae
25. Tamaricaceae.

Cohort -V **Guttiferales** : Sepals imbricate; stamens many; ovary multilocular; placentation axile; endosperm absent or fleshy; It includes 6 orders.

Orders (families):

26. Elatineae
27. Hypericineae
28. Guttiferae
29. Transtroemiaceae
30. Dipterocarpaceae
31. Chalanaceae

Cohort - VI: **Malvales:** Flowers bear definite number of perianth lobes (usually five), stamens indefinite or definite, monoadelphous, carpels definite in number and fused. Placentation axile, seeds non-endospermic or fleshy. Malvales have three orders.

Orders (families):

32. Malvaceae
33. Sterculiaceae
34. Tiliaceae.

Series - II: **Disciflorae**: Flowers hypogynous, perianth lobes definite in number, stamens usually definite and are arranged (inserted) within or upon around receptacle (disc). Ovaries free or embedded in the disc. Disciflorae include 4 cohorts, 21 orders in addition to 2 anamolous orders.

Cohort - VII: **Geraniales**: Disc usually as a ring between the stamens or adnate to the staminal tube or reduced to glands, alternate with petals. Gynoecium syncarpous or sub-apocarpous, 5-2 pendulous ovules in each locule with ventral raphae.

Orders (families):

35. Linaceae
36. Humariaceae
37. Malphigiaceae
38. Zygophyllaceae
39. Geraniaceae
40. Rutaceae
41. Simoroubeae
42. Ochnaceae
43. Burseraceae
44. Meliaceae
45. Chaillatiaceae

Cohort - VIII: **Olacales**: Disc cup like or circular bearing stamens and petals on its edge. Ovules 1-3 in unilocular ovaries, pendulous, dorsal raphae. It inlcudes 3 orders.

Orders (families):

46. Olacineae
47. Ilicineae
48. Cyrillae

Cohort -IX : **Celastrales**: Disc united with calyx at its base, stamens inserted round the disc. Ovules 2 in each locule, erect, ventral raphae, leaves simple or rarely compound. It includes 3 orders.

Orders (families):

53. Sapindaceae
54. Sabaceae
55. Anacardiaceae

Two anomolous orders are also included in this cohort. These are:

56. Coriarieae
56. Moringeae

Series-III: **Calyciflorae**: Flowers are perigynous or epigynous with enlarged saucer, cup or urn shaped thalamus which surrounds ovary and perianth lobes, Calyx tube usually surrounding ovary or adnate to it. Stamens definite or numerous, arranged along margins of thalamus. Calycilorae includes 5 cohorts and 27 orders.

Cohort- X : **Rosales**: Flowers perigynous or rarely epigynous. Gynoecium one to many, apocarpous, rarely unite at base. Styles usually free, rarely united into a column. It includes 9 orders.

Orders (families):

58. Connaraceae
59. Leguminosae
60. Rosaceae
61. Saxifragaceae
62. Crassulaceae
63. Droseraceae
64. Hamamelideae
65 Bruniaceae
66. Haloragaceae

Cohort-XI: **Myrtales:** Flowers epigynous or perigynous, definite number of carpels syncarpous, ovary 2-3 loculed, ovules 2 to many, placentation axile. It includes 6 orders.

Orders (families):

67. Rhizophoraceae
68. Combretaceae
69. Myrtaceae
70. Melastomataceae
71. Lythraceae
72. Onagraceae

Cohort-XII: **Passiflorales:** Flowers epigynous or perigynous, carpels united, ovary unilocular, placentation parietal, styles free and forked at the apex, It includes 7 orders.

Orders (families):

73. Samydaceae
74. Loasaceae
75. Turneraceae
76. Passifloraceae.
77. Cucurbitaceae
78. Begoniaceae
79. Datiscaceae

Cohort - XIII: **Ficoidales:** Flowers epigynous or hypogynous, carpels fused, placentation parietal or sub-basal, styles distinct or united at the base. Embryo curved or concentric, It includes 2 orders.

Orders (families):

80. Cactaceae
81. Ficoideae

Cohort-XIV: **Umbellales:** Flowers eipigynous, carpels united but styles are free arising on a well developed epigynous disc. One ovule in each chamber, pendulous. It includes 3 orders.

Orders (families):

82. Umbelliferae (Apiaceae)
83. Araliaceae
84. Coranaceae

Division (Sub-Class) - II: **Gamopetalae:** Flowers epigynous with inferior ovary. It includes 3 cohorts and 9 orders.

Cohort -I: **Rubiales:** Flowers regular or irregular, stamens epipetalous, gynoecium bicarpellary, syncarpous. Ovary 2-many locular with single ovule in each locule, axile-placentation. It includes 2 orders (families).

Orders (families):

85. Caprifoliaceae
86. Rubiaceae.

Cohort-II: **Asterales:** Flowers regular or irregular, stamens epipetalous, sympetalous or free, ovary one loculed with one ovule in basal placentation. Asterales include 4 orders.

Orders (families):

87. Valerianeae
88. Dipsaceae
89. Calycereae
90. Compositae (Asteraceae)

Cohort -III : **Companulales:** Flowers usually irregular, stamens not always epipetalous, ovary 2-6 loculed with many ovules in each locule. It includes 3 orders.

Orders (families):

91. Stylidiaceae
92. Goodeniaceae
93. Campanulaceae.

Series -II: **Heteromerae:** Flowers hypogynous, stamens epipetalous or free. Opposite or alternate to stamens. Carpels more than two. It includes 3 cohorts and 12 orders.

Cohort-IV: **Ericales:** Flowers regular, hypogynous, stamens as many as or twice as many as petals. Ovary 1-many locular with 1-many ovules in each locule. Seeds minute. It includes 6 orders.

Orders (families):

94. Vacciniaceae
95. Ericaceae
96. Monotropaceae
97. Epacrideae
98. Diapensiaceae
99. Lenaceae

Cohort-V : **Primulales:** Stamens as many as petals, opposite to them. Ovules 1-many in unilocular ovary. Placentation free central or basal, It includes 3 orders.

Orders (families):

100. Plumbagineae
101. Primulaceae
102. Myrrsineae

Cohort -VI: **Ebenales:** Stamens are as many as or twice as many as petals, epipetalous, opposite to corolla, ovary 2-many locular, placentation axile, seeds few and large, plants trees or shrubs.

Orders (families):

103. Sapotaceae
104. Ebenaceae
105. Styraceae

Series-III : **Bicarpellatae:** Flowers hypogynous, stamens as many petals or few, carpels two, rarely one or three, It includes 4 cohorts.

Cohort -VII: **Gentianales:** Corolla regular, stamens equal to lobes of corolla leaves usually opposite. It consists of 6 orders.

Orders (families):

106. Oleaceae
107. Salvadoraceae
108. Apocynaceae
109. Asclepiadaceae
110. Loganiaceae
111. Genetianeae

Cohort - VIII: **Polemoniales**: Corolla regular, stamens are as many as petals. Leaves generally alternate. It includes 5 orders.

Orders (families):

112. Polemoniaceae
113. Hydrophyllaceae
114. Boraginaceae
115. Convolvulaceae
116. Solanaceae.

Cohort -IX : **Personales**: Corolla irregular, oblique. Stamens less than the petals. Posterior stamen becomes a staminode or altogether absent. Ovary bilocular with numerous ovules, placentation axile. It includes 8 orders.

Orders (families):

117. Scrophulariaceae
118. Orbanchaceae
119. Lentibulariaceae
120. Columelliaceae
121. Gesnaraceae
122. Bignoniaceae
123. Pedaliaceae
124. Acanthaceae

Cohort -X: **Lamiales**: Flowers zygomorphic, corolla bilabiate, stamens didynamous and less than petals in number and posterior stamen is reduced to staminode or supressed. Ovary bilocular with collateral ovules. Fruit enclosed in the persistant calyx. It has 5 orders.

Orders (families):

125. Myoporineae
126. Selagineae
127. Verbenaceae
128. Labiatae (Lamiaceae)
129. Anomolous Plantigineae

Division (Sub-Class) -III : **Monochlamydeae**: (Incompletae or Apetalae): Flowers bisexual, rarely unisexual, petals lacking. There are no cohorts but classified into 8 series.

Series-1. **Curvembryae**: Flowers bisexual, rarely unisexual, stamens are as many as sepals, single ovule in each locule. Ovule compylotropous, seeds endospermic. It includes 7 orders.

Orders (families):

130. Nyctaginae
131. Illecebraceae
132. Amarantaceae
133. Chenopodiaceae
134. Phytolaccaceae
135. Batideae
136. Polygonaceae

Series-2: **Multiovulatae Aquaticae**: Aquatic submerged herbs. Gynoecium syncarpous and ovules many in each locule. It includes 1 order.

Orders (families):

137. Podostemaceae

Series - 3 **Multiovulatae terrestrae**: Usually terrestrial, trees or shrubs. Ovules many in each locule. It includes 3 orders.

Orders (families):

138. Nepanthaceae
139. Cystinaceae
140. Aristolochiaceae.

Series - 4. **Microembryae:** Gynoecium apocarpous or syncarpous. Single ovule in each locule. Seeds endospermic. It is divided into 4 orders.

Orders (families):

141. Piperaceae
141. Chloranthaceae
142. Myristicaceae
143. Monimiaceae

Series - 5. **Daphnales:** Flowers bisexual, perigynous, perianth in one or two whorls, sepaloid, stamens as many or double the number of perianth lobes. Gynoecium monocapellary, 2-4 ovules in each locule. It includes 5 orders.

Orders (families):

145. Lanrineae
146. Proseaceae
147. Thymelaceae
148. Penacaceae
149. Elaegnaceae

Series -6. **Achlamydosporae:** Perianth sepaloid or petaloid, ovary unilocular with 1-2 ovules, which are poorly developed, this series is further classified into 3 orders.

Orders (families):

150. Loranthaceae
151. Santalaceae
152. Balanophoraceae

Series - 7. **Unisexmales:** Perianth sepaloid or minute or absent. Flowers unisexual, gynoecium syncarpous, styles forked, ovules 1-2 in each locule, seeds endospermic. It includes 9 orders.

Orders (families):

153. Euphorbiaceae
154. Balanopseae
155. Urticaceae
156. Leitherieae
157. Plantanaceae
158. Juglandeae
159. Myricaceae
160. Casuarinaceae
161. Cupuliferae

Series -8 **Ordines anomoli (Anamolous orders)** It includes 4 orders.

162. Salicaceae
163. Lacistemaceae
164. Empetracaea
165. Ceratophyllaceae

Group -II **(Gymnospermae):** Shrubs or trees, Leaves needle like or scaly, Flowers in cones, spikes or clusters, unisexual, perianth absent. Ovules naked, not enclosed in ovary. Pollination through direct contact of the pollen grain with ovules due to the absence of stigma and style. It includes 3 orders.

Orders (families):

166. Gnetaceae
167. Coniferae
168. Cycadaceae

Group-III. **Monocotyledons (Endogenes) :** Herbs, more rarely shrubs, or arboreous as in palms. Leave large, parelled-veined, perianth generally trimerous, free-or united. Embryo with one cotyledon. Adventitious root system present. It is divided into 7 series and 34 orders.

Series- 1. **Microspermae**: Inner series of perianth petaloid, ovary inferior, parietal or axile placentation seeds minute, numerous, without endosperm. It includes 3 orders.

Orders (families):

169. Hydrocharitaceae
170. Burmanniaceae
171. Orchidaceae.

Series -2. **Epigynae**: Inner whorl of perianth petaloid, ovary inferior, endosperm copious. It includes 7 orders.

Orders (families):

172. Scitaminae
173. Bromeliaceae
174. Hoemadoraceae
175. Iridaceae.
176. Amaryllidae
177. Taccacae
178. Dioscoreaceae

Series -3. **Coronarieae**: Inner perianth lobes petaloid, ovary free, rarely slightly adnate at the base, endosperm copious. It includes 8 orders.

Orders (families):

179. Roxburghiaceae
180. Liliaceae
181 Pontederiaceae
182. Philydraceae
183. Xyrideae
184. Mayaceae
185. Commelinaceae
186. Rapataceae

Series - 4. **Calycineae**: Perianth sepaloid, inner series sub-petaloid, ovary free. It includes 3 orders.

Orders (families):

187. Flagellareae
188. Juncaceae
189. Palmae (Arecaceae)

Series -5 **Nudiflorae** : Perianth absent or reduced to hairs or scales, ovary superior, carpels generally single, 1-many ovuled. Endosperm present. Nudiflorae further divided into 5 orders.

Orders (families):

190. Pandanaceae
191. Cyclanthaceae
192. Typhaceae
193. Aroideae
194. Lemnaceae.

Series -6. **Apocarpae**: Perianth 1-2 seriate or some times absent ovary superior carpets, single if more than one apocarpous, endosperm absent, It includes 3 orders.

Orders (families):

195. Trivurideae
196. Alismaceae
197. Najadaceae

Series -7. **Glumaceae:** Flowers solitary, sessile, arranged in spikelets with bracts, segments of perianth small, scale like, glumaceous or absent. Ovary with 1 ovule, endosperm present. Glumaceae divided into 5 orders.

Orders (families):

198. Eioculeae
199. Centrolepideae
200. Cyperaceae
201. Gramineae (Poaceae)
202. Restiaceae

2.12.3 Salient Features of Bentham and Hooker's System of Classification

A perusal of Bentham and Hookers systems of classification reveals that the flowering plants have been divided into three (3) classes (Groups), three (3) sub-classes (Divisions), 21 series, 25 cohorts and 202 orders (Families). They included only 200 orders at first instance and each order was given a serial number. However, two more orders, viz., Vochysiaceae and Cyrillaceae were added making it to a total of 204 orders.

The Bentham and Hooker's system of classification was based on the classifications proposed by De Jussiue, de Condolle and Lindely. This system infact was a refined and modified version of de Condolle. A new series called Disciflorae is created and placed along with Thalamiflorae and Calyciflorae. Further, a new sub class Monochlamydeae was added where all the achlamydous groups were included.

Bentham and Hooker identified Gymnosperms as a separate group and placed them between Dicotyledons and Monocotyledons.

A total number of 202 families are included out of which 165 families belong to dicots, and 34 to monocotyledons and three families belong to gymnosperms.

2.12.3.1 Classification of Dicotyledons: As per the Bentham and Hooker's system, the Dicotyledons are spread over 3 sub-classes and 14 series which were further divided into cohorts and orders.

The classification into 3 sub-classes was based on the nature of perianth, i.e., presence of number of perianth whorls, whether petals are free or united, petaloid or sepaloid and the presence or absence of perianth lobes. The three sub-classes are:

1) Polypetalae (petals free)
2) Gamopetalae (petals united or fused)
3) Monochlamydeae (perianth in one or two whorls, sepaloid, rarely absent).

Polypetale is further divided into 3 series here, the nature of ovary and disc were considered.

i) Thalamiflorae: Ovary superior, torus or receptacle concave or convex shaped.
ii) Disciflorae: Thalamus in the form of disc, which is present below the ovary, ovary superior.
iii) Calyciflorae: Ovary superior, half-inferior, half-superior or inferior. calyx or thalamus cup shaped. If inferior,calyx tube surrounds or adnate to the ovary.

If one considers the evolution with regard to the position of gynoecium in polypetalae, it transformed from hypogynous to epigynous condition through perigyny (from thalamiflorae to calyciflorae via disciflorae).

The first family of dicotyledons being the Ranunculaceae, belongs to Ranales and is considered to be the most primitive order among the Angiosperms. Hallier, Arber and Parkin, the well-known taxonomists considered that the placement of Ranales at the beginning of Angiosperms by Bentham and Hooker is a right placement, since it is based on phylogeny. This view was well supported by other taxonomists also. As per the Ranalian school of thought, Ranalian flower is evolved from the cycedeioidea flower or cone and the so called hypothetical hemiangiosperms. The bracts of cycadeioidea cone evolved into perianth lobes, microsporophylls into stamens, megasporophylls to carpels and the cone axis got reduced and evolved into thalamus of Ranalian flowe. However, this concept was opposed by the taxonomists of Englarian school like Eichler, Engler-Prantil and Lawrence. Inspite of this opposition, Bentham and Hooker firmly believed in the Ranalian school of thought and placed Ranales at the beginning of docots. Evidences from floral anatomy, palynology and wood anatomy also support this view and considers that the Ranales are primitive angiosperms.

Reduced thalamus, cyclic arrngement of floral parts, gradual reduction in sepals, petals, stamens and carpels and transformation from apocarpous to syncarpous condition are some other features considered by Bentham and Hooker in their classification.

The sub-class Gamopetalae is divided into 3 series based on the position of ovary.

i) Inferae: with inferior ovary.

ii) Heteromerae: Superior orary with more than two carpels.

iii) Bicarpellatae: Ovary (superior) with two carpels.

The series inferae with epigynous flowers includes three orders namely, Rubiales, Asterales and Companulales. Asteraceae of asterales is considered as highly advanced inview of complex floral parts and placed at the beginning of gamopetalae. In the heteromerae, the ovary is superior and with more than two carpels. It has 3 orders namely Ericales, Primulales and Ebenales. The last series, bicarpellatae contain families with superior ovary and two carpels. It includes 4 orders starting from Gentianales and ending with Lamiales. Besides above characaters in Gamopetalae, a change from actinomorphic to zygomorphic flowers and so also a reduction in number of stamens to 4 or 2 in Personales and Lamiales are also considered.

Monochlamydeae characterized by the presence of a single whorled perianth includes 8 series staring with Curvembryae and ends with Ordines anomali. Inspite of its drawbacks, Monchlamydeae considered as an artificial group, the inclusion of families like Loranthaceae, Euphorbiaceae and Urticacaceae in unisexuales is well appreciated.

The gymnosperms with naked seeds were arbitarily placed between dicots and monocots and the phylogenetic importance of naked seeds is not recognised. This drawback however, corrected by Eichler in his classification and placed gymnosperms before angiosperms.

Bentham and Hooker gave emphasis on the relative position of the ovary and perianth characters of monocotyledons. The first two series, microspermae and epigynae have epigynous flowers Orchidaceae which has been grouped under microspermae is characaterized by minute seeds, complex floral structure and pollination mechanism. The third series Coronariae however, have families with hypogynous flowers including Lilliaceae, Calycinae and Nudiflorae.

The next two series of monocotyledons are characterized by the sepaloid or with the total absence of perianth lobes (naked), In the sixth series i.e. Apocarpae, perianth is either absent or uni or bi-seriate and the ovary is superior with free carpels. The last series of monocots is Glumaceae where, flowers are minute and are arranged in spikes or spikelets. The perianth if present is in the form of glumes and pollination is entamophilous. The glumaceae includes the well known famialy with abundant distribution, the Graminae or Poaceae.

2.12.4 Merits of Bentham and Hooker's system:

1. The Bentham and Hookers system of classification was accepted through out the British Empire and later in the United States of America, This was also adopted to a lesser extent by some continental botanists. They regarded the system as the most convenient and suitable for all practical purposes.
2. It is considered as the first great natural system of classification for many reasons.
3. Every genera and species were studied from the actual specimens available in the British and Continental herbaria and their descriptions were based on their detailed investigations.
4. Although, the system was largely a compilation of the earlier classification of de Candolle and Lindley, however, every effort was made to look it more natural and complete.
5. Unlike de Candolle system, where conifers (gymnosperms) were distributed among dicot toxa, Bentham and Hooker regarded gymnosperms as a third taxon and placed them between dictos and monocots.
6. Monocotyledons are discribed after dicotyledons.
7. For the sake of convenience, the genera were divided into sub-genera to which were assigned all the relevant and important species.
8. Placing of Cucurbitaceae in polypetalae inspite of having gamopetalous corolla. Its inclusion is a right step since as in other polypetalae families, the ovule in Cucurbitaceae are covered by two integuments and are crossinucellate.
9. Placing of Umbelliferae (Apiaceae) at the end of polypetalae due to advanced flower and fruit characters.
10. Heteromerae with more than two carpels given a position before bicarpellatae having two carpels.
11. The entire description and placement of different orders look natural. They were classified taking into account the shape of flowers (regular/irregular) and number of stamens (5 to less than five). In this context Gentianales and Polemoniales with regular flowers and with five stamens are placed before Personales and Lamiales having zygomorphic flowers and with 4 or 2 stamens. Hence, the transformation of actionomorphy to zygomorphy and reduced number of stamens considered as an evolutinary tendency. In the same way, reduction in carpel number (only two in bicarpellatae) is regarded as an advanced character.
12. Inspite of its artificial amalgamation of different groups in monochlamydeae, inclusion of Curvembryae series considered as natural.
13. While classifying monocots into different series, emphasis has been laid on the position of ovary and the nature of perianth. Hence, in series, Coronerae where perianth is petaloid is placed at the beginning, followed by Calycinae in which it is sepaloid. Further, the Nudiflorae in which reduced perianth considered as an advanced and classified at the end. Grouping of Poaceae as last family of monocots is also a positive aspect of this system.
14. Bentham and Hooker's system of classification was never claimed as a phylogenetic system by its authors. Hence, any criticism of the system on the basis of phylogeny is not justified.

2.12.5 Demerits of Bentham and Hooker's System

Inspite of its many positive features, this system of was subjected to criticism from various quarters due to some drawbacks. Some of these are outlined below.

1. Placing of gymnosperms between dicots and moncots considered as anomalous since dicots were considered to have evolved from gymnosperms and correspondingly monocots are found to be evolved from dicots.
2. The third sub-classs of dicotyledons, monochlamydae is not a definite group as it includes several families which show close affinity with those placed in polypetalae for example, all families of Curvembrayae are related to polypetalae. The Chloranthaceae, Myresticaceae and Monimiaceae are related to Cyclanthaceae and Magnoliaceae among Ranales.
3. Due to arbitrary selection of diagnostic characters, some of the related orders are widely separated from each other.
4. Orchidaceae and Scitaminae in which the flowers are hypogynous with a highly developed and complicated form of corolla are considered to be most primitive order of the monocots, on the other hand, orders like Gramineae and Cyperaceae in which the flower is hypogynous and simple are considered as most highly evolved.
5. Liliaceae is sperated form Iridaceae and Amaryllidaceae merely on the characters of ovary without taking into consideration of other similarities.
6. In Polypetalae, Bentham and Hooker considered epigyny as advanced character and placed myrtales, Passiflorales and Umbellales at the end under calyciflorae, on the contrary, in Gamopetalae, epigynous orders like Rubiales (Rubiaceae) Asterales (Asteraceae) were classified under Inferae, which is the first series of Gamopetalae. In the same way, in monocots, Microspermae (Hydrocharitaceae, Orchidaceae) Epigynae (Scitaminae) series were placed in the beginning. This makes it clear that Bentham and Hooker used hypogypous, epigyous flower characters only for grouping and failed to identify the evolutionary importance of these features.
7. The Monochlamydeae which is the last sub-class of dicots considered as an artificial amalgamation of distinct families. Chenopodiaceae belonging to monochlamydeae infact is closely resembled to Caryophyllaceae of polypetalae. Further, Podostemataceae have more common features with rosales than the other monochlamydeae. The different families found in monochlamydeae resemble just in one characters, i.e. nature of perianth.
8. Two series of monochlamydeae, viz., Multiovulatae aquaticae, multiovulatae terrestris are created considering the habitat of different genera. Hence, this grouping is also considered as an artificial.

Table 2.2 : Summerized data on number of families, genera and species incorporated in Bentham and Hooker's Genera Plantarum.

Taxonomic group	No. of families (Orders)	No. of genera	No. of species (estimated)
Polypetalae	84	2,610	31,874
Gamopetalae	45	2,619	34,556
Monochlamydeae	36	801	11,784
Gymnosperms	3	44	415
Monocotyledons	34	1,495	18,576
TOTAL :	202	7,569	97,205

QUESTIONS

A. Answer the following in about 25 lines.

1. What is the difference between artificial and nature system of classification. Briefly describe the artificial system of Linnaeus.
2. Give an account of the natural system of classification of angiosperms studied by you. Discuss its merits and demerits.
3. Give an account of Engler and Prantl classification and mention its merits and demerits.
4. Write an account on Hutchinson's system of classification and how does it differ from Bentham and Hooker's system.
5. Give an account of the system of classification of flowering plants by Bentham and Hooker.
6. Briefly discuss the salient features of Bentham and Hooker's system classification and mention the merits and demerits of the system.

B. Answer the following in about 10 lines.

1. Classifications based on habit.
2. Artificial systems
3. Sexual system of classification.
4. Classifications based on form-relationships
5. Englarian school
6. Ranalian school
7. Phylogenetic system
8. Classification of Polypetalae
9. Distinguishing features of inferae
10. Monochlamydeae is an antificial group, justify this statement.
11. Classification of Monocotyledons.

C. Answer the following in 2 or 3 lines.

1. Monandria
2. Syngensia
3. Tetradynamia
4. Ameniferae
5. Strobiloideae
6. General dicta
7. Lignosae
8. Herbacae
9. Mitra's system
10. Monopetalae
11. Disciflorae
12. Calyciflorae
13. Cohorts
14. Ranales
15. Ordines anomoli

3

PLANT TAXONOMY

BASIC RULES OF NOMENCLATURE

3.1 INTRODUCTION

Nomenclature may be defined as 'The system of naming objects, particularly those of biological origin'. In nature each and every object has to be named for its identification. Hence, living organisms like plants also need to be named and a system of nomenclature has to be applied. Nomenclature is one of the important aspects of plant taxonomy, which deals with correct name to be applied to a known taxon or to a known plant. Once the plant is identified, it has to be named and later classified into a group.

3.2 COMMON NAMES VIS-À-VIS SCIENTIFIC NAMES

Man has been interested in plants since pre-historic times. In all nations, regions and tribes names have been assigned to plants in their own language. These names are described as vernacular or local names. The vernacular name of a given plant differs in various languages and are mostly popular only when they are of some importance to the people of a specific region or tribe. Further, these local names vary with the place, language of the people. This thus results in the multiplicity of names for a singles plants. In a country like India, where several languages are spoken, nomenclature becomes still difficult. It is all interesting to know that the some plant may be called by several names in one language. For instance, *Viola tricolor* has about 50 names in English. In the same way *Mangifera indica* (mango) has more than 50 different names in Sanskrit.

In many cases, the same name may mean widely different plant in various languages sampangi in Tamil is *Zephranthus* as well as *Magnolia*, while in Telugu it refers to *Artabotrys*. In another instance the name pine is indiscriminately used for *Pinus, Agathis, Araucaria, Callitris* or *Casuarina* causing confusion about the identify of plants. A person from other part of the world cannot recognize these plants by their local vernacular names which are quite different form the region he hails. Hence, a plant should be known by a single name throught the world. This necessitated the need for assigning scientific names to the plants by adopting a universal or international nomenclature system to name plants. Since then, the scientists mooted the idea of scientific naming of plants.

Botanists world over recognized Latin language as the most suitable language for naming plants. Latin is known by scholars and students throught the world and does not belong to any particular country. Further, the Latin language is specific and exact in its meaning and the use of Latin words in plant names thus avoid confusion as against other languages. Therefore, the use of scientific names rather than common or vernacular names are international in scope and common to all tongues.

3.2.1 Polynomical systems : During the early 18th century, taxonomists gave names to plants which composed of several words or names giving on brief description of the plant. This type of nomenclature is known as polynomial nomenclature. Polynomial system was followed by many toxaonomists including Linneaus (1975). In polynomial system, the present *Sida cordlifolia* was earlier known as *Althia maderaspatina, Subrotondofolio molli* and *Hissuto multipilis.* Such a polynomial nomenclature was also adopted by Linneus. He first named *Plantago major* as *Plantago follies ovatus glabris.* Later, when he identified new species of *Plantago,* a new polynomial was given as *Plantago follies, Ovatis glabris, nudo scapo tereti spica flosculis imbricates.* However, inview of this complex nomenclature, it was found very difficult to remember and to prepare keys. Thus, the polynomial system of nomenclature was abandoned and replaced by a new system called the binomial nomenclature.

3.2.2 Binomial nomenclature : Gaspard Bauhin (1956) was the first taxonomists to suggest the binomial system. However, he failed to follow it constantly in his descriptions of plants. Later, Rivinus (1690) felt that the name of any plant should not contain more than two words. Atlast Linneus (1753) devised a system of giving two names to the plants in the first edition of his well known work 'Species plantarum'. According to the binomial system of nomenclature, the scientific name of a plant, consists of two Latin or Latinised words.

1) The first word representing the genus or generic name (generic epithet) and

2) The second, the species or specific name (specific epithet)

For example, the name of custard apple as per the bionomial nomenclature is *Annona squamosa.* Hence, a scientific name of any plant consists of two latinised words. The first or the generic name always starts with capital letter and or the specific name begins with small letter, ex. *Hibiscus rorasinensis.*

In a binomial system a complete scientific name must be followed by a third element, i.e, the name of the person or author (citation) who formally described the plant. For example, the complete scientific name for custard apple is *Annona squamosa* L. named after Linneus who described the plant for the first time. For neem it is *Azadirachta indicu* A de Juss. The author element of names is usually abbreviated and 'L'. and 'A de Juss' are normally used for authority in place of Linneus and A.L. de Jusse. Further, the scientific names are printed in italics and underlined while typed.

3.2.3 Rules and regulations for Generic epithets : The generic name is always a singular noun or a word treated as a noun. As said earlier, the generic epithet is always written with a capital letter in the beginning. Further, if a generic name has been spelled more than once, it may be abbreviated by using the initial capital letter. For example

Gossypium arboreum (Tree Cotton)

G. barbadense (Sea Island Cotton)

The generic name may be given after a botanist to honour him

eg: *Bauhinia* After a well known botanist Bauhin.

Thunbergia Named after Thumberg

Caesalpina after Caesalpino

Linnaea after Linneus.

The generic name may be given to describe a certain definite character.

eg: *Xanthoxylon* The colour of the wood is yellow

The generic names may be derived from the locality and language in which plants were discovered.

Ginko From China

Nelumbo : From Srilanka

Pandanus : From Malaya

The generic names may originate by the combination of two or more Greek or Latin words.

eg: *Leucodendron* (Silver tree)
Leuco : Silver
Dendron : tree

Phytelephas (Plant elephant)
Phyton : Plant
Elephas : Elephant

Hygrophila (Marsh loving plant)
Hygrow : Marshy)
Phily : loving

The generic names depict mythological or poetical origin

eg: *Nymphaea* (beautiful nymphs)

Theobroma (Chocolate plant; Gods food)

3.2.4. Specific epithat : The species name is often an adjective referring to a distinguishing feature of the species like colour, size, habit or habitat.

eg. *Rosa alba* (white rose)

Mikania scandense (climbing habit)

Calotropis gigantea (big size)

The specific epithet may indicate its place of occurrence (locality or country)

eg: *Argemone mexcicana* (Mexican origin)

Coccinea indica (Indian origin)

Agave americana (American origin)

Riccia pathankotensis (found at pathanket)

The specific epithet may be named after botanists.

eg: *dellinic* (after Dillon)

wildenowil (Willdenow)

wrightii (Wright)

devisii (Davis)

The specific epithet some times may have combination of two words.

eg: *latifolia* (broad leaved)

grandiflora (large flowers).

The species name generally in agreement with the gender of the generic name. It may be masculine, feminine or neuter. The names ending with us are masculine, *-a* are feminine and with *um* are neuter.

3.2.5. Advantages of Binomial nomenclature:

1. These names are more definite and precise than polynomial names.
2. Being generally in Latin, they have a universal acceptance by people of all languages.
3. Binomial names are usually descriptive.
4. They are more independent.
5. The binomials are more comprehensible and easier to remember than the long descriptions.

3.2.6. International Code of Botanical Nomenclature (ICBN)

With ever increasing number of newly identified plants (numbering about three lakhs) from the Linneus period to the first half of 19th century nomenclature became very difficult. Botanists from different countries gave different names to same genus in the absence of proper keys and idexes. This lead to disorder in the botanical nomenclature. At this juncture botanist all over the globe felt that there should be an international code for Botanical nomenclature so that uniformity can be achieved.

Linneus in his philosophia Botanica (1751) proposed certain principles of Botanical nomenclature. A significant contribution was also made by Augstin de Condolle (1867). A detailed account of rules and regulations was given by him in his book 'The therie elemantaire de la botanique'. However, the first organised efforts to develop a precise and simple system of nomenclature was made at the first International Botanical Congress held in Paris (1867).

In the Paris Congress, laws of botanical nomenclature was proposed by Alphonose de Candolle were adopted with some modifications. The nomenclature rules adopted at Paris are known as the de condolle rules (de candollean code) or as Paris code of 1867. The Paris code contained 3 chapters and 68 rules and was valid for a considerable period of time.

The International Botanical Congress of Paris was followed by the Vienna Congress in 1957. During this meeting, the first edition of the International Code of Botanical Nomenclature was published. In the subsequent congress held at Brussels (1910), some of the nomenclature problems were resolved and an internationally accepted rules were established.

The fifth Botanical Congress was held at Cambridge in 1930, where another set of International nomenclature rules were formulated. Subsequently International Botanical Congresses were held at Amsterdam (1935), Stolkholm (1950), Paris (1954), Montreal (1959) Edinborugh (1964), Seattle (1969) and Leningrad (1975). The current International Code of Botanical Nomenclature was adopted at 12th Congress of Leningrad.

3.2.7. Salient Features of ICBN :

The International code of Botanical nomenclature has following three components.

1. Principles
2. Rules
3. Recommendations.

3.2.7.1 Principles : The principles of ICBN are the basic points on which the whole code is based. These principles, however, do not give detailed rules about nomenclature but show the main ideas that should be kept in mind by any botanist who wish to name a new plant.

The six principles that included in the ICBN are :

1. Botanical nomenclature is independent of Zoological nomenclature.
2. The application of name to taxonomic groups is determined by means of nomenclatural types (holotype specimen)
3. The nomenclature of taxonomic group is based upon priority of publication
4. Each taxonomic group can bear only one correct name, the earliest, that is accordance with the rules except in specific cases.
5. The scientific names of taxonomic groups should be only in Latin regardless of their derivation. If they are from other than Latin, then they have to be translated into Latin or a Latin description should be given.

3.2.7.2. Rules : The rules or articles gave a detailed prescriptions on all the pointed connected with the naming of plants. Naming of plants against rules will be considered as illegitimate. There are 75 rules and meant for bringing about greater uniformity in futune nomenclature. Some of these rules are described in the following paragraphs.

3.2.7.3. Recommendations : The recommendations are practical application of rules. Their main object is to bring about greater uniformity and clarity.

The three important appendices are added to the ICBN code, these are,

Appendix I : Nomenclature of hybrids.

Appendix II : **Nomine Familiarum Conservanda**
(includes names of families concerned because of language)

Appendix III : **Nominal Generica Conservanda Etregciender**

(The names of genera which are against the principle of priority, however, conserved because of their long usage)

3.2.8. Some Important rules of ICBN :

3.2.8.1. Rank of taxa : In the Article — 1 of the ICBN, the term 'Taxon' (plural = Taxa) was introduced which refers to a taxonomic group of any rank. The 'species' is the basic taxon or species is the basic unit of taxon. The other important ranks are Genus, Family, Order, Class, Division and Kingdom. Hence, Kingdom will be in the top of the hierarchy and species at the bottom. It a taxon falls between those units mentioned above, then a taxon may be further sub-divided by adding the prefix — sub to the unit.

eg: Sub-species
Sub-genus
Sub-family

If needed, new units can be created and placed before the fixed units for example,

a) A unit called tribe can be placed between family and genus
b) The other taxonomic units called section and series are placed between genus and species.
c) The units like variety, form and local population are placed after the species.

3.2.8.2. Modes of ending in different taxonomic ranks : ICBN rules underline that the mode of ending of each taxonomic unit (grammatical endings) i.e., from the division to the sub-tribe level.

Table: 4.1. Modes of ending in different taxonomic ranks

Taxonomic Unit	Modes of ending (grammatical ending)	Example
Kingdom	a, ae,	Plantae, Mycetae (Fungi)
Division	-ta, ota	Embryophyta, Eumycota
Class	-opsida	Hepaticopsida
Order	-eles	Dicotyledoneae
Sub-order	-ineae	Rosineae
Family	-aceae	Rosaceae
Sub-family	-oideae	Minosoideae
Tribe	-ceae	Paniceae
Sub-tribe	-ineae	Malvinae

Article 18 of ICBN states that the name of family is formed by adding the suffix-aceae. eg., Rosaceae. However, the names of the following eight families however treated as expection to the rule, because of their long usage. Article 18.5 authores the use of alternative names for those families. The alternative names are shown in brackets along the generic holotype.

Palmae (Arecaceae) - *Areca*

Cruciferae (Brassicaceae) - *Brassica*

Gramineae (Poaceae) - *Poa*

Leguminosae (Fabaceae) - *Faba*

Guttiferae (Clusiaceae) - *Clusia*

Umbelliferae (Apiaceae) - *Apium*

Labiatae (Lamiacae) - *Lamium*

Compositae (Asteraceae) - *Aster*

3.2.8.3 Typification or type method: Typification of type method is found in the Article-7 the ICBN. It states that the application of names of taxa of the rank of family or below is determined by means of nomenclatural types. Further, the nomenclature type is that element with which the name is permanentlty associated or the preserve herbarium sheet of the plant which was used by the author to give a authentic description of the plant. The specimen designated as the type of the species (the orginal) on which the name of the species is based is always preserved and the same is called the nomenclatural type.

If the berbarium specimen can not be preserved, then an illustration or description or figure can serve as nomenclature type.

Article 9 of ICBN states that the type of genus is a specied and in the same way type of family is a genus. Several terms are used to different nomenclatural types.

a) ***Holotype (Article 7.3):-*** Holotype is the one, a specimen or the other element used by the author or designated by him as the nomenclatural type (the original type material or specimen).

b) ***Paratype (Article7.4):-*** A Paratype is a specimen cited along with the orginal descriptions other than holotype or isotype. The paratype usually is a specimen from a collection other than the type, but on which the description of new taxon has been based. Paratype is generally designated with different field method from that of holotype or isotype.

c) ***Lectotype (Article 7.5):-*** In some cases, when the author of the species does not designate the type specimen (holotype), in this case a competent authority selects a specimen from the original material of the author to serve as a nomenclature type. The term lectotype is used to designate for such a specimen.

d) ***Isotype(Article 7.6):*** A specimen exactly like the herlotype (from the original collection, with same locality, date and reference number) is known as Isotype for example, if several hubaceous plants (mounted on a separate sheet) or several branches of a tree are collected at the same time, one, is designated as holo type and the others become isotypes. These duplicate types will be like holotype.

e.g: *Andrographis nallamalayana* Ellin: Isotype Ellin 25546

e) ***Syntype: (Article 7.7):*** Syntype is any one of two or more specimens cited by the author where no holotype was designated or any one of two or more specimens simultaneously designated as types. Some times, it so happens that when an author describes a new species and cites two or more species, these become syntypes.

f) ***Neotype: (Article 7.8):*** It is a specimen or other element selected as nomenclatural type. When all the original material is missing, lost or destroyed, then it can be substituted by a newly collected specimen, such a specimen is called the neotype. The neotype is collected from the same locality or place where original type was collected.

g) ***Topotype:*** If an additional nomenclature type is collected from the same locality (type locality) from where holotype was also collected such a type is known is known as topotype.

3.2.8.4 Principles of Priority: With a few exceptions, where alternate names are permitted each of plants can heave only one correct name (The correct name is the earliest legitimate one and validly published) In the principle of priority, if there are several names for a taxon, the earliest name of which is published is regarded as the legitimate name, published is regarded as the legitimate name. The valid publication of names is usually taken as beginning (cut off date) in May 1753, i.e., the date of publication of Linneus,'The species plantarum'. However, the principles of priority and typification does not affect the form of names of taxa above the rank of family.

3.2.8.5 Citation of Author's name: Article 46 of ICBN states the name of a taxon is considered valid only when the author (The person who first described the taxon) name is cited. The authors name is abbreviated and follows the species name.

e.g: DC = De Candolle

Linn=Linnaeus

Retz=Retzius

Roxb=Roxburgh

i) However, if authors name is short, then full name is cited but not the abbreviated form

eg:Vahe, Nees.

ii) When both father and his son are the authors of a taxon, the letter f=filial;son) follow the name of the son.

eg: *Zanthoxylum tomentellus.*Hook, f.,
(Where JD Hooker is son of William Hooker)

iii) When a taxon's name is published jointly by two authors, then two letters 'et' or and are used.

eg: *Polycarpia auria* white *et* Arnot

P.auria white *and* A

iv) When a taxon is described by more than two authors, then the name of the first author and the name of the taxon is followed by etal.

eg: *Arthraxon raizadae* Jain *etal*

iv) When a plant name is proposed by one author but could not publish, but same was published by another author, then the botanical name of the taxon is followed by name of the first author. This is followed ex and then the name of the present author (s) who validly published the name.

eg: 1) Clemne apera Koenig ex DC

2) *Portulaca whitiana* Wall ex While and Arnot.

3.2.8.7. Conditions for effective and valid publication: Articles 29 and 32 of ICBN state that, the publication of new names and descriptions are effective only when the printed matter is distributed to the general public or to atleast to ten botanical institutions with libraries accessible to botanists. The date of effective publication is the date on which the printed matter become available.

The published taxon becomes effective and valid only when it is accompanied by a description or diagnosis or a reference to previously and effectively published description or diagnosis.

The newly published taxon must be accompanied by a Latin description or diagnosis or by a reference to a previously published Latin description or diagnosis.

3.2.8.8. Retention of names of taxa which are divided: When a genus is divided into two or more genera, the original generic name must be retained for the genus including the type species. In the same way, when a species is divided into two or more species, the original specific name must be retained for the species containing the type.

a. eg: When *Euphorbia* genus was devided into two genera the names should be *Euphorbia chamescyce Lyctinis diorca* was divided into two species which were named as *L.dioorca Linn. Emend pull*, and L. alba mill

3.2.8.9. Names of plants in cultivation: Plants brought from the wild habitat into cultivation will have their original wild names. For example, retention of name of wild plant *Chrysanthemum parthenium* brought into cultivation but not with cultivated form name, i.e. *Matricaria eximaia.*

3.2.8.10. Legitimate and illegitimate names: When a taxon is named strictly by following the rules of ICBN,it is considered as illegitimate name.

A name is considered illegitimate even though it is published validly but failed to follow one or more rules if ICBN.

eg: i) Superflous names: Superflous names come under illegitimate names because a particular name given to a plant which has already has another validly published name.

ii) Nomen dubium: A name that is used in different senses and so has become a source of error or confusion.

1) *Synonym:* If a new name is given to a taxon which already has a valid name, such a name is called synonym. There are two types of synonyms.

a) *Nomenclatural or Homotypic synonym:* It is a different name based on the same nomenclatural type, the second name is synonym of the first one.

eg: *Milensa tomentosa* (Ruxburgh) Sinclair has two other names.

1) *Saccopetalum tomentosum* (Roxbugh) Hook.f. & Thorns

2) Uveria tomentosa (Roxburgh)

The above three name are based on same nomenclature types.

b) Taxonomic or Heterotypic synonym: It is a name given to taxon based on two different nomenclatural types.

eg: *Linneus* named a plant as *Boerhaavia erracta* in 1753 and the same plant was given another name by Saha and Murthy (1962)

As *Boerhaavia punarniva:* Later, Nair (1967), Rao and Hemadri (1970) opined that as per the priority of publication, the second name must be rejected.

b) *Basionym*: A specific or intraspecific word or epithet that has priority and retained due its legitimacy even when transferred to a new or different taxon.

eg: *Cypripedium pubescens* Willd.

Changed to *C.calceolus* var. pubescence (willd). Correl, as per the basionym rules of ICBN.

c) Two or more identical names based on different types out of which only one is legitimate. The first name is valid and the second name becomes Homonym.

eg: *Jatropha heterophylla* Studeyl is the first name and valid and it homonym is *Jatropha heterophylla* Hooker.f.

d) *Hoponym*: It is a name not based on any type species. It is rejected due to lack of identified type.

e) *Tautonym:* It is an illegitimate bionomial in which the generic name and specific epithet is the same.

eg: *Malus malus*

Strobus strobus

But *Cajanus cajan, Samenea saman* are not tanuonyms, since the specific epithet is not exactly similar as that of genus.

QUESTIONS

A. Answer the following in about 25 lines.

1. **Discuss the important principles of International code of Botanical Nomenclature.**
2. **Write an essay on binomial nomenclature.**
3. **What is the type method? Give in detail the terminology of type methods.**

B. Answer the following in about 10 lines.

1. **International code of Botanical Nomenclautre**
2. **Type method**
3. **Principles of priority**
4. **Valid publications.**
5. **Polynomiales**

C. Answer the following in 2 or 3 lines.

1. **Specific epithet**
2. **Index Kewens**
3. **Paris code**
4. **Isotype**
5. **Neotype**
6. **Author citations**
7. **Tautonym**

4

PALNT TAXONOMY

ANGIOSPERMS TAXONOMY

4.1 ANNONACEAE (The Curstard Apple Family)

Systematic Position

Dicotyledonae
Polypetalae
Thalamiflorae
Ranales
Annonaceae

4.1.1 Distribution : Annonaceae is a large family consisting of aromatic trees, shrubs and lianes. Lawrence(1952) recognised 62 genera and 820 species, while Takhtajan (1966) included 120 genera and 2,100 species in this family. In India, it is represented by 26 genera and 129 species, which are mostly distributed in the states of Andhra Pradesh, Assam, West Bengal, Tamilnadu, Uttar Pradesh and Madhya Pradesh. *Annona squamosa* (Custard apple) and *A. reticulata* (Bullock's heart) are extensively cultivated in India for their edible fruits. About 60% of Indian species are endemic.

4.1.2. Important Plants

1. *Annona squamosa* L. (Custard Apple) Telugu : Sitaphalam
2. *A. reticulata* L. (Batard Apple or Bullock's heart) Tel. : Ramaphalam
3. *A. muricata* L. (Sour sop) Tel: Laxmana phalam
4. *Artabotrys odoratissimus* R.Pr. (climbing ylang - ylang) Tel: Teega sampenga.
5. *Cananga odorata* Hook and Thoms. (Ylang - Yland) Tel: Chettu Sampenga
6. *Desmos chinensis* Lour.
7. *Monodora myristica* Dunal.
8. *Polyalthia longifolia* Benth & Hook f. (Mast tree) Tel: Naramamidi - Ashoka
9. *Unona discolor* Vahl. (Lavender champa)
10. *Unaria cersoides* L. Tel. Chitti madduga
11. *Xylopia aromatica* Bail

4.1.3. Habit and Habitat: :

Trees (*Polyalhia, Cananga*), shrubs (*Annona*) or lianes (*Unona, Unaria* and *Oxymitra*), *Artabotrys* is a hook climber. Plants mostly terrestrial and perennial, ever green cultivated as well as wild. Oil ducts are present in the bark, leaves and perianth.

4.1.4. Vegetative Characters

Root System : Tap root system. Extensively branched, surface feeders or deep feeders are present.

Stem : Erect or climbing, branched, monopodial (*Polyalthia*) or sympodial (*Artabotrys*) branching due to modification of terminal bud into inflorescence.

Leaf : Simple, exstipulate, alternate and arranged in two longitudinal series (distichous). Leaf lanceolate (*Polyalthia*), ovate (*Annona*) or obovate, margin entire or wavy, apex acute, obtuse or acuminate. Mucronate in *A. mucronata.* Leaf lamina coriaceous, glabrous or glaucous, venation reticulate, unicostate, leaves aromatic and gland dotted.

4.1.5. Floral Characters

Inflorescence : Solitary, some times cluster of flowers arise in the axils of leaves or on old stem (Cauliflory) in *Polyalthia fragrans.*

Flower : Bracteate or ebracteate, pedicellate, pedicel hooked (*Artabotrys*) and helps in climbing, ebracteolate, actinomorphic or zygomorphic due to presence of irregular corolla (*Monodora*) bisexual, rarely unisexual (*Stelechocarpus*), polygamous (*Popowia*) hypogynous, trimerous, thalamus swollen, cone shaped (*Annona*), convex (*Artabotrys*) hemispherical (*Alphonsea*) floral parts arranged in a cyclic manner (*Perianth lobes*) and spiral manner (androecium and gynoecium) on the thalamus. Flowers variously coloured, usually large and conspicuous (fig 4.1.1 & 4.1.2).

Perianth : Perianth distinguishable into calyx and corolla, indistinguishable in *Monodora*, absent in *Eupomatia.*

Sepals six, some times three or rarely two, polysepalous, slightly petaloid, valvate or imbricate aestivation.

Corolla lobes six, arranged in two whorls (3+3), free, valvate, fleshy (fig. 4.1.1C) and aromatic.

Androecium : Stamens indefinite, six in *Orophea zeylanica*, spirally arranged on thalamus. Filaments short and thick. Connectives broad and thick, produced above anther lobes to form a hood like structure at the top. Hood is truncate (*Annona*), (fig 4.1.1D) Coneshaped (*Artabotrys*), leaf shaped (*Uvaria*) anthers dithecous, adnate, extrose. In *Polyalthia* and *Fissistigma*, pollen chambers are completely sunken, pollen uniaperturate, monocolpate with one germpore.

Gynoecium : Polycarpellary apocarpous, monocarpellary (*Kingstonia*), ovary superior, unilocular, some times carpels slightly fused at their lower ends (*Monodora*). This condition is described as paracarpus. Carpels inserted spirally on thalamus. Placentation usually parietal or basal each locule with many anatropous ovules. Styles separate, stigma simple or tri-lobed or tetralobed (*Meranthus*).

Fruit : An etaerio of berries. In *Annona squamosa* berries unite into a receptacle forming a single fruit (fig 4.1.1H).

Seed : Seeds large with small embryo, endosperm ruminate (fig 4.1.1L).

Pollination : Entomophilous, insects are attracted due to the presence of scented glands and pollinate flowers.

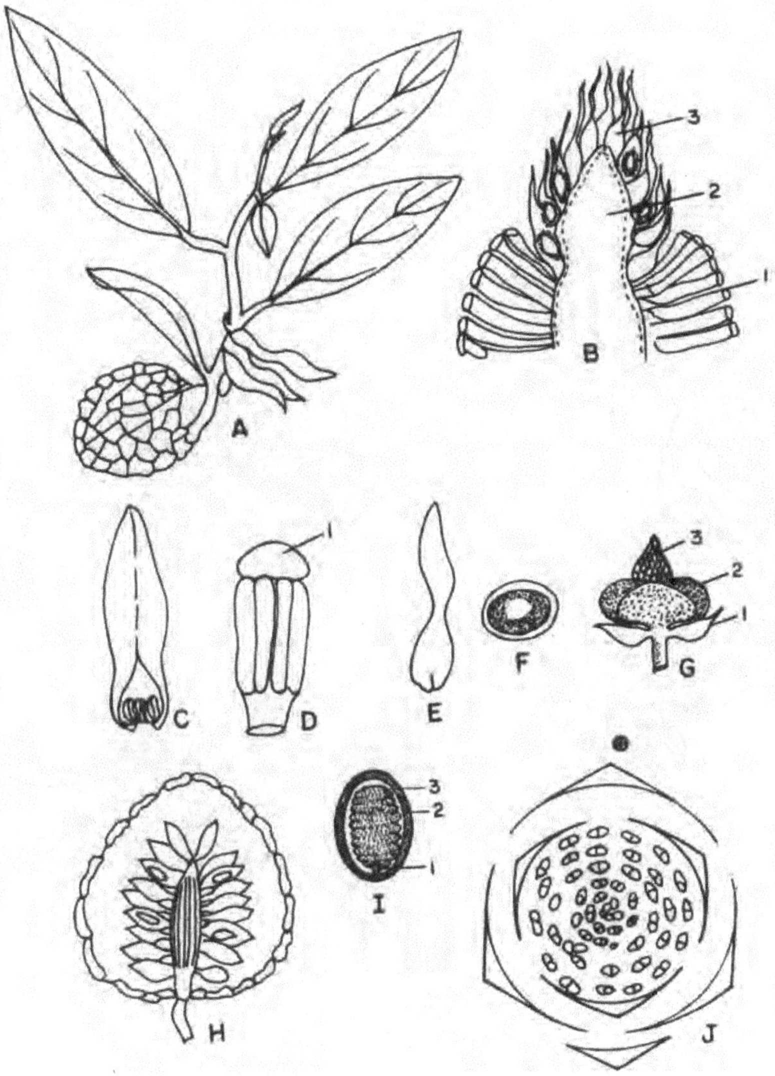

Fig 4.1.1 *Annona squamosa* (Custard Apple)
A. Flowering twig; B. L.S. Flower 1. Stames, 2. Thalamus, 3. Carpels; C. Fleshy Petal; D. Stamen 1 hood; E. Gynoecium; F. T.S. Ovary; G. Opened flower 1. Sepal 2. Androecium, 3. Carpels; H. Agregale fruit L.S. I. L.S. Seed with ruminate Endosperm ; J. Floral diagram

4.1.6 Distinguishing Features

1) Mostly trees or lianes climbing by means of hooks
2) Exstipulate, leaves simple
3) Distichous arrangement of leaves
4) Presence of oil passages in stems, leaves and flowers
5) Presence of long, swollen cone shaped thalamus
6) Trimerous and cyclic perianth
7) Indefinite and free stamens and carpels, spirally arranged on thalamus
8) Stamens with short filament with hooded connective
9) Monoaperturate and monocolpate pollen
10) Multiple aggregate fruit
11) Ruminate endosperm

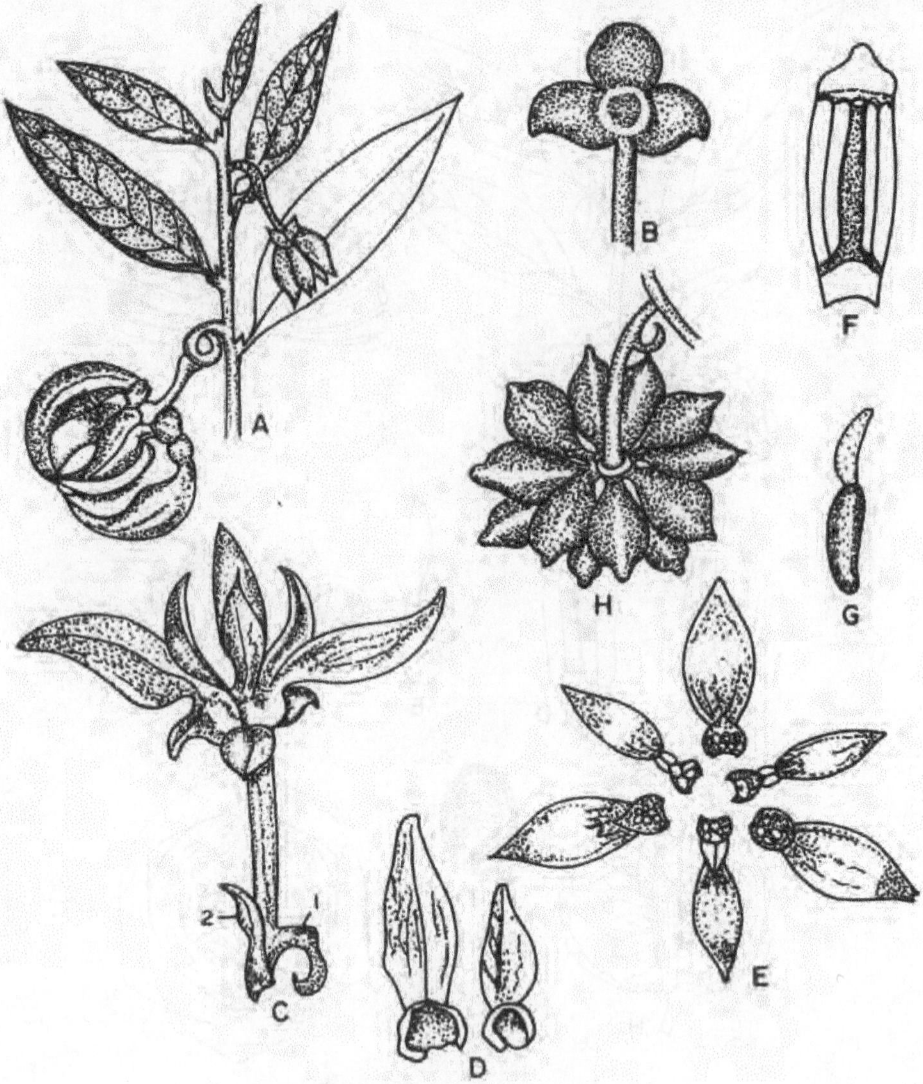

Fig 4.1.2 *Artabotrys odoratissimus*
A. Flowering twig; B. Sepals; C. Flower 1. Hook, 2. Bract; D. Fleshy Petals ; E. Petal arrangement; F. Stamen; G. Carpel; H. Fruits

4.1.7. Economic Importance

4.1.7.1 Edible fruits : The large fleshy delicious fruits of *Annona squamosa, A. reticulata, A. muricata* and *A. cherimalia are edible.*

4.1.7.2 Ornamental Plants : *Polyalthia longifolia, P. pendula* and *Desmos chinensis* are grown as ornamental plants.

Artabotryis odoratissimus, Cananga odorata and *Unona discolor* are also grown in gardens for their sweet scented flowers.

4.1.7.3 Perfumes: Ylang-Ylang or macassor oil is produced from the flowers of Cananga and scented oil is obtained from the flowers of *Artabotrys.*

4.1.7.4 Other Miscellaneous uses : Leaves of *Milusa velutina* are used as fodder. The wood of *Xylopia* used in the manufacture of light furniture, whereas the wood of *Polyalthia* is used in the preparation of boxes and pencils.

4.1.7. Primitive Characters and Affinities

1) Majority of genera are trees and shrubs, a few are lianas. Herbs absent.
2) Flowers are larger in size and sweet scented.
3) Flowers solitary or in cauliflorous clusters.
4) Flowers hypogynous, bisexual and spirocyclic.
5) Perianth lobes usually six in each, arranged in two whorls.
6) Stamens numerous, polyandrous and arranged spirally on the cone shaped or convex thalamus, stamens characteristically laminate with elongated connective extended beyond the anthers to form a hood.
7) Pollen monocolpate and uniaperturate.
8) Gynoecium multicarpellary and apocarpous
9) Parietal placentation with many anatropous ovules.
10) Fruits large and aggregate.
11) Seeds large with small embryo.
12) Endosperm ruminate
13) Basic chromosome number is seven.

The family Annonaceae is closely related to Magnoliaceae and considered to have been evolved from Magnoliaceous stock. There are resemblences between two families in number of vegatative and reproductive characters. This view is accepted by number of taxonomists. Takhtajan (1968) felt that Annonaceae is related to Magnoliaceae but more advanced than the latter. Their relationship is more pronounced by the presence of ethreal oil and these are different from other advanced families like Rosaceae and Zingiberaceae. Cronquist (1968) related Magnoliaceae, Coirinteraceae, Degeneriaceae, Himatandraceae with Annonaceae which are characterized by the absence of vessels in xylem, with regard to karyology. Annonaceae is related to Myrsticaceae as their basic chromosome number is also seven.

4.2. MALVACEAE (Cotton or Mallow family)

Systematic position

Dicotyledonae

Polypetalae

Malvales

Malvaceae

4.2.1. Distrbution

Malvaceae or the cotton or mallow family contains about 82 genera and 1,500 species, many of which are distributed in the tropical and sub-tropical areas of the world. In India, the family is represented by 22 genera and about 110 species. They comprise annual and perennial herbs together with shrubs and trees, many with large and beautiful flowers. Example for perennial herbs are *Abutilon, Abelmoschus, Pavonia and Sida malvaviscus,* shrubs *(Gossypium, Hibiscus)* trees *(Thespeia populnea, Hibiscus tileaceous)* other trees include *Adansonia, Bombax, Ceiba, Durea* and *Chorisea.*

4.2.2 Important Plants

1. *Abutilon indicum* (Linn) Sweet, Telugu; tutturubenda
2. *Abelmoschus esculentus* (L) Moench; (Okra), Tel :benda
3. *Adansonia digitata* Linn (Monkey bread tree)
4. *Althea rosea* Cav. (Holly hock)
5. *Bombax ceiba Linn* (Silk cotton) Tel, erra buruga
6. *Ceiba pentandra Gaertn* : Tel: tella buruga
7. *Grossypium arboreum Linn* (Cotton tree), Tel: erra puvula patti
8. *Gossypium herbaceum Linn* (Asiatic cotton) Tel:patti
9. *Malva sylvesris Linn* (Mallow)
10. *Malveviscus arboreus*, Cav Tel: mirapa mandara
11. *Hibiscus cannabinus Linn*. (Deccan temp) Tel: tella gongura
12. *Hibiscuss sabdariffa Linn* (hemp) (jamaica sorell) Tel: erra gongura,
13. *Hibiscus rosasinensis Linn*. (China rose or Shoe flower), Tel; mandara
14. *Hibiscus schizopetalous* (Mast) HookTel: Juka Mandara
15. *Hibiscus micrarthus Linn*. Tel; Nityamalle
16. *Pavonia zeylanica* (L) Cav.
17. *Sida cordifolia* (L).
18. *Thespesia populnea* (L) Soland. Ex Corr (Portia tree) Indian tulip tree

4.2.3. Habit and Habitat

Many genera are cultivated and a large number of them also grow as wild plants in the plains and the hills. Mucilage occurs in all parts of the plant body and vegetative plants are covered by stellate or tuffed hairs.

4.2.4 Vegetative Characters

Root system: Tap roots are profusely branched and with the surface feeders or deep feeders.

Stem: Aerial, erect, scandent or semiprostrate in some species of *Malva*. herbaceous or herbaceous below and woody above stem cyllindrical or angular or ribbed, mostly hairy and glabrous. In *Abutilion* the branches are covered wirh the minute glandular hairs.

Leaf : Leaves simple rarely compound, petiolate or subsessile, stipulate, free lateral in *Hibiscus*. In *Abutilon sp*, the stipules are often coloured, leaves some times palmately lobed (*Gossypium*), alternate phyllotaxy, shape of lamina cordate, orbiculor, rhomboidal ovate or obovate, margin dentate or crenate, apex obtuse, acute or acuminate, surface glabrous, venation reticulate multicostate divergent (*Gossypium*) or unicostate. Leaves of *Abutilon* megapotamium are yellow and variegated.

4.2.5 Floral Characters

Inflorescence Solitary axillary (*Hibiscus, Thespesia*) or Cymose (*Althea, Pavonia*) cauliflory is seen in *Goethea*.

Flower: Bracteate or ebracteate, bracteolate or ebracteolate, regular, complete, bisexual, rarely unisexual (*Napea*) male and female flowers in *Kydia*, hypogynous, pentamerous, cyclic, large and attractive in china rose (fig 4.2.1A).

Epicalyx: A Whorl of bracteoles, known as epicalyx is present above the calyx. Epicalyx lobes range in 3 to 12 lobes free (*Hibiscus*), or united (*Althaea*) and green, brightly coloured in *Goethea*. In *Gossypium* and *Malope* however, the bracteoles are present in the form of large leaf like structures (fig 4.2.3E).

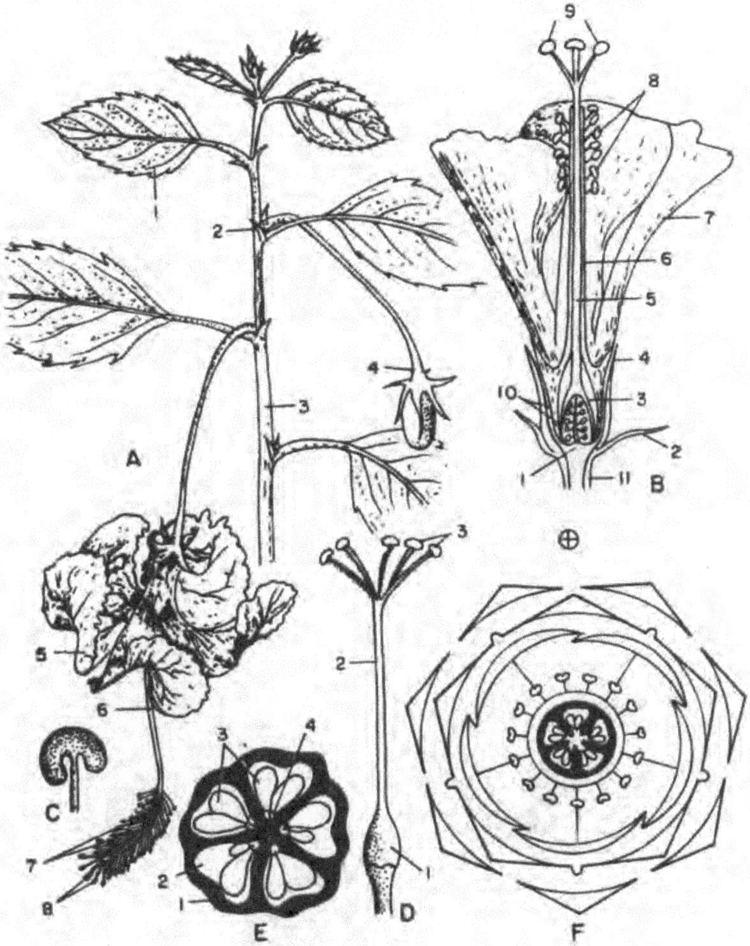

Fig 4.2.1 *Hibiscus rosa - sinensis* (China Rose)
A: A flowering twig 1 Leaf; 2. Stipule; 3. Stem; 4. Flower bud; 5. Petals; 6. Staminal tube; 7. Monoadelphous stamens; 8. Stigmos
B: L.S. Flower 1. Thalamus; 2. Epicalyx; 3. Ovary; 4. Sepals; 5. Style; 6. Staminal tube; 7. Petal; 8. Stamens; 9. Stigmas; 10. Ovules; 11. Pedicel
C. Stamen; D. 1. Ovary, 2. Style, 3. Stigmas;
E. 1. Ovary Wall, 2. Locules, 3. Ovules, 4. Placenta; F. Floral diagram

Calyx : Five, gamosepelous, cup shaped, compound or tubular, some times persistent (*Abelmoschus*) green, dark red in some *Abutilon* spp, valvate. The odd speal is posterior where the epicalyx is absent (*Sida*) or numerous (*Hibiscus*). However, where the epicalyx members are few in number the odd sepal becomes anterior (*Malvastrum*). In the genera belonging to tribe Bombaceae sepals are leathery. Trichiferous nectories are present below the calyx lobes.

Corolla: Five, polypetalous or appears slightly gamapetalous at the base on account of its being fused with staminal tube (fig 4.2.1B). Corolla mucilaginous in china rose and okra, twisted, usually large and attractive colours variable. In *Althaea* **rosea** different colured flowers (white to nearly black) occurs in the same plant. Corolla totally absent in *Cullenia*.

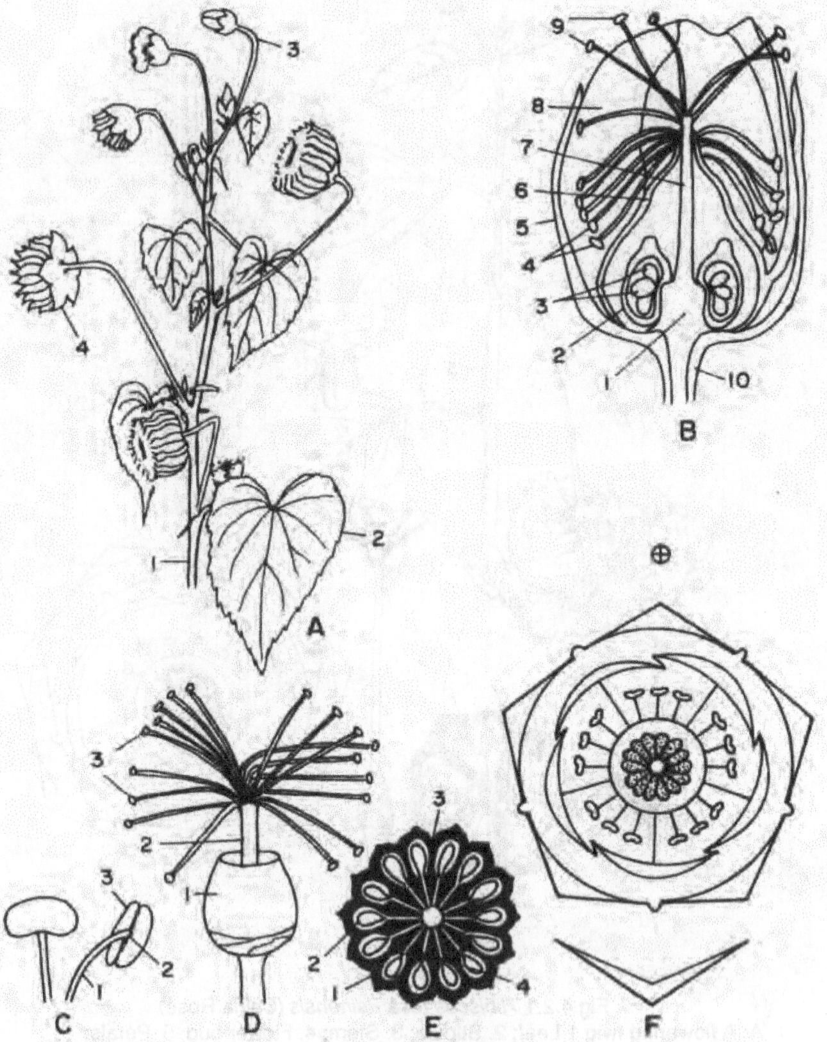

Fig 4.2.2 *Abutilon indicum*
A: A flowering twig 1 Stem, 2. leaf, 3. Flower bud, 4. Flower
B: L.S. Flower 1. Thalamus; 2. Ovary; 3. Ovules, 4. Stamens, 4. Sepal,
6. Staminal tube, 7. Style, 8. Petal, 9. Stigmas. 10. Pedicel
C: Stamens : 1. Filament, 2. Connective, 3. Antherlobes ;
D:Gynoecium: 1. Ovary, 2. Style, 3. Stigmas;
E: T.S. Ovary 1. Ovule, 2. Locule, 3. Ovary wall, 4. Placenta F: Floral diagram

Androecium :- Stamens numerous, the filaments of the numerous stamens are unified to form a staminal tube around the gynoecium. Hence, stamens are monoadelphous, further the staminal tube is fused at the base with petals, hence stamens are epipetalous. In Bombax tree, however, the stamens are polyadelphous. Anther reniform (kidney shaped), variously coloured, monothecous, extrose, transverstly attached to the filaments (fig 4.2.1B&C 4.2.2 B&C).

The tip of the staminal tube is dentate (*Hibiscus*), and these teeth represent the anti-sepalous staminodes.

Pollen grains big spherical with spiny exine, Multiporate (*Althea, Malva, Gossypium*) Pentaporate and pentacolpate (*Hibiscus*) triporate (*Thespesia*) arranged in circular manner (zonocolpate)

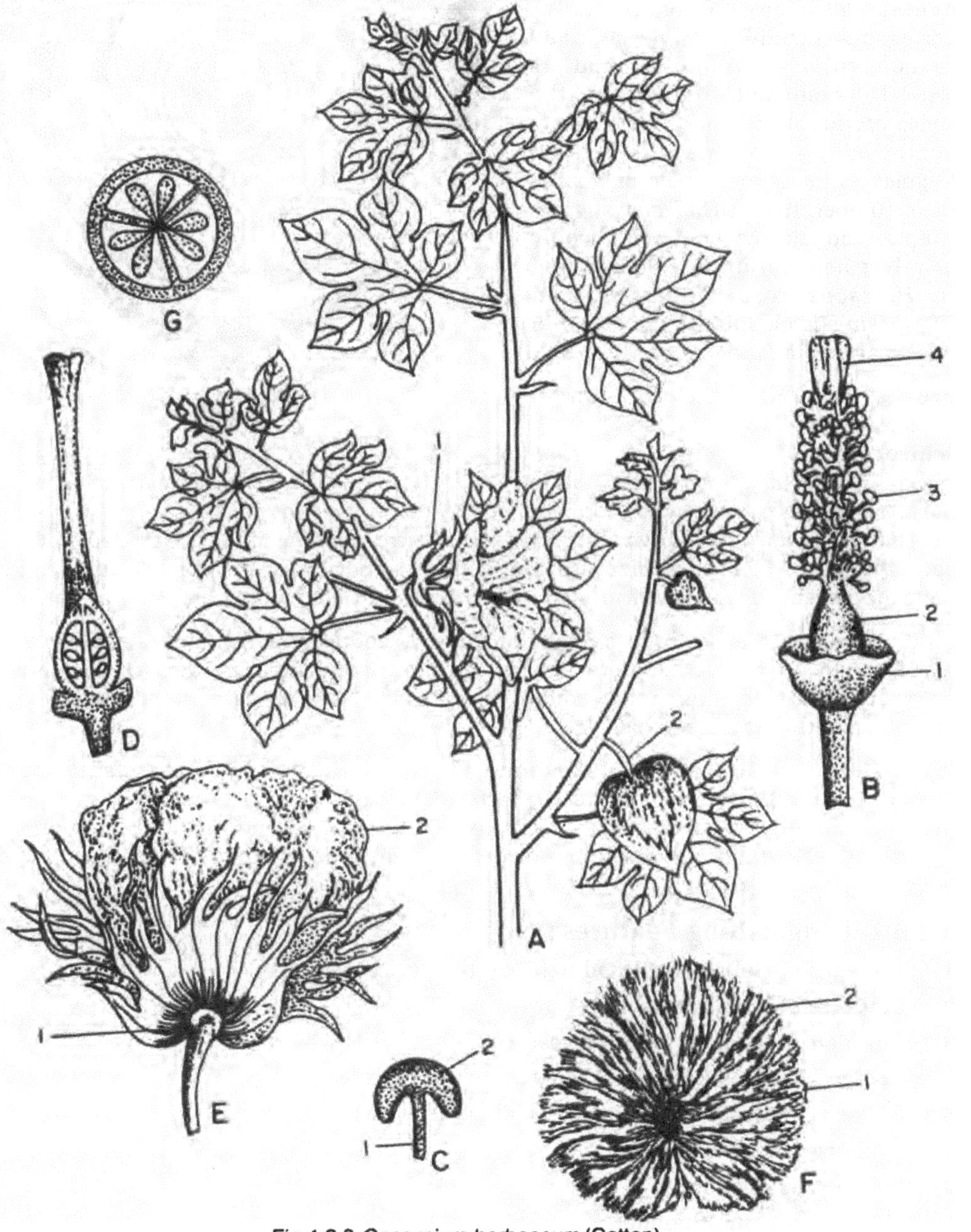

Fig 4.2.3 *Gossypium herbaceum* (Cotton)
A: Flowering twig: 1. Flower, 2, Fruit; B: Flower: 1. Calyx, 2. Ovary, 3. Staminal tube, 4. Stigma; C: Stamen: 1. Filament, 2. Reniform anther; D: Gynoecium; E: Fruit : 1. Epicalyx, 2. Cotton, F: Seed: 1. Testa, 2. Lint fibres G. Floral diagram

Gynoecium:- The number of carpels varies from one to many (*Abutilon*) (fig 2.2) pentacarpellary condition is very common (*Hibscus, Malva*) Syncarpous ovary, superior, bi-to multilocular ovary, placentation axile, one to may ovules in each locule. Ovules anatropous, pendulous, styles as many as the carpels, fused into one and are arpically branched. In *Althea*, the styles are fused only at the base, style encircled by the staminal tube. Stigmas as many as carpels or double their number, five in *Hibiscus*, stigmas simple, capitate, covered with wooly or velvet hair, variously coloured (red in *Hibiscus rosa sinemsis*). In *Thespesia* stigma spirally curved 3 - 5 fused carpels in *Gossypium*. (fig 2.3).

G: Floral diagram

Fruit : Generally it is loculicidal capsule (*Hibiscus, Gossypium*), schizocarpic (Carcerulus) in *Althea rosea*, In *Hibiscus rosasinensis* it is mostly abortive, capsule when formed. In *Abutilon*, *Sida* and *Malva*, the fruit schizocarp breaking into 1-3 seeded parts called cocci (fig 2.2). Hooked schizocarp found in *Pavonia*, Berry in *Malvaviscus*. Epidermal hairs are produced from the fruit cap in *Bombax* and *Ceiba* which are used as cotton.

Seed : Endospermic, embryo straight or curved, endosperm contains abundant proteins and fats. In *Gossypium*, epidermal hairs are produced which are densely tomentose, the hairs on the seeds furnish the cotton fibres, Each fibre or hair is the extension of an epidermal cell of the seed coat.

Pollination: The flowers are protandrous and are pollinated by insects (entomophilous) which visit the flowers for the nectar secreted by the petals and stored in basal pits and pollen. The flowers have special mechanism to prevent self-pollination. Humming birds pollinate the *Abutilon* and *Gossysium* flowers. In genera belonging to Bombaceae, pollination is done by bats.

4.2.6 Distinguishing Features

1. Presence of stellate or tuffted hairs on the vegatative parts.
2. Presence of mucilage cavities.
3. Stipulate, palmately lobed leaves.
4. Epicalyx is present in most genera.
5. Stamens monoadelphous or polyadelphous.
6. Anthers monothecous, reniform.
7. Stamens epipetalous.
8. Pollen spiny, multiporate.
9. Axtile placentation
10. Fruit capsule,schizocarpic or carcerules.

4.2.7 Economic Importance

The family is of great economic importance. It includes many important fibre yielding, medicinal, vegetable and ornamental species.

4.2.7.1 Cotton: Many species of *Gossypium* are grown for the hairy epidermal outgrowths of the seeds which constitute the cotton wool or cotton of commerce. The important species are *G. barbadense* (sea island cotton) *G. hirsutum, G.* arborium (Indian cotton) and *G. herbaceum.* Cotton is used in the manufacture of textiles, cellulose and other products. Cotton treated with a strong alkali is strong and more or less silky and is known as mercirized cotton. Gun cotton is an explosive is made by treating the cotton lint with nitric and sulphuric acids. Cotton seeds are used to extract oil which is hydrogenated to produce vanaspathi ghee, oil cake is used as fodder and manure. *G. indicum, G. neglectum, G. sanguineum* are other common species of desi cotton that are grown in India.

4.2.7.2 Fibres: *Hibiscus cannabinus* (Deccan hemp) is another important species, is grown mainly as a fibre crop. The cortical fibre is used for cordage, ropes and floor covering. It is also employed in the manufacture of paper. The seed yield a fatty oil which is used a lubricant and in the manufacture of linoleum, paints and varnishes. Fibres inferior to that of Deccan hemp also obtained from several species of *Sida, Abutilon, Urnea, Thespesia* and *Kydia*, wooly cotton is obtained from the silk cotton tree *Bombax malabarcium.*

4.2.7.3. Vegetables: *Abelmoschus esculentus (Hibiscus esculentus,* Lady's finger, or Okra) used as vegetable and in soups. Leaves of *H.cannabinus* (Deccan hemp or Bimili) and *H. sabdarifa* are used as vegetable (gongura) particularly in Andhra Pradesh. Petals of latter being acidic are used for making pickles and jellies.

4.2.7.4 Ornamental plants: Several genera of Malvaceae are grown as ornamental plants for their large showy flowers. These include *Hibiscus rosasinensis, Malva sp, Malvaviscus, Althaea, H. schizopetalous, Adansonia sp, Kydia, Thespesia, Bombax* and *Ceiba.*

4.2.7.5. Medicinal Plants: The roots of *Sida acuta* are astraingent, cooling tonic and are useful in nervous and urinary diseases and also in the disorders of the blood and bile. In Ayurveda, *Sida* species are known as 'bala' (strength), preparations from *S. cordifolia* (Atibala), *S. spinosa* (Nagabala), *S. cordata* (Bala) are found to increase vigour, energy to the body.

Roots and bark of *Urena lobata* possess anti-hydrophobic properties. Petals of china rose are used for cough, fever and also as hair tonic, flowers of *Althaea rosea*, are used in colouring wines.

4.2.7.6. Timber yielding plants: *Kydia calycina, Bombax* ceiba and *Thespesia populnea* provide timber, used for making agricultural implements, toys, pencils and also in match industry.

4.2.8. Affinities of the Family

The family is closely related to Sterculiaceae, Tiliaceae and Bombacaceae. The affinities are 1) Presence of stellate hairs 2) Presence of mucilag cells. 3) The monothecous anthers and monadelphous stamens are the special characters of Malvaceae. Further, the pollen of most members are large and spinous. The presence of epicalyx and contorted corolla are also other diagnostic characters of the family and considered advanced over the other families. Malvaceae subdivided into two groups and three tribes.

1. Group - A a) Tribe - Malopeae
2. Group - B b) Malveae c) Ureneae

4.3 RUTACEAE *(The Orange family)*

Systematic Position :

Dicotyledonae

Polypetalae

Disciflorae

Geraniales

Rutaceae

4.3.1 Distribution

Rutaceae commonly known as the orange family comprising about 150 genera and 1,300 species. Mostly distributed in tropical and sub-tropical regions. They are abundant in Australia and South Africa. In India the family is represented by 23 genera and over 80 species distributed mainly in sub-tropical Himalayas and Western peninsular India (eg. *Citrus, Aegle, Atlantia, Feronia and Toddalia* etc,).

4.3.2 Important Plants

1. *Aegle marmelos correa. (Bacl)* Tel: maredu or Bilva
2. *Atlantia monophylla L.* Tel: Karunimma
3. *Citrus aurantium L* (Bitter Orange) Tel: Narinja
4. *C. aurantifolia Swingle.* (Lime) Tel : Nimma
5. *Citrus limon Burn.f* (Lemon) Tel : Dabba or gaza nimma
6. *Citrus medica L.* Tel : Madiphalam
7. *C. sinensis Osbeck* (Sweet orange) Tel: Battai
8. *C. reticulata Blanco* (Loose skinned orange, Santra) Tel: Kamalaphalam
9. *C. decumana L* (Pomelo) Tel: Pampara panasa
10. *Feronia elephantum Correa:* (Wood apple) Tel: Velaga
11. *Murraya koenigii Spring* (Curry leaves) Tel: karivepaku
12. *M. paniculata (L) Jacq.* or (Orange Jasmine) Tel: poovelaga
13. *Naringi crenulata L.* Tel: Torra Velaga
14. *Ruta graveolens L* (Common Rue) Tel: sadapaku
15. *Xanthoxylum piperata L.* (Japanese pepper)

4.3.3 Habit and Habitat The members are mostly shrubs or small trees, Herbs are rare (*Boenning hausevia*) Mostly mesophytic, *Diosma* is a xerophyte.

4.3.4 Vegetative Characters

Root system: Tap root system, branched, often infected with mycorrhizae.

Stem: Erect, woody, branched, armed with spines (*Aegle, Citrus and Xanthoxylum*) or unarmed *(Evodia, Murraya)* cylindrical or angular or ribbed. Young shoots gland dotted, aromatic, pubscent or tomentose, climbing stragler in *Toddalia*.

Leaves: Leaves simple or compound, pinnately compound in *Murraya*. In *Citrus* sp. leaves are unifoliate with a joint at the junction of leaf blade and winged petiole (fig 4.3.1) exstipulate, Leaves opposite, or alternate, gland dotted, In many genera the first leaf of axillary bud modifies into spine, venation reticulate unicostate.

4.3.5 Floral Characters

Inflorescence: Usually axillary (*Citrus, Murraya*) or in terminal cymes (*Clausena, Skimmia*) or panicles (*Murraya exotica*). In some species of *Citrus* and *Correas* flowers are axillary solitary or fasicled, *Feronia* shows cauliflory.

Flower : Bracteate, bracteolate, pedicellate or subsessile, complete, actinomorphic or zygomorphic (*Dictamnus*) bisexual or unisexual (*Xanthoxylum*), polygamous (*Feronia*) hypogynous, pentamerous. *Toddalia* shows bi, tri or tetra-merous condition. A fleshy

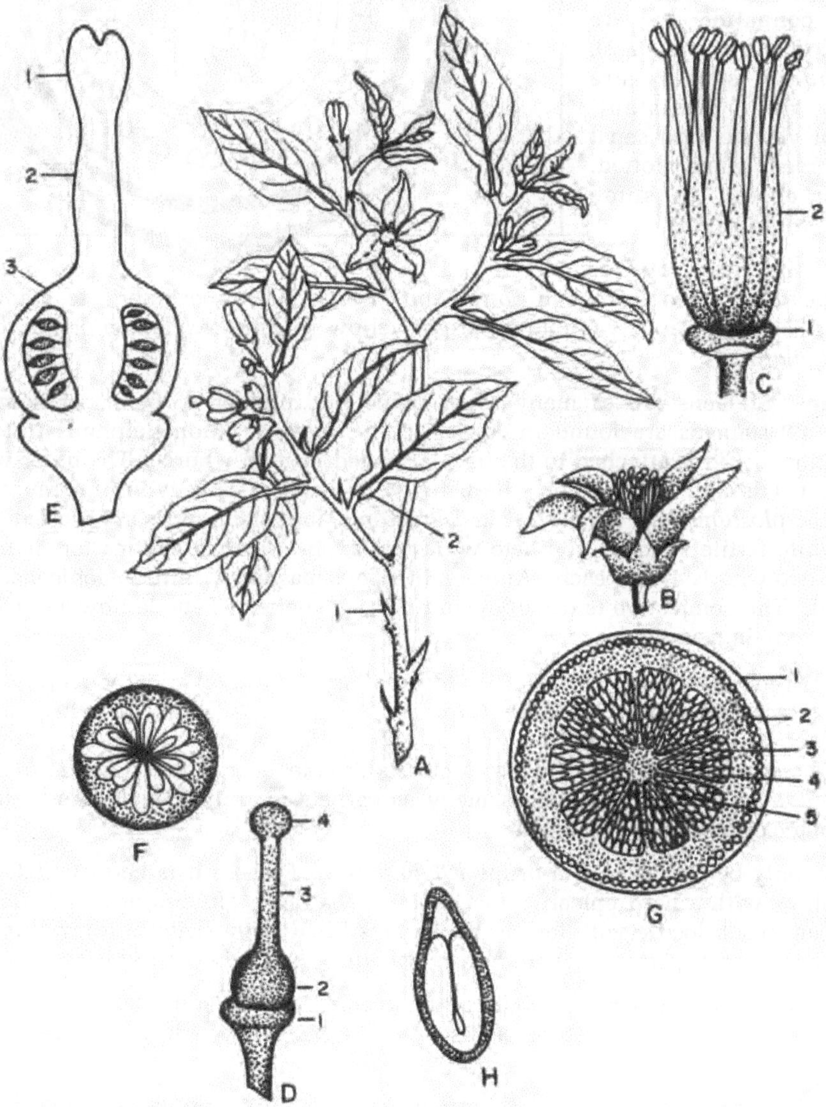

Fig 4.3.1 *Citrus limon*

A: Floweirng twig 1. Spine; 2. Unifoliate leaf; B. Flower; C. Androecium: 1. Disc, 2. Stamen, D: Gynoecium: 1. Disc, 2. Ovary, 3. Style, 4. Stigma; E: L.S. of Gynoecium 1. Stigma, 2. Style, 3. Ovary, F: T.S. Ovary; G. T.S. Fruit: 1. Pericarp, 2. Mesocarp, 3. Endocarp, 4. Juicy Hairs, H. L.S. Seed, I. Floral diagram

distinct nectariferous disc (Crinate or lobed) is present below the ovary. In *Dictamnus* the thalamus modifies into a androphore. Sepals and petals are also gland dotted and aromatic or fragrant or even some times pungent. (fig 4.3.1)

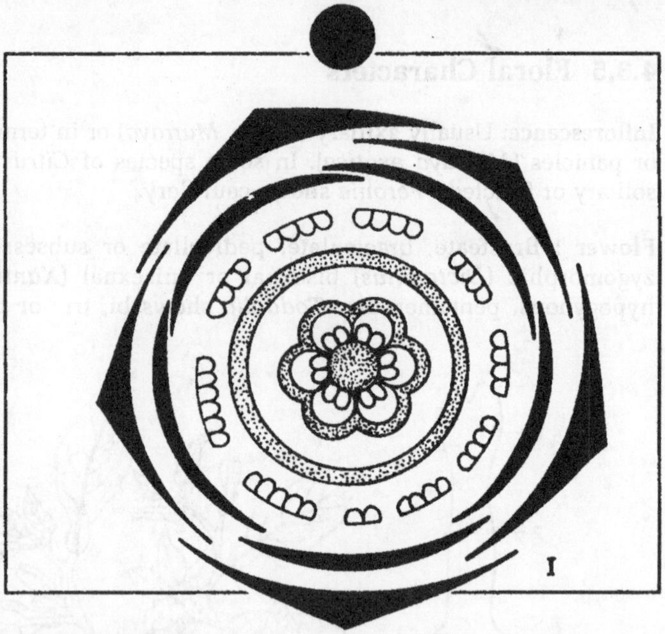

I

Calyx: Sepals four or five, free or more often show various degrees of connation and are cup shaped in *Citrus* (fig 4.3.2). In *Peganum*, sepals are foliaceous and persistant, imbricate or valvate, odd sepal posterior, green, gland dotted. sepals are absent in some species of *Xanthoxylum*.

Corolla: Petals five, four or three, polypetalous, gamopetalous in *Correa and Ticorea*. Valvate or imbricate, gland dotted and sweet smelling, variously coloured, bright yellow in *Xanthoxylum piperata*, however absent in *X. alatum*.

Androccium: Stemens are as many as or twice the number of petals or some times numerous. 60 stamens are found in *Aegle*, 10 in *Murraya*. Monadelphous in *Atlantia*, Polyadelphous in *Citrus* attached to the disc, arranged in two whorls. The outer whorl lie opposite to petals and inner whorl alternates with petals. This type of arrangement is known as *obdiplostamenous* (fig 4.3.2). In *Tetractonia*, four staminodes are present opposite to petals while fertile stamens alternate with petals. In *Murraya* exotica ten stamens are arranged in two whorls of five each. Anthers dithecous, basifixed, introse, dehicence usually longitudinal. The connective is usually glandular at the apex. Female flowers of *Toddilia* and *Evodia* contain staminodes.

Gynoecium: Monocarpellary (*Emplurum*) bicarpellary (*Murraya*) tricarpellary (*Triphasia*) Pentacarpellary (*Ruta*) or polycarpellary (*Citrus*) and syncarpous (fig 4.3.2D). In some genera ovaries are free whereas styles are fused at top. Ovary bilocular to multilocular, or unilocular (*Feronia*) Placentation usually axile, rarely parietal, ovules anatropous. In *Xanthoxylum* a small gynophore is present. Styles as many as carpels or united, style short, thick and fleshy, stigma capitate, entire or lobed.

Fruit: Commonly berry (*Ruta*) or a drupe (*Skimmia*). In *Feronia* it is berry which is large, globose, yellow with a hard epicarp. In *Citrus* fruit is large, globose or ovoid, fleshy and many loculed. Each locule with few seeds, locules filled with fusiform juicy hairs arising from the inner layer of pericarp. This fruit is known as *hesperidium* (fig 4.3.1G).

Seed: Endosperm with a large straight or bent embryo. Polyembryony (with about 13 embroys) is a very common feature of Rutaceae. Folded cotyledons are present in the embryo.

Pollination: Flowers are protandrous and it prevents self pollination. Pollination is brought about by insects (entomophily). The strong or sweet smelled nectar secreting disc below the ovary attracts insects. In *Ruta* if cross pollination fails, the protandrous stamens stay upright and shed the left-out pollen on the stigma of same flower and brings about self-pollination. This is known as *safety mechanism*.

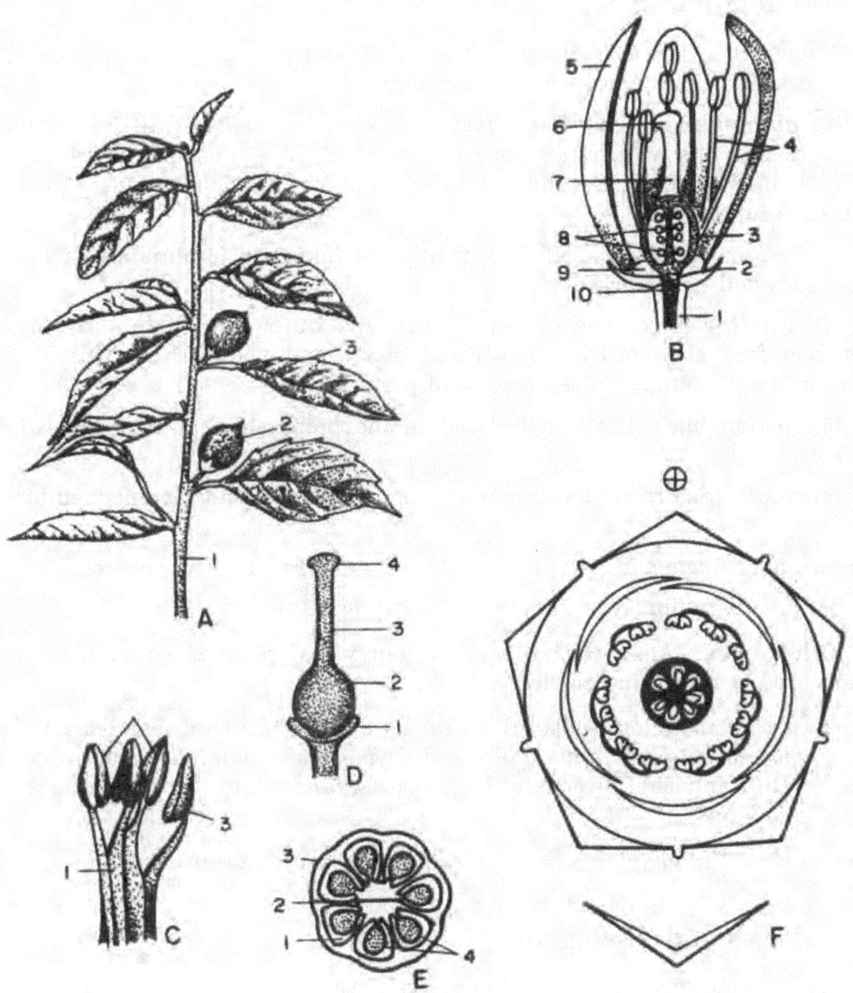

Fig 4.3.2 *Citrus medica*

A: Floweirng twig 1. Stem, 2. Fruit, 3. Leaf; B. L.S. Flwoer 1. Pedicel, 2. Sepal, 3. Ovary, 4. Stamens, 5. Petal, 6. Stigma, 7. Style, 8. Ovules, 9. Disc, 10. Thalamus
C: Stamens: 1. Filaments, 2. Anthers, 3. Connectives; D: Gynoecium 1. Disc, 2. Ovary, 3. Style, 4. Stigma
E: T.S. Ovary 1. Locule, 2. Placenta, 3. Ovary wall, 4. Ovule ; F: Floral diagram

4.3.6 Distinguishing Features

1. Plants predominently trees or shrubs.
2. Leaves are existipulate, with winged petiole and gland dotted.
3. The sepals and petals are also gland dotted.
4. Flowers bisexual, regular with a large honey disc present below the ovary.
5. Androccium obdiplostamenous, polyandrous polyadelphous or monadelphous.
6. Gynoecium penta or multicarpellary, multilocular with axile placentation.
7. Fruit, a characteristic hesperidium.
8. Polyembryony is a common feature.

4.3.7 Economic Importance

4.3.7.1 Edible fruits: The most important genus of the family *Citrus* provides various edible fruits like oranges, limes and lemons. These are,

a) *Citrus aurantifolia* (lime): is cultivated all over India. The fruit juice contains citric acid, volatile oils with citral, limonene, linalool, linalyl acetate, turpinol and cymen. The oils distilled from peel of the fruit is mainly used in confectionary, pharmaceutical and soap industry.

b) *C. aurantium* (sour orange): The fruit is of immense importance in confectionery, cosmetics and perfumery.

c) *C. sinensis* (Sweet orange or mausambi): The fruits are edible. The essential oils obtained from flowers (Neroli oil) and leaves and young shoots (petil grain oil) is used in the preparation of essences and perfumes.

d) *C. limon* (lemon); It is widely used in the preparation of lemonate squashes and pickles.

e) *C. reticulata* (Santra or loose skinned orange) is the most valued edible orange of India.

2. The fruits of *Aegle marmelos* are edible and are also useful in controlling dysentry.

3. *Feronia elephantum* (Wood apple) fruit pulp is edible.

4.3.7.2. Other Uses : *Murraya koenigii,* the curry leaf plant is cultivated for its leaves which are used in flavouring curries.

Murraya paniculata (orange jasmine), Ruta graveolens, Limonis acidisma, Calodendrum, *Skimmia arborescens, Dictamnus albus* and *Glycosmis pentaphylla* are grown in gardens for their beautiful foliage, flowers and fragrance. *Xanthoxylum, Toddalia and Atlantia* have medical applications.

The wood of *Chloroxylon swietenia* is known as Indian satin wood and used in making furniture.

4.3.8 Affinities and Phylogeny

Rutaceae is closely related to Malvaceae, Anacardiaceae and Sapindaceae due to following features.

a) Obdiplostamenous conditionsof androecium
b) Presence of a disc below the ovary
c) Axile placentation and 1-many ovules for each locule
d) Morphology of leaf
e) Shrub or tree like habit.

Engler, However, believed that Rutaceae has affinities with Zygophyllaceae, Meliaceae, Burseraceae and Simarulaceae. All these families resemble each other in the presence of disc below the ovary.

Hallier (1905) opines that Rutaceae may have originated from some Ranalian ancestors like Lardizabaceae with families like Capparidaceae as an intermediate. His observations were based on the presence of gynophore in *Xanthoxylum* of Rutaceae.

Some botanists relate Rutaceae to Euphorbiaceae due to the presence of ventral raphae in ovule in some genera.

4.4. FABACEAE (The Pea family)

Systematic position

Dicotyledonae
Polypetalae
Calyciflora
Rosales
Fabaceae

4.4.1 Distribution Fabaceae includes about 490 genera and 1200 species which are mostly distributed in the tropical and sub-tropical regions of the world. In India, Fabaceae is represented by over 100 genera and 750 species, with variety of habitats like xerophytes (*Alhagi*), mesophytes and hydorphytic (*Aechynomene*). The members of the family are distributed in plains as well as hilly areas. A large number of them grow as halophytes (*Derris scandens, Desmodium latifolius*).

4.4.2 Important Plants

1. *Aeschynomene aspera* L (Pith plant) Tel: Jeluga.
2. *Abrus precatorius* L (Rati) Tel: Gurivinda
3. *Arachis hypogea* L. (Ground nut) Tel: Verusanaga
4. *Butea frondosa* (Syn: *B. monosperma*) (Lamk) Tamb. (Flame of the forest) Tel: Moduga
5. *Cicer arietinum* L. (Chickpea or bengal gram) Tel: Sanaga
6. *Cajanus cajan* (L) Millsp (Red gram or pigeon pea) Tel,kandi
7. *Clitoria ternatea* L. (butterfly pea) Tel: Sankupushapam
8. *Canavalia ensiformis* L (Sword or Jack bean) Tel: Chemmakaya
9. *Crotolaria juncea* L. (Sun hemp) Tel: Janumu
10. *Cyampsis tetrogonoloba* L. Tanbert (Cluster bean) Tel: Goruchikkudu
11. *Dalbergia latifolia Roxb.* (Indian rose wood) Tel: Iruguducheva
12. *D. sissoo* Roxb. (Indian red wood)
13. *Dolichos biflorus* L (Horse gram) Tel: Ulavalu
14. *D. lablab* L (bean) Tel: Chikkudu
15. *Erythrina indica* (Coral tree) Tel: baditha
16. *Glycine max* (L) Merr. (Soybean) Tel: Soya chikkudu
17. *Lathyrus odoratus* L (Sweet pea) Tel: Teepi batani
18. *Lens esculenta* Moench. (Lentil) Tel: Mysore pappu
19. *Indigofera tinctoria* L (Indigo) Tel; neelimandu
20. *Pisum sativum* L (Garden pea) Tel: Batani
21. *Phaseolu trilobus* Wall. Tel: Pillipesara
22. *P. vulgaris* L (French bean)
23. *Pongamia pinnata* (L) Pierre (Pongam oil tree) Tel: Kanuga
24. *Psoroalea corylofolia* L Tel: Bavanchalu
25. *Psophocarpus tetragonobobus* DC (Winged or Goa bean)
26. *Ptorocarpus sanatalinus* (L) f. (Red sanders) Tel: Raktha chandanam
27. *P. marsupium* Roxb. (Indian Kino) Tel: Aegisa
28. *Sesbania grandiflora* (L) Merr. (Sesban) Tel: Ausa or Agastya

29. *Vigna mungo* (L) Hepper (black gram) Tel: Minumulu
30. *V radiata* (L) wilczek. (green gram) Tel: Pesalu.

4.4.3 Habit and Habitat The family is characterized by variety of habits, these include, herbs (*Medicago, Melilotus*), shrubs (*Tephrosia*), woody climbers (*Mucuna*), herbaceous climbers (*Lathyrus, Pisum, Viccia*), twinners (*Dolichos, Vigna*), and Trees (*Dalbergia, Erythrina,* and *Pongamia*). The members of the family grow in different habitats like xeric (*Althagi*), hydrophytic (*Aeschynomene*) and halophytic (*Desmoidium*) besides number of species grow in mesophytic environment.

4.4.4 Vegetative Characters

Root system: A much branched tap root that bears nodules containg nitrogen fixing bacateria (symbiotic) like *Bacillus, radicicola* and *Rhizobium* sp. These bacteria convert the atmospheric nitrogen into nitrogenous compounds, which will be absorbed by the host plant. Tubular roots are seen in *Dolichos falcatus.*

Stem : Aerial and erect, Twinning and climbing stems also very common. Climbing being achieved by leaf tendrils (*Lathyrus*) or leaf-let tendrils (*Pisum*) twinner (*Dolichos, Clitoria, Canavalia* and *Atylosia*). Stem branched, angular or cyllindrical, glabrous, tomentose and hairy (*Mucuna*). In *Mucuna* and *Rhynchosia* anomolous secondary growth of stem is observed.

Leaf: Stipulate, leaf base pulvinate in many genera, alternate. Leaves simple (Heylandia) trifoliate, compound (*Medicago, Trigonella, Cajanus, Phaseolus, Vigna*) bifoliate (*Zornia diphylla*). In *Astragalus*, the rachis of the compound leaf ends in a spine, modified into tendril in *Lathyrus aphaca.* Stipules become large and leafy to perform photosynthesis (*Lathyrus* and *Pisum*) (fig 4.4.2). The leaf-lets become tendrillar and help in climbing (*Pisum* and *Lathyrus*). In *Desmodium gyrans* the two leaf-lets show autonomus movements. Leaf let modify into spines (*Ulex*).

4.4.5 Floral Characters

Inflorescence: Usually axillary or terminal (*Crotolaria*) a raceme or spike or panicle (*Dalbergia*): rarely condensed into a head (*Trifolium*) or solitary (*Heylandia*).

Flower: Flowers pedicillate or subsesile, bracteate, bracteolate, complete, pentamerous, bisexual, partially actinomorphic or zygomorphic due to the presence of unequal corolla lobes (papilionaceous): hypogynous or mostly perigynous, Thalamus is flat, slightly concave or mostly cup shaped, Flowers showy in different genera like *Clitoria, Erythrina, Crotoloria and Sophora.*

Calyx: Five and gamospelous, companulate or cup shaped, toothed, calyx lobes some times unequal, tubular (*Astragalus* and *Cyamopsis*) odd sepal anterior, greeen, hairy, persistent in some species. Aestivation ascending imbricate (*Viccia*) and valvate (*Tephrosia*). (fig 4.4.2)

Corolla: Five petals, papilionaceous (butterfly appearance) polypetalous (however, keel petals fused at the base). The posterior petal is large and conspicuous known as standard or vexillum. It encloses two posterio-lateral part of petals (medium sized) called wings or alae, which in turn enclose another two anterio-lateral petals (small sized) which partially fuse at the anterior margins forming a boat shaped structure called the keel or carina. The carina in turn encloses the essential organs (Androecium and gynoecium). In *Phaseolus* spp. keel petals are prolonged into a long beak like structure and spirally twisted. In *Dumasia* the wings and keel are adherent and have long claws. However, wings and keel petals are absent in *Amorpha.* The aestivation is descending imbricate or vexillary and is

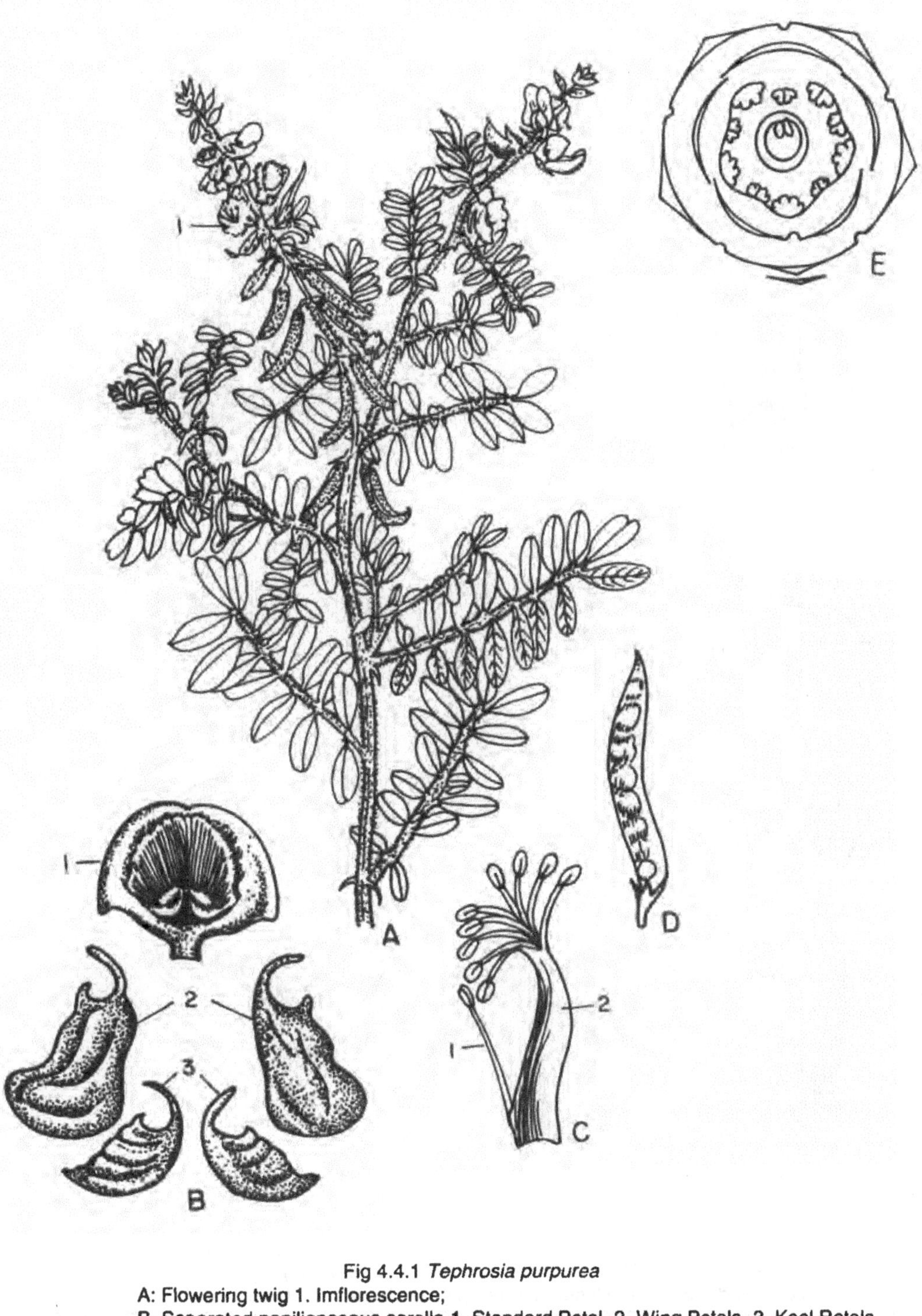

Fig 4.4.1 *Tephrosia purpurea*

A: Flowering twig 1. Imflorescence;

B. Separated papilionceous corolla 1. Standard Petal, 2. Wing Petals, 3. Keel Petals

C: Diadelphous stamens (9+1) 1. Single stamen, 2. 9 petals in one bundle

D: Legume or pod ; E. Floral diagram

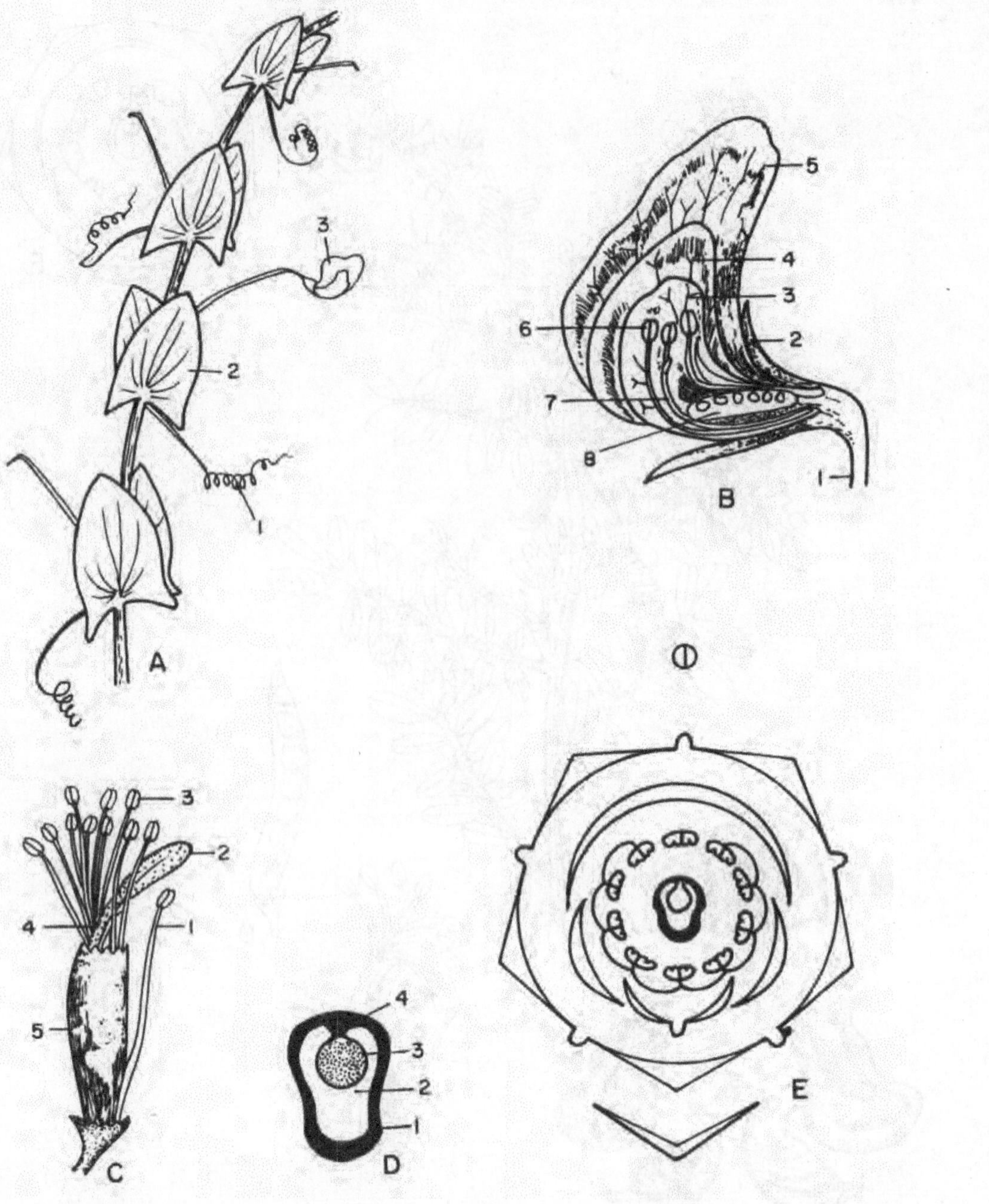

Fig 4.4.2 *Lathyrus odaratus* (Sweet Pea)
A: Flowering twig 1. Modified leaf into tendrils, 2. Foliaceous stipules, 3. Flower;
B. L.S. Flower 1. Pedicel, 2. Sepal, 3. Keel Petal, 4. Wing Petal, 5. Standard Petal, 6. Stamen, 7. Style, 8. Ovary,
C: Essential organs of flower 1. Stamen, 2. Stigma, 4. Stamens, 5. Style, 6. Ovary;
D: Ovary T.S. 1. Ovary wall 2. Locule, 3. Ovule, 4. Placenta; E. Floral diagram

a characteristic feature of the family. In *Tephrosia* petals are green in young flowers and become violet in fully developed flowers (fig 4.4.1B).

Madhusudhana Rao and Bir Bahadur (1981) identified different types of vexillary aestivations based on the twisting (folding either towards right or left directions) of standard, wings and keel petals. Accordingly, following vexillary aestivations types are recognised,

1. *Butea* type (*Butea frondosa*)
2. *Cajanus* type (*Cajanus cajan*)
3. *Phaseolus* type (*Phaseolus vulgaris*)
4. *Glyricidia* type (*Glyricidia, Pongamia*)

Androecium: Mostly ten in number, inserted on a disc below calyx. Nine in *Abrus* and *Dalbergia*. Arranged in two bundles (5+5) in *Smithia* and *Aeschynomene*. However mostly (9)+1, where the posterior stamen is free and filaments of the rest nine are fused to form a tube like structure (staminal tube) surrounding the ovary. The free stamen lies opposite to standard petal (eg. *Tephrosia*). (fig 4.4.1C & 4.4.2C.)

Stamens monadelphous in many genera eg. *Dalbergia, Abrus, Arachis, Desmodium* and *Glycine*. However, they are free in *Sophora, Ormosia* and *Myroxylon*.

Stamens in majority of genera are arranged in two different heights i.e., 5 long and 5 short. Rao (1983) while studying the pollination biology of Fabaceae has observed that only pollen form short stamen bring about self-pollination, while pollen of long stamens used for cross-pollination. In *Arachis hypogea* there are two posterior sterile anthers, two globose monothecous anthers and three oblong monothecous and three dithecous anthers. However, in other genera anthers are dithecous. Pollen grains tricolpate, porate and exine reticulate (See Rao, 1983).

Gynoecium : Monocarpellary, ovary superior, or half inferior-half superior, stalked or sessile, unilocular, ovules one (*Medicago*) to many, ovules anatropous or campylotropus, placentation marginal, style long and filiform (*Arachis*), Stigma simple, spathulate or papillate (*Vigna*). It is observed that the style is twisted towards left side in Asiatic species (*Vigna radaita*), while it is right handed in American species (*Phaseolus multiflorous*). Smart (1977) opined that it may be due the geomagnetic force.

Fruit: Legume or pod dehisces along both ventral and dorsal sutures (fig 4.4.1D). It may be indehicent (*Dalbergia, Arachis and Pongamia*). Fruit spirally twisted in *Medicago*, and winged in *Psophocarpus* (winged bean). Fruit is a samara in *Pterocarpus* and indehiscent lomentum (*Desmodium*), Geocarpic in *Arachis*.

Seed: Non-endospermic, food stored in thick cotyledons. Embryo large and curved.

4.4.6 Pollination Mechanism

Pollination in most of Fabaceae it is entomophilous. In *Erythrina*, pollination is carried out by bats, while in *Trifolium* and *Viccia* flowers are self-pollinated due to cleistogamy.

Pollination in Fabaceae is carried out by piston mechanism. The flowers are well adopted for insect pollination because of papilionaceous corolla and the stamens and pistil are enclosed by keel petals.

The standard or vexillum petal being the largest and colourful stands perpendicularly and attracts the pollinating bees. The nectar is secreted on the inner sides of stamen bases. The nectar acumulates around the base of the ovary and is protected by the staminal tube and cup shaped thalamus.

As described earlier, the bees are attracted by the conspicuous corolla and are easily accessible to nectar. However, only insects with long probosis can reach the honey secreting place. When bees visit flowers, they alight on the wing petals which serve as platform or landing place. Once the bee lands on wings and when tries to get nectar in this process the wings together with keel petals are pressed down. At this juncture, the pollen present in the anther tubes of stamens and papillate stigma are pushed as a piston, the pollen come in contact with bee body. When this bee insect visit another flower, the stigma comes into contact with insect and the pollen present on the insect body is shed and thus cross-pollination is achieved. Piston mechanism of pollination is observed in many genera including *Lotus corniculatus, Lupinus and Crotalaria.*

4.4.7 Distinguishing Features

1. Herbs, shrubs or trees growing in different habitats.
2. Presence of root nodules.
3. Usually trifoliate compound leaves.
4. Stipules and leaflets modified into tendrils.
5. Pulvinate leaf base.
6. Flowers bisexual, zygomorphic and perigynous.
7. Papilanaceous corolla with vexillary aestivation.
8. Stamens 10, monadelphous or diadelphous.
9. Monocarpellary gynoecium
10. Ovary unilocular with marginal placentation
11. Fruit, a legume
12. Piston mechanism of pollination.

4.4.8. Economic Importance

The family is of great economic importance and ranks second after Poaceae. The members of Fabaceae provide many important pulses, vegetables, fodder, fatty oils, fibres, timber, dyes, gums and also several horticulturally important plants.

4.4.8.1 Pulses: The seeds (cotyledons) of several genera are rich source of proteins and form an important part of diet. Some commonly cultivated pulses plants are given below:

Cajanus cajan (Pigeon pea); *Cicer arietinum* (chickpea); *Vigna mungo* (black gram) *Vigna radiata* (green gram), *Lens culinaris* (Lentil) *Vigna sinensis* (Cow pea) and *Dolichos biflorus* (horse gram).

4.4.8.2 Vegetables: Following are the vegetable yielding plants.

Dolichos lablab (Beans); *Cyamopsis tetragonoloba* (Clustured beans); *phaseolus vulgaris* (French beans); *Glycine max* (Soy bean); *Canavalia ensiformis* (Sword bean);

4.4.8.3 Fodder : Many plants of family yield excellent fodder, these include, *Phaseolus trilobus, Trifolium* (Clover); *Midicago sativa.*

4.4.8.4. Edible Oil: Edible oil is obtained from the seeds of *Arachis hypogea* (Groundnut).

4.4.8.5 Fibres: Following plants are the source of fibres which are used in rope making, *Crotalaria juncea* (Indian hemp); *Sesbania grandiflora* and *Viccia faba* (broad bean).

4.4.8.6 Timber: Quality wood is obtained from *Dalbergia latifolia* (Indian rose-wood), *Pterocarpus santalinus* (Red sanders); and *P. marsupium* (Indian kino) which are used in making furniture and dolls.

4.4.8.7 Dyes: *Indigofera tinctoria* (Indigo) yields a blue dye useful in textile industry.

4.4.8.8 Green Manure: Many plants like *Trifolium, Lupinus, Medicago* and *Crotalaria* are used as green manure to improve nitrogen content of agricultural fields.

4.4.8.9. Medicine: Roots of *Glycirrhiza glabra* useful as diuretic agent in controlling throat problems. *Psarolia corilifolia* yields a drug, which is useful in controlling leprosy. The flowers of *Sesbania grandiflora* are applied to cure night blindness.

4.4.8.10 Ornamentals : Species of *Glyricidia, Clitoria, Butea, Erythrina, Lathyrus* and many others are grown in gardens for their beautiful flowers. Others of horticultural importance are *Robinia* (black locust) *Sophora japonica* (Japanese pagoda tree).

4.4.8.11 Gums: *Astragalus gumifera* is the source of commercially important gum known as trogacanth which is used as emulsifier in confectionary and cosmetic preparations. It is also used in textile industry as a sizing material. The seed material of cluster bean is the source of a commercial gum, known as guar gum, is used extensively in food, paper and textile industries.

4.4.8.12 Other uses: The bark of serveral species of *Erythrina* have bacteriostatic properties. The seed and roots of *Spatholobus* and *Millettia* possess insecticidal and pesticidal properties.

The seed of *Physostigma* (Calabar bean) are extremely poisonous and cause paralysis.

Leaves and seed of *Trigonella foenum-graceum* used as green vegetable and also as flavouring agent.

The oil cake obtained from the seeds of groundnut is a valuable cattle feed and a good organic nitrogenous material. A synthetic fibre called ardil is also manufactured from ground nut seed.

4.5 CAESALPINACEAE (The Cassia family)

Systematic Position

Dicotyledonae

Polypetalae

Calyciflorae

Rosales

Caesalpinaceae

4.5.1 Distribution

The family Caesalpinaceae includes about 152 genera and 2,800 species largely distributed in tropical and sub-tropical regions of the world. In India, it is represented by over 23 genera and 80 species chiefly distributed in the western peninsular India and also in tropical Eastern Himalyas.

4.5.2 Important Plants

1) *Bauhinia vahli* Wild. Tel : Addakulu
2) *B. purpurea* (Camels's boot) Tel : Devakanchanam.
3) *B.anguina* Roxb. (Snake climber)
4) *Cassia auriculata* L.Tel : Nela tangedu
5) *C.fistula* L. (Indian Laburnum) Tel: Rela.

6) *C. occidentalis* L. Tel : Kasinta

7) *C. samea* L. Tel : Chettu Tangedu

8) *Caesalpina bonducella* L. Tel. Gacha poda

9) *C. pulcherrima* Switz (Poinciana or Peacok flower) Tel. Ratna gandhi.

10) *C. cristata* L. (Fever nut)

11) *Delonix regia* (Boj). Raf. (Royal Poinciana or Gulmohar) Tel. Erraturai

12) *D.alata* L. (white gulmohar) Tel. Tella turai

13) *Haematoxylon campechianum* L. (Leg wood or haematoxylon).

14) *Hardwickia binata* Roxb. Tel; Yepi

15) *Kingiodendron pinnatum* (Roxb) Harms.

16) *Parkinsonia aculeata* L. (Jerusalem thorn) Tel: Seema tumma or Jeluga.

17) *Peltophorum ferrugineum* (DC) Backer

18) *Pterolobium hexapetalum* L. Tel: Pariki

19) *Saraca indica* Roxb. Tel: Asokamu

20) *Tamarindus indica* L. (Tamarind) Tel: Chinta.

4.5.3 Habit and Habitat

The members of Caesalpinaceae are mostly trees or rarely herbs (*Cassia tora* and *C.pumila*) *C.occidentalis* is an undershrub, woody climbers (*Bauhinia vauhlii* and *Caesalpina bonducella*) and they climb with the help of stem tendrils or sharp prickles. *Cassia fistula* is a medium sized tree, *Peltophorum ferrugineum* is a ever green large tree. Plants are mostly xerophytes. However, *Parkinsonia* is a xerophyte where, stipules and primary rachis are reduced to spines and secondary rachis is modified into phyllodes.

4.5.4 Vegetative Characters

Root system : A well branched tap root system.

Stem : Erect, cylindrical, woody, branched or some times weak and climb by means of spines or prickles (*Pterolobium*) or by modified tendril in *Bauhinia vauhilii.* In many species tanin sacs and gum passages are found.

Leaf : Leaves usually alternate, simple and bilobed (*Bauhinia*). But more commonly pinnately compound (*Cassia, Tamarindus*) Leaf base often swollen (Pulvinous leaf base) leaves stipulate, stipules auriculate (*Cassia auriculata*) (fig. 4.5.1A) spiny (*Parkinsonia*) or foliaceous (*Delonix regia)* (fig 4.5.2) venation reticulate. Leaflets ovate or obovate. In xerophytic *Parkinsonia* leaf rachis is flat and ends in a sharp point where minute leaf-lets fall off or do not develop, rachis resembles a phyllode.

4.5.5 Floral Characters

Flowers usually arranged in racemose inflorescence. Inflorescence pendulous in *Cassia fistula*, corymb (*Delonix* and *Cassia auriculata)* corymbose panicle (*Saraca*) and compound corymb (*Cassia siamia*). In *C.tora* a group of 2-3 flowers arise from leaf axils.

Flower : Flowers bracteate, bracteolate and pedicillate, complete, bisexual, slightly zygomorphic, perigynous and pentamerous. In many genera flowers are large and showy (*Delonix* and *Peltophorum*). (fig 4.5.2.A)

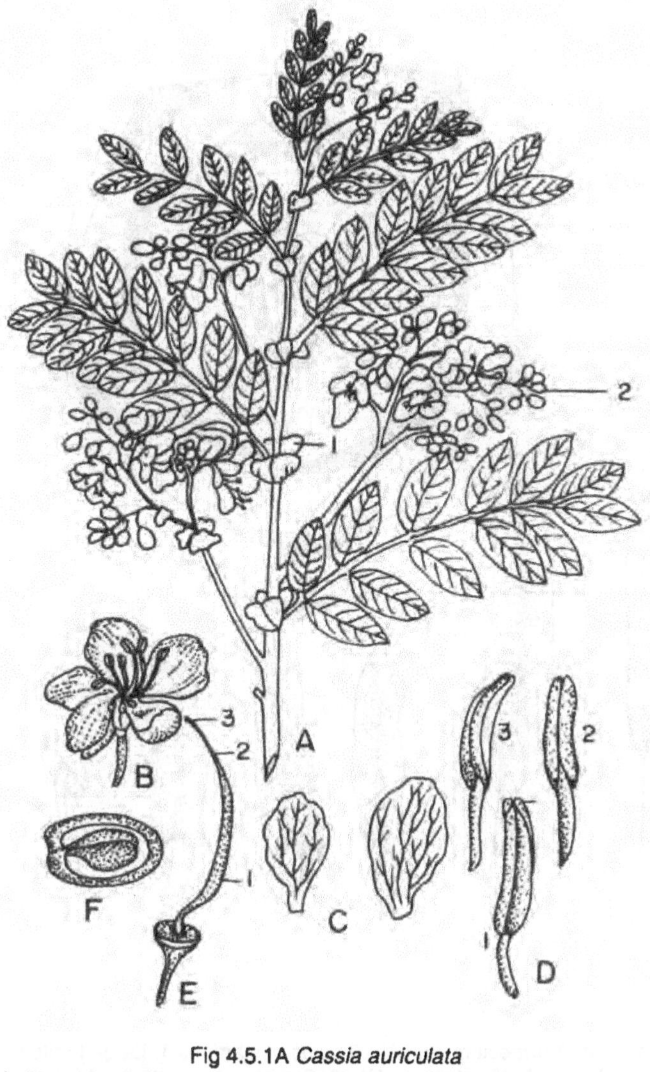

Fig 4.5.1A *Cassia auriculata*
A: Flowering twig 1. auriculate stipules, 2. Inflorescence
B: Flower; C. Corolla (Petals); D. Stamens showing various faces (1, 2, 3);
E. Gynoecium 1. Ovary, 2. Style, 3. Stigma; F. T.S.Ovary

Calyx :- Sepals five or four as in *Amherstia* three in *Tamarindus polysepalous* (*Cassia, Caesalpina* and *Delonix*) or fused (*Bauhinia, Acerocarpus*) aestivation imbricate or rarely valvate (*Cercis*) green or petaloid (*Saraca, Dialium*) usually the odd sepal lies in the anterior side.

Corolla : Petals five in *Cassia, Delonix* and *Bauhinia*, three in *Tamarindus* and only one in *Afzelia*. In Tamarindus the two anterior petals are suppressed and represented by scaly structures. Petals are altogether absent in *Saraca*, while *Cercis* shows papilionaceous corolla. Petals polypetalous, aestivation ascendingly imbricate. The odd petal is posterior and is different from others in size and colour. All petals are clawed and with spreading limbs. In most of the genera petals brightly coloured and attractive.

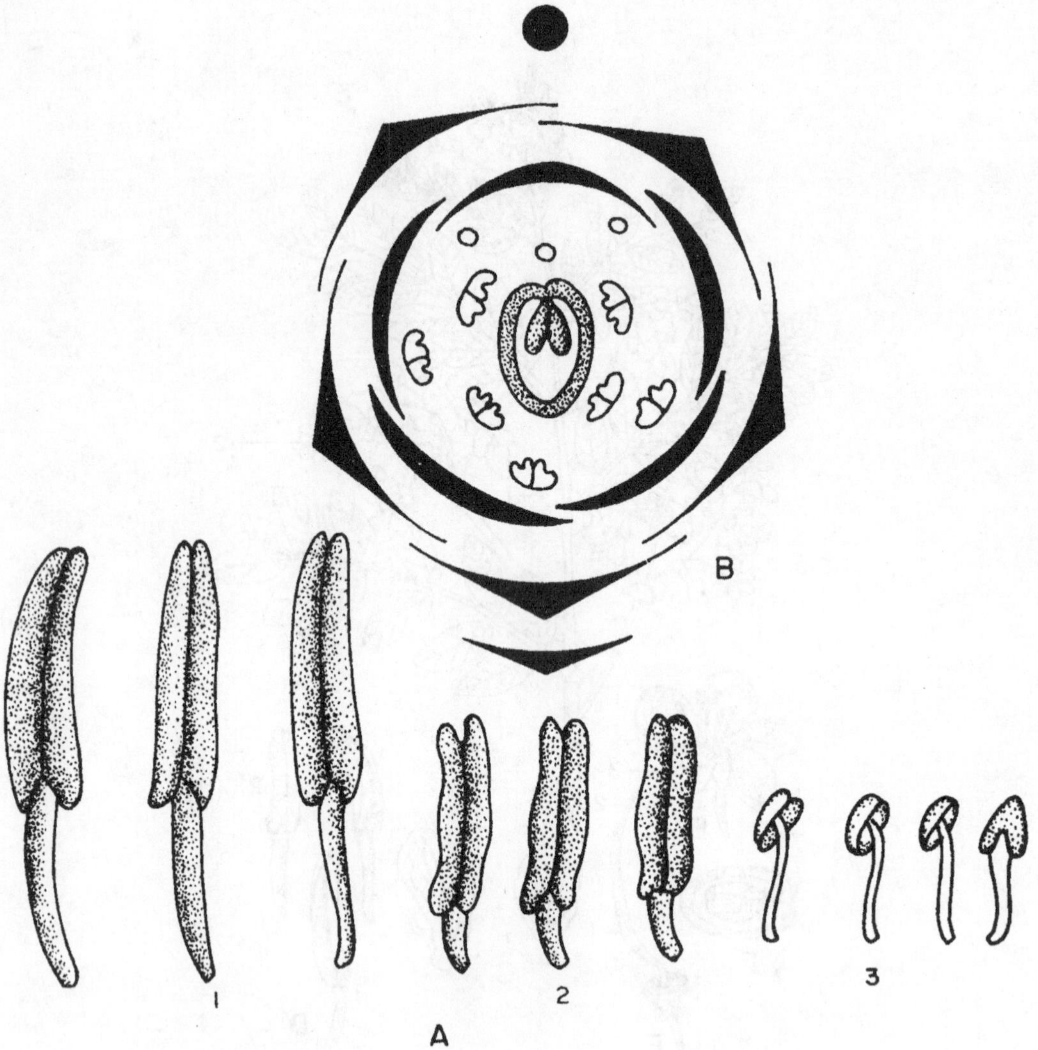

Fig 4.5.1B : A. Androecium showing 3 types of stamens 1. Large fertile stamens (3) 2. Medium fertile stamens (3); 3. Staminodes or sterile stamens (4) B. Floral diagram

Androecium : Stamens usually 10, diplostemanous with two alternate whorls of five stamens each and are free. However, diadelphous in *Amperstia* and monadelphous in *Tamarindus* where, 3 fertile stamens and 4 staminodes form a bundle. In *Cassia* spp. three smaller stamens are sterile, four are medium sized and three large and sickle shaped, arranged in three whorls. (fig 4.5.1.B). In *Parkinsonia* and *Delonix* all 10 stamens are fertile while in *Bauhinia monandra* only one stamen is fertile. (fig 4.5.2.B)

Stamens dithecous, basifixed or versatile, dehiscing by longitudinal slits or by pores as in *Cassia* spp.,

Gynoecium : Usually monocarpellary, rarely bicarpellary and apocarpous in *Saraca*. Ovary half- inferior half superior, unilocular with two rows of ovules in marginal placentation.

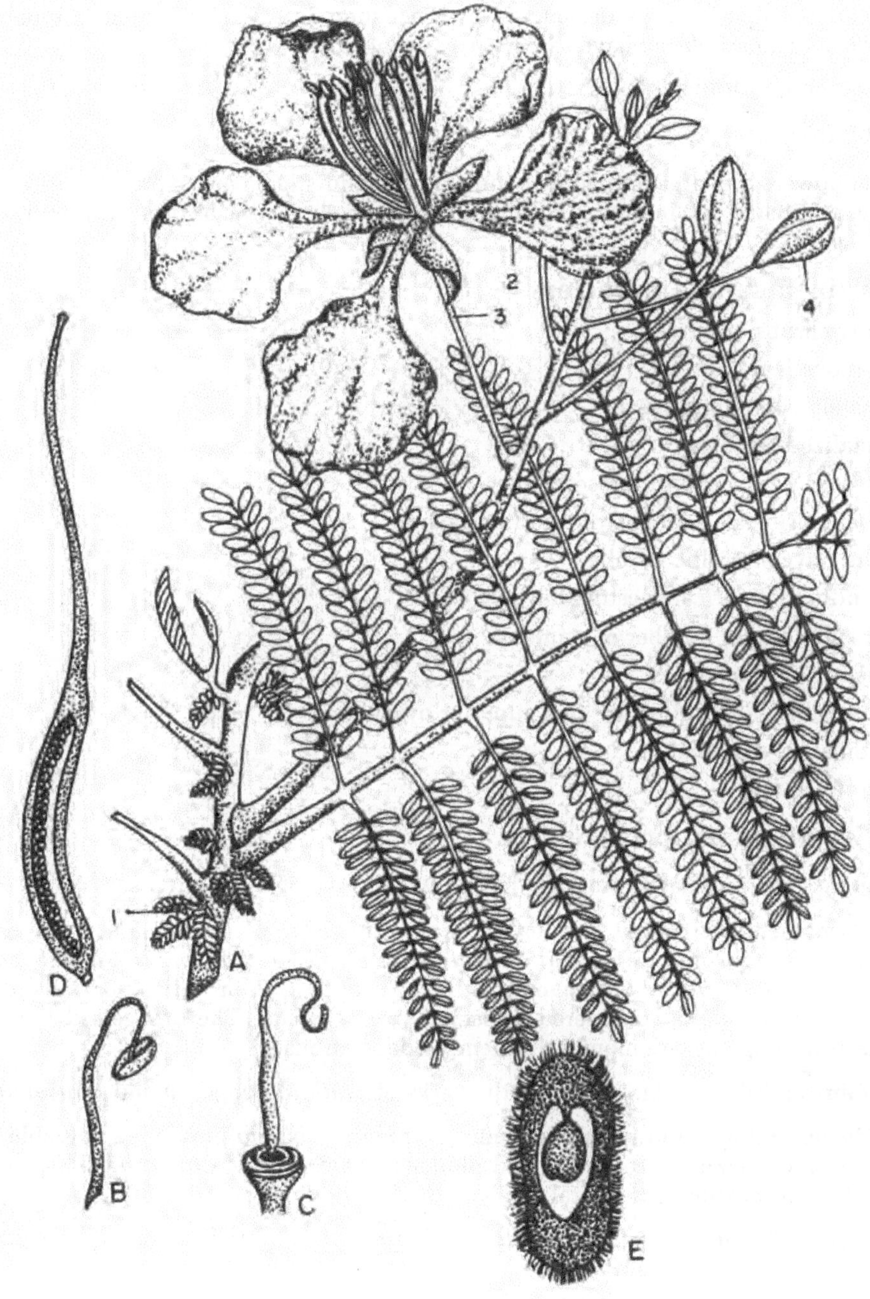

Fig 4.5.2 *Delonix regia*
A. Flowering twig 1. Leafy stipules, 2. Odd petal, 3. Flower 4. Floral bud
B. Stamen; C. Glynoecium; D. L.S. gynoecium; E. T.S. Ovary

In some genera, ovary curved or sickle shaped, hairy or tomentose. Style long, filiform, stigma capitate or simple. Nectary glands presnt at the base of the ovary (fig 4.5.1A).

Fruit : Legume or pod and dehiscent. In *Cassia fistula* and *Tamarindus* it is indehiscent. *Hardwickia* and *Pterolobium* have samara fruit (winged).

Seed : Non-endospermic.

4.5.6 Pollination

Since flowers are large and showy and stamens and stigma are exposed, hence flowers are easily pollinated by insects. Pollination is carried out by birds (ornithophilous) in *Delonix regia*.

4.5.7 Distinguishing Characters

1. Mostly shrubs and trees.
2. Pulvinous leaf base
3. Pinnately compound leaves
4. Specialized stipules
5. Racemose or corymb inflorescence.
6. Flowers bisexual, pentamerous and perigynous.
7. Slightly zygomorphic corolla
8. Ascendingly imbricate aestivation
9. Free and definite number of stamens
10. Presence of staminodes
11. Unicarpellary, half inferior - half superior ovary
12. Placentation marginal
13. Fruit, a legume.

4.5.8. Economic Importance

Different genera of Caesalpinaceae have a wide range of economic uses.

4.5.8.1. Ornamentals : Many species of *Cassia* (*C.fistula, C.nodosa, C. marginata), Caesalpina pulcherrima, Bauhinia racemosa, Delonix regia, Parkinsonia aculeate, Peltophorum* are grown as ornamentals which produce golden yellow or pink flowers. *Saraca indica* (Ashoka) is considered as sacred tree of India. The orange-red flowers are widely used in religious ceremonies and temple decorations.

4.5.8.2. Fibres : Fibre is obtained from the barks of *Bauhinia anguina* and *B.racemosa*.

4.5.8.3. Timber and agricultural implements : *Hardwickia binata* provides a valuable timber which is used for making cart wheels and ploughs. The wood of *Cassia fistula* provides a useful timber for making agricultural implements.

4.5.8.4. Dyes : The heartwood of *Caesalpina, Sappan* yields a red orange dye which finds use in dying of cotton, silk and wool.

4.5.8.5. Medicines : Leaves of *Cassia fistula* are applied to control ring worm and a decoction of plant parts is administered in acute bronchitis.

Fruits of *Tamarindus* yield tamarind which contains acidic pulp around the seed is extensively used for souring curries. Tartaric acid is obtained from unripe fruits, which is employed extensively in chemical and pharmaceutical preparations.

4.5.8.6. Other uses : The bark of some *Bauhinia* species yield tannins of commercial importance. The bark and pods of *Caesalpinia coriaria* also yield tanning material.

Seed material of *Caratonia siliqua* used for sizing fabrics and also in printing, tanning and paper industry.

Haematoxylin is extracted from the heart wood of *Haematoxylon compechianum* (log wood) is used as nuclear stain in the biological studies.

4.5.9 Evolutionary Tendencies The family Caesalpinaceae shows number of advanced characters when compared to other families of calyciflorae.

These includes

1. Reduction in the number of petals form 5 to 3 (*Tamarindus*) to one *Atzelia*) and ultimately their total absence in *Saraca.*
2. The stamen number is also reduced from 10 to 3 or 8 to 2
3. Monadelphous condition of Androcium in *Tamarindus.*
4. Leaves stipulate and also some times exstipulate.

4.6 MIMOSACEAE (The Acacia family)

Systematic position :

Dicotyledonae
Polypetalae
Calyciflorae
Rosales
Mimosaceae

4.6.1 Distribution Mimosaceae is the smallest family among three sub-families of erst-while family Leguminosae. It comprises about 58 genera and 2800 species distributed mainly in tropical and sub-tropical areas of the world. About 15 genera and 72 species are found in India.

4.6.2 Important Plants

1. *Acacia catechu*, Wild, (Cutch) Tel: Catachu or Kaviri.
2. *A. concina* DC (Soap pod) Tel: Seekaya
3. *A. leucocephala* Willd, Tel: Tella tumma
4. *A. nilotica* (L) Del. (Acacia) Tel: Nalla tumma
5. *A. melanoxylon* R.Br. (Australian acacia)
6. *Albizzia lebbeck* L. (Silk flower) Tel: Dirisena
7. *Dichrostachys cinera* L Tel: Veluturu chettu
8. *Adenanrthera pavonia* L (Bead tree or Coral tree)
9. *Enterolobium saman* L. Tel: Nidraganneru
10. *Entada rheedii* L. Tel: Gila teega
11. *Inga dulce* Benth. (*Pithecolobium dulce*) (Manila tamarind) Tel: Seema chinta.
12. *Leucaena glauca* L (*L.leucocephala*) (White popinae) Tel: Pedda nidrakantha
13. *Mimosa pudica* L (Touch me not) Tel: Nidrabhangi or Attipatti
14. *Neptunia oleracea* (Lamk)Baill. Tel: Neeru talavapu

15. *Prosopis julifera* L. Tel: Sarcar tumma
16. *P.spicigera* L. Tel: Jammi
17. *Samnea saman* (Jacq) Merr. (Rain tree)
18. *Xylia xylocarpa* L. Tel: Konda tangedu

4.6.3. Habit and Habitat

Herbs (*Mimosa pudica*) shrubs (*M.rubicalis*) and trees (*Acacia nilotica Xylia* and *Albizzia*). *Entada* is a woody climber, *Neptunia oleracae* is an aquatic plant. However, most members are thorny and show xerophytic characters.

4.6.4 Vegetative Characters:

Root System: Branched tap root system, some times modify into nodular roots. In *Neptunia* adventitious root develop from the nodes of vegetative branches.

Stem: Aerial, erect, woodystem, weak stems climb by means of prickles (*Acacia sinuata*) or by tendrils (*Entada*)

Leaf: Leaf alternate, leaf base pulvinous, bipinnately compound (unipinnate in *Inga*) leaflets oval, unicostate reticulate venation.

Leaves stipulate, stipules modified into spines (*A.nilotica*) (fig. 4.6.1) and *Prosopis julifera*) or into myremecophilous spines (*Acacia sphaerocephala*). The petiole is flattened into a phyllode and leaflets are suppressed in *Acacia melanoxylon* and *A auriculata*. In *Entuda* the terminal part of petiole modify into tendrils. Leaves show sensitive movements in *Mimosa pudica* and *Neptunia oleracae*.

4.6.5 Floral Characters

Inflorescence: Usually racemose (*Edenonthera*) however, it is an elongated spike (*Prosopis*) or a condensed cymose head (*Albizzia, Acacia*) and *Dichrostachys*) (see fig. 4.6.2).

Flower: Bracteate, ebracteate, sub-sessile complete, bisexual, actinomorphic, Pentamerous, or tetramerous (*Mimosa*), trimerous (*Acacia* spp). Perigynous, polygamous (*Dichrostachys*) (fig 6.2).

Calyx: Four to five sepals, gamosepalous, Polysepalous in (*Parkia*) valvate or quincuntial, Calyx toothed (*Mimosa*).

Corolla: Petals five, gamopetalous (*Samania* and *Albizzia*) four and polypetalous (*Mimosa, Prosopis*), valvate aestivation.

Androccium: Stamens indefinite in number, monadelphous (*Samania Albizzia Acacia* (fig.6.1) and *Pithecolobium*), stamens 10, free and present in two whorls 5 each (*Prosopis, Entada* and *Adenanthera*) 4-8 and free (*Mimosa* spp).

Anthers usually minute, introse, dithecous, filaments of various colours and attractive, pollen grains tricolpate and porate, present in compound polyads.

Gynoecium: Ovary half-interior half-superior, monocarpellary, style long, filiform, stigma minute, ovary with one locule and plancentation marginal. (Fig. 4.6.2)

Fruit: Usually characteristic lomentum.

Seed: Non-endospermic. In *Pithecolobium* dulce, seeds are enveloped by a fleshy edible aril. In Inga naked embryo falls from the fruit and directly germinates.

Pollination: Pollination entomophilous, however it is Chiropterophyly in Parkia where pollination is done by the bats.

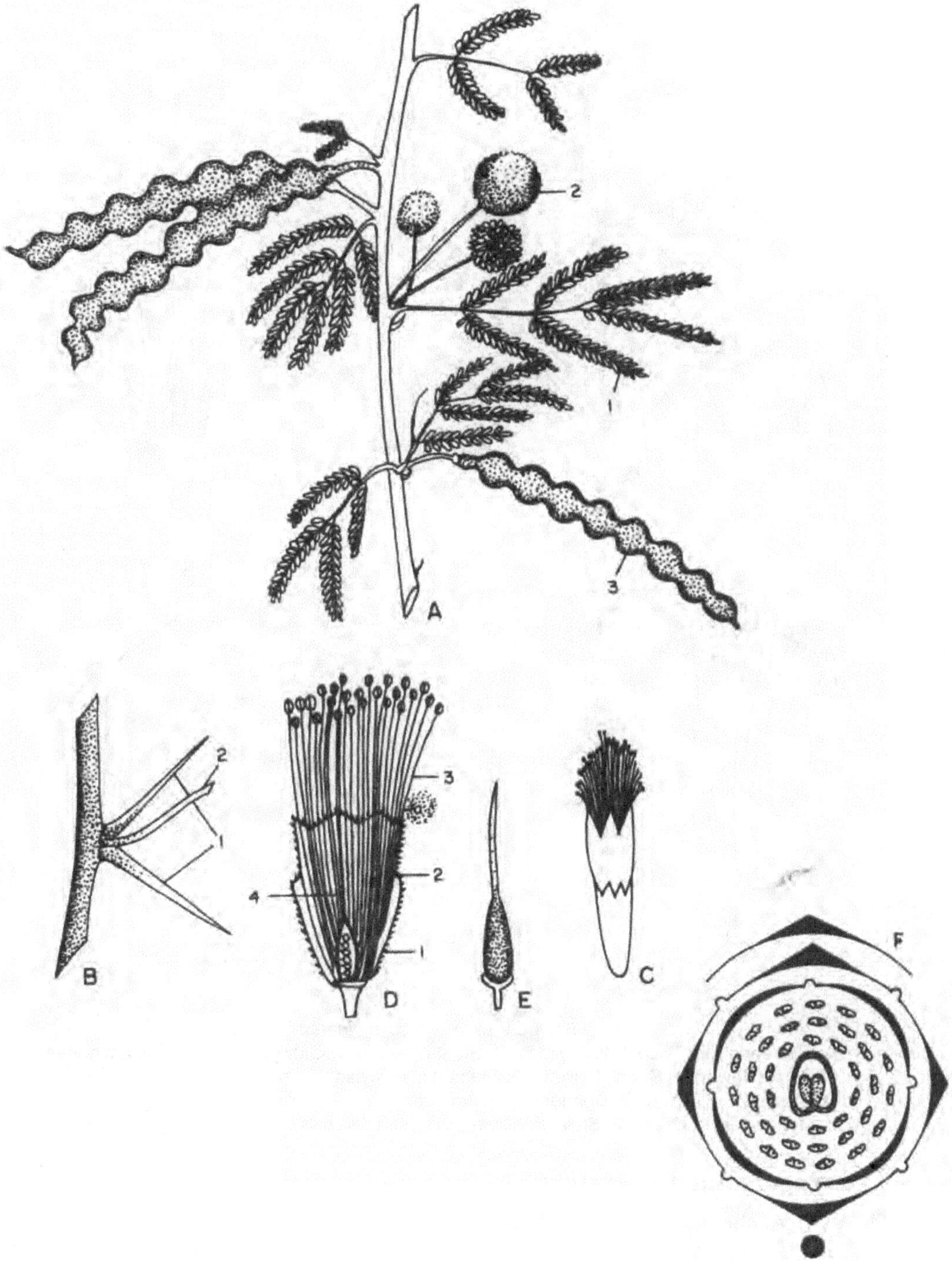

Fig 4.6.1 *Acacia nilotica*

A: Flowering twig 1. Compound leaf, 2. Inflorescence, 3. Lomentum
B.Shoot region 1. Spines, 2. Petiole, C: Flower D. L.S. Flower 1. Calyx, 2. Corolla, 3. Stamens, 4. Gynoecium, E: Gynoecium, F: Floral diagram

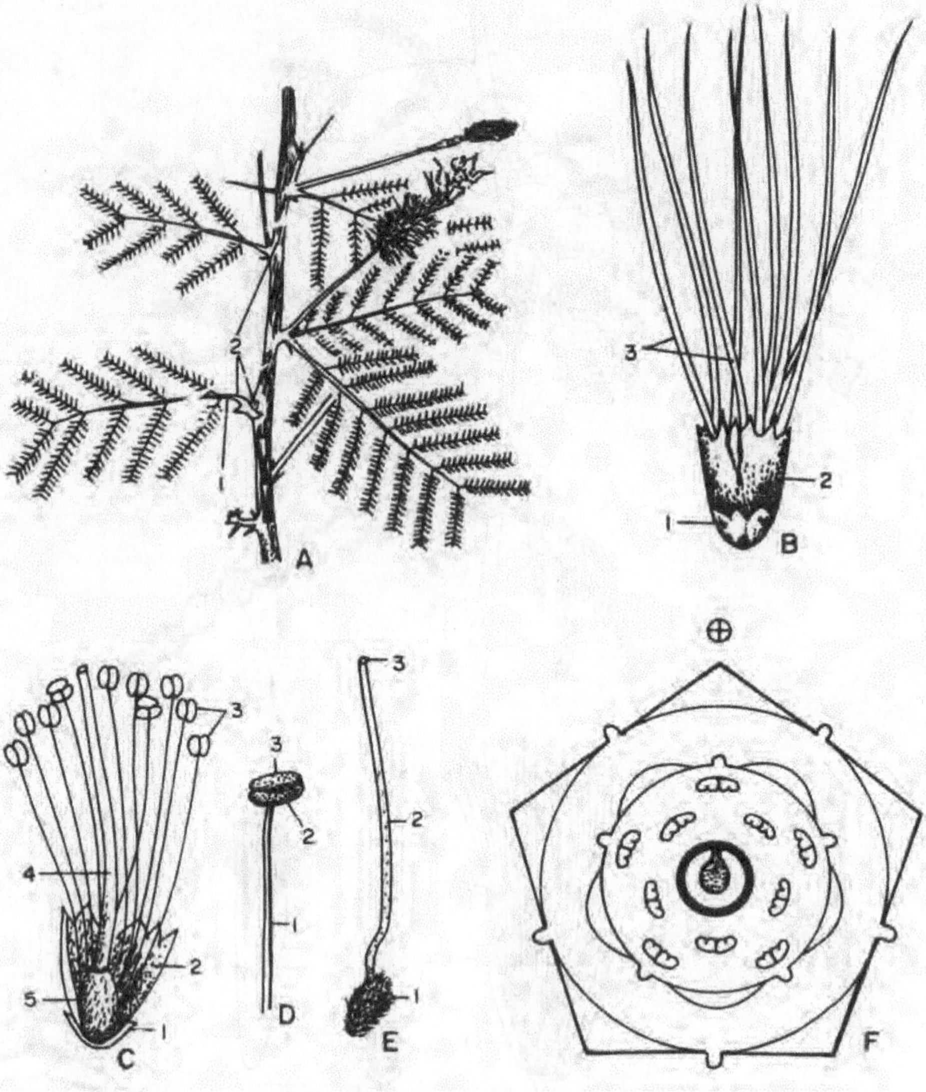

Fig 4.6.2 *Dichrostachys cinarea*
A: Flowering twig 1. Leaf, 2. Stipules, B. Sterile flowers, 1. Calyx 2. Corolla, 3. Staminodes;
C. Fertile flowers 1. Sepal, 2. Petal, 3. Stamens, 4. Styles, 5. Ovary;
D: Stamen 1. Filament, 2. Connective, 3. Anthers
E: Gynoecium 1. Ovary, 2. Style, 3. Stigma; F: Floral diagram

4.6.6 Distinguishing Characters

1. Mostly trees showing xerophytic habit
2. Pulvinous leaf base.
3. Stipulate, pinnately compound leaves.
4. Cymose head inflorescence
5. Actinomorphic, penta or tetramerous flowers.

6. Perigynous condition.
7. Fused petals with valvate aestivation.
8. Stamens indefinite in number, monadelphous with showy filaments.
9. Monocarpellary unilocular ovary
10. Marginal plancentaion
11. Legume or lomentum fruits.
12. Non-endospermic seed.

4.6.7 Economic Importance

4.6.7.1 Timber: *Acacia nilotica, A catechu, A.ferruginea* and *A.leucocephala* yield important timber for a variety of purposes like agricultural implements, tool handles and cart wheels. They are also a valuable source of fuel.

Wood of *Xylia xylocarpa* used for railway sleepers, boat building and also for producing pulp from which wrapping paper is produced. Species of *Adenanthera* yield red wood which is useful in making furniture. The wood of *Acacia catechu* yield catechu of commerce. Catechu has several medicinal and industrial uses besides being an important ingredient of betel.

4.6.7.2. Avenue trees: Species of *Albizzia, Adenanthera* and *Samanea* are grown as avenue trees.

Pithecolobium dulce, Prosopis specigera and *P. julifera* are grown as hedge plants and these also have considerable value in afforestation and reclaimation of waste lands.

4.6.7.3. Tannins and dyes: Bark of *Acacia nilotica, A.decurrens,* yields tannins and dyes.

4.6.7.4. Other uses: Fruits of *Acacia concina* are used as soap nuts.

Flowers of *Acacia farensiana* are rich source of a perfume called the cassia perfume.

The tree trunks of *A.nilotica, A scorpoides* and *A. senegal* yield gum acacia, gum ghati and gum arabic respectively.

4.6.8 Phylogeny of Leguminosae

Bentham and Hooker placed the familay Leguminosae in order rosales and series calyciflorae. They recognised three sub-families in Leguminosae. These are Papilionatae, Caesalpinoidae and Mimosoidae. Engler and Prantl also classified Leguminosae into three sub families and also included under order Rosales.

Hutchinson (1965) however, raised the rank of above three sub-families and gave the status of family and renamed them as Papilionaceae (Fabaceae), Caesalpinaceae and Mimosaceae. Among these three families Papilionaceae is considered to be advanced while Caesalpinaceae as primitive.

4.7 CUCURBITACEAE (The Pumpkin family)

Systematic position:

Dicotyledonae
Polypetalae
Calyciflorae
Passiflorales
Cucurbitaceae

4.7.1. Distribution

Cucurbitaceae is an essentially a tropical and sub-tropical family with 110 genera and 650 species. In India, it is represented by 37 genera and 97 species several of which are being cultivated in many parts of India.

4.7.2. Important Plants

1. *Benincasa hispida* (Thumb) cogn (White or Ash gourd) Tel: Budida gummadi
2. *Coccinia cordifolia* L cogn (Ivy guard) Tel: Donda
3. *Citrullus vulgaris* Schrad (Water melon) Tel: Pucha kaya
4. *Cucurbita maxima* Duch: (Red pumpkin) Tel: Theeya gummadi
5. *C. moschata* Duch: (Pumpkin)
6. *C.pepo* L (Field pumpkin) Tel: Gummadi
7. *Cucumis melo* L (Musk melon) Tel: Karbuja
8. *C. sativus* L. (Cucumber) Tel: Dosa
9. *Ecballium eletarium* L (Squirting cucumber)
10. *Lageneria vulgaris* Serv: Syn *L. siceraria* (Mol) Strdl. (Bottle Gourd) Tel: Sorakaya
11. *Luffa acutangula* (L) Raxb. (ribbed gourd) Tel: Beerakaya
12. *L.cylindrica* (L) M.J. Roen. (Sponge gourd) Tel: Neti beera
13. *Momordica charantia* L. (bitter gourd) Tel: Kakara kaya
14. *Methria madaraspatana* (L) Roem. Tel: Potti budama
15. *Momordica divica Roxb* (Pointed gourd) Tella adavi kakara
16. *Sechium edule* (Jacq) Sw. (cho-cho) Tel: Bangalore vankaya
17. *Trichoranthes anguina* L (Snake gourd) Tel: Potlakaya
18. *T.cucumerina* L Tel: Eedu potla.

4.7.3. Habit and Habitat

Members of Cucurbitaceae are mostly annuals or perennial herbs, ususlly climbing by means of tendrils, some genera are succulents, trailing or decumbent. *Acanthosicyos horridus* is a tropical leafless xerophyte armed with spines. *Socotra* is a small tree.

4.7.4. Vegetative Characters

Root system: Branched tap root system. Large tuberous roots (almost equivalent to a human body) are found in *Echinocystis lobata* and also (about 12 meters) in *Acanthociryos.*

Stem: Aerial, weak, climbing by means of simple or branched tendrils. Tendrils generally extra-axillary (fig 4.7.1). However, morphology of tendrils is a much disputed subject, rarely stem erect (*Dendrosisys*) stem five angled, hollow and green in colour with hairs.

Leaf: Alternate, petioles long, cordate, simple but often palmately or pinnately lobed or divided, leaves ex-stipulate, however in *Acanthosistys* stipules modify into spines, hairy, palmately reticulate venation, multi-costate.

4.7.5. Floral Characters

Inflorescence: Usually female flowers are solitary and male flowers arranged in racames or corymbs (fig 4.7.1 & 4.7.2).

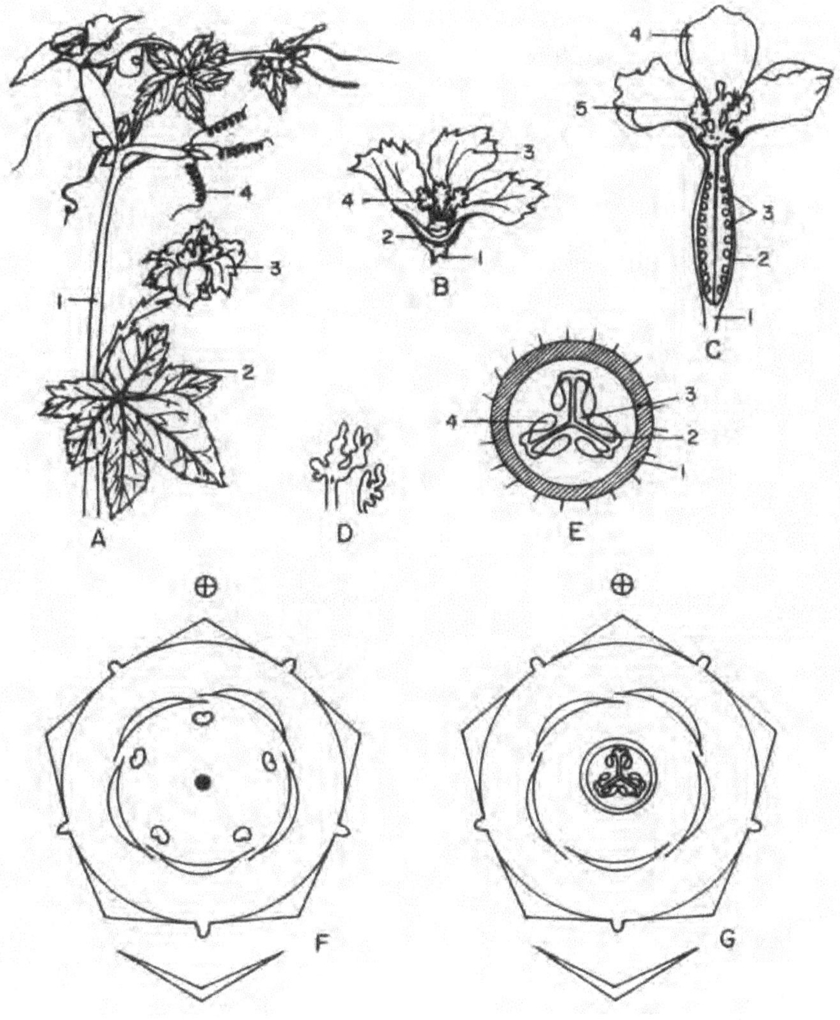

Fig 4.7.1 *Luffa cyllindrica*
A: Flowering twig 1. Stem, 2. Leaf, 3. Flower, 4. Tendrils
B: L.S. Male Flwoer 1. Pedicel, 2. Sepal, 3. Petal, 4. Synandrous stamens
C: L.S. Female Flower 1. Pedicel, 2. Ovary, 3. Ovules, 4. Corolla, 5. Stigma
D: Stamens ; E: T.S. Ovary 1. Ovary wall, 2. Locule , 3. Placenta 4. Ovule
F: Floral diagram (Male flower) G. Floral diagrame (female flower)

Flower: Bracteate, incomplete, unisexual usually monoceious (*Coccinia* and *Momordica*) or some times dioecious (*Bryonia*). Rarely, bisexual (*Schizopepon*), epigynous; rarely perigynous (*Actinostemma*). Actinomorphic. Flowers pentamerous. (Fig 7.1)

Calyx: Sepals five, fuse at the base forming a cup shaped structure which is adnate to gynoecium. Aestivation valvate or imbricate and generally hairy.

Corolla: Petals five, gamopetalous forming a companulate (*Coccinia* and *Cucurbita*), rotate (*Cucumis*) or salver form in *Trichosanthes*, however nearly free (*Fevillea*, *Luffa* and *Lageneria*), valvate or imbricate aestivation.

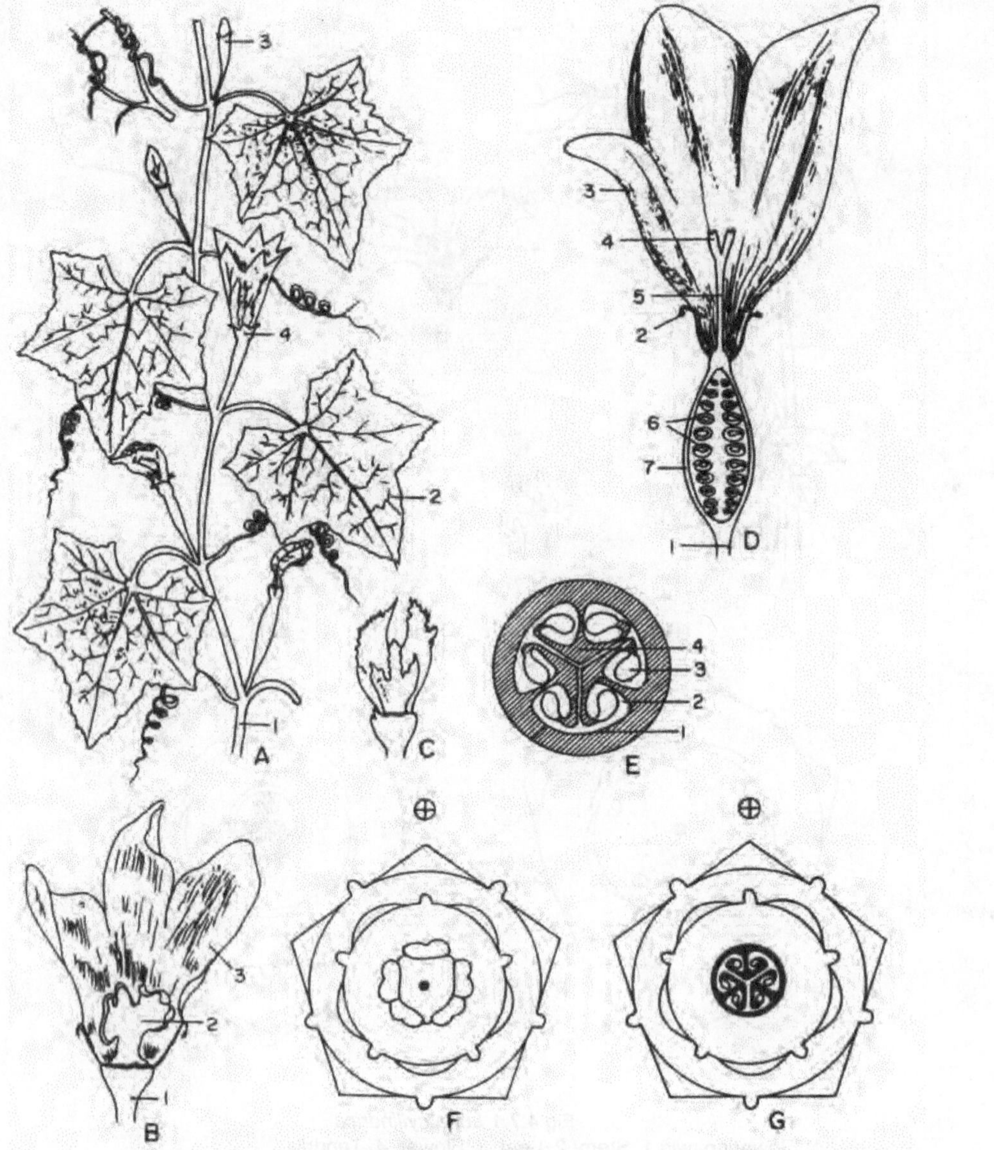

Fig 4.7.2 *Coccinia cordifolia*
A: Flowering twig 1. Stem, 2. Leaf, 3. Flower bud, 4. Flower
B: L.S. Male Flower 1. Pedicel, 2. Stamens, 3. Petal C: Androecium ;
D: L.S. Female Flower 1. Pedicel, 2. Sepal, 3. Petal, 4. Style, 5. Stigma, 6. Ovules, 7. Ovary ; E: T.S. ovary
1. Ovary wall 2. Locule 3. Ovule, 4. Placenta; F: Floral diagram (male flower), G: Floral diagram (female flower)

Androecium: five stamens, show different types of cohesion (Synandrous condition). They are however free and alternate with petals (*Fevillea* and *Luffa*) (fig. 4.7.1). Various cohesion types are given below.

1. In *Sicyos*, the filaments of all stamens are united into a column bearing anthers which are remarkably curved.
2. In *Thaldianatha* 4 stamens form 2 pairs and the fifth remains separate.

3. In *Sycydium*, out of 5 stamens 4 form 2 pairs and the fifth remains separte as in Thaldiantha but the paired 4 stamens become united at the base but the anthers remain free.
4. In *Momordica* and *Bryonia* stamens form pairs of (2+2+1) and are completely united forming a synandrous androecium. Due to this fusion, they look like possessing three stamens only.
5. In *Cucurbita, Lageneria* and *Bryonia* also five stamens totally unite so that only 3 stamens can be observed, of these two stamens show 4 loculed anthers and the third show dithecous anthers. The anthers are twisted and folded, such highly folded anthers are described as conduplicated condition.
6. In *Cyclanthera* all the stamens are united into a central column with two ring like pollen chambers at the top.
7. In. *Sechium* also stamens show synandrous condition. Here, one stamen bears monthecous anthers and others bear dithecous anthers.

Anthers basifixed, mostly dithecous, Anthers 'S' shaped and extrose. Staminodes are present in female flowers.

Gynoecium: Absent in male flowers, tricarpellary syncarpous, tetracarpellary in some species of *Cucurbita*, ovary inferior, unilocular, however, appears trilocular due to the growth of placenta towards the centre and thus placenta incompletely divide ovary into the three chambers.(figs 7.1&7.2) Numerous anatropous ovules are found on three parietal placentae, single ovule which is pendulous present in Sicyos and *Sicyosperma*. Stigmas as many as number of carpels, very thick and forked. In *Cucurbita* a dark brown penta-lobed ring like disc encircling the base of the style is present.

Fruit: A fleshy berry like fruit called pepo is present. In this, the pericarp may be soft (*Trichoanthes*) or hard (*Benincasa*) pepo may be indehiscent (*Cucurbita maxima, Luffa, Cucumis* etc. or dehiscent (*Momordica Cyclanthera* etc.) In *Cucurbita maxima, Citrullus vulgaris* (Watermelon) pepo attains large size. The club-shaped pepos are coated with a layer of wax on their hairy skin (*Benincasa hispida*). Fruits of *Sechium* are single seeded berries.

Seed: Non-endospermic, winged (*Macrozonia macrocarpa*), Embryos straight with large oily cotyledons. In *Sechium* edible seeds are large and germinate before fruit drops from the plant (vivipary). Seeds have hard seed coat.

Pollination: Cross pollination is a rule since flowers are unisexual and it is entomophily. In many genera flowers are large and conspicous. They are visited by insects for nectar which is secreted by a cup shaped disc at the base of ovary.

4.7.6 Distinguishing Characters:

1. Mostly herbaceous climbers, climb by means of tendrils.
2. Presence of bicollateral vascular bundles in the stem.
3. Stems hollow.
4. Simple, palmately lobed leaves.
5. Unisexual and epigynous flowers
6. Synandrous androecium
7. Tricarpellary syncarpous gynoecium with inferior ovary
8. Placentation parietal
9. Fruit, a pepo
10. Vivipary germination in some genera.

4.7.7 Economic Importance

4.7.7.1 Vegetables: The family is economically important as it provides many vegetable yielding plants. Some of these include,

Trichosanthes anguina (Snake gourd)

Lageneria vulgaris (Bottle gourd0

Cucumis sativus (cucumber)

Cucurbita pepo (field pumpkin)

Luffa acutangula (ribbed gourd)

Sechium edule (cho-cho)

Momordica charantia (bitter gourd)

Coccinia indica (Ivy gourd) and

Benincasa hispida (Ash gourd).

4.7.7.2 Edible fruits: Fruits of *Cucumis melo* (muskmelon), *Citrullus vulgaris* (watermelon) are edible.

4.7.7.3 Medicinal Plants: Some members of the family have medicinal value. A drug called 'Bryony' is obtained from *Bryonia dioca*. The dried pulp of *Citrullus colocynthis* yields a powerful laxative called colocynth. A poisonous drug eleterium is extracted from the fruits of *Ecballium eleterium*. *Luffa echinata* produces a bitter alkaloid useful in curing dropsy. Seeds of *Bryonia callosa, B. rostratum* are employed in treating piles and asthama.

4.7.7.4 Musical instruments: The hard dried shells or woody pericarps of pumpkin, pepo and bottle gourd are used in prepraing musical instruments by the villagers.

4.7.7.5 Ornamental plants: Some plants of the family like *Corallocarups, Ibervillea* and *Xerosicyos* are grown as ornamental plants.

4.7.8 Morphology of Tendrils

Majority of genera in this family possess branched or unbranched tendrils. The tendrils are useful in climbing. However, the morphology of tendrils is variously interpreted by different systematic botanists. It is regarded as an extra-axillary body, an axillary structure, an axis, a leaf, a stipule and even as a bracteole and also as a organ of suigeneris.

Braun - Eichler (1876) : Braun and Eichler, opined that tendrils are modified bracteoloes by observing the origin of tendrils in *Cucumis sativus and Cucurbita pepo*. Accordingly, they felt that the single flower is formed by the modification of an axillary bud of the leaf. The two bracteoles present on the lateral sides of the flower are modified into tendrils. The inflorescence is interpreted as an axillary shoot of one of the tendrils, so also the vegetgative branch as the axillary shoot of the second tendril. The full compliment of parts is not commonly seen and in many cases, one tendril is seen and the other is suppressed.

Muller (1887) and Hagerup (1930): According to Muller and Hagerup, tendrils are considered to be the modified leaf. To support their view, they observed that the lower stiff part of the tendril is the stem and the twining part is the leaf. This observation was based on the transitional forms between the leaf and tendrils that were observed in *Cucurbita pepo*. Hence, the petiole, the leaf apex or lateral ribs may be tendrillar in function. It thus makes clear that the lower part of tendril is modified from stem, and the twining part is a modified leaf. Though, Muller and Hagerup made their observations separately, the views expressed by Muller on the tendril morphology was fully supported by Hagerup, todate, this observation that the tendril is a modification of a branch and a highly modified foliage leaf is still gains wide support.

Engler (1904) believed that tendrils are modified stipules. This assumption is based on the observations in *Kedrostis spinosa* where, a tendril is found in place of a stipule on one side and on other side it is replaced by a spine.

Sen Sarma (1955) opines that the bracts are modified into tendrils. This view is supported by anatomical observations where, vascular supply to bracts and the tendrils is similar. Besides above opinions, following are some other assumptions on the morphology of tendrils advanced by various workers.

1. Tendril is a modification of the peduncle.
2. It is an extra-axillary shoot.
3. Tendril is developed as a result of the splitting of the petiole.

4.7.9 Affinities and Phylogeny

There is a considerable difference of opinion regarding the position of the family on account of 1) The epigynous nature of flower.

2. Presence of gamopetalous corolla
3. Tendency to form epipetalous stamens and
4. The general nature of the calyx.

Eichler and other taxonomists place Cucurbitaceae in gamopetalae near Campanulaceae. But, studies on floral morphology and embryology clearly show its affinities with Polypetalous families. These include,

1) Presence of two integuments in the ovule
2) Large persistent nucellus and
3) Presence of extensive tapetum.

Bentham and Hookers, also placed Cucurbitaceae in the order Passiflorales of polypetalae near Passifloraceae because of

1. Presence of bitegmic crossinucellate ovules.
2. Parietal placentation.

On the other hand, Engler and Prantl place Cucurbitaceae in the order cucurbitales of Sympetalae (Gamopetalae) due to the presence of gamopetalous corolla.

However, the close relationship of Cucurbitaceae with gamopetalous members like Campanulaceae has been confirmed by sero-taxonomical studies (Serum diagnostic tests) besides their tenency to form synandrium and similar nature of calyx and corolla in both the families.

4.8 APIACEAE (Umbelliferae) (The carrot family)

Systematic position :

Dicotyledonae
Polypetalae
Calyciflorae
Umbellales
Apiaceae

4.8.1 Distribution Apiaceae or Umbelliferae commonly known as the carrot family, is a large one with over 3,000 genera and 2,900 species. In India, it is represented by 53 genera and 200 species. Apiaceae members are commonly found in north-temparate regions. In tropical and sub-tropical areas they are distributed in the hilly areas or cultivated in the winter season.

4.8.2 Important Plants

1) *Apium graveolens* L. (Ajumud or Celery).
2) *Carum carvi* L (Caraway) Tel: Vaamu.
3) *C. copticum* Koch. (Ajowan)
4) *Centella asiatica* ((L) Urban. (Syn.*Hydrocotyle asiatica* (Penny wort) Tel: Brahmi.
5) *Daucus carota* (L)DC. (carrot)
6) *Coriandrum sativum* L (Coriandrum) Tel: Dhaniyalu
7) *Cuminum cyminum* L (Cumin) Tel: Jeelakarra
8) *Ferula asafoetida* Sprang. (Asafoetida) Tel: Inguva
9) *Foeniculum vulgare* Mill, (Fennel) Tel: Sopu
10) *Oenanthe javanica* (Bl) DC.

4.8.3 Habit and Habitat

Plants are mostly annual or perennial herbs (*Coriandrum*, *Carum* and *Hydrocotyle*) biannuals (*Daucus*) shrubs (*Ferula*, *Eryngium*) with aromatic odour due to the presence of schizogenous oil ducts called vittae in stems, leaves and mostly in fruits. Climbing shrubs (*Pseudocarum*) and rarely trees (*Aengelica* sp). *Centella* species are small prostrate herbs rooting at the nodes. The plants are mostly mesophytes.

4.8.4 Vegetative Characters

Root system: Branched taproot, tuberous roots (Conical) with stored food material are found in *Daucus carota*. Nodular roots present in *Centella*.

Stem: Aerial, erect, prostrate runner (*Hydrocotyle*), hollow, with swollen nodes some times ridged, internodes long, stem fistular and glacous.

Leaf: Radicle leaves in *Daucus*. Alternate phyllotaxy, *Apiastrum* shows opposite leaves, exstipulate, simple (*Hydrocotyle*) or usually much divided several times as in *Coriandrum* (fig 4.8.1A) and *Daucus*, petiole generally sheathing at the base. Certain species of *Eryngium*, *Aciphylla* have narrow paralled veined lamina with broad sheathing base like those of monocotyledons. Leaf reniform (*Hydrocotyle*) or palmately lobed (*Sanicula*).

4.8.5 Floral Characters As the name of the family indicates, the inflorescence is characteristically an umbel. It may be

1) a simple umbel (*Hydrocotyle*) or
2) A compound umbel surrounded by an involucre of bracts (*Foeniculum*, *Coriandrum*, *Daucus*) (fig 4.8.1A).

Each umbel also contains (where, flowers originate) an involuvre of bracts. The involucre is usually petaloid.

Umbel is reduced to a single flower (*Azorella*) while in *Eryngium giganteum* a compact head (cymose umbel) is found.

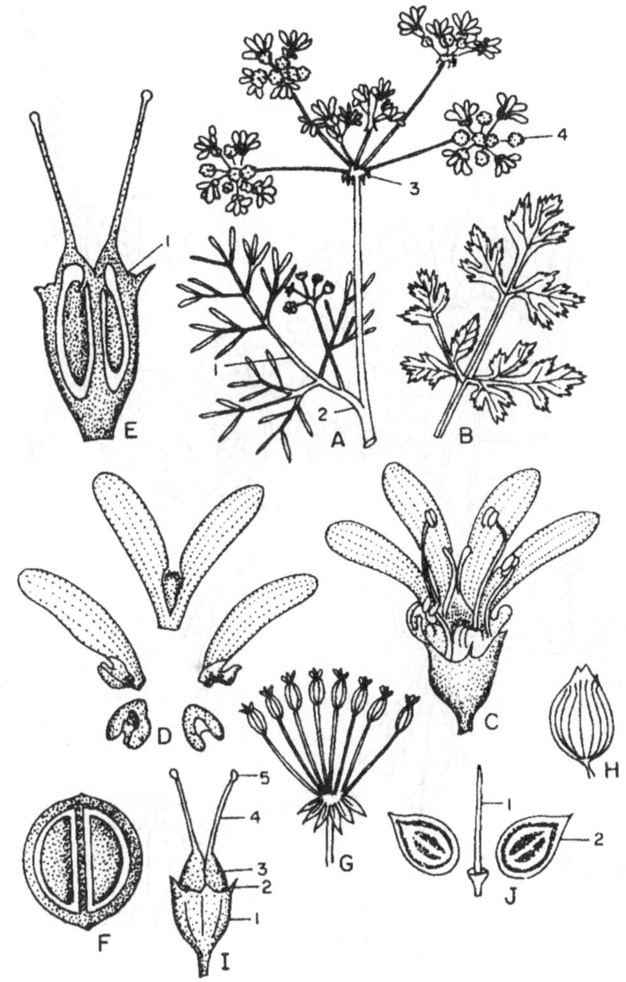

Fig 4.8.1 *Coriandrum sativum*

A: Flowering twig 1. Decmpound leaf, 2. Shealhing leaf base, 3. Involucre bracts, 4. Umbel
B: Lower leaf, C: Outer flower, D: Corolla of outer flower; E: L.S. Gynoecium 1. Disc, G: Umbel
H: Fruit, I. Gynoecium 1. Ovary, 2. Calyx,
3. Stylopodicm, 4. Style, 5. Stigma; J. Cremocarp 1. Carpophore, 2. Mericarp

Flower: Bracteate, pedicellate, complete, bisexual, small, usually perfect, white or yellow, rarely blue, showy, monoecious in *Echinophora*, dioecious (*Arctopus*), actionomorphic, In *Coriandrum* the outer flowers (peripheral) of umbel are usually zygomorphic. Flowers pentamerous and epigynous. (fig 4.8.2 C.)

Calyx: Sepals five, polysepalous, or gamosepalous as in *Hydrocotyle*, absent in *Apium*. If present, they are in the form of small teeth or scales, some times sepals adnate to the ovary. Aestivation valvate or imbricate.

Corolla: Petals five, polypetalous, In *Coriandrum* the flowers (peripheral flowers of umbel) have two small petals and are bifid which are posterior in position (fig 4.8.1C). The anterior and the lateral petals on the other hand are large and dissected unequally. Petals

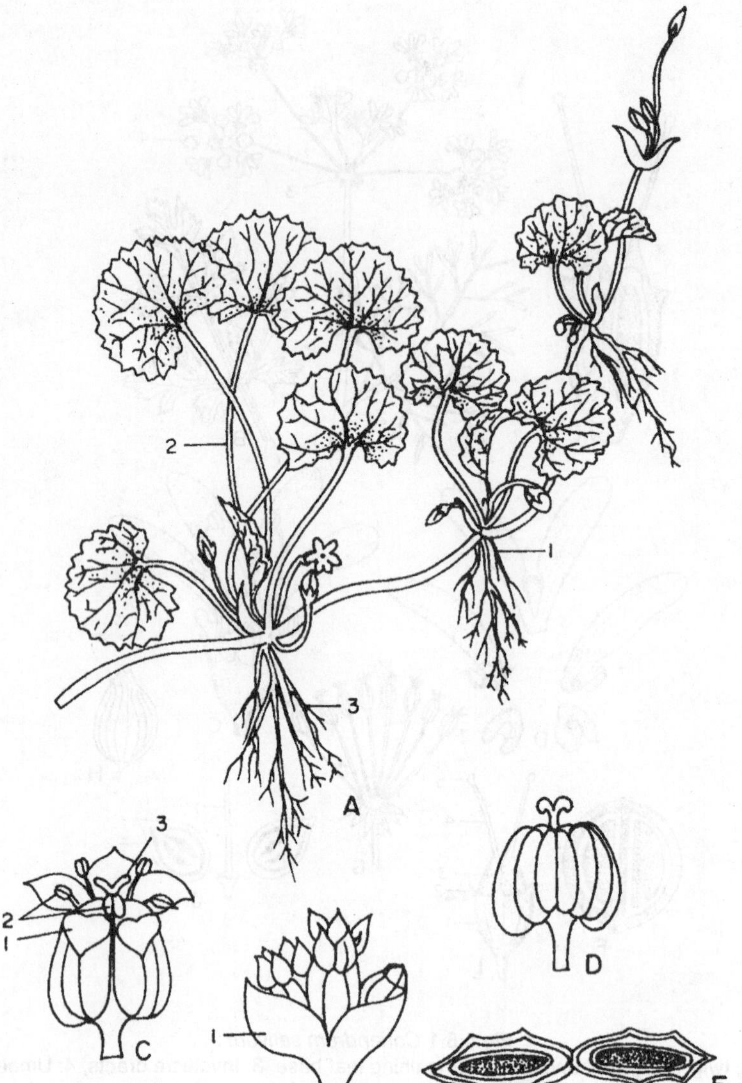

Fig 4.8.2 *Hydrocotyle asiatica*

A: Flowering twig 1. Stem, 2. Leaf, 3. Rootsystem; B: Simple umbel 1. Involucre of bracts. C: Flower 1. Petal, 2. Stamens, 3. Gynoecium, D: Fruit, E: T.S. of Fruit

usually white (*Coriandrum*) or yellow (*Foeniculum*). The tips of petals are inflexed in bud condition and fall off soon. Aestivation imbricate or valvate.

Androecium: Stamens five, free, alterni-petalous, inflexed in bud, anthers dithecous, introse, dehisce longitudinally, basifixed or dorsifixed.

Gynoecium: Bicarpellary, syncarpous, inferior, carpels placed anterio-posteriorly, ovary bilocular with single unitegemic, anatropous, pendulous ovule in each locule, placentation axile. Styles two, stigmas two and capitate, surrounded by a bi-lobed nectar secreting disc called stylopodium (fig. 4.8.1H).

Fruit: Schizocarpic cremocarp and splits into two one seeded indehiscent mericarps which remain attached to a slender forked axis called the carpophore. Each mericarp is usually marked by five longitudinal ridges called costae. They contain the vascular bundles. Between these ridges furrows are present which are known as valleculae under which oil ducts, the vittae are present (fig 4.8.1I, J & 4.8.2 E).

Seed: Endospermic with oil ducts. Embryo small. The mericarps being light weight are dispersed by wind. The edges of mericarps develop into wings (*Angelica)* or spinous outgrowths (*Daucus)* which facilitate either wind or animal dispersal.

Pollination: Entomophilous. The umbel inflorescence with aggregation of flowers attract large number of insect pollinators. The epigynous honey secreting disc (stylopodium)is easily accessible to short-tongued insects, flies and beetles which effect cross-pollination.

4.8.6 Distinguishing Characters

1. Herbaceous habit
2. Presence of decompound leaves
3. Sheathing leaf base
4. Hollow and ribbed stem
5. Simple or Compound umbel
6. Epigynous bisexual flowers
7. Inflexed petals and stamens in bud condition
8. Bicarpellary, syncarpous gynoecium
9. Presence of stylopodium
10. Schizocarpic cremocarp

4.8.7 Economic Importance

4.8.7.1 Spices and condiments: The seeds and leaves of *Coriandrum sativum (*Coriander) and seeds of F*oeniculum vulgare* (Saump), *Carum capticum* (Azwain*)*, *Cuminum cyminum* (Cumin), *Carum carvi* (Caraway) P*ipinella anisum* (Aniseed) and *Apium gravelolens* (Ajmud) are used as spice and condiments for culinary purposes. They are also used for flavouring bread, biscuits, cakes, cheese, confectionery, and liquors. The oil obtained from the seeds is used in perfumery.

Some species of *Ferula* namely *F. asafoetida* is an importance source of oleo-gum resins. Asafoetida (hing) is obtained from the dried latex after making incisions in living root stocks and roots. Asafoetida is extensively used as a condiment for flavering curries, sauces, pickles. It also has medicinal properties and finds use in curing indigestion problems and cardiac diseases.

*Centella asiatica (*syn. *Hydrocotyle asiatica* Brahmi) is an important source of medicine in the Indian system of medicine. It is used as hair oil and is a good tonic for improving mental abilities. It is also employed in the treatment of leprosy and possesses insecticidal properties.

4.8.7.2 Other Uses: The storage roots of *Daucus carota* are used as vegetables. Species of *Pimpinella* (Coweparsnip); *Heracieum, Angelica, Trachymene* (bluelac flower) and *Eryngium* (Sea holly) are grown as ornamental plants.

Conium (Hemlock), *Cicuta* and *Aethusa* are the poisonous plants. The fresh leaves and green fruits of *Conium* contain a volatile, oily alkali called conine which is poisonous. However, it is valuable in small doses in treatment of cancer and nervous disorders.

4.8.8 Systematic Position and Phylogeny

Bentham and Hooker placed Apiaceae as the last order of polypetalae i.e. umbellales. In the same way, Engler and Prantl included it in the umbellales of archichlymydae. Hutchinson classified the family in Herbaceae due to its hebaceous character.

Among the polypetalous dicotyledons, Apiaceae is a fairly advanced family because of number of attributes. Some of these include, 1) Arrangement of flowers in clusters forming simple or compound umbels which to some extent resemble head inflorescence of Asteraceae 2) The outer flowers in the umbel are sterile or male and having a tendency towards zygomorphy due to variations in size of the petals 3) Reduction in number of sepals 4) Bicarpellary inferior ovary 5) Presence of two distinct styles 6) Presence of honey disc and 7) Cremocarpic fruit.

4.9 RUBIACEAE (The Coffee Family)

Systematic position:

Dicotyledonae

Gamopetalae

Inferae

Rubiales

Rubiaceae

4.9.1 Distribution

Rubiaceae is a very large family comprising of 500 genera and 7,800 species. Members of the family are mostly distributed in tropical regions and some grow in temperate zones. Species of *Galium* occur in arctic regions. In India, Rubiaceae is represented by about 76 genera and 274 species distributed in the mountains of southern and western India. Many species are cultivated.

4.9.2 Important Plants

1. *Adina cordifolia* Hook f. (Saffron teak) Tel: Bandaru or Kamba.
2. *Anthoceptalus cadamba* Miq. (Kadam) Tel: Kadambam.
3. *Borreria hispida* L Tel: Madana budama
4. *Cephalis ipecactuanha* Rich. (ipecac)
5. *Cinchona officinalis* L. (quinine tree) or Cinchona
6. *Coffea arabica* L. (Coffee)
7. *C.liberica* Hievn (Liberican Coffee)
8. *C robusta pierre*. (Congo coffee)
9. *Gardinia jasminoides* Ellis. (Cape jasmine) Tel: Scentumalle.
10. *Hamelia patens* Jacq
11. *Ixora coccinea* L. Tel: Ramabhanam
12. *I. bractiate* L. Tel: Koravi
13. *Morinda tinctoria* L. Tel: Togaru
14. *Oldenlandia umbellata* L. Tel: Verrinelavemu
15. *O.corymbosa* L.
16. *Rubia tinctoria* L. Tel. Mangista
17. *Randia spinosa* L (Comon emectic nut) Tel: chinnamanga.

4.9.3 Habit and Habitat Mostly mesoplytes some species of *Randia* are xerophytes. Myrmecophily is common in *Hydnophytum myrmecodia* and *Nauclea* which are epithytes.

Majority of Rubiaceae are trees (*Anthocephalus, Adina*) shrubs (*Ixora, Hamiltonia)* and climbers (*Gallium*) which climb by hooks or stiff hairs *Uncaria* is a liane.

4.9.4 Vegetative Characters

Root system; Much branched tap root system.

Stem: Aerial, erect, runner in *Geophyla* twinners (*Manettia)*; hairy or smooth.

Leaf: Simple, opposite decussate, (*Gallium)* Stipulate, stipules are characteristically interpetiolar (The stipules stand between the petioles (*Ixora*) (fig 4.9.1) or intrapetiolar (The stipules stand between the petiole and the axis) as in *Gardenia* and *Coffea* sp. The two stipules one from each leaf form a pair and often are united as in *Gardenia* sp and related genera, the four stipules are united into a conical cap which is thrown off as the leaf bud opens. In *Gallium* and others, the stipules become foliar and are indistinguishable from leaves. The stipules are reduced to glandular setae (*Pentas*) or bristle like (*Borreria*) leafy nodules with symbiotic bacteria are found in *Pavetta*.

4.9.5 Floral Characters

Inflorescence: Dichasial cyme in number of species, (fig 4.9.2) In a few cases (*Anthocephalus* and *Adina)* small flowered cymes aggregate into dense heads. Inflorescence solitary in *Gardenia*, in *Coffea* one to three flowers are present in the axils of a leaf. Panicle inflorescence found in *Heydatis paniculata*, corymb (*H corymbosa*) and umbel *(H umbellata*).

Flower: Bracteate, pedicellate, complete, bisexual, rarely unisexual (*Anthospermum*) actinomorphic or zygomorphic (*Henriquezia*) Tetaramerous, epigynous or rarely perigynous or hypogynous. (fig 4.9.1 & 4.9.2)

Calyx: Sepals four, polysepalous, sepals usually small or reduced to a ring like rim (*Morinda*), totally supressed in *Rubia, Galium*. In *Mussaenda* one or more sepals are enlarged and petaloid. These sepals are described as advertisement flags. Persistent in *Anthocphalus*.

Corolla: Usually four, gamopetalous, valvate (*Ixora)* twisted (*Heydotis*) rarely twisted (*Guettarda*) tubular (*Hemelia)* or may be rotate or funnel shaped. *Henriquenzia* shows bilipped corolla.

Androecium: The stamens are as many as the number of petals and alternate with them, epipetalous, inserted in or present at the mouth of corolla tube. Anthers dithecous, basifixed and introse (fig 4.9.1) .

Gynoecium: Ovary inferior or rarely half-inferior (*Gaertinera*) bicarpellary, syncarpous ovary bilocular, with axile placentaion. Ovule one (*Coffee*) or numerous (*Cinchona*) ovules in each locule. Sometimes placenta is 'T' shaped (*Mussaenda*) or stalked (*Oldenlandia*) In *Gardenia* ovary is unilocular with basal placentation.

Style long, smiple, stigma bifid. Characteristic heterostyly is observed in many genera (eg. *Oldenlandia*) with pin and thrum forms. A nectariferous disc is presenct at the base of style. (fig 4.9.2)

Fruit: A berry (*Mussaenda Coffea* and *Randia*) capsule (*Cinchona*), Drupe (*Gardenia*) and compound fruit (*Morinda).*

Seed: Small, endospermic, straight or curved embryo. Embryo cartilaginous in *Coffea* and winged in *Cinchona* and *Henriquezia*.

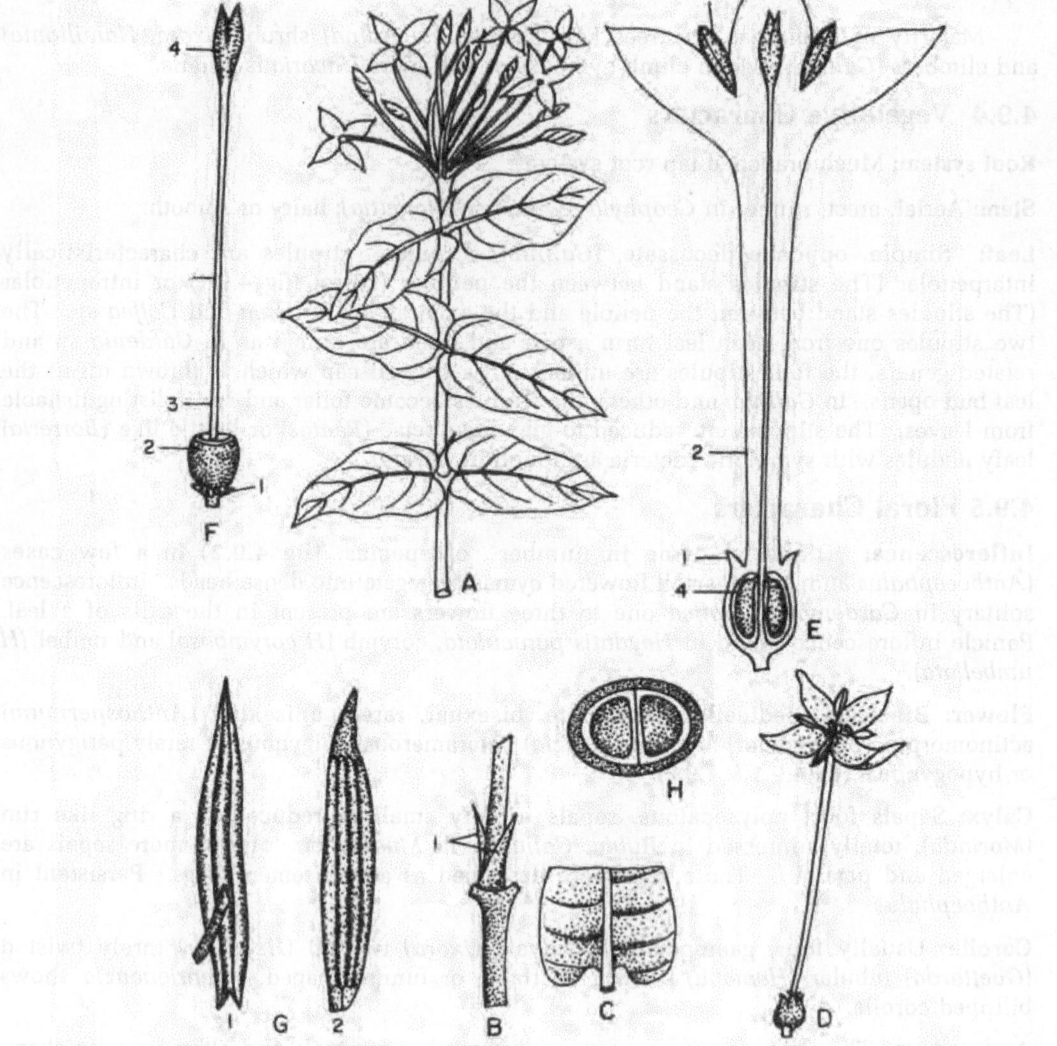

Fig 4.9.1. *Ixora coccinia*

A: Flowering twig ; B: Stem portion 1. Interpetiolar stipules; C: Leaf lower part; D: Flower; E: L.S. Flower, 1. Calvx, 2. Corolla tube, 3. Stamen (epipetalous) 4. Ovary; F: Gynoecium 1. Fallen leaf scar, 2. Ovary, 3. style, 4. Stigma; G: 1&2 Dorsal and ventral faces of stame, H: T.S. Ovary; I. Floral diagram

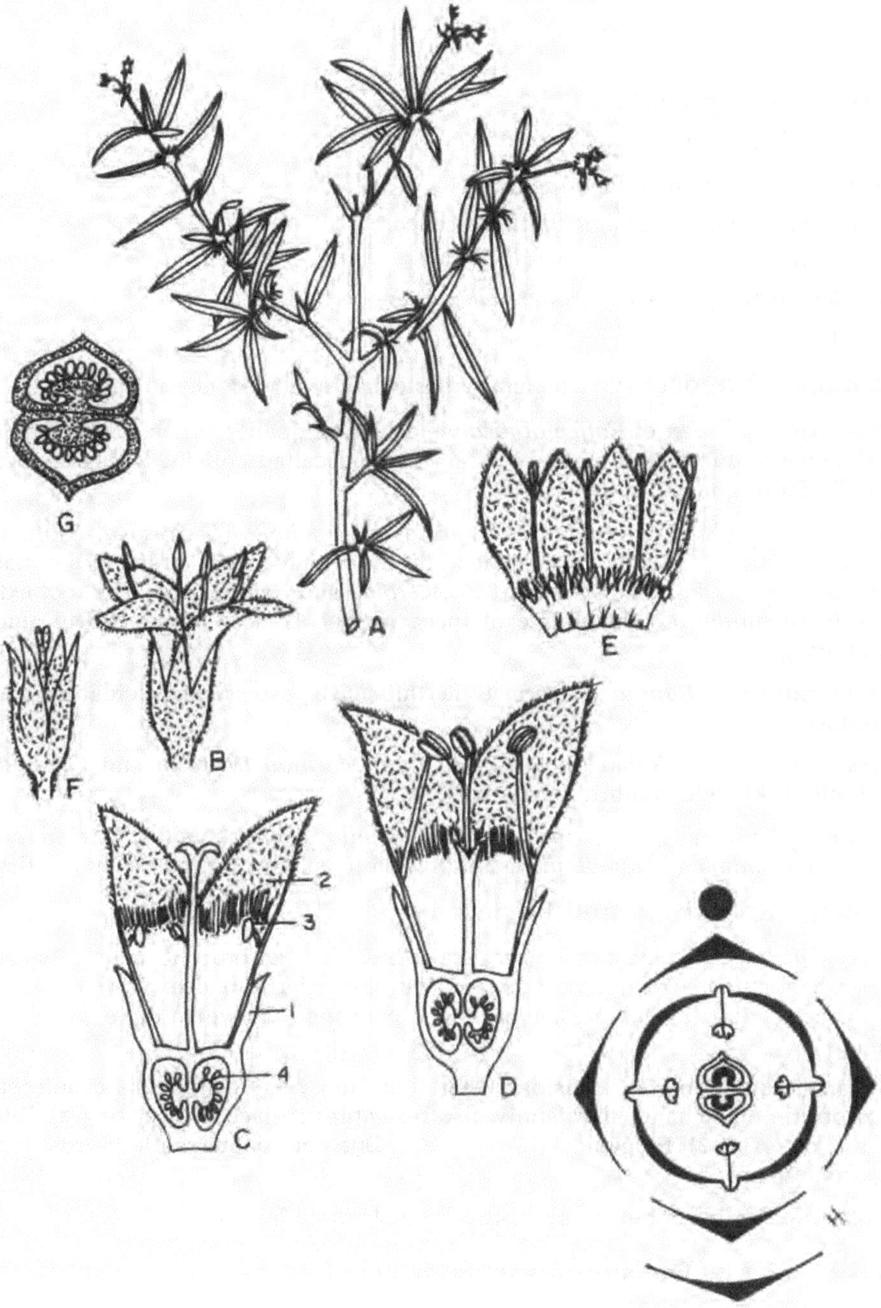

Fig 4.9.2. *Oldenlandia umbellata*
A: Flowering twig ; B: Flower with short stigma; C&D L.S. Flower, 1 Sepal, 2. Petal, 3. Stamen, 4. Ovary, E: Corolla opened showing epipetalous stamens; F: Gynoecium, G. TS Ovary; H: Floral diagram

Pollination: Pollination is brought about by insects which visit flowers for attractive aggregated infloressences and nectar. The wide accurance of heterostyly and protandry in several genera favour cross-pollination. Chirapterophyly (pollination by bats) is seen in *Anthocephalus.*

4.9.5. Distinguishing Features

1) Opposite decussate phyllotaxy.
2) Interpetiolar and intrapetiolar stipules.
3) Cymose, dichasial cyme or umbel inflorescence.
4) Epigynous flowers.
5) Bicarpellary syncarpous ovary, axile palcenation
6) Endospermic seeds.

4.9.6 Economic Importance The family has considerable economic value.

4.9.6.1 Beverages: The seed of *Coffea arabca* yield highest quality coffee. The stimulating property of coffee is due to the presence of an alkaloid called coffein. *C. liberica* and *C. robusta* yield inferior quality coffee.

4.9.6.2 Drugs: The bark of roots of several species of *Cinchona* (*C calisaya, C officinalis* and *C succurubra)* yield a valuable drug called quinine which acts as a remedy for malaria. Quinine is also obtained from *Exostemma, Ladenberbion* and *Remigia* spp. The root extracts of *Cephalis ipecachunha* are the source of ipecac which is used to control of amoebic dysentry and cough.

The fruit extract of *Randia tinctoria* (Emerthic nut) possess insecticidal and insect repellent properties.

4.9.6.3 Dyes: The roots of *Rubia tinctoria* (red dye) *Morinda tinctoria* and *Oldenlandia umbellata* (dull pink) yield colouring material.

4.9.6.4 Ornamental plants: Various species of *Mussaenda, Anthocephalus, Morinda, Ixora, Pentas, Hamelia, Galium* and *Hamiltonia* are cultivated in gardens for their beautiful flowers.

4.9.7 Systematic Position and Phylogeny

Rubiaceae according to most taxonomists just preceeds the order to which Asteraceae belongs. Bentham and Hooker placed this family in the order Rubiales. Engler and Prantl placed Rubiaceae also in Rubiales, which is positioned between Plantaginales and Cucurbitales.

Takhtajan (1980) however, considers that this family is related to Loganiaceae of Gentianales on the other hand, this family also resembles Apiaceae in having 1) Tetra or pentamerous flowers ; 2) Epigynous flowers ; 3) One anatropous ovule in each locule ; 4) Endospermic seeds.

These features indicate the probable origin of Rubiaceae from Umbellales.

4.10 ASTERACEAE (The sun-flower family)

Systematic position:

Dicotyledonae
Gramopetalae
Inferae
Asterales
Asteraceae

4.10.1 Distribution

Asteraceae is the second largest family in the Angiosperms with over 900 genera and 15,000 species which comprise more than 10 percent of flowering plants. The members of the family have worldwide distribution in various habitats. In India, the family is represented by 140 genera and 700 species.

4.10.2 Important Plants

1. *Artemisia vulgaris* L. Tel. Machapatri
2. *Ageratum conyzoides* L (Goat weed)
3. *Acanthospermum hipidum* L.
4. *Aster amellus* L.
5. *Blumea solidaginoids* L. Druce. Tel. Kukka Pogaku
6. *Calendula officinalics* L. (pot marigold)
7. *Cichorium intybus* L (Chikori)
8. *Centaurea moschata* (Sweet william)
9. *Centrantherm anthelmintium* L.
10. *Chrysanthemum coccineum* L. Tel. Adavi jeelakarra
11. *C. parthenium* Tel: Chamanthi
12. *Cosmos bipinnata L* Tel. Damidi chamanti
13. *Dahlia pinnata* L. (Dahlia)
14. *Echinopsis echinatus* Roxb.
15. *Eclipta alba* L. Tel. Gunta kalagara or Bhringaraja
16. *Elephantopus scaber L.* Tel: Nela marri
17. *Helianthus annus L.* (Sunflower): Tel. Poddutirugudu or Suryakanthi
18. *H. tuberosus* L. (Jerusalem artichoke)
19. *Lactuca sativa* L. (Lettuce)
19. *Parthenim histerophorus* L. (Congress weed) Tel. Vayyaribhama
21. *Pyrethrum cineraraefolium* L. Tel. Kumkuma puvu
22. *Solidago virgaurea* L. (Golden rod)
23. *Sphaeranthus indicus* L. Tel. Bodasaram
24. *Tagetus patula* L. (Marigold) Tel: Banti
25. *Vernonia cinerea (*L). Less
26. *Tridax procumbens* L. Tel: Gaddi Chamanti
27. *Xanthium stumarium* L. (Cockelbur)
28. *Zinnia elegans* Jacob. (Garden Zinnia) Tel: Bangala banthi.

4.10.3 Habit and Habitat

Mostly herbs, either annual or perennial *(Helianthus, Tagetus, Calendula)* some are shrubs (*Inula, Vernonia)* rarely trees (*Vernonia arborea, Senecia johnstorui*) Climbing shrubs (*Vernonia illignifolia), Senecio macroglossus* and *Mutisia.* Many genera are mesophytes. Few genera xerophytes (*Senecio articularis, Halophyllum, Proustia)* and Hydrophytes *(Bidens*

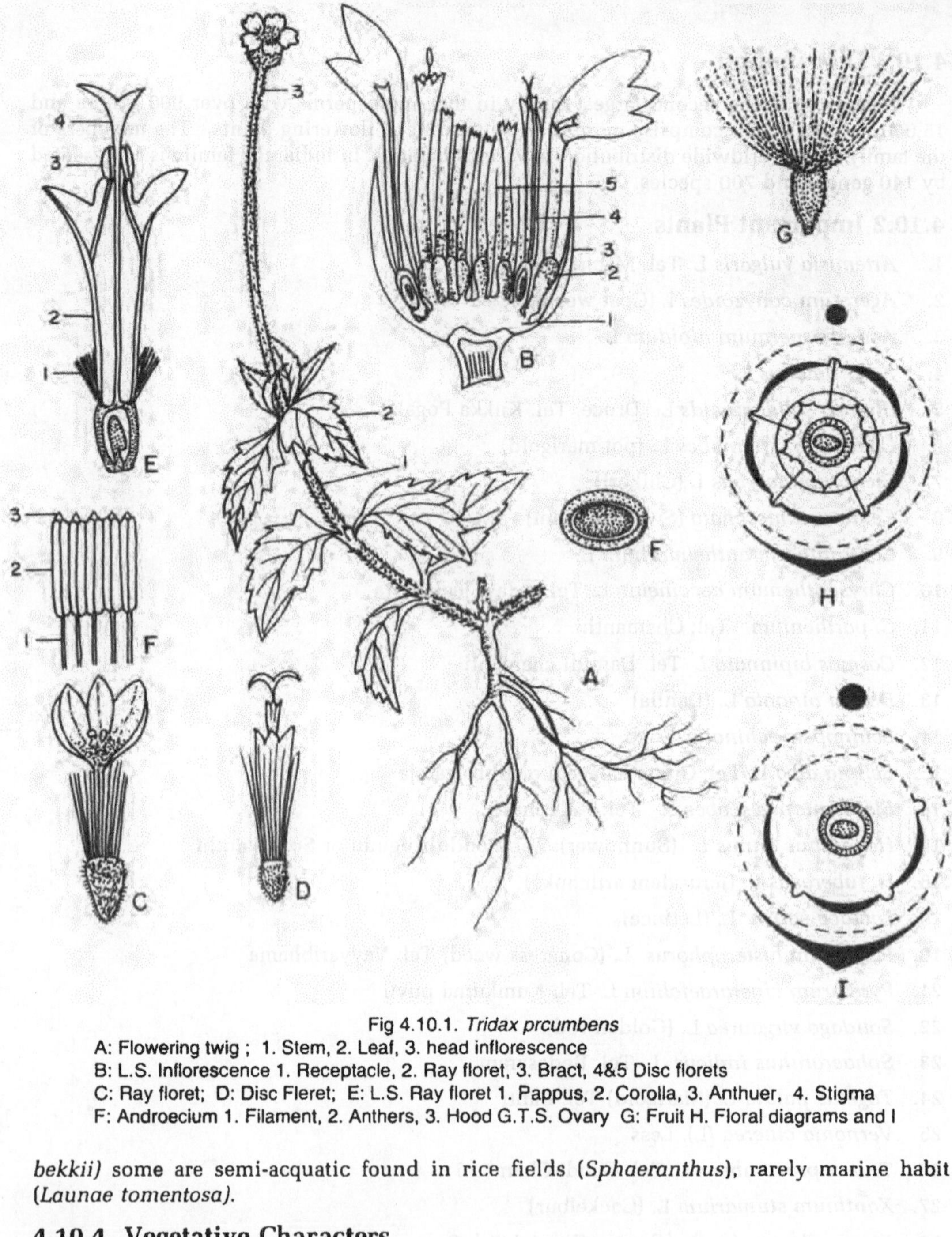

Fig 4.10.1. *Tridax prcumbens*

A: Flowering twig ; 1. Stem, 2. Leaf, 3. head inflorescence
B: L.S. Inflorescence 1. Receptacle, 2. Ray floret, 3. Bract, 4&5 Disc florets
C: Ray floret; D: Disc Fleret; E: L.S. Ray floret 1. Pappus, 2. Corolla, 3. Anthuer, 4. Stigma
F: Androecium 1. Filament, 2. Anthers, 3. Hood G.T.S. Ovary G: Fruit H. Floral diagrams and I

bekkii) some are semi-acquatic found in rice fields (*Sphaeranthus*), rarely marine habit (*Launae tomentosa)*.

4.10.4 Vegetative Characters

Root system:- Usually tap root, root-tubers are found in *Helianthus tuberosus*,and *Dahlia* Roots store food in the form of inulin.

Stem: Aerial, errect or prostrate (fig. 4.10.1A), rarely climbing, some times woody below (*Artemsia*) runner (*Launea)* underground tuberous stem (*Helianthus tuberosus)* stem hollow, hairy or smooth. Succulent in *Senecio*, some stems contain milky latex (*Taraxacum)*.

Leaf: Radical or cauline or both types on the same plant, usually exstipulate, pedicellate, sessile (*Inula, Zinnia)*; leaf simple or many dissected (*Chrysanthemum, Parathenium)* rarely pinnate compound (*Dahlia*), Ovate, serrate, hairy or smooth, leaf surface spiny (*Acanthospermum, Echinops*) succulent (*Notonia*). In *Nutisea* the mid vein extends and modified into a tendril. *Biders bekki* shows heterophylly. Venation pinnately reticulate, however, *Corymbirum* shows parallel venation, phyllotaxy usually opposite decussate, alternate (*Vernonia*) rarely whorled (*Zinnia verticillata*).

4.10.5 Foral Characters

The inflorescence typically a head or capitulum which consists of a large number of sessile flowers and are commonly described as florets. The florets are grouped on a common disc like (*Helianthus)* (fig 10.2) convex (*Tagetus*) or cup shaped (*Sphaeranthes*) axis called the receptacle. However, a single flower is present in the capitulum. eg: *Echinops*.

The capitula are surrounded by a whorl of bracts called the involucre thus producing a flower cluster resembling a single flower. A true involucre is absent in *Wilkesia*. The involucre if present may be in one whorl (*Emilia*) or in two to three whorls (*Tridax*). Involure looks like a leaf (*Helianthus*),(fig 4.10.2B) leathery (*Scinera*) and petaloid and persistent in *Helichrysum*.

The heads are solitary (*Tridax*) (fig 4.10.1) solitary or axillary terminal in *Sphaeranthus* the heads are aggreagated into a compound head or look like an umbel (*Achillea*) or a compound panicle as seen in *Solidago*.

The heads or capitula are recognised into two groups.

1. Homogeneous heads
2. Heterogeneous heads

If the flowers are of same type in a capitulum such capitula are described as homogenous head or if flowers are of two different types, it is known as heterogeneous head. In a homogeneous head all the flowers may be bisexual, actinomorphic and tubular (eg: *Vernonia*) or all are bisexual. In a heterogeneous head, the arrangement of disc and ray florets may vary. In *Tridax*, the ray florets are present on the periphery and the disc florets at the centre of the receptacle (fig 4.10.1) on the other hand, in *Blumea*, the capitulum consists of only tubular florets, where, the flowers present at the periphery are unisexual female flowers where as the disc florets are bisexual. In *Xanthium*, male and female capitula are present on two different plants.

Flower : Usually bracteate, sessile, the tubular florets are complete, actinomorphic and bisexual, the ray florets are zygomorphic, ligulate, unisexual female flowers, flowers epigynous.

Calyx: Modifies into pappus, persistent. In some genera (*Heleanthus, Ageratum*) calyx modifies into a scaly structure. Calyx absent in *Ambrosia siegesbeckia* and in *Xanthium*,

Corolla: Petals five or three tubular in disc florets or ligulate (*Helianthus*) in ray florets some times bilabiate (*Mutisea*) gamopetalous, aestivation valvate. (Fig 4.10.2)

Androecium: Five stamens, syngenecious (anthers united and filaments free) (fig 4.10.1F) epipetalous, the connective extends beyond anther lobes and forms a hood like structure, dithecous, introse *Xanthium*, however, has free stamens.

Gynoecium: Bicarpellary syncarpous, ovary inferior with single locule. A anatropous ovule is present at the base of every, on basal placentation. Stigmas two and bifid, A tuft of haris are present at the base of stigma. A nectariferous disc is present at the base of style.

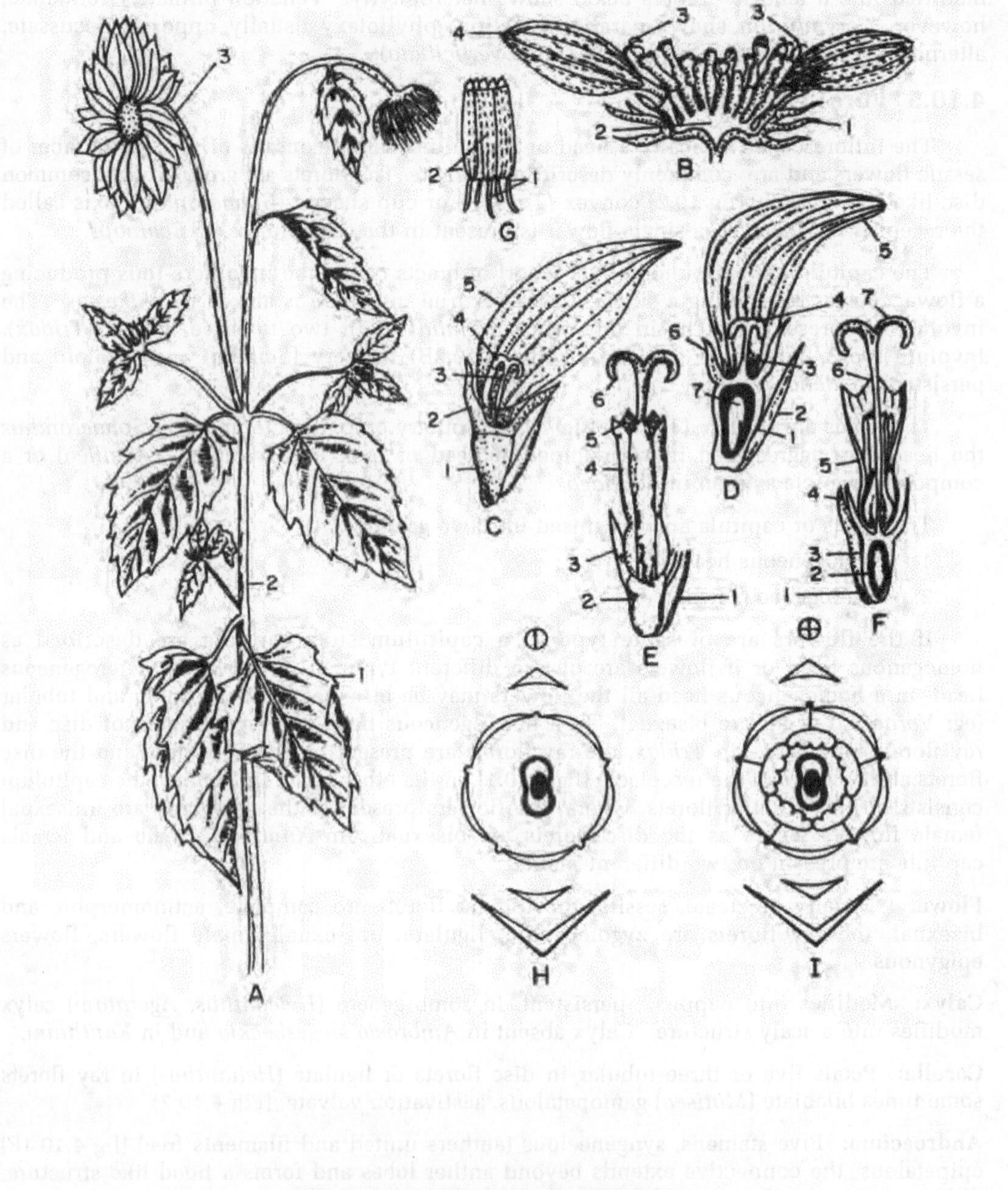

Fig 4.10.2. *Helianthus annus* (Sunflower)

A: Flowering twig ; 1. Leaf, 2. Stem, 3. Inflorescence; B: L.S. Head inflorescence 1. Involucre of bracts 2. Fleshy receptacle, 3. Disc florets, 4. Ray floret; C: Ray floret; 1. Ovary, 2. Calyx, 3. Style, 4. Stigma, 5. Petal; D: L.S. Ray floret 1. ovary, 2. Bract, 3. style, 4. Stigma, 5. Petal, 6. Sepal, 7. Ovule (basal); E: Disc Floret: 1. Bract, 2. ovary, 3. Sepal, 4. ract, 5. Stamen, 6. Style; F: L.S. Disc Floret: 1. Bract, 2. Ovary, 3. Sepal, 4. Bract, 5. Stamen, 6. Style, 7. Stigma

Fruit: A indehicent cypsela. (fig. 4.10.1G). Due to the presence of persistent pappus, the fruits are carried to long distances by parachute mechanism through wind (eg: *Tridax*). In *Xanthium*, the fruits are dispersed by animals due tǒ the presence of hooks. Fruit dispersal by means of animals is also seen in *Siegesbeckia* (due to the presence of mucilage hairs) (Fig 4.10.1 & 4.10.2)

Pollination: In many genera protandry is a common feature, which prevents sefl-pollination. Though the flowers are small, the presence of capitulum due to their aggregation makes them atractive for the insects, the nectar secreted by the flowers also available, thus, paving the way for entamophily. In view of this, a single insect can pollinate a number of flowers in a single visit. Further, the flowers are also adopted to a great variety of insects which is the chief cause of their wide distribution under varied habitats.

The narrow anthers are united to form a hollow cyllinder into which the pollen is liberated.

4.10.6 Distinguishing Features

1. Plants mostly herbs, some are climbers.
2. Leaves usually exstipulate
3. Homogamous or heterogamous head inflorescence
4. Flowers epigynous, actinomorphic or zygomorphic, unisexual or bi-sexual
5. Calyx is reduced to pappus or scales
6. Stamens epipetalous with syngenecious anthers.
7. The connective of anthers elongated into a hood
8. Ovary inferior, bicarpellary syncarpous, unilocular
9. Basal placentation
10. Ovules anatropous
11. Stigma bifid
12. Cypsela Fruit
13. Perfect mechanism for fruit and seed dipsersal with parachute mechanism

4.10.7 Economic Importance:-

Asteraceae provides variety of economically important plants,

4.107.1 Ornamentals:- The family Asteraceae includes several ornamentals. They are cultivated for their beautiful flowers, These include,

Calandula spp; *Chrysanthemum indicum, C. parthenium, Cosmos bipinnata, Tagetus, petula, Aster* spp, *Helianthus annus, Senecio* spp, *Solidago vellutina* and *Tithonia* spp.

4.10.7.2 Edible and Medicinal oils:- Edible oil is extracted from the seeds of *Helianthus annus Carthamus tinctorius* and *Giegotia abiciniam* the seeds of *Eclipta frostrata* yield an oil known as bringharaia. *Arnica montana* seed yield arnica oil which is extensively used in nervous disorders.

4.10.7.3 Medicines and Insecticides:- The plants of *Grindelia, camphorum* are useful in controlling chronic cough and *Artemisia vulgaris* in renal diseases. The well known bio-insecticide pyrethrin is obtained from *Pyrethrum cinerarifolium* flow flower heads.

4.10.7.4 Others:- The corms of *Chicorium intybus* yield the chickory which is mixed with coffee for flavour. The tubers of *Helianthus, tuberosum* are edible, leaves of *Lactuca sativa* used as vegetable.

The rag weed (*Ambrosia*), golden rods (*Solidago*) are causes of hay-fever, certain species of *Xanthium* and *Solidago* are poisonous to live stock.

4.10.8 Advanced Features and Phylogeny

Bentham and Hooker classified the family under gamopetalae and placed in series inferae and order Asterales after Rubiales. Hutchinson put this family under 67th series of Asterales far below the Lamiales and did not consider it to be an advanced family. However, Engler and Prantl considered Asteraceae as an advanced family and placed as the last family of Dicotyledens and grouped under order Companulatae. Rendle also opined that the Asteraceae is the highly evolved family of Dicotyledons.

Following are the advanced features of the family Asteraceae,

1. Herbaceous nature of the majority taxa
2. Aggregation of flowers into heads or capitula with attractive ligulate florets and dense tubular disc florets
3. The capitula are surrounded by conspicous involucral bracts
4. The construction of capitula enables a single visit of insect can help in pollinating several florets, there is no necessity for the insect to fly from one flower to another.
5. The capitula exhibit many adaptations in the flowers for cross-pollination by insects. The nectar is secreted at the base of a narrow tubular corolla and accessible to very short tongued insects and is protected by the anthers from rain, The nectar may also be sucked by bees and also by most specialised type of insects.
6. Presence of syngenecious anthers, protandry of androecium ensures cross pollination
7. Occurrence of self-pollination, in the event when cross-pollination fails is yet an advanced feature.
8. Calyx modified into pappus, reduction in number of carpels, presence of a single ovule also confirms highly evolved stage in floral morphology.
9. Highly conducive mechanism for the dispersal and distribution of fruits and seeds due to the presence of parachute mechanism. In some cases, the pappus is hygroscopic and spread out in dry air. In some genera the pappus is glandular and sticky and the fruits are carried away by animals, In some other plants it is in the form of barbed bristles and fruit stick to the body of the animals and are dispersed effectively.

The family shows affinity with Rubiaceae, Dispaceae and Fabaceae taking into consideration the polyphyletic origin of sympetalous families from polypetalous ancestors, hence, the origin of Rubiales and Asterales started from Apiaceae which shows a tendency of aggregation of flowers and reduction of calyx lobes, carpels and ovules.

4.11 SAPOTACEAE (The Sapodilla family)

Systematic Position

Dicotyledonae

Gamopetalae

Heteromerae

Ebenales

Sapotaceae

4.11.1 Distribution Sapotaceae is represented by 40 genera and 800 species mainly distributed in the tropical and sub-tropical regions of the world. About 12 genera and 23 species are found in the North Eastern and southern parts of India.

4.11.2 Important Plants

1. *Achras sapota L.* (Syn. Manilkara achras (Mill) Forberg. (Sapodilla Pulm) Tel: Sapota
2. *Chrysophyllum roxburghii L.* (star apple)
3. *Diploknema butgracea* (Roxb) H.J. Lamb.
4. *Eucuma nervosa L.* (caristel)
5. *Madhuca indica Gmel.* (Mohwa tree) Tel: ippa.
6. *Manilkara hexandra* (Roxb.) Dub. Tel: Pala
7. *Mimusops elengi L.* (Bullet wood or Indian medlar) Tel. Pogada
8. *Palaquium ellipticum* Burck. Tel: gutta percha
9. *Sideroxylon tomentosum L.* (syn. *Xantolis tormentosa* (L)).
10. *Sarcosperma arboscum* Hook. F.

4.11.3 Habit and Habitat Mostly woody trees or shrubs and generally ever green with abundant milky latex in laticiferous ducts. The leaves, flowers and fruits are often covered with unicellular brown coloured tomentose hairs. Plants usually mesophytes.

4.11.4. Vegetative Characters

Root system: Tap root system. Butresses roots are found in some genera.

Stem : Aerial, erect, branched and woody, covered with hairs.

Leaf : Simple, alternate, rarely opposite (*Sarcosperma*) exstipulate, leaf surface is coriaceous or leathery covered by silky tomentose hairs (*Madhuca*). Pinnately reticulate venation. The leaves often clustered towards the ends of the branches (*Achras*) (fig 4.11.1A).

4.11.5 Floral Characters

Inflorescence : The inflorescence some times arises on old branches (cauliflory) as in *Dichopsis* and *Madhuca*. The flowers may be solitary or arranged in cymose clusters in the leaf axils. In *Sarcosperma* the inflorescence is either a simple or compound panicle.

Flower : Bracteate, pedicillate, complete, bisexual, actinomorphic, trimerous (*Achras*) tetramerous (*Manilkera hexandra*) or pentamerous (*Sideroxylon*), hypogynous.

Calyx : Four to eight sepals, usually connate at the base arranged in two whorls of 4 + 4, (*Manilkara hexandra*); 3+3 (*Achras*); 2+2 (*Madhuca*) or in one whorl of 5 sepals (*Chrysophyllum*), outer whorl of sepals persistent in *Mimusops*; Aestivation imbricate but when the calyx is biseriate, the outer whorl is valvate.

Corolla : Gamopetalous, number of petals are equivalent or double to the number of sepals but however, arranged in one whorl only. Alternipetalous. In *Mimusops* each petal (8 in number) show a pair of out-growth on its dorsal side. Aestivation twisted. In *Achras* aestivation is imbricate. The corolla (10-12 in number) is thick, fleshy and deciduous. (fig 4.11.1).

Androecium : Definite or indefinite in number and usually arranged in two to three whols inserted upon the corolla tube (epipetalous). In *Madhuca* the stamens are in two whorls of 3 each which are fertile and 6 stamens present in the outer whorl are petaloid staminodes. In *Achras* the stamens are 6 + 6 of which outer whorl are sterile. (fig 4.11.1 C) Androecium

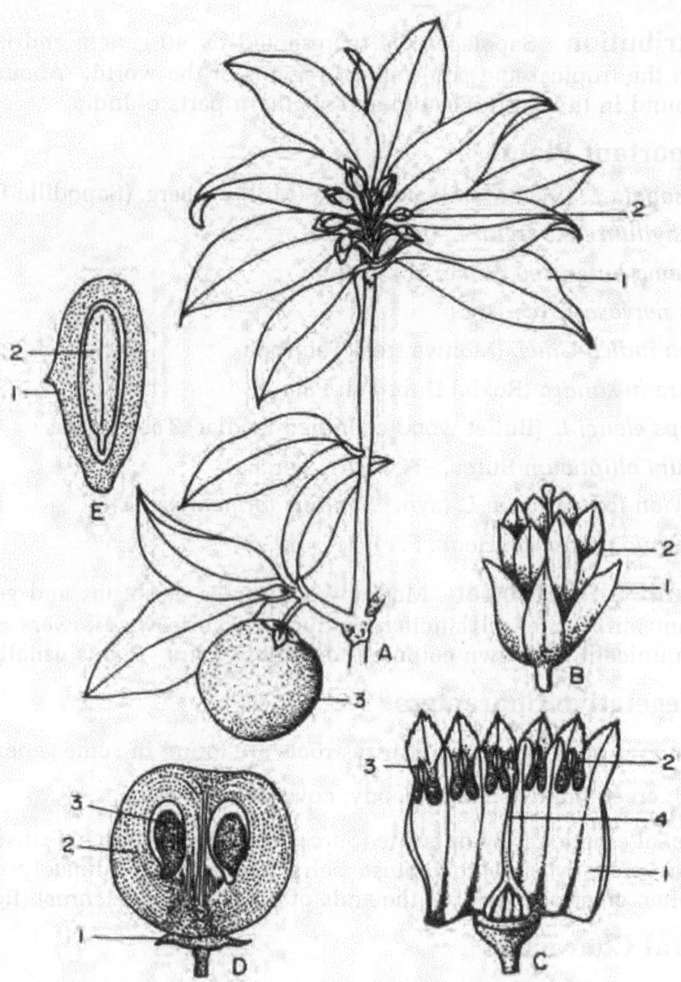

Fig 4.11.1. Achras sapota
A: Flowering twig ; 1. Leaf, 2. Flowers, 3. Fruit; B: Flower 1. Calyx, 2. Corolla, C.L.S. Flower, 1. Petal, 2. Fertile anthers, 3. Staminode, 4. Gynoecium; D: Fruit L.S. 1. Calyx, 2. Freshy fruit wall, 3. Seed ; E: Seed L.S. 1. testa, 2. Embryo, F: Flower L.S. G Floral diagram

is represented by 8 fertile stamens alternating with 8 sub-petaloid staminodes in *Mimusops* in a single whorl (fig 4.11.2C). *Palaquium* shows 18 stamens arranged in 6 per whorl and all are usually fertile.

The filaments are usually short and connective emerges protruding the anthers. Anthers oblong-lanceolate in shape, dithecous extrose, dehiscing by longitudinal slits (fig 4.11.2 D).

Gynoecium : Two - eight or some times carpels in indefinite number, (*Madhuca*) syncarpous, ovary superior with as many locules as the number of carpels. Each locule consists of a solitary, erect, anatropous, unitegmic ovule on axile placentation. In *Restonia* placentation, however, is basal. Stigma inconspicuous, secretes mucilage.

Fruit : The fruit is a one to eight seeded berry (fig 4.11.1D). It is surrounded by sclerenchymatous outer layer and the seeds are embedded in the fleshy pulp. Latex sacs are also present.

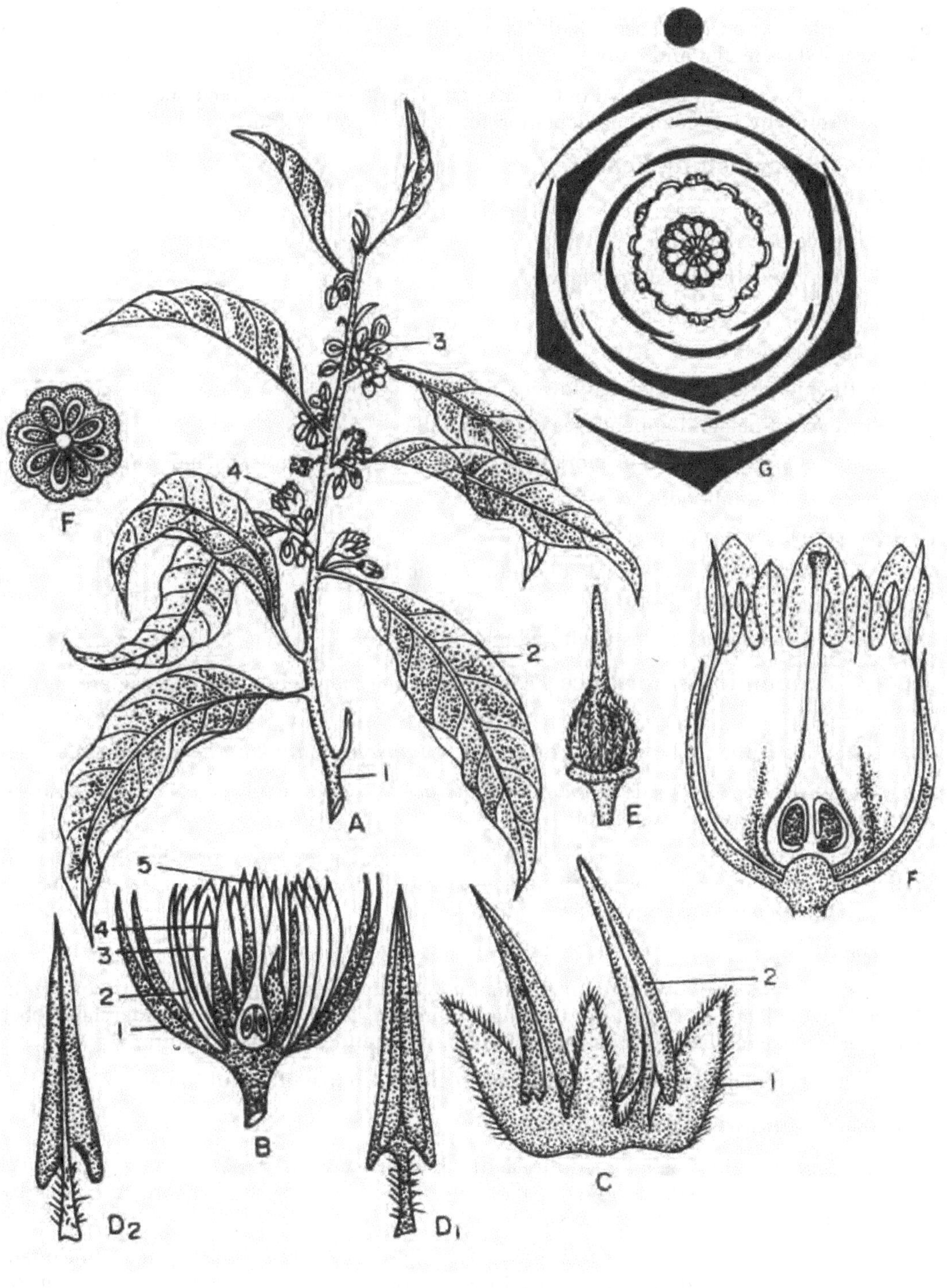

Fig 4.11.2. *Mimusops elengi*

A: Flowering twig 1. Stem, 2. Leaf, 3. Inflorescence, 4. Flower; B: Flower L.S. 1. Sepal, 2. Petal, 3. Staminode, 4. Fertile Stamen, 5. Gynoecium, C: Staminal tube 1. Staminode, 2. Fertile Stamen D: Stamen D1 Dorsal view D2 Ventral view E: Gynoecium, F: T.S. Ovary

Seed : Endospermic, oily. seed with hard seed coat and shiny testa with broad white hilum. Cotyledons flat and embryo straight.

Pollination : Pollination is carried out by insects. The flowers are scented and with abundant nectar. Protogyny is a common adaptation for cross — pollination.

4.11.6 Distinguishing Features

1. Mostly trees
2. Presence of milky latex
3. Leathery, exstipulate simple leaves.
4. Solitary or cymose inflorescence
5. Many whorled flower parts
6. Actinomorphic, hypogynous flowers
7. Presence of petaloid appendages on the corolla
8. Stamens are as many as or more than the number of corolla lobes and epipetalous
9. Presence of staminodes.
10. 2-8 carpellary, syncarpous ovary
11. Axile placentation with single ovule in each locule
12. Fruit many seeded berry
13. Seeds with hard stony testa with a broad hilum.

4.11.7 Economic Importance Economically the members of Sapotaceae are useful in several ways.

4.11.7.1 Edible fruits : Fleshy fruits of *Achras, Mimusops, Manilkara hexandra* are edible.

4.11.7.2 Other uses: The seeds of *Madhuca butyracea* produce a substance, called ' phulwa' used as a cold cream and luminant.

The latex of *Achras sapota* yields chickle or chewing gum. An aromatic oil is obtained from the flowers of *Mimusops elengi* finds use in the manufacture of perfumes. *M. globosa* is the source of 'balata' a commercially important latex.

The flowers of *Madhuca indica* are used for distilling spirit.

The milky latex of *Palaquium ellipticum* is used as a cement for water proofing. In the same way, latex of *P. guttu* is the chief source of gutta percha of commercial rubber which is extensively used for insulation, cables, pipes and dolls manufacture.

Some species of *Sideroxylon* and *Bumelia* are the source of useful timber.

4.11.8 Systematic and Phylogeny

Bentham and Hooker included the family Sapotaceae in the order Ebenales along with Ebenaceae. This classification was also followed by Engler and Prantl and later by Hutchinson, Takhtajan and Cronquist.

Sapotaceae is a loosely knit family with loosely defined genera and genera of uncertain phylogentic position. It shows distinct line of evolution from polypetalae and it can be taken to be evolved from Malvaceae, Rutaceae in which diplostamenous condition and axile placentation are found.

4.12 ASCLEPIADACEAE (The Milk weed family)

Systematic position:

Dicotyledonae

Gamopetalae

Bicarpellatae

Gentianales

Asclepiadaceae

4.12.1 Distribution The family Asclepiadaceae consists of 175 genera and 2,000 species and is distributed mostly in the tropical and sub-tripical regions of the world and only a few species are distributed in the temperate zone. In India, the family is represented by 53 genera and 250 species ocurring mostly in the tropical Himalayas and Southern and Western India.

4.12.2 Important Plants

1) *Asclepias curassavica* L. (Silk weed)
2) *Calotropis gigantea* (L) R.Br.ex.Ait. (Madar) Tel: Jilledu
3) *Cryoptostegia grandiflora* (Roxb) R.Br. (Rubber wine) Tel: Rubber teega.
4) *Dischidia rafflesiana* Wall Tel: Dudipala
5) *Gymnema sylvestra* L. Tel: podapatri
6) *Hemidesmus indicus* (L) Schult (Indian sarasparilla) Tel: Sugandhipala
7) *Leptodenia reticulata* L (Patatuge) Tel: Pala teega
8) *Oxystelma esculentum* (L) Karst.
9) *Pergularia daemia* (Forsk) chivo. (Syn. *Daemia extensa)* Tel: Peddamandu or Dushtapu teega
10) *Periploca aphylla* L. (Silk wine)
11) *Sarcostemma acidum* (Roxb) Viol. Tel: Aaku jemudu
12) *Sarcolobus carinatus* Wall , Tel:Palabudda teega.
13) *Tylophora indica* (Burm.f) Merr. Tel: Goripala or Mekameyaniaaku
14) *Wattakaka vulubilis* L. Tel: Bandigurivinda
15) *Hoya parasitica* Wall (Wax plant).

4.12.3 Habit and Habitat Most of the genera are xerophytic, perennial herbs (*Asclepias*) or shrubs (*Calotropis),* woody climbers or lianes (*Perigularia, Leptadaenia* and *Tylophora).* Some are succulents (*Hoya)* or totally xerophytic showing cactoid habit (*Stapelia), Dischidia* is an epiphytic climber and *Sarcolobus* is a mangrove plant.

4.12.4 Vegetative Characters

Root system: A deep tap root, adventitious in *Dischidia.* A perennial rootstock is commonly present and they may be fleshy or tuberous. The roots of succulents usually show endotropic mycorrhizae.

Stem: Aerial, erect eg. *Calotropis*, weak and twining (*Leptadaenia, Tylophora, Pergularia* and *Cryptostegia).* In several xerophytic forms, the stem is green, phylloclade, succulents covered by wax or hairs (*Hoya, Stapelia* and *Sarcostemma*). The stem is tuberous in

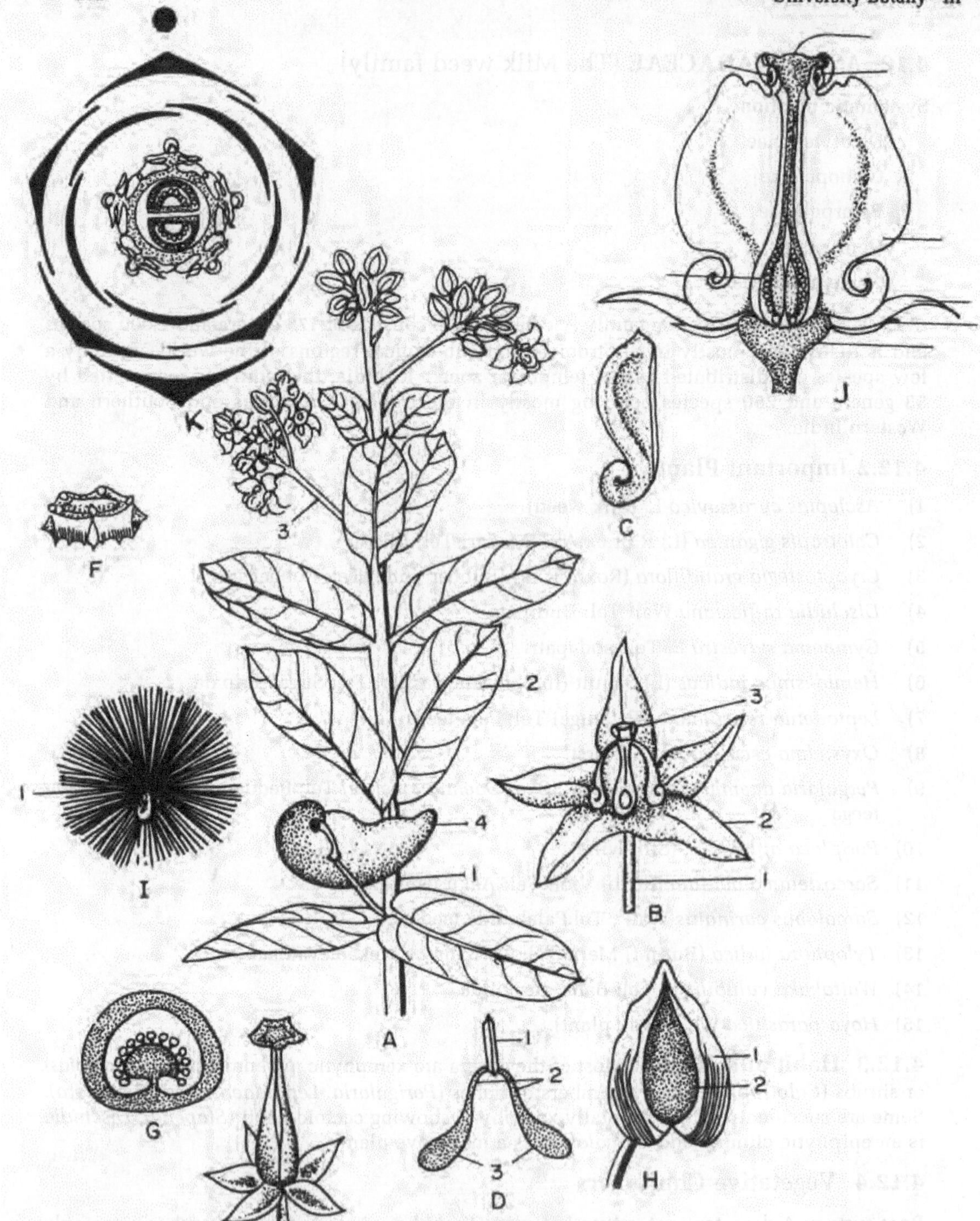

Fig 4.12.1. *Calotropis gigantia*

A: Flowering twig ; 1. Stem, 2. Opposite Leaf, 3. Inflorescence, 4. Fruit,
B: Flower, 1. Pedicel, 2. Petal, 3. Gynoestegium, C: 'S'-shaped corona, D: Translator 1. Corpusculem,
2. Retinacula, 3. Pollinia, E: Gynoecium; F: Pentagonal Gynoestegium G: T.S. ovary, H: Fruit 1. fruit wall, 2. seed,
I: Seed 1. comose hairs, J: L.S. Flower, K: Floral diagram

Ceropedia. Internally the stem shows bicollateral vascular bundles. Abundant milky latex is present in stems.

Leaves: Simple, show characteristic opposite decussate phyllotaxy, rarely whorled. Usually leaves exstipulate, linear to orbicular, obovate, leaf lamina usually coated by wax. Leaves in xerophytic species are small, scaly or deciduous. In *Stapelia* leaves are reduced to spines, while in *Dischidia*, leaves are transformed into pitchers.

4.12.5 Floral Characters

The inflorascence is terminal or axillary or some times extra-axillary (Calotropis) due to the adnation of inflorescence with the axis throughout the internodal region (fig 4.12.1A). Inflorescence may be a raceme or an condensed umbel (*Asclepias, Calotrpis*). Dichasial cyme (*Cryptostegia, Cryptolepis)* with a tendency to monochasial condition in the ultimate branches, such an inflorescence is described as Cincinnus. Flowers solitary in *Stapelia grandiflora*.

Flower: Pedicellate, bracteate, bracteolate, ebracteate in *Stapelia,* complete, bisexual, actionomorphic, partially actinomorphic in *Ceropegia*. Pentamerous, hypogynous, large flowers (about 15 cm diameter) are found in *Stapelia gigantea*.

Calyx: Five, gamopetalous, corolla tube usually short, aestivation twisted.

Corolla: Corolla are of different shapes, i.e., salver-shaped in *Stephanotis,* pitcher like *(Ceropegia)* campanulate (*Calotropis*), tubular (*Daemia*); funnel shaped (*Cryptostegia*). (fig 4.12.1B) Aestivation twisted or valvate.

Corona: The Asclepiadaceae flower is characterised by the presence of specialized appendages called corona.

The corona may be scaly or hairy arises from petals or the staminal tube. In the former case it is called as the *corolline cororna*, while in the latter case it is called *staminal corona*. The corolline corona is present in *Cryptostegia*, (fig 4.12.2C) *Cryptolepia, Gymnema* and *Brachylepis.* The staminal corona is found in many genera including *Calotropis, Asclepias, Tylophora, Periploca* and *Hoya* (fig. 4.12.1).

In *Calotropis* the corona lobes arise near the base of the filaments and are more or less free with coiled basal spurs (fig 4.12.1B). In *Hoya*, on the other hand, five corona lobes are convex, conical and star shaped.

In *Dischidia*, five staminal coronary lobes are found and each lobe is double hooked. *Marsdenia* has the corona in three whorls the outer whorl is corolline where as the inner two are staminal. However, in number of other species corona may be absent.

Corona is primarily meant for secreting and storing nectar besides providing additional attraction to the flowers.

Androecium: Five stamens, united, monadelphous, (*Calotropis*) or free (*Cryptostegia*), (fig 4.12.1) epipetalous, filaments short, anthers dithecous, introse, unite with stigmas to form a gynostegium. The pollen grains are dispersed by insects by a special translator mechanism.

Based on the nature of translators, the family is divided into two sub-families these are:

1) Periplocoideae

2) Cynanchoideae

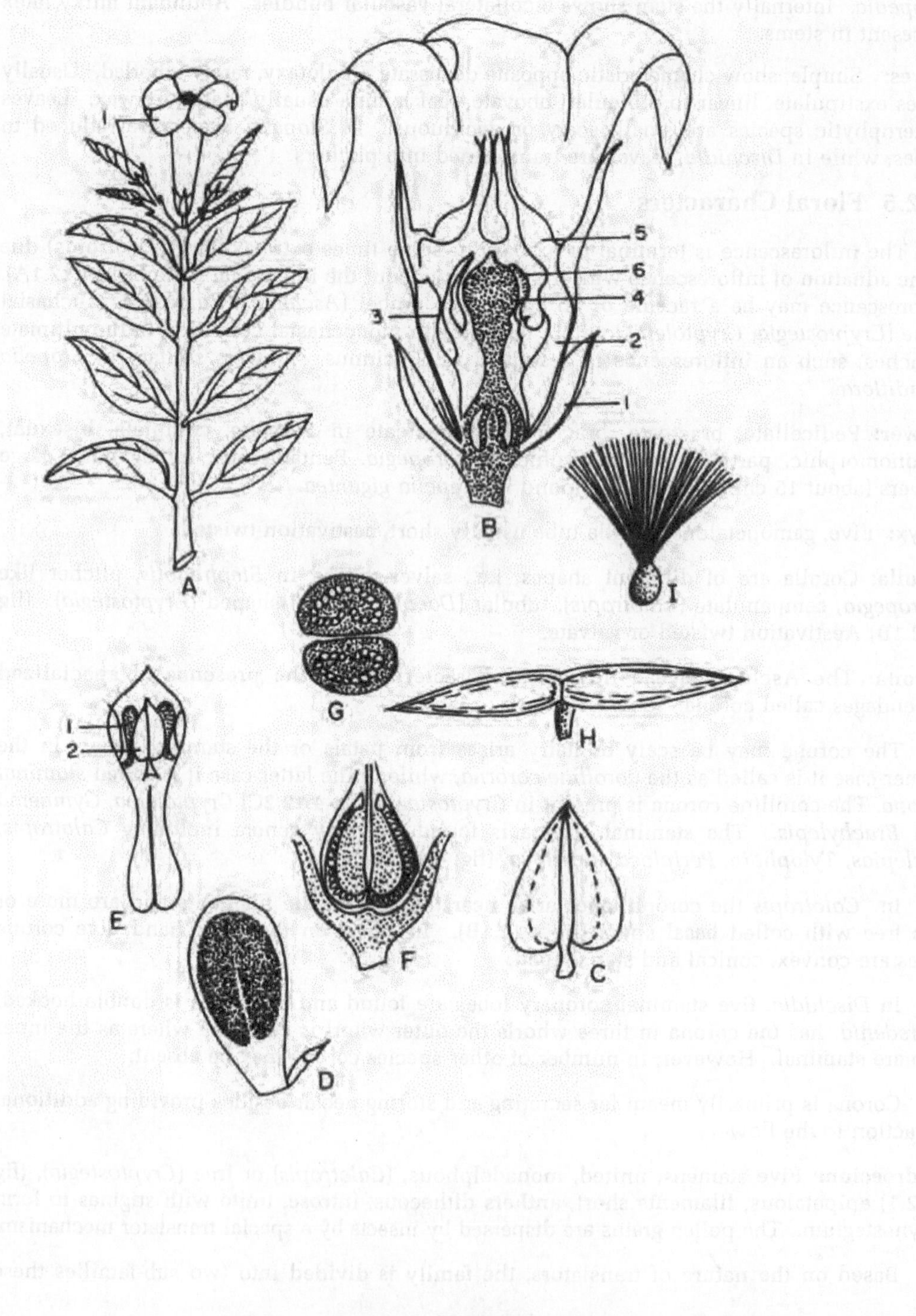

Fig 4.12.2. *Cryptostegia grandiflora*

A: Flowering twig ; 1. Dichasial Cyme inflorescence; B: L.S flower 1. Calyx, 2. Corolla, 3. Stamen, 4. Stigma 5. Corona, 6. Translator; C: Stamen, D. Translator (spoon shaped); E: Stamen with translator 1. Stamen, 2. translator, F: Ovary L.S. G: Ovary T.S. H: Fruit ; I. Comose seed

Periplocoideae: In this sub-family the filaments are free but the anthers are adherent to the expanded stylar head. The pollen grains are granular and are united into tetrads. Five spoon shaped (spathulate) translators arise from the stylar head alternating with the anthers. The pollen from one half of two adjacent anthers are discharged on to this spathulate translator. Each translator has a adhesive disc at the base with which translator easily attached to the body of the insect. (fig 4.12.2D)

Cynanchoideae: In this sub-family (which is also known as Euasclepiadaceae) the filaments of five stamens unite forming a short fleshy column. The anthers present at the tip of the filaments are joined to the pentagonal disc through their connectives. The structure thus formed is called as the gynostegium. The pollen in each lobe of the anther aggregate to form a mass called pollinium. Thus each anther has two pollinia. The pollinia of adjacent anthers are united by means of short stalks or caudicles to form a translator. The translator is λ (lambda) shaped and consists of two parts a) The corpusculum or gland like body attached to the margins of the stigmatic head b) a pair of diverging arms called conncectives or retinacula by means of which the pollinia of adjacent anther halves are attached to the corpusculum (*Calotripis, Daemia* and *Hoya)* (fig 4.12.1D).

Gynoecium: Bicarpellary, sub-apocarpous ovaries and styles are free while stigmas connate to form a pentangular stigmatic disc with which the anthers are coherent to form a gynostegium (fig 4.12.1F). Ovary superior, unilocular with many ovules on marignal placentation. Ovules are anatropous and bitegmic.

Fruit: A pair of follicles in many genera (fig. 4.12.2H).

Seed: Albuminous, numerous, small, covered with a tuft of hygroscopic hairs (commose hairs) which help in seed dispersal.

4.12.6 Pollination

Entomophilous, usually carried out by bees. Pollination mechanism in this family is unique in many ways.

In the sub-family *cynanchoideae*, when insect visit the flowers for honey, soon gets its legs trapped in the slit present between the anthers. When it tries to come out, the corpusculum of translator gets attached to the leg of the insect, with it the pollinia are drawn out of the anther and are carried. Due to the hygroscopic action of retinaculum, the pollinial masses come together and firmly clasp around the leg so that they are not lost during the flight of the insect. When the insect visits another flower for honey, the pollinial masses are deposited on the receptive surface thus effecting cross-pollination.

In the sub-family *periplocoideae*, the stamens are stalked, pollen grains are granular and united into tetrads. The spoon shaped translator present between each anther and terminating below in an adehesive disc. The sticky gynostegium disc is turned upwards so shat it can be easily attached to the head of the insect which is in search of honey. When this insect visits another flower, the translator gets deposited on the stigmatic surface thus assuring cross-pollination.

4.12.7 Distiguishing Features

1) Plants herbs, shrubs or mostly twinners with milky latex.
2) Leaves with opposite-decussate phyllotaxy.
3) Presence of bicollateral vascular bundles.

4) Exstipulate leaves.

5) Inflorescence usually in axillary umbels.

6) Flowers regular, bisexual and hypogynous.

7) Presence of corallary or staminal corona.

8) Anthers coherent with stigmatic disc to form a pentangular disc called gynostegium.

9) Presence of spoon shaped or lambola shaped translators containing pollen mass in pollinia.

10) Bicarpellary, sub-apocarpous pistil.

11) Ovary unilocular, placentation marginal

12) Fruit, a pair of follicles.

13) Seeds surrounded by commose hairs.

4.12.8 Economic Importance

4.12.8.1 Ornamentals: A few species of *Asclepias, Hoya, Stapelia, Periploca* and *Cryptostegia* are well known garden plants.

4.12.8.2 Medicine: The roots of *Hemidesmus indicus* yield 'Sugandhipalu' which is extensively used in the preparation of drugs. Roots of *Oxystelma* are useful in the treatment of jaundice, while leaves of *Jamnea, sylvestrae* are employed to cure excessive urination and that of *Pergularea* are used in controlling pulmonary diseases and also as a insect repellant. Latex of *Matelea* is an arrow poison.

4.12.8.3 Rubber: Latex of *Cryoptostegia* (rubber wine) is used for the manufacture of cheap quality rubber.

4.12.8.4 Others: The seed material of several species is useful as an insulating material. Some genera like *Calotropis, Marsdenia* yield silky fibres employed for making ropes. Some species of *Asclepias* act as livestock poisons. *Marsdenia tinctoria* yields an indigo like dye. *Oxystelma recamone* yields drinkable milk. The latex of *Gymnema lectiferum* is used as a substitute for milk and also contain hydrocarbons

4.12.9 Affinities and Systematic Position

Bentham-Hooker's placed the family Asclepiadaceae along with Apocynaceae and Loganiaceae in the series Bicarpellatae of Gamopetalae. Engler and Prantl included in contortae of Sympetalae. Hutchinson (1969) treated the sub-family periplocoideae as distinct family, Periplocaceae. He classified Asclepiadaceae along with Apocynaceae in the order Apocynales. This shows that Asclepiadaceae and Apocynaceae are closely related families by having exstipulate leaves, presence of corona, epipetalous stamens, partially syncarpous ovary, marginal placentation, nuclear endosperm, winged and hairy seeds. However, these two families differ in that in Apocynanceae pollen grains do not form pollinia, absence of translator mechanism and gynostegium.

4.13 CONVOLVULACEAE (The Ipomea family)

Systematic position:

Dicotyledonae
Gamopetalae
Bicarpellatae
Polemniales
Convolvulaceae

4.13.1 Distribution

The family contains about 56 genera and 1,200 species widely distributed all over the world particularly in the tropics and sub-tropics. In India, Convolvulaceae is represented by 20 genera and 150 species chiefly distributed in southern and western parts.

4.13.2 Important Plants

1) *Argyreia setosa* (Vahl) chois Tel: Maya teega or Verriboddi tega.
2) *Convolvulus arevensis* L.
3) *Cuscuta reflexa* (Roxb). Tel: Seetamma savaram
4) *Evolvulus alsinoides* (L) .E Tel: Vishnu kranthi
5) *Ipomoea cornea* Jacq. (Syn *I. fistulosa* Mart.) Tel: Tutaku or Palasamudram
6) *I. nederifolia* L. (syn. *Quamoclit pinnata* (Desr) Boj) (Needle creeper)
Tel: Kasiratnam or Malathi Madhavam
7) *I. batatas* L (Lamk) sweet potato Tel: Ratnapuri gadda or chilagada dumpa.
8) *I. cairica* (L) Sweet (Morning glory)
9) *I. biloba* Forks. (Goats foot creeper) Tel:Bedaku
10) *I. aquatica* Forsk. (Water bind weed) Tel: Tooti kura
11) *I. palmata* L. (Railway creeper)
12) *Merremia emerginata* (Burm.f) Hall.f. Tel: Elakachevi
13) *Porana paniculata* Roxb. (Bridal creeper)
14) *Stictocardia tilifolia* (Dess) Hall.
15) *Rivea hypocreteriformis* (Desv.) Chois. Tel: bodditeega.

4.13.3 Habit and Habitat

The plants are mostly annual or perennial twiners. They are either herbs (*Ipomoea* spp*)* or shrubs (*I. fistulosa, Convolvulus arvensis*) may be woody climbers (*Lettusomia*). They grow in variety of habitats. *Cressa cretica* grows in saline soils. *Ipomoea biloba, Calystegia soldanella* are sand binders and are prominent members of sand-dune flora. *Ipomoea hastata* is a floating aquatic plant, while *Hildebrandtia* grows in Western African deserts. *Cuscuta reflexa* (dodder) is a leafless, rootless total stem parasite.

4.13.4 Vegetative Characters

Root system: Mostly tap root system. In *I. batatas* (sweet potato) however, adventitious roots become fusiform tubers due to storage of food material. In *Cuscuta*, special parasitic roots known as haustoria are developed to absorb nutrition from the host plant.

Stem: Herbaceous, often twining, usually covered by hairs. Latex usually present. Thick tuberous rhizomes are found in *Ipomoea scammonias.* The stem in *Cuscuta* is filiform

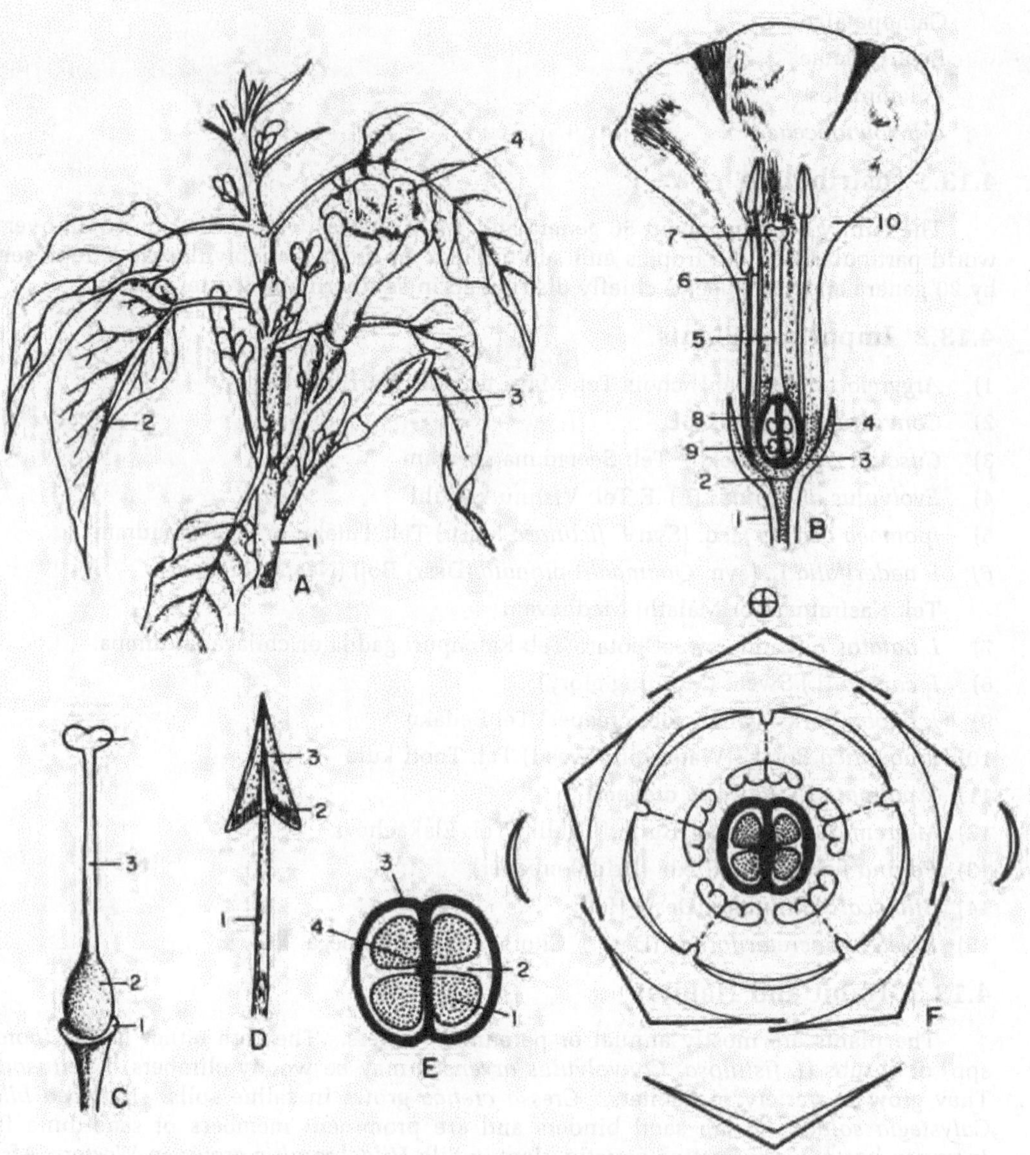

Fig 4.13.1. *Ipomoea fistulosa*

A: Flowering twig ; 1. Stem, 2. Leaf, 3. Flower bud, 4. Flower

B: L.S. Flower 1. Pedical, 2. Thalamus, 3. Disc, 4. Sepal, 5. Style, 6. Stamens, 7. Stigma, 8. Ovary, 9. Ovule, 10. Corolla C: Gynoecium 1. Disc, 2. Ovary, 3. Style, 4. Stigma; D: Stamen 1. Filament, 2. Connective, 3. anther lobe E: T.S. Ovary 1. Ovules 2. Locule 3. Ovary wall 4. Placenta; F: Floral diagarm

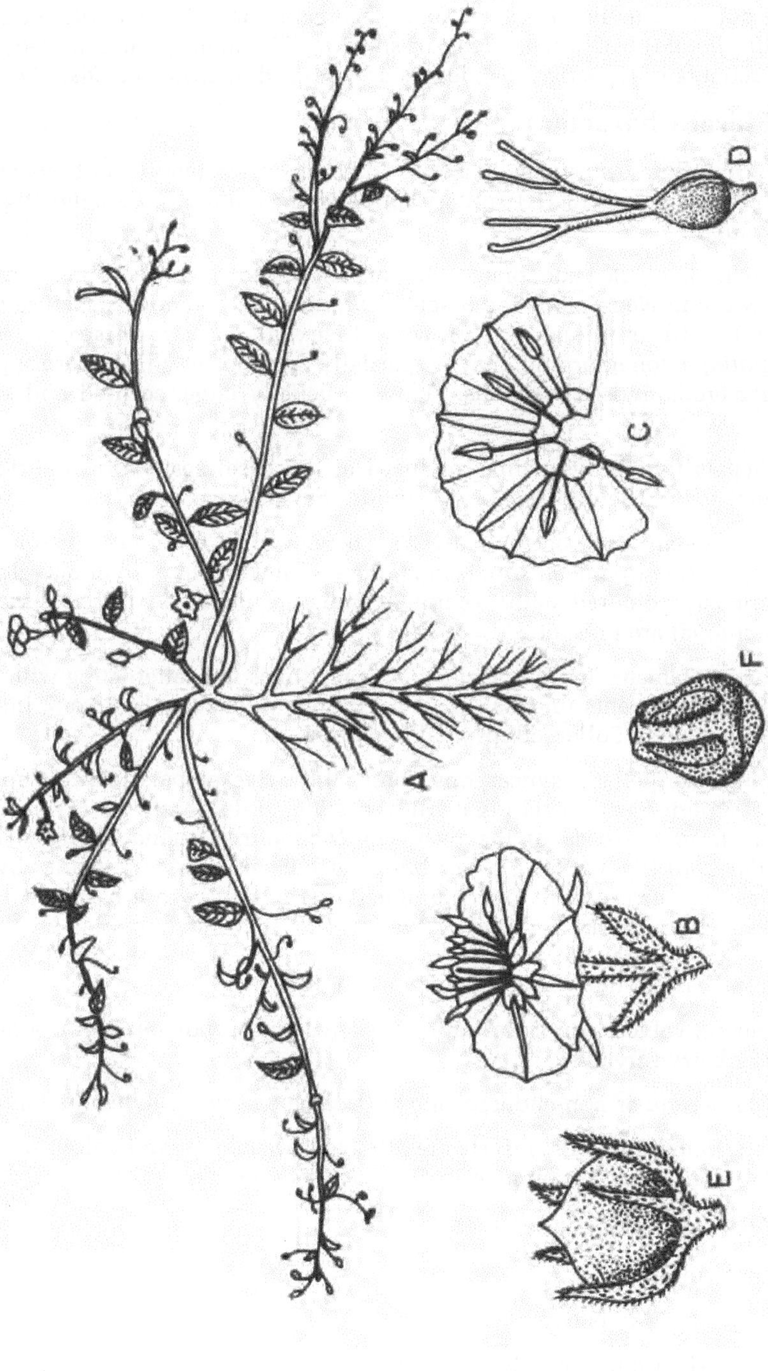

Fig 4.13.2. Evolvulus alsenoides

A: Flowering twig ; B. Flower; C. Opened Corolla showing epipetalous condition; D: Gynoecium; E: Fruit; F: Seed

and is pale greenish white or orange yellow without leaves. Runner in *Merremia.* Internally stem shows bicollateral vascular bundles.

Leaf: Exstipulate, usually simple, petiolate, palmately lobed (*I.palmata*) or pinnately lobed (*Quamoclit pinnata*) leaf surface hairy or smooth. Phyllotaxy alternate, reticulate venation, pinnately reticulate (*C.pestigradis*), leaves are reduced to scales in *Cuscuta.*

4.13.5 Floral Characters

Inflorescence: Inflorescence usually axillary dichasial cyme tending to become a monochasial (*Ipomoea fistula*) (fig. 4.13.1A) solitary axillary (*Evolvulus, Calystegia*) (fig 4.13.2A) large panicle (*Porana*).

Flower: Pedicillate, bracteate, bracteolate. In *Calystegia* the bracteoles are large and brightly coloured, while in *Neuropeltis* the bracteoles form a membranous structure during the fruit formation to aid in dispersal. Flowers bisexual, rarely unisexual and dioecious (*Hildebrandtia*) actonomorphic, may be partially zygomorphic (*Hembertia*) pentamerous or rarely tetramerous and hypogynous. Flowers usually large, scented and showy in many genera.

Calyx: Sepals five, polysepalous or fused at the base, aestivation imbricate or valvate, usually persistent. The odd sepal is posterior. Sepals have awn like tips in *Quamoclit.*

Corolla: Five petals, gamopetalous, twisted, however in *Cuscuta* aestivation is quincuntial, corolla shape rotate (*Ipomoea spp*) campanulate (*Convolvulus.spp*) or sub-globose (*Cuscuta*) corolla variously coloured with prominent radiating lines. In *Cuscuta* scaly appendages (Corallary corona) are present.

Androecium: Stamens five, epipetalous, or some times free, alternating with petals, inserted on corolla tube, filaments unequal in size. Anthers dithecous, basifixed, introse, dehiscing by longitudinal slits. Pollen ellopsoidal and spinous.

Gynoecium: Bicarpellary, syncarpous, rarely three-five carpels (*Erycibe* spp) and *Ipomea spp*) ovary superior with a terminal filiform style, style absent in *Erycibe.* Two distinct styles arise from the base of the ovary in *Dichondra,* stigma capitate (*Ipomoea)* bifid (*Convolvulus* spp) (fig 4.13.1C), ovary bilocular but becomes tetralocular by the formation of a false septum or replum. Ovary unilocular (*Erycibe)* with a basal ovule. However in others, placentation is axile. One or two sessile anatropous ovules are present in each locule. Raphae ventral, micropyle bents downwards. A cup or ring like nectary disc is present at the base of the ovary. (See fig 4.13.1 and 4.13.2)

Fruit : Four-six valved loculicidal capsule *(Convolvulus, Ipomoea).* A dry or fleshy capsule (*Cuscuta)* or berry (*Erycibe*).

Seed: Endospermic or non-endospermic with large embryo, cotyledons folded or plicate.

Pollination: Entomophilous, insects are attracted by the size and colour of the flowers and nectar. Self-pollination is common in *Convolvulus arvensis.*

4.13.6 Distinguishing Features

1) Mostly twiners.

2) Presence of milky latex

3) Bicollateral vascular bundles in stem.

4) Parasitic habit of some taxa

5) Exstipulate, alternate simple leaves.

6) Mostly cymose or dichasial cymose inflorescence.

7) Large flowers, complete, actinomorphic, bisexual and hypogynous.

8) Persistent calyx lobes.

9) Companulate or rotate corolla.

10) Stamens epilpetalous and alternipetalous

11) Bicarpellary syncarpous ovary bi-to tetralocular.

12) Placentation axile with anatropous ovules.

13) Filiform style.

14) Capitate or bifid stigma

15) Presence of a nectariferous disc at the base of the ovary.

16) Fruit, a loculicidal capsule.

17) Seeds endospermic with large curved embryo and cotyledons folded.

4.13.7 Economic Importance

4.13.7.1 Food: Starch and sugar rich root tubers of *Ipomoea batatas* (Sweet potato) are edible. The root stocks of *Calystegia sepium* are cooked and eaten. The young stems and leaves of *I. aquatica* and *Rivea* spp are edible and also used as fodder.

4.13.7.2 Ornamentals: Following species are cultivated in gardens for their beautiful flowers.

Ipomoea purpurea (Morning glory).

I.violacea (heavenly blue)

I.coccinia (Star ipomoea)

I.tuberosa (Wood rose)

Quamoclit pinnata (Needle creeper)

Porana paniculata (bridal creeper)

4.13.7.3 Medicinal Plants: The dried latex of *Exogynum purga, Convolvulus scammonia* and *Ipomoea purga* are used as medicine on account of their purging properties. *Cuscuta reflexa* (dodder) stems are used as blood purifier and for bilous disorders. Roots of *Argyreia specosa* are used in rheumatic infections and leaves for curing skin diseases and wounds.

4.13.7.4 Soil binders: Species of *Ipomoea* (*I. pes-capre*) *Convolvulus, Calstigia* are grown on the shores of seas to act as soil binders.

4.13.8 Affinities and Systematic Position

Bentham and Hooker regarded Convolvulaceae being related to Boraginaceae and Solanaceae in many respects, which are grouped under Polemoniales. Engler-Prantl classified this family in Tubiflorae along with Boraginaceae, Polymoniaceae, Hydrophyllaceae and Solanaceae. Hutchinson however included Convolvulaceae in Solanales along with Solanaceae.

Some taxonomists (Hutchinson, Airyshaw, Takthajan and Cronquist) prefered to place the parasitic member *Cuscuta* in a different family called Cuscutataceae on account of 1) parasitic habit 2) Imbricate corolla 3) Coronary structures on corolla and 4) filiform spiral embryo.

4.14 LAMIACEAE (The mint family)

Systematic position:

Dicotyledonae

Gamopetalae

Bicarpellatae

Lamiales

Lamiaceae (Labiatae)

4.14.1 Distribution

Lamiaceae family includes about 200 genera and 3500 species, is cosmsopolitan in distribution, growing under variety of climatic and soil conditions. In India, over 65 genera and 4,000 species occur mostly in South and North Western India. About 260 species are reported to be endemic to Indian sub-continent. Mostly they are restricted to moist and hilly areas.

4.14.2 Important Plants

1) *Anisomeles ovarta* (L) O.ktze. Tel: Chinna ranabheri
2) *Coleus aromaticus* (Benth.) Tel: Vamu aku
3) *C. blumei* L. Tel: Nemali pincham
4) *Hyptis suaveolens* (L) Poit Tel: Seema tulasi
5) *Lavendula vera* L (Lavender plant)
6) *Leonotis nepaetaefolia* (L) R.Br. Tel: Magabeera
7) *Lamium album* L (Wild) (dead nettle).
8) *Leucas aspera* (Willd.). Spreng. Tel: Tummi.
9) *Mentha arvensis* L. (Menthol)
10) *M. piperata* L. (Pepper mint)
11) *M. sylvestris* L. (Mint)
12) *M. viridis* L. (Spear mint) Tel: Pudina
13) *Ocimum basilicum* L. (basal) Tel: Rudrajada or sabji mokka
14) *O. sanctum* L. (Sacred basil) Tel: Krishna tulasi or Laxmi tulasi
15) *O. canum* Sims Tel: Kukka tulasi
16) *O. gratissimum* L. Tel: Rama tulasi
17) *Origanum majorana* L. Tel: Maruvam
18) *Nepeta compestris* L. (Ground Ivy)
19) *Salvia plebija* (R. Br)
20) *S. officinalis* L. (Sage)
21) *Pogostemon bengalensis* (Burm.f) Oktza. Tel: Pedda tulasi
22) *Rosmarinus officinalis* L (Rose mary)
23) *Tinnea aethiopica* L.
24) *Thymus serphyllum* L.

4.14.3 Habit and Habitat

Members of Lamiaceae mostly are annual or perennial aromatic herbs. Species of *Hyptis* and *Leucosceptrum* are arborescent. Few species of *Scutellaria* are lianes. A few species are shrubs (*Rosmarinus, Lavendula),* trees (*Leuosceptrum).*

Species of *Mentha* and *Lycopus* are found in marshy areas. *Lycosus* species grow in muddy soil also.

4.14.4 Vegetative Characters

Root system: Commonly a tap root system . In few cases adventitious roots are present.

Stem: Herbaceous, quadrangular, aerial or underground (*Hederacia*), stolon, or suckers (*Mentha*). Stem hairy, vegetatively propagated by suckers, stolons and runners. In *Hedoma*, green stem performs assimilatory function due the reduction of leaves.

Leaf:- Simple, exstipulate, phyllotaxy opposite decussate, rarely whorled (*Ocimum*) pedicellate or sessile. Venation pinnately reticulate, leaf surface glabrous, margin entire or toothed or some times finely disected (*Salvia* sp), leaves variegated in *Coleus*. Leaves aromatic due to the presence of oil glands secreting volatile oils.

4.14.5 Floral Characters

Inflorescence:- Terminal or axillary usually verticellaster (eg: *Leucas, Leonotis*) it is a type of inflorescence where flowers in axillary cymose clusters arranged in a racemose manner. Each axillary cluster is at first dischasial but becomes monochasial on the flowers appear in clusters around the stem because of shortening of peduncles. The two opposite clusters together is known as veriticillaster.

In *Ocimum* spp. It is a mixed type of inflorescence called thyrsus where, an indeterminate peduncule (racemose) contained three flowered cymes at each node (fig 4.14.1)

In some cases, the inflorescence looks like a spicate head (*Anitochilus*), panicle in *Hyptis* and solitary in *Scutellaria*.

Flower: Pedicellate or sesile, bracteate, bracteolate, complete, partially zygomorphic, actinomorphic (*Mentha*), bisexual or rarely unisexual (*Nepeta, Thymus*) with pistillate flowers only. Flowers pentamerous. In *Monarda*, the bracts are highly specialised. Flowers characteristically are bilabiate

Calyx : Sepals five, gamosepalous, tubular (*Lavendula, Nepeta*) companulate (*Tucriun*) or bilabiate (*Salvia, Ocimum*). In *Ocimum* upper lip has one sepal and lower lip with four sepals (1+4),(fig 14.1C) while in *Salvia* it is 3+2 calyx green and succulent (*Hoslundia*). Usually persistent (*Salvia, Ocimum*) (fig 4.14.1 & 4.14.2).

Corolla: Petals five, gamopetalous, usually irregular, however, regular in *Mentha* due to the reduction of the upper lip to a small round tubule which is almost indistinguishable from the other three lobes of the lower lip. In many genera it is of 2/3 arrangement, where, the upper lip consists of 2 petals while the lower lip has 3 petals (*Salvia, Leucas*). (fig 14.1D) In other genera like *Ocimum, Hyptis* and *Coleus* it is 4/1 arrangement (upper lip of 4 petals, lower lip with 1 petal) the corolla lacks upper lip and the lower lip has all 5 petals (0/5) (*Tencrium, Rosmarinus*). Aestivation imbricate, corolla attractive, variously coloured and scented.

Androecium: Stamens usually four in number (*Ocimum, Coleus, Mentha, Leucas)* only two stamens are present in *Salvia* and *Monarda*. If stamens are four, the posterior (fifth one) is completely supressed or represented by a staminode. The stamens are didynamous (The anterior pair of stamens have long filaments and the posterior one have short filaments). In *Salvia* only two fertile stamens are present and the posterior pair are reduced (fig. 4.14.2.B). The stamens are monadalphous in *Coleus*.

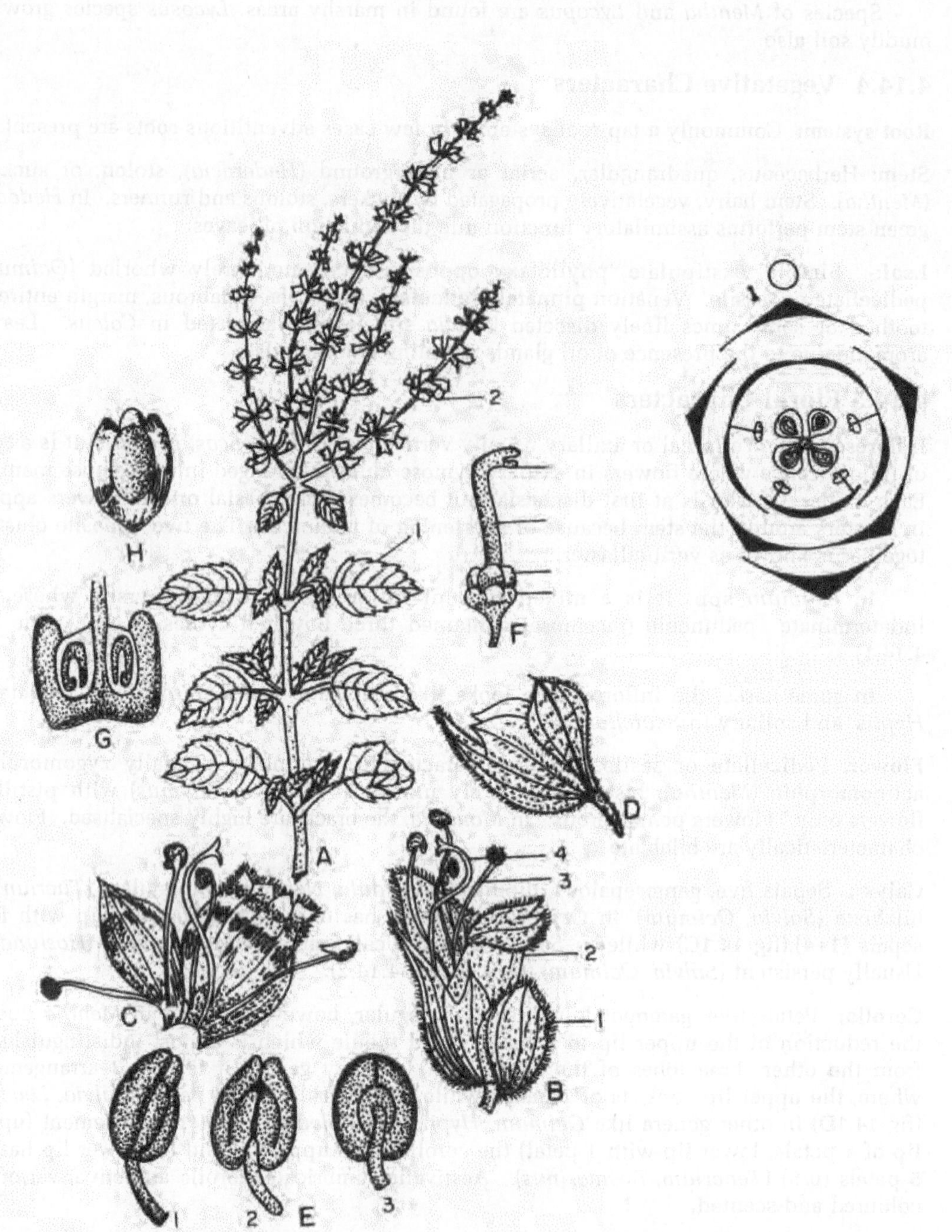

Fig 4.14.1. *Ocimum sanctum*

A: Flowering twig ; 1. Leaf, 2. Inflorescence; B: Flower 1. Calyx, 2. Corolla, 3. Androcium, 4. Gynoecium; C: Corolla opened; D: Calyx; E: Different faces of stamen 1. Lateral, 2. Ventral, 3. Dorsal face; F: Floral diagram

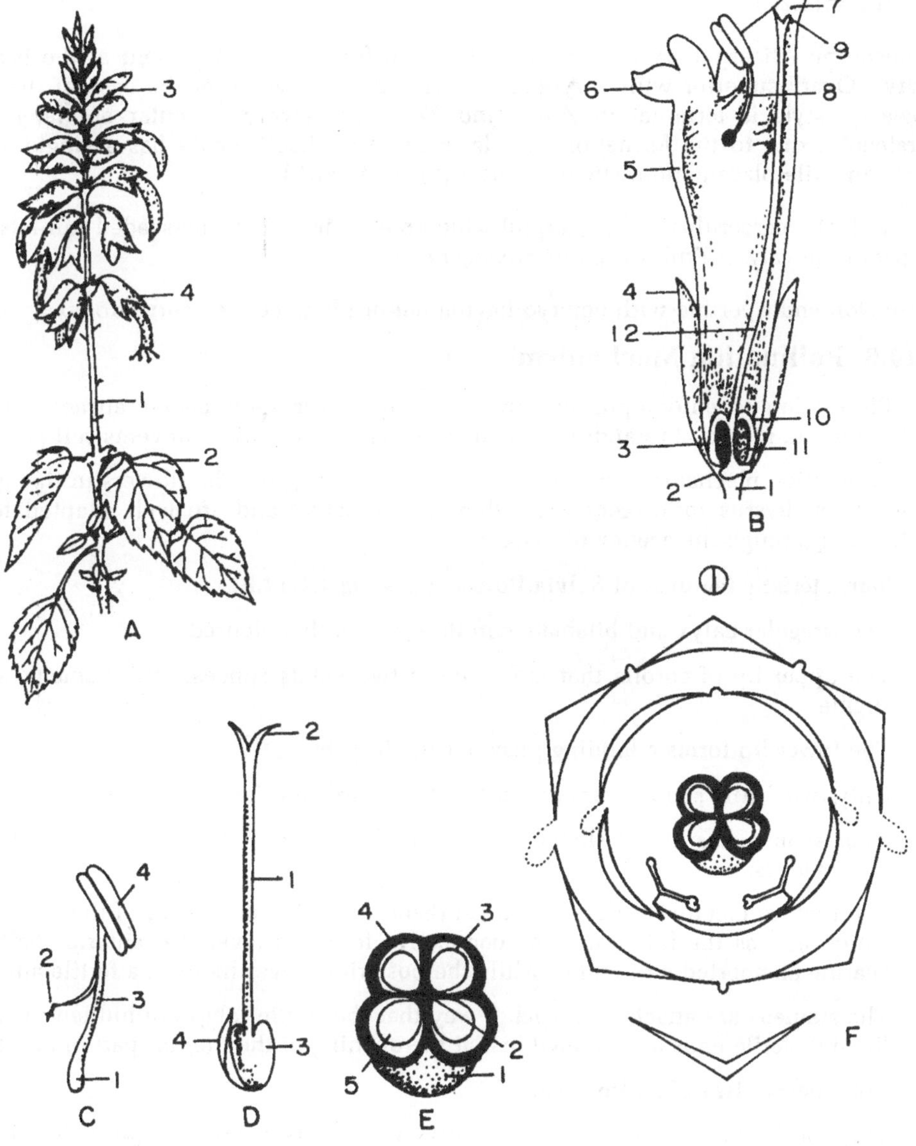

Fig 4.14.2. *Salvia spendens*
A: Flowering twig ; 1. Stem, 2. Leaf, 3. Flower bud, 4. Flower, B: L.S. Flower 1. Pedicel, 2. Thalamus, 3. Nectary, 4. Calyx, 5. Corolla, 6. Outer lip, 7. Posterior lip, 8. Stamen, 9. Stigms, 10. Ovary, 11. Ovule, 12. Gynobasic style; C: Stamen 1. Sterile Lobe, 2. Filaments, 3. Connective, 4. Fertile Lobe D: Gynoecium, E: T.S. Ovary, F: Floral diagram

Anthers dithecous and rarely monothecous as seen in *Geniosporum* and *Lycopus*. Stamens highly reduced in *Scutellaria*.

In *Salvia* the two anther lobes are separated by a long connective, of the two lobes the anterior one is reduced to a knob like structure. Anthers introse, dehisce by longitudinal slits (fig 4.14.2C).

Gynoecium: Bicarpellary syncarpous. A nectariferous disc is present at the base of the ovary. Ovary superior with a gynobasic style. (style arises from the base of the carpels). However, style is terminal in *Ajuga* and *Teucrium*. Ovary bilocular but later becomes tetralocular due to the formation of a false septum. Each locule has a single anatropus ovule on axile placentation, stigma is bifid (fig 4.14.1E&F).

Fruit: Mostly carcerulus (schizocarpic) which breaks into four one-seeded nut-lets covered by persistent calyx. Fruit drupe in *Stenogyne*.

Seed: Non-endospermic with embryo having flat or plano-convex cotyledons.

4.14.6 Pollination Mechanism

Pollination is entomophilous, fragrance of flowers and nectar attract insects like butterflies and moths. Protandry is a common adaptation which prevents self-pollination.

Pollination mechanism in *Salvia* is a unique lever type. The flowers in this genus are arranged in thysrus inflorscence and they are attractive and are well adapted for cross-pollination through the agency of insects.

A) Characteristic features of Salvia flower : (see fig 4.4.14.B).

1. The irregular calyx and bilabiate corolla are brightly coloured
2. The upper lip of corolla that is formed of two petals conceals the stamens, style and stigma.
3. The lower lip forms a landing place for the insects.
4. Only two fertile stamens are present and they are anterior in position.
5. Each stamen has a "T" shape due to, the elongation of the connective between the anther lobes.
6. The long connective of the versatile anthers can swing on the tip of the short filament which acts as the fulcrum. The connective forms as lever and anterior half is short bearing an aborted half anther while the posterior longer half has a fertile anther.
7. The stamens are attached in such a way that the fertile lobes are hidden in the upper lip and sterile ends are situated just at the opening of the tubular part of corolla,.

B) Lever mechanism of pollination:

The corolla has a spreading upper lip, which acts a landing place for the insect. When the insect lands on the lower lip of corolla and in its endeavour to obtain nectar that is secreted by the hypogynous disc, pushes the sterile lobes of stamens inwards. Due to this, the upper half of the connective with the fertile lobes acts as a lever. Then, the anther lobes come down and shed the pollen on the back of the insect.

As the flowers are characteristically protandrous, the immature stigma is concealed in the upper lip in the early stage. In the next stage, the style of ripe gynoecium bends down from the upper lip and places the stigmatic lobes up in such a way that it can receive the foreign pollen from the back of the visiting insect body.

4.14.7 Distinguishing Features

1. Herbaceous habit
2. Quandrangular stem
3. Aromatic leaves
4. Opposite decussate phyllotaxy
5. Thyrsus or vertilcillaster inflorescence
6. Irregular and persistant calyx
7. Bilabiate corolla (2/3, 4/1, 0/5)
8. Didynamous, epipetalous stamens
9. Presence of staminodes
10. Bicarpellary, syncarpous tetralocular ovary
11. Gynobasic style
12. Bifid stigma
13. Nectariferous disc at the base of ovary
14. Carcerulus fruit
15. Lever mechanism of pollination.

4.14.8 Economic Importance

Economically, the members of Lamiaceae are of wide importance as the source of volatile aromatic oils and several are cultivated as ornamentals.

4.14.8.1 Volatile oils and perfumes:- An essential oil is extracted from the dried leaves and tops of *Hedeoma pulegioides* is used as an insect repellent. *Lavendula veera* yields 'Lavender oil' which is an important constituent of Eaude cologne. *Thymus serphyllum* gives 'thyme oil' which is also an important ingredient of tooth paste. *Mentha piperata* yields peppermint oil, and menthol is obtained from *M. viridis* (spear mint) and *M. crispers* (Curley mint). The peppermint contains 50-55% of menthol which is widely used in flavouring foods, manufacturing of mouth washes, medicines, chewing gum and also in confectionary industry. Dried leaves of *Pogostemon cablin* yield 'patchouli oil' by distillation. The oil is commonly used in the preparation of high quality scents. *Rosmarinus officinalis* flowers, leaves young stems produce 'rosmary oil' which is finds use in the preperation of scents, soaps and hair oils. Dried leaves of *Salvia officinalis* yield 'sage oil', commonly applied as an insect repellent.

Some herbs are in great demand for the flavour or aroma imparted to foods, these include *Majorana hortensis, Melissa officinalis*; and *Orignanum vulgare.*

4.14.8.2 Medicines: Medicinal drugs like menthol and thymol are obatined from various species. The leaves of *Ocimum sanctum* are used for treating cold, cough and bronchitis, leaves of *Ocimum, Kilmands charicum* contain camphor. Flowers of *O. basilicum* have medicinal properties are employed in curing renal and digestary diseases. In the same way, vegetative parts of *Leucas* are used in controlling skin diseases, leaves of *Coleus amboinicus* useful in urinary and pulmonary disorders.

4.14.8.3 Ornamental Plants:- Some species of *Salvia* such as *S. coccinia, S. officinalis, S. spendens, S. leucontha* are cultivated for their beautiful scarlet red flowers. *Coleus blumei* for their variegated coloured leaves. Various species of *Dracocephalum Lamium, Melissa, Molucella, Nepeta* and *Origanum* are grown as ornamentals.

4.14.8.4 Food: Tubers of *Stachys, Coleus* are edible. The tubers of *Coleus rotundifolia* are consumed as a potato substitute. Leaves of *Mentha viridis, Melissa officinalis* and *Lamium* spp. are used as condiments for flavouring foods. The leaves of *Lamium album* contain carotene and ascorbic acid.

4.14.8.5 Dyes: Red dye is obtained from the fruits of *Lycopus europeas*.

4.14.9 Affinities and Systematic Position

Bentham and Hooker included family Lamianceae in the order Lamiales along with Verbenaceae, Selaginaceae and Mayoporaceae. Engler Prantl however, placed the family in Tubeflorae along with above families except Selaginaceae. Hallier, Wettstein and Rendle accepted the veiws of Engler and treated Lamiaceae as a part of Tubiflorae.

The Lamiaceae bears relationship with Boraginanceae but set apart by the character of the ovule. On the other hand, Lamiaceae is also closely related to Acanthaceae, but differs from it by the presence of verticillaster inflorescence, gyno basic style and the nature of ovary as well as fruit.

Among the gamopetalous families, the Lamiaceae is considered as highly advanced. This is due to the elaboration of floral structure and adaptation for insect pollination.

4.15 EUPHORBIACEAE (The Rubber family OR The spurge family)

Systetmatic position:

Dicotyledonae

Monochlamydeae

Unisexuales

Euphorbiaceae

4.15.1 Distribution

Euphorbiaceae is considered as fourth largest family of Angiosperms after Asteraceae, Fabaceae and Orchidaceae, nearly equivalent to Poaceae in distribution world over.

The family constitutes about 300 genera and 5,000 species, cosmopolitan in distribution. However, majority of them occur in tropical regions including Africa and South America and rarely seen in Arctic regions. In India, the family is represented by 70 genera and 450 species distributed in different habitats like deserts and also as mangroves. The members are quite common in the tropical and sub-tropical Himalayas and the mountains of South India.

4.15.2 Important Plants

1. *Acalypha indica* (L) Willd. (Copper leaf).
2. *Baliospermum micranthum* Mudl. Tel: Muripinda.
3. *Codiacum variegatum* Blume (Croton).
4. *Croton bonplandianum* Bailon. Tel:galivana mokka or verrimirapa
5. *Cicca acida* (L) Merril. Tel: Racha usiri
6. *Emblica officinalis* Gaern. (Indian goose berry) Tel: Pedda usiri.
7. *Euphorbia hirta* L. Tel: pacha botlu.
8. *E. pulcherrima* L. Willd. (Poinsettia).

9. *E. antiquorum* L . Tel: Pedda jamudu
10. *E. nerifolia* (L) (Schund) Tel; Akujemudu
11. *E. tirucalli* L (Milk bush) Tel: Kadajemudu
12. *E. milli* Des.Moulins
13. *Fluggia leucopyrus* L: (Crown of thorns) Tel: Siripala chettu
14. *Hevea brasiliensis* Muell. (Para-rubber tree)
15. *Hura crepitans* L. (Sand box tree)
16. *Hemicyclia elata* Bedd.
17. *Jatropha curcas* L (purging nut) Tel: Pedda nepalam
18. *J. pandurifolia* Andr. (Garden jatropha).
19. *Manihot utilissima* Pohl. (Cassava) Tel: Karra pendalam
20. *M. glaziovii* Muell (Arg) (Manicoba rubber) Tel: Sindhuram
21. *Mallotus philippinensis* (Lamk) Muell. (Kamela tree).
22. *Phyllanthus niruria* L. Tel: Nela usiri
23. *Putranjiva roxburghii* Wall. (Putijiva)
24. *Pedilanthus tithymaloides* (L) Poit. (Jew's slipper).
25. *Ricinus communis* L (Castor bean) Tel: Amudam
26. *Tragia involucrata* L (forget me not) Tel: Duradagondi
27. *Trewia nudiflora* L (False white teak)

4.15.3 Habit and Habitat

The plants may be small herbs (*Phyllanthus* spp., *Euphorbia hirta, Acalyptha indica)* shrubs *(Jatropha glandulifera, Ricinus communis, Euphorbia splendens, E. pulcherrima)* Trees *(Hevea brasiliensis, Emblica officinalis, Putranjiva roxburghii)* climbing shrubs *(Bridelia stipularis, Dalechampia indica, Phyllanthus reticulatus, Tregia involucrata)* *Excoecaria* is a mangrove plant. Several species of *Euphorbia (E. royleana, E. nerrifolia* and *E. trigona)* are cactus like in habit with thick and fleshy stems and the leaves are reduced to spines. The plants often contain a milky latex in special laticiferous vessels.

4.15.4 Vegetative Characters

Root system: Tap root system. In *Manihot utlissma* tap root becomes tuberous by the storage of starch.

Stem: Aerial, usually erect, weak and prostrate in *Euphorbia prostrata,* however, usually woody and in few species it is hollow. Stem, spinous (*E. splendens*) or smooth (*Ricinus)* hairy (*E. hispida*) cyllindircal (*E. tirucalli*) or angular (*E. antiquorum*). Many genera show xerophytic adaptations like leafless or reduced leaf character. Stems thick and succulent. In some species of *Euphorbia* (*E.trigona, E.roylena*) stem is green, cactus like and performs photosynthesis like a phylloclade. The leaves are reduced and modify into spines.

Leaf: Leaves usually exstipulate, in some species (*Jatropha glandulifera*) they are modified into hairy glands or spines (*E. splendens*). Leaves pedicellate or sub-sessile. Leaves caducous in *E. roylena* and ever green (*Buzus* and *Ricinus)*, phyllotoxy alternate (*Ricinus*) (fig 4.15.1), opposite (*Euphorbia* spp) or whorled (*Mischodon*) In *Croton bonplandianum* the upper leaves are arranged in whorled phyllotaxy while lower leaves are opposite (fig 4.15.1A). Leaves usually simple, palmately lobed (*Ricinus, Jatropha*) or trifoliate compound (*Hevia, Bischofia*). Leaves variegated in *Codieum, E. milii*).

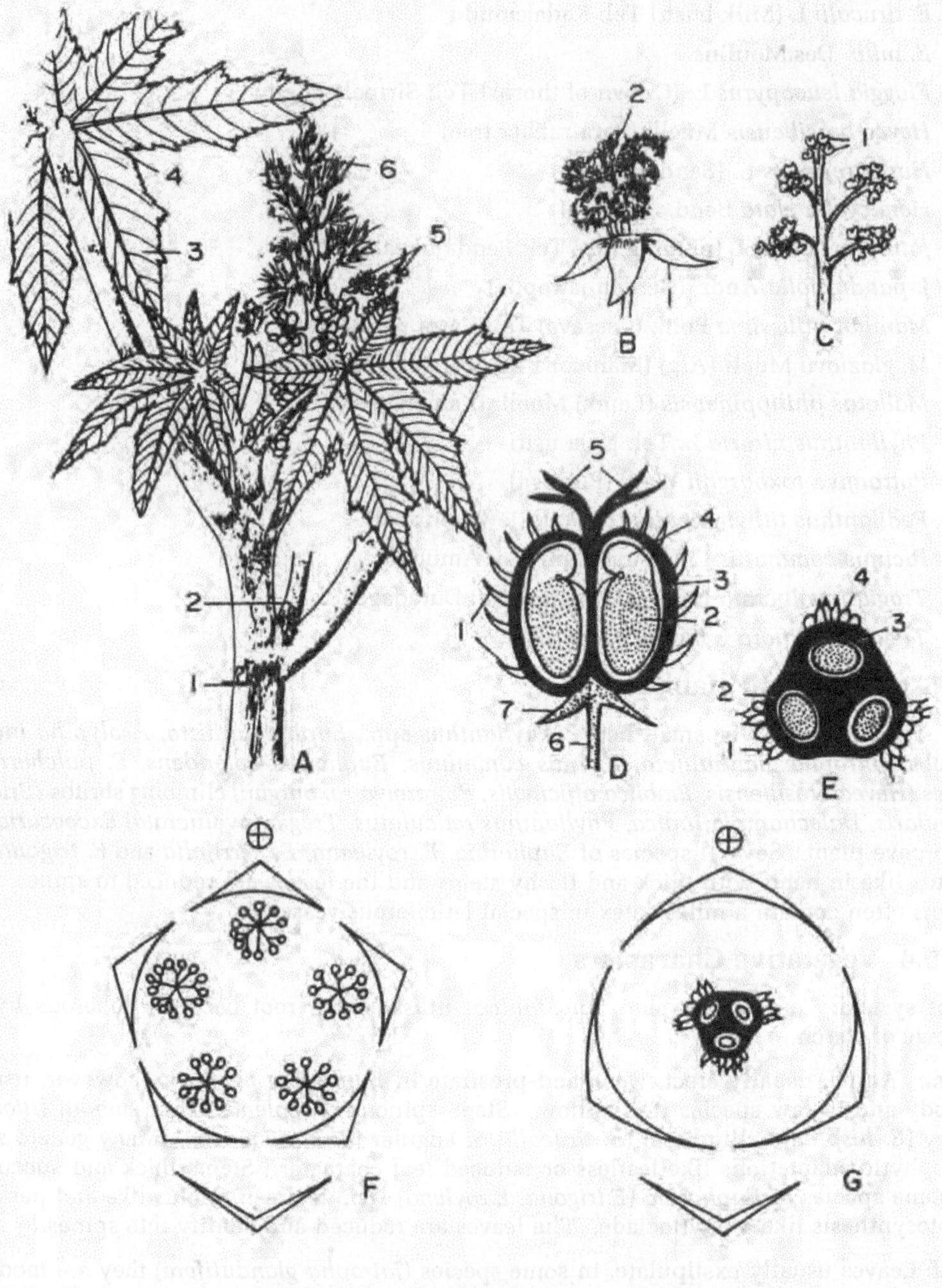

Fig 4.15.1. *Ricinus communis* (Euphorbiaceae)

A: Flowering twig ; 1. Stem, 2. Axillary Branch, 3. leaf, 4. Gland, 5. Male Flowers, 6. Female Flowers; B: Male Flower 1. Tepal, 2. Branched Stamens, C: Branchlet of stamen 1.anthers D: L.S. Female flower 1. Appendages 2, Ovule, 3. Ovary, 4. Carencle, 5. Stigmas, 6. Pedicel, 7. Tepal; E: T.S. Ovary 1. Ovary wall, 2. Locule, 3. Ovule 4. Appendages F: Floral diagram male flower; G: Floral diagram (female flower)

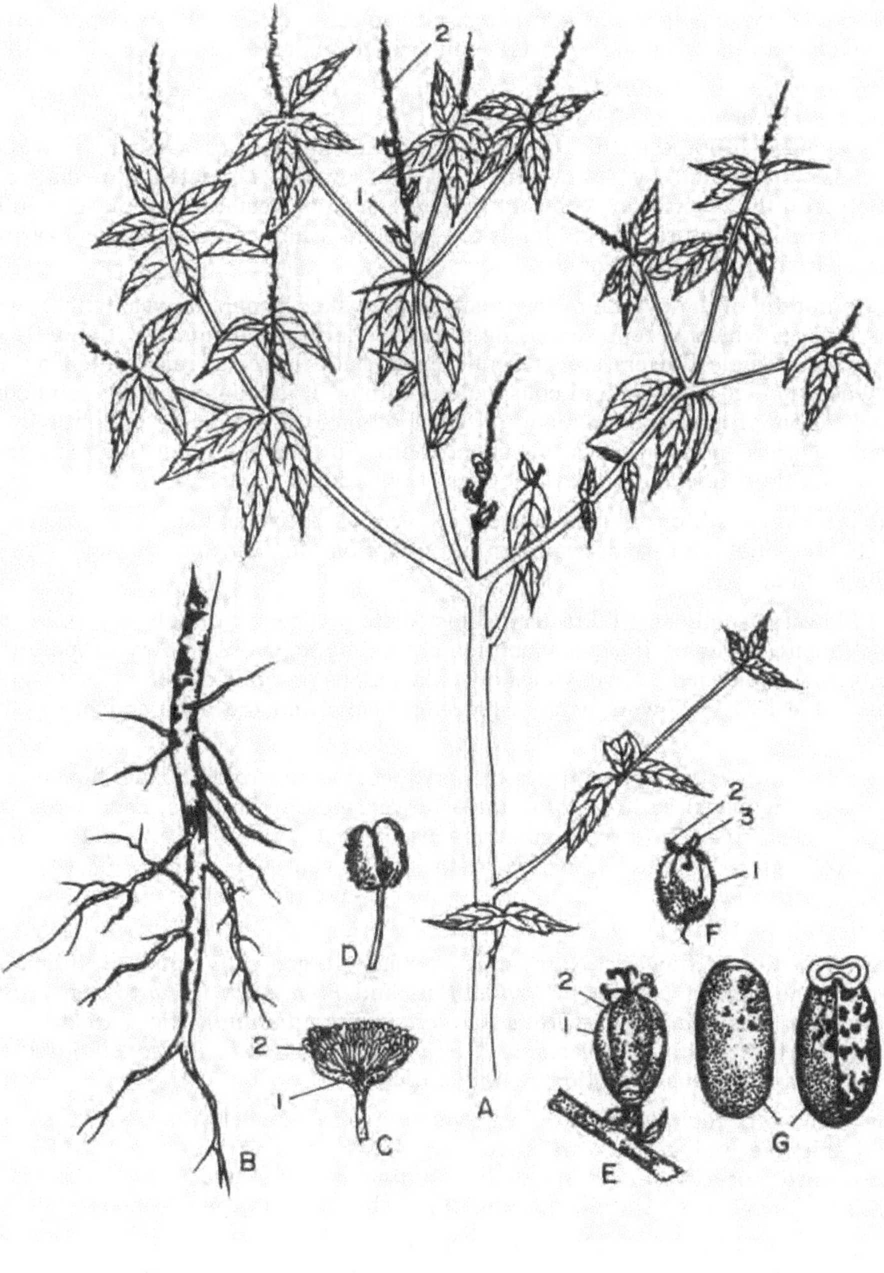

Fig 4.15.2. *Croton bonplandianum*

A: Flowering twig ; 1. Female Flower, 2. Male Flower, B: Root; C: Male Flower 1. Perianth, 2. Stamens; D: Stamen; E: Female Flower 1. Perianth, 2. Gynoecium; F: Gynoecium 1. Ovary, 2. Style, 3. Stigma; G: Seeds with Caruncle.

4.15.5 Floral Characters

Inflorescence : In Euphorbiaceae considerable variations are observed in the inflorescence types.

A dichasial cyme is found in *Jatropha*, a raceme (*Croton*) (fig 4.15.2A), spike (*Acalypha, Antidesma)* Catkin in *Acalypha*. Axillary solitary flowers are observed in *Phyllanthus*.

In some genera like *Euphorbia, Anthostemma Poeinsettia* and *Pedilanthus* a specialized inflorescence characteristic of Euphorbiaceae called the *Cyathium* is present which is a modified cymose inflorescence. Infact, a cyathium appears as a single flower. Each cyathium is surrounded by an involucre of four or five connated cup-shaped bracts alternating with the bracts 4 or 5 crecent shaped glands or nectaries are present on the rim of the cup-shaped involure. These glands are large and often coloured which bear petaloid appendages in *Euphorbia splendens*.

In the middle of the cyathium there is a single naked female flower (without perianth lobes) is present which is represented by a tricarpellary gynoecium. In the axils of each bract number of male flowers are present. Each male flower is represented by a single stamen with a jointed filament and conspicuous anther. It is opined that the joint condition is an indication of supressed perianth. The presence of a tubular perianth in similar position in *Anthostemma* justifies this conclusion. Further, the older flower is nearest to the centre and thus the maturation of flowers is centrifugal.

The bracts of cyathium in many genera are very attractive and substitute for the petals. The bracts present at the base of cyathium in *Euphoribia pulcherrima* are large red coloured and quite distinct.

Flower : Flowers usually small, bracteate, bracteolate pedicellate or sub-sessile, incomplete, unisexual, monoeceous or dioecious, actinomorphic, hypogynous, however, perigynous in *Bridelia*, pentamerous, monochlamydous or achlaymdous or rarely dichlamydous. In *Ricinus* and *Croton* the female flowers being aggregated below and the male above. (fig 4.15.1 & 4.15.2).

Perianth : Perianth in one whorl (monochlamydeous) as in *Croton* female flower or in two whorls (dichlamydeous) as in *Croton* male flower and *Phyllanthus*. Each whorl contain three tepals each. If perianth is dichlamydeous the inner whorl being petaloid (*Jatropha*) or scaly (*Actiphila*). If inflorescence is cyathium, flowers lack perianth (Achlamydeous). Perianth lobes free (*Croton, Phyllanthus*) or united (*Ricinus)*. Aestivation may be imbricate (*Jatropha*), valvate (*Ricinus*) or twisted in the whorl (*Jatropha pondurifolia*).

Male flower (staminal flower) : Number of stamens are generally equal to number tepals. Rarely only one stamen present in *Pedilanthus* and *Poinsettia*. Five to many (*Ricinus*), eight (*Jatropha gossipifolia*), numerous (*Croton bonplandianum*). (fig 4.15.2C) Stemens monadelphous (*Jatropha*), however united by the filaments only (*Kirganelia*) and the five stamens show synandrous condition in *Phyllanthus*.

The stamens in *Ricinus communis* are numerous in number and are characteristically branched tree-like and anthers are borne on short branches (fig 4.15.1 B&C). Anthers dithecous, introse or extrose, dehisce by longitudinal or transverse slits or apical pores (fig 4.15.2D). In many genera a rudimentary pistil may be present in male flower.

Female flower (Pistillate flower) : Female flowers usually bigger thatn the male flowers, tricarpellary syncarpous, superior, however, bicarpellary (*Mercurialis*) pentacarpellary (*Wielandia*), when tricarpellary ovary is trilocular, each locule contains one or two pendulous ovules are present on axile placentation. Ovules anatropous with ventral raphae and obturator, styles free or may be fused at the base. Each stigma bifid, simple or feathery as in *Ricinus*. A nectariferous disc is situated below the ovary (fig 4.15.1D). Staminodes may be present in the female flower.

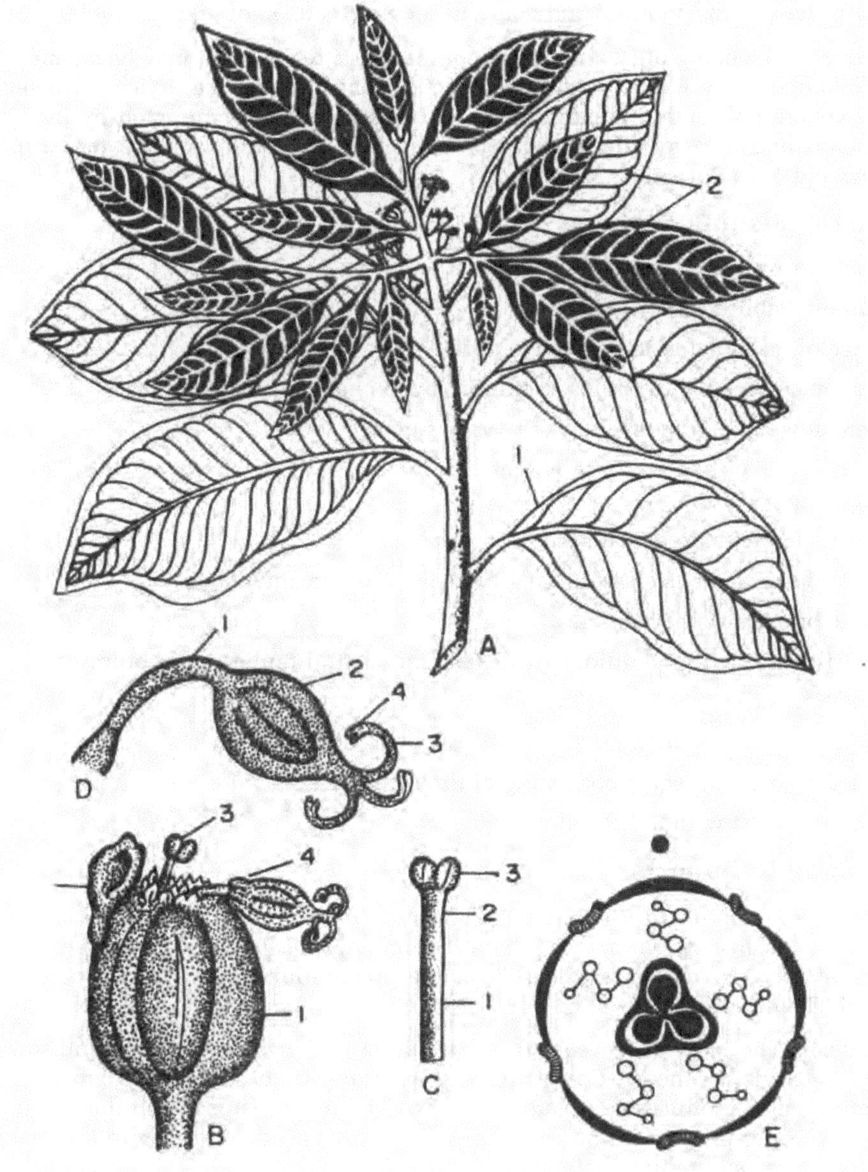

Fig 4.15.3 *Euphorbia pulcherima*

A: Flowering twig ; 1. leaf, 2. Thick coloured leaves, B: Cyathium inflorescence 1. Involucre of bracts, 2. Nectary, 3. Anthers, 4. Gynoecium, C: Male Flower, 1. Pedicel, 2. Filament, 3. Anthers, D: Female flower 1. Pedicel, 2. Ovary, 3. Style, 4. Stigmas, E: Floral diagram

Fruit : Schizocarpic capsule (*Jatropha*), regma (*Ricinus*) which is dehiscent and splits into three cocci each with one seed. Fruit may be drupe (*Travia*) or berry (*Bischofia*).

Seed: Endospermic, oily (*Ricinus*) and the seeds are oblong as in castor with smooth testa which is crustaceous and mottled and contains a conspicous knob like caruncle.

Pollination: Flowers being unisexual, cross-pollination is the rule. It may be anemopholous (*Mercurialis* and *Ricinus*) due to the presence of long thread like styles. However, in majority of genera pollination is carried out by insects which are attracted by the brightly coloured bracts and also for the honey secreted by extra- floral nectaries (eg. *Euphorbia pulcherrima* and *Manihot spp.)*

4.15.6 Distinguishing Features

1. Plants mainly shrubs, trees and few herbs.
2. Xerophytic habit of many genera.
3. Presence of milky latex in vegetative parts.
4. Inflorescence variable, complex cyathuim is the characteristic feature.
5. Flowers unisexual, hypogynous and ovary superior.
6. Perianth in one or two whols or absent. If present, may or may not show differentiation into calyx and corolla.
7. Stamens one-many, free or united
8. In a cyathium, stamens show jointed filament and achlamydeous.
9. Presence of extra-floral nectaries.
10. Ovary trilocular with pendulous ovules having ventral raphae and obturator.
11. Placentation axile.
12. Styles three, stigmas bifid
13. Fruit a schizocarpic regma consisting of three cocci.
14. Seed albuminous with a distinct caruncle.

4.15.7 Economic Importance

4.15.7.1 Food: *Manihot utilissima* (tapioca) and *M. palmata* yield starch from their tuberous roots. Fruits of *Antidesma acuminatum*, *A. ghesambilla* are edible. *Emblica officinalis* and *Cicca acida* fruits are used for the preparation pickles which contain high quantities of 'C'-vitamin.

4.15.7.2 Rubber: The latex of *Hevea brasiliensis* (Para rubber of commerce) and few other species of *Hevea* yield rubber. Ceara rubber is obtained from *Manihot glaziovii*. *Sapium indicum* latex is also produces commercial rubber. The rubber is useful in the manufacture of tyres, tubes, water-proof clothing and various electrical goods and also in the making of sports goods, gloves and tennis balls.

4.15.7.3 Oils: Castor oil is extracted from the seeds of *Ricinus communis*, which is used as lubricant, illumination oil and also as a purgative medicine. Castor oil is insoluble in benzene and has a very low melting point. It is ideally suited to the lubrication of aeroplane engines. It is also finds use in the preparation of detergent washing powder and also in the manufacture of hydraulic brake fluids.

The oil obtained from the seeds of *Aleutes foridii* (tung oil) is used in the manufacture of varnishes and paints. Oil extracted from the *Croton* species is employed as a purgative agent.

4.15.7.4 Medicine: *Emblica officinalis* (amla) has an immense medicinal value. Its fruits are rich in vitamin-C and are used in the treatment of scurvy, asthama and bronchitis. The dried fruits are employed in the control of haemorrhage, diarrohoea, dysentry and in combination with iron is used for treating anaemia and jaundice.

Two alkaloids, namely ricinine and taxalbumin ricin are obtained from the seeds of *Ricinus communis* which have high medicinal properties. Leaves and latex of *Euphorbia nivulia* are employed in the control of cough and other pulmonary diseases. *Acalypha indica* has pesticidal properties and also to control skin diseases. The species of *Phyllanthus* find their application in the control of jaundice, insect repellant and urinary troubles.

4.15.7.5 Ornamentals: Because of the brilliant colour of the leaves, bracts and cyathium inflorescence of many species are grown as ornamentals, some of these include, *Excoecaria bicolor, Codiacum variegatum, Pedilanthus tithymeloides, Jatropha pandurifolia, J. hastata, Acalypha hispida, Croton eluteria, Euphorbia pulcherrima, E. milii* and *Phyllanthus distichus.*

4.15.7.6 Dyes: *Chrozophora tinctoria* and *Sapium* spp. yield dyes. Fruits of *Kirganelia reticulata* are source of ink.

4.15.7.7 Other Uses: Strong wood is obtained from *Bridelia retusa* and *Bischofia javanica* which is used in the construction of bridges and boat making. various species of *Cleistanthus* yield wood and used in the manufacture of agricultural implements, poles and others. Box-wood is obtained from *Buxus* spp., which is handy in wood engraving. The wood of *Trewia nudiflora* is used for making drums. Gum exuded from the trunk of *Croton xalapensis* is used as a substitute for tooth paste. *Malotus phillipinensis* and *Glochidion velutinum* are used for tanning.

4.15.8 Systematic Position and Affinities

Bentham and Hooker placed Euphorbiaceae in the sub-division Monochlamydeae and series Unisexuales along with eight other families viz., Urticaceae, Moraceae, Plantanaceae, Lytheraceae, Juglandaceae, Balanopsaceae, Myricaceae and Casuarinaceae. Engler-Prantl however, placed Euphobiaceae in sub-division Archichlamydeae and order Geraniales. In Hutchinson's system, the family is placed under Euphorbiales.

The affinities of Euphorbiaceae with other families have been interpreted differently. Formon (1952) opined that it is closely related to Pandanaceae. According to Cronquist, this family has originated from the order Celastrales. However, the family is found to be related to the families of order Malvales and particularly, Sterculiaceae due to the presence of monadelphous stamens, presence of rudimentary stamens and pistil, presence of mucilaginous fibrous stem, caruncle and nature of embryo. The family also bears relationship with Geraniales and Sapindales on account of the nature of disc and ovules.

4.16 ORCHIDACEAE (Orchid family)

Systematic Position :

Angiosperms

Monocotyledonae

Microspermeae

Orchidaceae

4.16.1 Distribution The Orchidaceae is one of the largest families of the monocots and second largest among the flowering plants. The family comprises about 735 genera and about 17,000 to 20,000 species. The members of the family are cosmopolitan in distribution, but abundant in tropics and rare in arctic regions. This is the second largest family of the Indian flora, represented by 128 genera and about 881 species growing mostly in the Himalayas, Western Ghats and Khasia Hills but, they are abundant in the Eastern Himalayas.

4.16.2 Important Plants

1. *Cymbodium grandiflorum* Swartz.
2. *Cypripedium californicum* L.
3. *Dendrobium ovatum* (Willd.)
4. *Epipactis royleana* L.
5. *Eulophia nuda* Lindl.
6. *Habenaria roxburghii* (Pers.) R. Br.
7. *Neottia listeroides* L.
8. *Paphiopedilum lawrenceamun* L.
9. *Orchis* latifolia L.
10. *Spiranthes sinensis* L.
11. *Vanda spathulata* (L.) Spr. Tel: veduru badanika
12. *Vanilla wightiana* Liindl. Tel: Gundu nalleru
13. *Zeuxine membranacea* L.

This is the largest family of the monocots and second largest of all the families of Angiosperms.

4.16.3 Habit and Habitat

In general, members of Orchidaceae are terrestrial (*Orchis, Epipactis*), saprophytic (*Neottia*) or epiphytic (*Cuttleya, Cymbidium Cypripedium*) perennial herbs. The tropical species are mostly epiphytes while those occurring in temperate zones are largely terrestrial.

4.16.4 Vegetative Characters

Root system: Tap root is short lived; it is replaced by adventitious roots that may be clinging, absorbing, aerial or photosynthetic (*Taeniophyllum, Gunonia*, Polyrhiza). In *Orchis epipactis* the adventitions roots become swollen and tuberous which helps the plant overwinter. In *Vanda, Cypripedium, Cuttleya* and *Cymbidium* the aerial roots are with outer layer of water absorbing tissue, the velamen. Their internal tissue is green which helps in photosynthesis. The saprophytic forms (*Neottia*) produce a much branched fleshy rhizome with roots. *Corallorhiza* is without roots. They derive their nutrition from the humus with the aid of mycorrhizae (*Epipogan*).

Stem: The epiphytic forms are adapted to xerophytic conditions. They shed the leaves in dry weather and tide over this period by storing water and food in thick internodes called pseudobulbs or fleshy leaves (eg. *Bulbopyllum). Vanilla* has no pseudobulb but only thick fleshy leaves. Many others have no leaves and photosynthesis takes place through green stem or aerial roots. In *Taeniophyllum*, the aerial roots, which are green, flattened and ribbon like look as foliage leaves.

Leaves: The leaves are simple, alternate, rarely opposite or whorled, entire, oval or linear with parallel venation. Leaf base is sheathing which envelops the stem. In saprophytes,

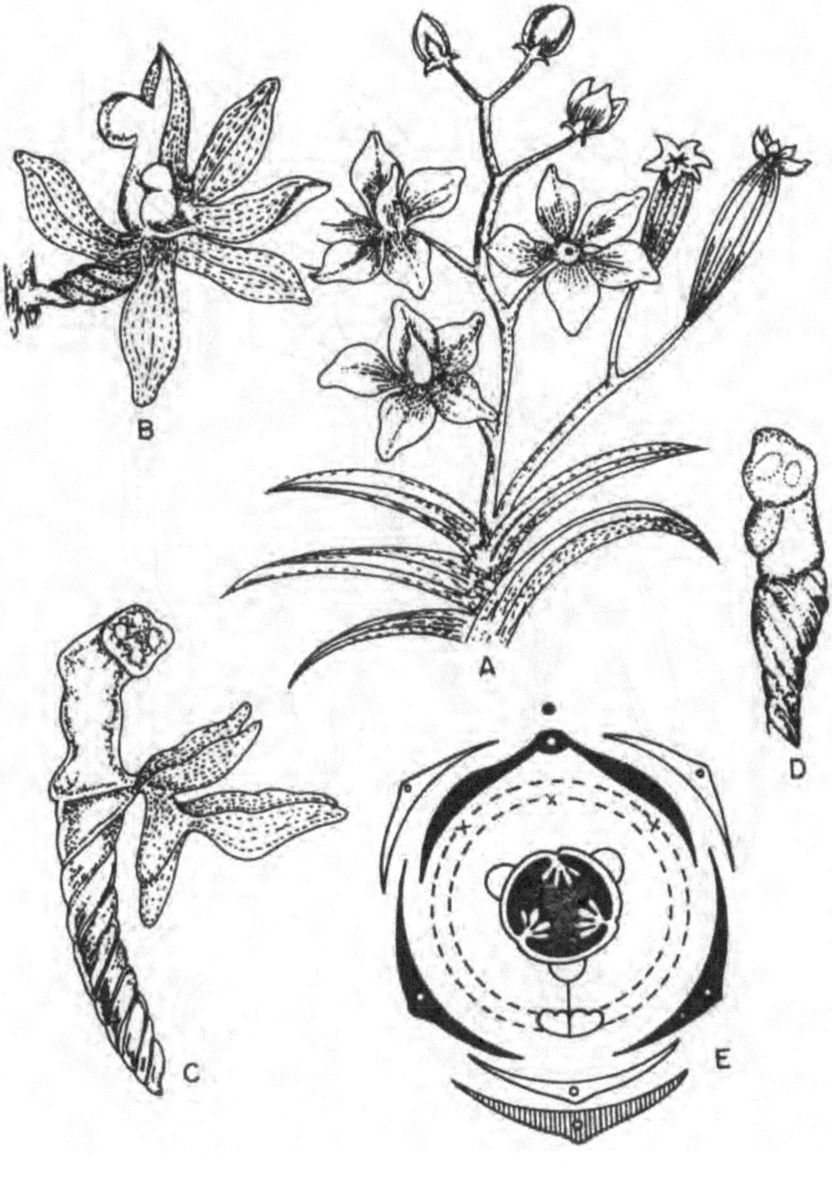

Fig 4.16.1. *Vanda roxburghii*

A: Flowering twig; B: Flower; C: Dissected flower showing resupinate ovary, Column, stigma anther and labellum, D: Ovary with column stigmas and anther; E: Floral diagram

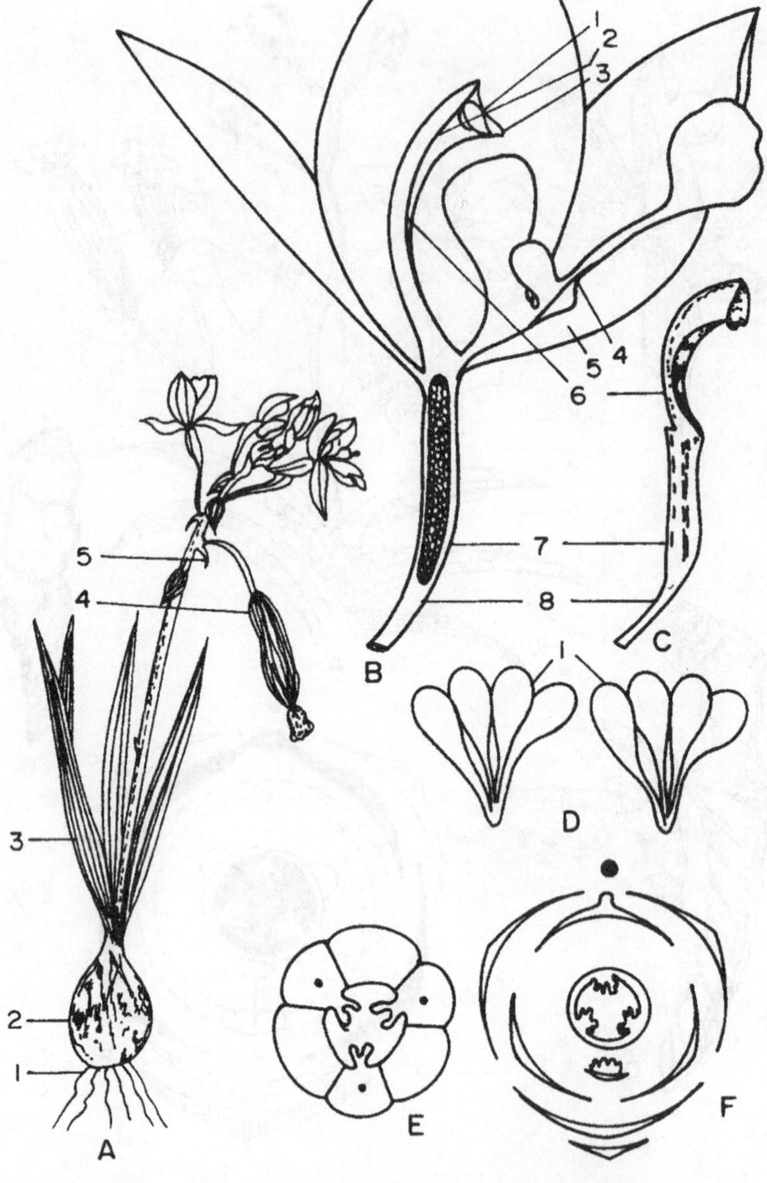

Fig 4.16.2 *Spathoglottia sp.*

A: Flowering twig ; 1. adventitious roots, 2. Bulb, 3. Leaves, 4. Fruit, 5. Bract, B and C : Flower L.S. 1. Pollonia, 2. Fertile stigma, 3. Rostellum, 4. Labellum, 5. Perianth, 6. Column, 7. Ovary, 8. Pedicel; D: Pollinial masses 1. Pollinia; E: T.S. Ovary; F: Floral diagram

leaves are reduced to scales. *Epipogon* is a leafless terrestrial herb. Occasionally there is only one leaf in *Bulbophyllum* or two leaves in *Dendrobium*.

4.16.5 Floral Characters

Inflorescence: Flowers are solitary in many species of *Cypripedium*, or borne in racemes or spikes (*Rhynchostylis, Cattleya*) or panicles (*Oncidium*).

Flower: The flowers are large and showy brightly coloured and scented. They are bracteate, bisexual or rarely unisexual, zygomorphic, trimerous and epigynous. The flowers are usually resupinate at anthesis as the pedical or ovary twists at an angle of 180º during development (fig 4.16.1C).

Perianth : Perianth is in two trimerous whorls, free or fused, fleshy and the outer whorl is sepal like. The median sepal becomes somewhat larger in size. The aestivation is imbricate or sub-valvate.

The inner whorl corresponds to petals and all are dissimilar. The posterior petal is larger and modified into a lip or labellum (fig 4.16.2 B&C). It is posterior in bud condition but it comes to lie in the anterior position due to resupination. It is in different shapes and colours. Funnel like in *Cattleya*, slipper like (*Cypripedium*), bucket like (*Corysanthes*) or curved upwards and spreading as in *odontoglossum*. This labellum contains nectar secreting tissue. The two lateral petals or wings are simple and alike.

Resupination : The long inferior ovary externally looks like the pedicle of a flower. This ovary undergoes a clock wise twisting through 180º which bring the posterior side of the flower in the bud towards the anterior side and vice-versa. This phenonmenon of reverse orientation is called resupination. The anterior side of the labellum is the landing place for the insects during pollination (fig 4.16.1C).

Androecium : The androecium is represented by one or two sterile anthers. Located at the tip of gynostegium, or column, which is highly complex structure formed by the union of the stamens, style and stigma. It is the most characteristic structure of the orchid flower. The orchids are distinquished into two groups namely Monandrae and Diandrae. In Monandrous forms there is only single fertile stamen, terminal on the column. This stamen is considered to represent the anterior member of the outer whorl. The lateral pair of the inner whorl (*Orchis, Epipactis*) or the posterior lateral pair of the outer whorl (*Diuris*) are reduced to staminodes. The other stamens are completely suppressed (fig 4.16.1D).

In Diandrous forms (*Cypripedium* and allied genera) there are two fertile stamens belonging to the inner whorl and they are lateral in position. The anterior stamen of the outer whorl is represented by a large staminode. The other staminal members are completely absent. The anthers are dithecous, introse and opening by longitudinal slits. The pollen grains are either granular or agglutinated into waxy pollinia (fig 4.16.2D). One end of the pollinium is extended into a sterile structure, the caudicle. There are 2-8 pollinia per anther and they are free or more or less losely united.

***Gynoecium*:** The gynoecium is tricarpellary syncarpous, unilocular, with inferior ovary with parital placentation. In *Apostasia* the ovary is trilocular with axile placentation. There are numerous ovules on each placenta (fig 4.16.1E)

In monandrous form the column has two fertile stigmas and a specialized organ, the rostetllum represents the third stigma. It lies in the centre of the column between the anther and fertile stigmas.

In diandrous forms all the three stigmatic lobes are fertile and there is no rostellum.

Fruit: Mostly opening laterally by five-six hygroscopically sensitive valves.

Seeds : The seeds are very large in number, minute, fusiform and rarely winged. They are non-endospermic and the embryo is undifferentiated.

4.16.6 Pollination Mechanism

The flowers of Orchidaceae are admirably adapted to insect pollination. The flowers are usually large, brightly coloured beautiful in form and conspicuous by the presence of inflorescence. The flowers are sweet scented. The flowers secrete nectar in the spur of the labellum or sacuate structure developed by the rostellum or at the base of the column. Due to the resupination of flowers, the labellum is brought into a position, where it makes an effective landing place for the insect. The labellum shows variation in size, structure and colour to attract the insects. The pollen grains are mostly compound and are united to form 2-8 pollinia. These pollinia have slender caudicles which are attached to the rostellum through a gelatinous disc. An insect vistor lands upon the labellum and then probes into the flower for nectar in doing so, the insect head strikes the sticky caudicle, which attaches to the body of insert. As the insect withdraws, the pollinium is taken out of the flower when the same insect visits another flower, the pollinia touch the stigma. The flowers of *Ophyrus* lack spurs and nectaries. The flower resemble in a remarkable manner the female species of the bees. The flowers are pollinated only by male insects. They are attracted by the flowers of *Ophyrus* due to their female bee appearence. On reaching the flower, the male bee gets excited, carry out mating movements, in doing so their bodies come into contact with the pollinia.

4.16.8 Distinquishing Features

1. Herbs, terrestrial or epiphytic or saprophytic forms
2. Bracteate, bisexual, epigynous, zygomorphic flowers
3. Perianth is in two trimerous whorls. The tepals are free 3 + 3 or connate (3 + 3)
4. Posterior segment of inner whorl developed into lobellum or lip
5. Stamens one or two
6. Pollen grains united into pollinia
7. Tricarpellary syncarpus unilocular ovary with parietal placentation
8. Formation of gynostegium or column.
9. Stamens and stigmas together and form a column
10. Fruit, Capsule with non-endospermic seed.
11. Seeds minute and light weight

4.16.9 Economic Importance

1. ***Cymbodium:*** The plant is emetic and purgative.

   C. grandiflorum, C. eburneum, C. insigne are ornamentals.

2. ***Cypripedium spp:*** ***Paphiopedilum, Orchis, Vanda, Vanilla, Dendrobium and Catleya*** grown as garden ornamentals.
3. ***Orchis:*** root tubers yield glucosides, loroglossin and salep
4. ***Vanda:*** the roots of vanda are used in rheumatism and antidote for scorpion stings
5. ***Eulopia:*** tubers are good vermifuse
6. ***Geodorum — G. densiflora*** root stocks are crushed and rubbed on cattle to kill flies.
7. ***Vanilla*** — The capsules of *V. fragrans, V. planifolia* are the source of Vanilla extract of commerce, which is extensively used for flavouring various food products

8. ***Dendrobium*** : The fibres from the stem cortex of *D. crumenatum* are used as a baiting material for hats.
9. ***Platenthera*** : The tubers of *P. sursannae* are edible.

4.16.10 Primitive and Advanced Characters or Orchidaceae

PRIMITIVE CHARACTERS:

1. The plants are perennial and perennate by means of pseudobulbs or root tubers. Perennial habit is primitive as compared to annual.
2. The simple, alternate leaves.
3. Presence of large, attractive and scented flowers.
4. Presence of solitary terminal flowers.
5. Flowers are mostly bisexual.
6. The perianth is biseriate and free in many species.
7. Insect pollination.
8. Ovules are numerous and anatropous.
9. Embryos without a suspensor
10. A feebly developed nuclear type of endosperm.
11. Presence of monocolpate pollen grains.

ADVANCED CHARACTERS :

1. The family includes only herbs.
2. Presence of variously modified advertitious roots.
3. Saprophytic and epiphytic habit.
4. Exstipulate leaves.
5. The flowers are arranged in definite inflorescence types.
6. In some orchids (*Habenaria*) the flowers are small, inconspicuous and green or brown in colour, non-scented.
7. Presence of gamophyllous perianth, modification of a perianth lobe into labellum.
8. Zygomorphic and epigynous flowers.
9. Reduction in the number of stamens to one or two.
10. Adhesion of stamens with the style and stigma to form a structure called as gynostegium or column.
11. Tricarpellary syncarpous unilocular inferior ovary.
12. Reduction in the number of fertile lobes of stigma to two and modification of third sterile lobe into rostellum.
13. Pollinia formation.
14. Parietal placentation.
15. Fruit is a simple capsule.

4.17 PALMAE (ARECACEAE) (The Date Family)

Systematic Position:

Angiosperms

Monocotyledonae

Calycinae

Palmae (Arecaceae)

4.17.1 Distribution

The Arecaceàe family contains 217 genera and 2500 species, widely distributed in the tropical and subtropical regions of the world. In India, the family is represented by about 28 genera and 93 species ocurring in the Eastern Himalayas and Assam.

4.17.2 Important Plants

1. *Areca catechu* L. (Betel nut plant) Tel: Vakkalu
2. *Borassus flabellifea* L. (Palmyra Palm) Tel: Thadi chettu
3. *Caryota urens* L. (Bastard sagopalm) Tel: Jeeluga
4. *Calaromus rotang* L. (care palm)
5. *Cocos nucifera* L. (Coconut palm) Tel: Kobbari chettu or narikelamu
6. *Elaeis guinensis* (Oil palm)
7. *Hyphaene thebaica* L. (Egyptian date palm)
8. *Metroxylon sago* L. (Sago palm)
9. *Phoenix dactylifera* L. (Date Palm) Tel: Kharjura
10. *Phoenix sylvestris* (L.) Roxb. (Wild Date) Tel: Eetha
11. *Phoenix humilis* Becc.Tel: Chitti Eetha
12. *Oreodoxa regia* L. (Royal Palm)

4.17.3 Habit and Habitat In habit the palms range from small herbaecous form in *Eleuthero petalurm* and *Reinharditia* to large unbranched trees that may range in height from 100-200 feet (*Ceroxylon, Borassus, Cocos, Roystonea, Calamus, Plectocomia). Plectocomiopsis* and *Daemonorops* are the climbers.

4.17.4 Vegetative Characters

Root system: The primary root of the seedling is replaced by adventitious roots arising from the lower internodes of the stem. Thick aerial roots are found in some species of *Manicaria*. Thick aerial prop roots support the slender stems of some species of *Iriartea*. In the species of *Phoenix* negatively geotropic swollen roots are developed known as pneumathodes.

Stem: Mostly aerial, unbranched columnar woody stems (caudex) in *Borrassus, Phoenix*. Creeping and horizontal in *Serenoa*, Creeping and rhizomatous in *Sabal minor*, Dichotamousely branched in *Hyphaene*, monopodial in *Phoenix*.Climbing in *Colamus* by means of recurved spines, herbaceous in *Eteutheropetalum*. The older parts of the stem are usually covered by persistant leaf bases (*Phoenix sylvestris*) or by scars left by fallen leaves (*Caryota urens*), stem internodes are usually short, long in species of *Colamus, bactris. Ancistrophyllis, Desmancus* and some others (fig. 4. 17.1).

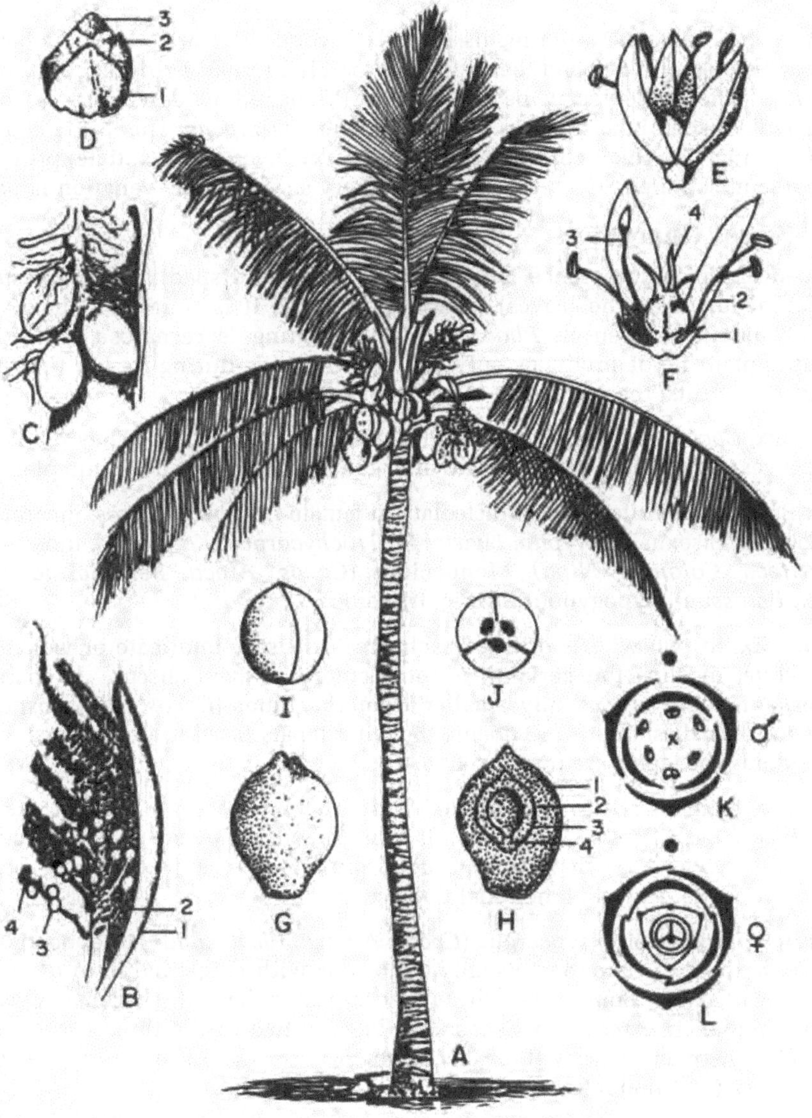

Fig 4.17.1. *Cocos nucifera* (Coconut)

A: Habit, B: Spadix, 1. Spathe, 2. Axis, 3. Female flower, 4. Male flower, C: Fruits, D. Female flower, 1&2 Perianth, 3. Pistil, EF Male flowrs 1&2 Perianth 3. Stamen, 4. Pistilode G: Fruit, H: L.S. Fruit 1. Pericarp, 2. Mesocarp, 3. Endocarp, 4. Endosperm, I. Fruit, J. Ovary T.S. K&L Floral diagram.

Leaf: Exstipulate, petiolate, simple in *Coccoturinax, Corypha*, Simple bifid in *Bactris*, mostly palamte in *Borarsus, paurotis, Crysophila*, pinnately compound in-*Phoenix, Cocos*, bipinnate in *Caryota*, Leaves vary in size from 5 inches long (*Chamedorea tuerckheinii*) to 50 feet long (*Corypha umbraculifera)* with lamina 10-12 feet in diameter; in some species the leaves are 1-2 feet long ex. *Reinhardtia. chamaedorea*. Shape of the pinnae is variable i.e., triangular in *Caryota*, sigmoid in *Bactris*, rhomboidal in *Ceratolobus* and ovate in *Desmonchus*.

In the climbing genera the rachis extends beyond the leaflet or pinna bearing portion and becomes flagelliform and bears spines or cirrhi that help in climbing (*Colamus, Kosthalsia*). The leaflets may be induplicate (V-shaped in a section) as in *Phoenix* or reduplicate (L-shaped in section) as in *Chamaedorea eurmpens.* The leaves show xerophytic characaters in having thick cuticle, presence of waxy layer above cuticle, presence of motor cells, abundant schlerenchyma, rough and leathery surface. The venation is parallel.

4.17.5 Floral Characters

Inflorescence: The inflorescence is spadix or a compound spadix. In *Borassus* the spadix is interfofliar (amongst the leaves) where as in *Areca* it is intrafoliar (below the leaves). Compund spadix in seen in *Phoenix*. The branching is recemose. *Plactocomia* is a monocarpic plant i.e. it produces only one inflorescence during its life time and after the matauration of fruits, the plant dies (*Corypha, Caryota).*

Phoenix, Cocos, Linistonia, Areca and *Borassus* are polycarpic. They produce numerous inflorescences during their life time and after the maturation of fruits the plant dies.

Flowers: Bracteate, sessile, small, bracteolate in female flowers of *Cocos*, incomplete, regular, actinomorphic, bisexual (*Corypha, Livistonia, Trachycarpus, Korthalsia*, unisexual (*Phoenix, Cocos, Areca, Colomus, Nipa).* Monoecious (*Cocos, Areca, Bentincksia*, or dioecious (*Phoenix, Borassua*), hypogynous, cyclic, trimerous.

Perianth: Six, in two whorls of three each, free or fused, imbricate or valvate, persistant, usually white, in *Nipa* the perianth is rudimentary or even absent. In many palms the outer perianth lobes are usually smaller from the inner (*Cocos, Didymosperma*) Tepals may be similar or dissimilar. In *Phoenix* the outer tepals fused to from cup-shaped structure where as the inner tepals are free (fig 4.18.1E).

Androecium: Stamens six, in two whorls of three each (*Phoenix, Cocos, Areca*) opposite to perianth lobes. 6-12 stamens, monoadelphous in the male flowers of *Plectoconia*, numerous in *Arenga, Caryota*, three in *Nipa, Areca triandra*, free or epipetalous in *Calaca* anthers dithecous, versatile, introse, longitudal dehiscence.

Gynoecium: Tricarpellary, syncarpus (*Cocos, Arenga, Plectocomia*) or apocarpous (*Phoenix, Rhapis*). The ovary is superior, usually, trilocular with single ovule in each locule. The placentation is axile. Some times the ovary is unilocular with parietal (*Oncosperma*) or basal placantation in *Areca*. In *Cocos* and *Phoenix* two of the three carpels abort during maturation. There are usually three sterile stigmas. In male flowers, gynoecium is absent or rudimentary (pistillode) (fig 4.17.1.F).

Fruit: A drupe (*Cocos, Areca, Licuba*) (fig 4.17.1.G) berry in *Rhapis, Phoenix* and *Pritchardia.*

Seed: Endospermic, ruminate in *Areca*, large in *Cocos*, embryo small and monocotyledonous.

Pollination: Pollination is chiefly effected by wind. Cross - pollination is favoured due to protandrous nature. It is entomophilous in some genera, by wasps in *Cocos nucifera.* In *Eupritchardia* pollination is brought about by birds.

4.17.6 Distinguishing Features

1. Small herbaceous forms are present
2. Leaves are exstipulete and compound.
3. Inflorescence is a spadix
4. Periants is present
5. Flowers are usually unisexual (*Phoenix, Cocos*)
6. Flowers trimerous.
7. Stamens many
8. Gynoecium tricorpelloaly, syncorpous rarely unilocular.
9. Style very short or absent.
10. Axile placentation.

4.17.7 Economic Importance

4.17.7.1 General Uses : The nuts of *A. catechu* are chewed along with betel leaves. The endosperm is the source of betel nut powder.

B. Flabellifer yields several products of commercial importance. It yields palmyra fibre which is used to prepare brushes and brooms. Its leaves are used for the manufacture of hand fans, umbrellas, baskets and mats. The fibre from petioles are used to make writing and printing paper. The fruits and seeds are edible. The sugary sap obtained from the young inflorescence is fermented to yield toddy.

The stems of *Calamus rotang* and *C.tenius* are used as cane in making chairs, baskets, hand bags, cables and umbrella rips.

4.17.7.2 Edibile oil: Leaves are used vegetable. Fibre obtained from the leaf sheeths is used for ropes, baskets and soft brushes.

4.17.7.3 Other Uses: The sugary milk of coconut makes a refreshing drink, tho endosperm is eaten raw or can be preserved dry. The dried kernel or endosperm is used for coconut oil extraction and the residue left behind is used as cattle feed. Its fibres are used in the manufacture of mats, brushes, brooms, strings and ropes. The hard shells left over are used as fuel and for making good quality charcoal.

E.quinensis an oil yielding palm with its seeds having 30-70% oil contents.

The young leaves of *Nipa fruticans* are used as cigarette paper.

Oredora sp., an ornamental plant, fruits yield oil and its stem yields sago.

Metroodylon sagu and *M.rumphii* yield sago or sagudana.

Fruits of *P. dactylifera* are edible dates that are rich in sugar. The leaves of *P.sylvestris* are made into mats and baskets.

4.18 GRAMINEAE (Poaceae) (The grass family)

Systematic Position :

Angiosperms

Monocotyledonae

Glumaceae

Gramineae (Poaceae)

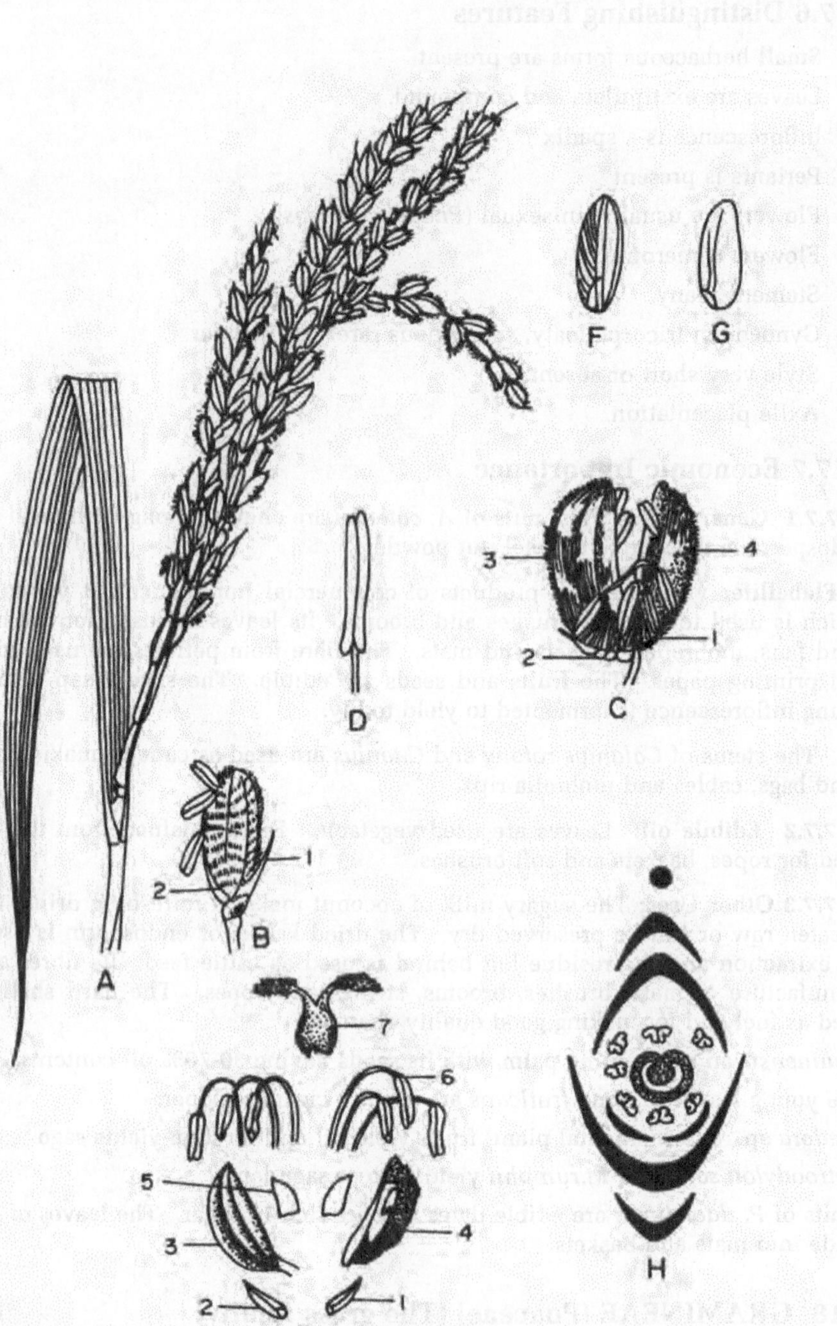

Fig 4.18.1. *Oryza sativa*

A: Plant with inflorescence; B: Spikelet 1&2 Glumes, C: Flower 1&2 Glumes 3. Upper (Superior) Palea 4. Lemne; D: Stamen; E: Spikelet 1&2 Glumes, 3. Palea, 4. Lemna, 5. Lodicule, 6. Stamens, E: Gynoecium; F: Fruit

4.18.1 Distribution Gramineae (Poaceae), the grass family with 620 genera and 10,000 species is the fourth largest among the flowering plants after the Asteraceae, Orchidaceae and Leguminosae. Many of them have world wide distribution. They grow under a wide range of climatic conditions from the warm tropics to within the Arctic circle.

4.18.2 Important Plants

1. *Andropogon odoratus* L. (Ginger grass)
2. *Aristida depressa* L. Tel: cheepuru gaddi
3. *Avena sativa* L. (Oat)
4. *Bambusa arundinacea* Retz.Tel: Veduru
5. *Chloris barbata* Sw.
6. *Cymbopogon citratus* Stapt (lemon grass) Tel: Nimmagaddi
7. *Cynodon dactylon* Rich. (Pers) (Dubgrass) Tel: Garika gaddi
8. *Dendrocalomus strictus* Roxb. (giant bamboo) Tel: sadanapu veduru
9. *Echinochloa colona* (Japanese Millet)
10. *Eleusine coracana* (L.) Gaertn. (Finger millet) Tel: Ragulu,
11. *Hordeum vulgare* (Barley, Jao)
12. *Oryza sativa* L. (Rice,) Tel: vari
13. *Panicum milliaceum* L. (Indian Millet) Tel: Varigalu
14. *Pennisetum typhoides* (Burn. f) (pearl Millet) Tel: Sajjalu
15. *Sacharum officinarum* L. (sugarcane) Tel: Cheruku
16. *Secale cereale* Bearun. (Rye)
17. *Setaria italica* (L.) Baeuv. Tel: korralu
18. *Sorghum valgare* (L.) Pers. Tel: Jonna
19. *Spinifex littoreus* (Burn. F) Tel: Ravanasurini meesalu or Gatti romala gaddi
20. *Triticum vulgare* L. (wheat) Tel: Godhuma.
21. *Zea mays* L. (Maize) Tel: Mokka jonna

4.18.3 Habit and Habitat

Majority of the genera are herbaeceious annuals (*Avena, Hordeum, Triticum, Secale, Zea* and *Cymbopogon)* many are perennials (*Saccharum, Arundodonax, Bromus, Agropyron*), many are shruby (*Bambusa, Arundinaria, Phyllostachys*) Tree like (*Bambusa nutans, B. khasina*). some species of *Bambusa, Dinochloa, Teinostachyum* and *Melocalomus compactiflorus* are climbing. Genera grow under mesophytic, xerophytic, halophytic, epiphytic and hydrophytic conditions. Many grow at higher altitudes and beyond tree limits in the alpine regions (*Agropyron repens*). The life span of *Bambuosa* varies from 15-20 years.

4.18.4 Vegetative Characters

Root System: Adventitious, branched, fibrous. In *Sorghum* and *Zea mays* stilt roots arise from the lower nodes of the stems.

Stem: Underground rhizome in all perennial grasses (*Cynodon*, *Agropyrum*, *Bambusa*, *Saccharum*, *Arundinaria*). Vegetative shoots that arise in clusters from the base or from the underground stems are called tillers or innovations. The aerial stems are always cylindrical and distinctly jointed, mostly hollow in *Triticum*, *Avena*, *Bambusa*, solid in *Zea*, *Saccharum*, herbaceous or woody in *Bambusa*, runners in *Cynodon*, stolons *in Hygroryza* species. Intercalary meeistem is present in most of the herbaceous grasses.

Leaf : leaves are cauline and ramal. Exstipulate, ligulate, mostly sesile, Leaf base sheathing, alternate and distichous, simple, entire, glabrous or hairy, lamina narrowed ribbon shapped, the epidermal cells are impregnated with silica.

4.18.5 Floral Characters

Inflorescence : The inflorescence is either axillary or terminal compound spike. The flowers are arranged in spikelets, which may be sesile (*Triticum*) or stalked (*Avena*). The spikelets are borne on the peduncle. Spikelet consists of a very much shortened and pointed axis called the rachilla. On the rachis are borne the sessile flowers. The number of flowers in a spikelet varies from one to many, one in *Oryza* (fig 4.18.1A) and more flowers in *Bambusa*. The flowers are usually arranged in an alternate manner or in two rows. At the base of rachilla are two sterile scales called the glumes. They lie one above the other and on the opposite sides. The glumes overlap and enclose all the flowers in the spikelets till maturity. (fig 4.18.1A)

Flower : Each grass flower is usually called a floret. It arises in the axil of a sterile scale or a bract called the *lemma*. The lemma frequently bears a long, stiff hair called as *awn*. Their number may vary from one to few. Opposite the lemma on the posterior side of the rachilla is another scaly bract called the *Palea* is present. The florets lie between the two membranous and variously nerved bracts and bracteoles are called the lemma and palea respectively. Flowers are bracteate, bracteolate, sessile, incomplete, bisexual (*Triticum*, *Avena*, *Hordeum*, *Oryza*) (fig 4.18.1C) or sometimes unisexual (*Zea*, *Coix*) (fig 4.18.2 D,E) irregular, zygomorphic, hypogynous, cylic, unisexual flowers may be borne on the same plant i.e. monoecious (*Zea mays*) or dioecious (*Spinifex littoreus*).

Perianth : Periant absent in *Dendrocalamus* and *Dinochloa*. It is represented by small scale like structures called as *lodicules*. There may be two lodicules in (*Triticum*, *Avena*, *Oryza*), (fig 4.18.1E) three in *Arundinaria* and many in *Bambusa*, two-twelve in various species of *Ochlandra*, They are hygroscopic and force apart the palea and lemma on absorption of water, there by exposing the essential organs for pollination.

Androecium : Stamens three in *Triticum*, *Zea mays*; (fig 4.18.1E) two-three in *Ergrostin*, six in two whorls of various species of *Bambusa*, *Oryza*, *Dendrocalomus* and six-twelve in species of *Ochlandra*. Anthers dithecous, versatile, dehiscence by longitudinal slits.

Gynoecium: Monocarpellary, superior, unilocular, single ovule on basal placenta, usually styles two or absent in *Avena* and *Triticum*. Generally, the stigmas are feathary. In *Zea mays* the stigmas are long, thread like and silky, they project out of the green bracts covering the female inflorescence (fig 4.18.2.D).

Fruit : Caryopsis, (fig 4.18.1E&G) rarely berry or nut in *Dendrocolamus*.

Seed : Endospermic, abundant endosperm, mealy with peripheral embryo. The plumule and the radicle are surrounded by distinct sheaths called coleoptile and colerorhiza respectively. The single cotyledon is called the scutellum. The scutellum is present in endosperm and does not come above the ground. It mainly acts as an absorbing organ. (fig 4.18.1 and 4.18.2)

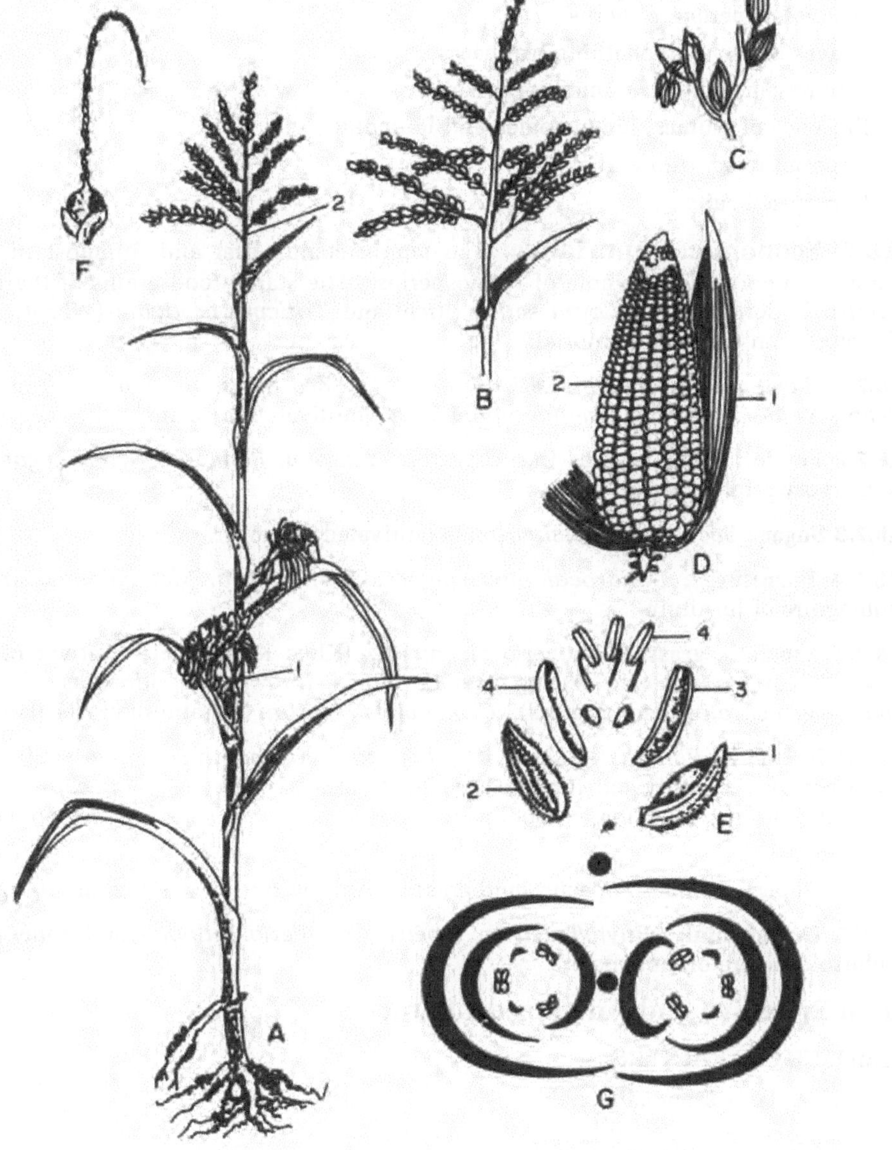

Fig 4.18.2. *Zea mays* (Corn)
A: Habit 1. Female Inflorescence, 2. male tassel, B: Male inflorescence, C: Flowers enalrged; D: Spike 1. Spathe, 2. Fruit, E: Parts of a male flower 1. Sterile glumes, 2. Lemma, 3. Upper Palea, 5. Lodicules, 6. Stamens, F: Gynoecium, G: Floral diagrams

Pollination : The flowers are small and without any frangrance and nectar. They are pollinated by wind (anemophilous). To enable cross pollination the flowers are either protandrous (*Zea mays*) or protogynous (*Anthoxanthum, Alopecurum*) cleistogamy occurs in *Hordeum*, and *Learsia*.

4.18.6 Distinguishing Features

1. Annual or perennial herbs
2. Leaves linear with shathing leaf bases
3. Inflorescence few to numerous spikelets
4. Presence of glumes, lemma,palea and lodicules
5. Superior ovary, unilocular
6. Caryopsis Fruits .

4.18.7 Economic Importance: The family stands first and foremost in respect of economic importance in whole of Angiosperms. The staple food grains of the population of world is derived from *Oryza sativa* (Rice) and *Triticum aestivum* (Wheat). They are cultivated from times immemorial.

4.18.7.1 Food : *Oryza sativa, Triticum aestivum, Zea mays, Hordeum vulgare, Sorghum vulgare, Avena sativa, Pennisetum typhoides* are cultivated for cereals and food grains.

4.18.7.2 Fodder: Many grasses like *Cynodom dactylon, Panicum, Cymbopogon, Agrostis, Poa* are gorwn for fodder.

4.18.7.3 Sugar : *Saccharum officinarum* is cultivated for sugar.

4.18.7.4 Furniture : *Dendrocalamus* species, *Arundinaria, Melocalamus* are used in manufacture of furniture.

4.18.7.5 Aromatic grasses: *Vetiveria zizanioides* (Khus Khus) yields vetiver oil from the roots. The roots are also woven into curtains. *Andropogon odoratus* (Ginger grass) *Cymbopogon citrates* (Lemon grass), *Cymbopogon martini* (Geranium grass) also yield oil.

4.18.7.6 Medicine: *Phragmites karka, Cymbopogon schoenanthus* etc are medicinal plants *Secale cereale* is cultivated for infection of its inflorescense by *Claviceps purpurea* for production of ergot and for extraction of ergotine. It is an excellent remedy for uterine contraction.

4.18.7.7 Paper: *Bambusa* species, and grasses are used for the manufacture of paper.

4.18.7.8 Ornamental: *Rhynchelytrum repens, Cortaderia selloana* and some species of Bambusoideae are ornamentals.

4.18.8 Primitive and Advanced Characters

PRIMITIVE CHARACTERS

1. A few plants arboreal in habit.
2. All florets in a spikelet are fertile.
3. Glumes are persistent.
4. Lemmas are herbaceous and leafy.
5. Stigmas are three
6. Leaves are simple and alternate.
7. Flowers are hypogynous and hermaphrodite.
8. Seeds are endospermic.

ADVANCED CHARACTERS

1. Plants are mostly herbaceous, annuals and perennials.
2. Leaves are exstipulate.
3. Flowers are arranged in distnince inflorescence.
4. Flowers are small, inconspicuous and zygomorphic
5. Perianth is represented by lodicules.
6. Stamens are reduced to 3
7. Gynoecium is monocarpellary and unilocular
8. Basal placentation.
9. Fruit caryopsis.
10. Seeds are small sized.

QUESTIONS

A. Answer the following in about 25 lines

1. Write the systematic position, general characters of Annonaceae.
2. Annonaceae is the primitive family among polypetalae, justify the statement.
3. Describe the floral characters of Malvaceae and mention the economic importance of the family.
4. Write on distinguishing features of Rutaceae with special reference to floral characters.
5. Write an essay on floral characters of family Fabaceae and describe the pollination mechanism.
6. Compare the floral characters of Fabaceae, caesalpinaceae and Mimosaceae.
7. Write the distinguishing features of family Apiaceae and mention its economic importance with special reference to spices and condiments.
8. Write an essay on the advanced characters of Asteraceae giving suitable examples.
9. Write an account on systematic position, vegetative and floral character of Asclepiadaceae.
10. Enumerate floral characters of family lamiaceae and add a note on pollination mechanism.
11. Describe the distinguishing features and economic importance of Euphorbiaceae.
12. Mention the floral structure and pollination mechanism in orchidaceae.
13. Describe the advanced features of orchidaceae.
14. Which family is considered as the most advanced among monocotyledons and underline important features of the family.
15. Give an account of important characters of Palmaceae (Arecaceae)
16. Write an account of economic importance of Arecaceae.
17. Give a critical account of the family Poaceae and its economic importance.
18. Describe an orchid flower and explain the mechanism of pollination in orchidaceae.

B. Answer the following in about 10 lines.

1. Flower of Malvaceae.
2. Various fruit types of Malvaceae
3. Fruit of Rutaceae.
4. Vexillary aestivation
5. Pollination in Fabaceae
6. Androecium in caesalpinaceae.
7. Morphology of tendrils in cucurbitaceae.
8. Androecium of cucurbitaceae.
9. Fruit in Apiaceae.
10. Economic importance of cucurbitaceae.
11. Infloresence of Asteraceae.
12. Structure of Male flower in Asteraceae.
13. Economic importance of Sapotaceae.
14. Asclepiadaceae flower.
15. Verticillaster inflorescence
16. Flower of Lamiaceae
17. Cyathium
18. Gynoecium in Euphorbiaceae
19. Orchidaceae habit.
20. Flower of orchidaceae.
21. Advanced characters of orchidaceae
22. Inflorescence in Arecaceae
23. Economic importance of Arecaceae.
24. Spikelet of Poaceae
25. Advanced characters of Poacae.
26. Gynostegium

C. Answer the following in 2 or 3 lines.

1. Ruminate endosperm.
2. Monadelphous condition
3. Epicalyx
4. Monothecous anther
5. Obdiplostemonous
6. Hesperidium
7. Pulvinus leaf base
8. Descending imbricate
9. Protandrous
10. Piston mechanism
11. Marginal Placentation
12. Myrmecophilous condition
13. Synandrium
14. Organ suigeneris
15. Cremocarp
16. Epigyny
17. Interpetiolar stipules
18. Syngenesious condition
19. Heterogamous head
20. Disc floret
21. Cypsela
22. Translators
23. Corona
24. Gynostegium
25. Follicle
26. Corpusculum
27. Periplocoideae
28. Thyrsus
29. Tetradynamous condition
30. Gynobasic style
31. Lever mechanism
32. Carcerulus fruit
33. Perianth
34. Cyathium
35. Regma
36. Epiphyte
37. Resupination
38. Labellum
39. Rostellum
40. Pollinia
41. Spadix
42. Caryopsis
43. Rachis
44. Glumes
45. Anemophily

5

PLANT TAXONOMY

MODERN TRENDS IN TAXONOMY

5.1. INTRODUCTION:

Till recently plant taxonomy or systematic botany was mainly based on the variation and form of morphological attributes for studying inter- relationships, phylogeny, diagnosis of taxa and finally for producing a system of classification. Turrill (1938) termed this taxonomy based on usually descriptive methods as Alpha or Classical or Orthodox taxonomy.

Maheshwari and Kapil (1963), while reviewing the progress of plant taxonomy in India for the last 50 years opined that the past Indian work in plant taxonomy has been purely observational and descriptive. They further said that substantial help could be taken from other branches of botany to solve some of the taxonomic problems. Stace (1980) also believes that the data from as many sources as possible should be employed in plant classification so that-inter relationships can be better judged. It is now felt that for a comprehensive plant systematics evidences from other disciplines like cytology, cytogenetics, anatomy, palynology, embryology, physiology, paleobotany and bio-chemistry are to be taken into account.

Data from some inter-disciplinary areas like mathematics and statistics are also needed. At present, the classification of plants based on data from various other disciplines of botany and other related sciences which is known as omega taxonomy.

An attempt has been made to describe various aspects of cytology and cytogenetics (cytotaxonomy), bio-chemistry (chemotaxonomy) serum analysis (sero-taxonomy) and application of different numerical methods and computers (numerical taxonomy) in the classification of plants.

5.2 CYTOTAXONOMY : Application of cytological data in the elucidation of taxonomic problems is known as cytotaxoanomy. Cytological characters such as chromosome number, size, morphology, their behaviour at meiosis are used for understanding relationships and evolution of different taxonomic groups. Some of these aspects are given below;

5.2.1 Chromosome number: The chromosome number is usually constant in a genus but varies from species to species. This makes it an important taxonomic character. The lowest number of chromosomes (4) are observed in; the somatic cells of *Haplopappus gracilis* (Asteraceae) whereas in *Poa* littorosa (Poaceae) there are 530 chromosomes, *Echieveria* (Crassulaceae) has 500 to 520 and in *Morus nigra* (Moraceae) it is 508. The highest number of chromosomes are found in a pteridophyte *Ophioglossum* (1240). This large variation in chromosome number paves way for the grouping of plants into different units. The chromosome number in whole group is constant. In *Quercus* and other members of Fagaceae, have the same basic number (n=12) showing homology among different genera of the family.

In Arecaceae two basic series exist which correspond to leaf types. In general, species with n=16 chromosome number, posesses pinnate leaves while those with chromosome number (n=18) have palmate leaves. The variation in chromosome number present in other members of the family may have taken place in the course of evolution.

A polyploid series is found in various members of a group. For example, the basic chromosome number in *Salix* is 19. The different species of this genus have multiples of basic chromosome number. For example, in *Salix virminalis* (diploid 2n=38); *S. atrocinera* (tetraploid 2n=76), *S. phylicifolia* (Hexaploid, 2n=152). Another interesting example of polpyloid series is found in the genus *Taraxacum* with 2n=16, 24, 32, 40 and 48. In *Rubus* (n=7) different multiples of basic chromosome number are found in polyploids i.e., 21, 28, 35, 42 and 49. Taxa differing in chromosome number are to be regarded as distinct. Further, a constant chromosome number of a genus is regarded as basic chromosome number based on this number, different polyplolids can be identified and also demarcation of taxa is also possible by considering the basic chromosome number.

5.2.2 Chromosome morphology:- Based on the form, size, volume and the distribution of chromatin material (Euchromatin and heterochromatin), chromosomes are arranged in a distinctive sequence called the karyotype or idiogram. Karyotype analysis provides a definite basis for classification of a taxonomic group.

Karyotype studies have shown that the monocotyledons have usually larger chromosomes than the dicotyledons. Similarly, woody plants have smaller chromosomes than the herbaceous plants. Sharma (1964) opines that large chromosomes, less chromosome number and symmetrical karyotype (A Karyotype consisting of chromosomes similar to each other in size and with medium or sub-median centromere is known as symmetrical karyotpe) represent a primitive character, on the other hand small chromosomes, high number and assymtrical kartytotype indicate advancement. Karyotype studies provide an interesting inter-relationship among various genera of Pandanaceae, Alismataceae, Hydrocharitaceae, Liliaceae, Amaryllidaceae and Dioscoreaceae. Like wise, creating of a separate family Basellaceae with *Besella* from the Chenopodiaceae also has been done on the basis of karyotype.

The genus *Yucca* with superior ovary is traditionally placed in Liliaceae and *Agave* with inferior ovary in Amaryllidaceae. A cytological examination revealed that their diploid chromosome number (2n=30) and karyotype morphology is also similar. In the light of above evidence and also because of similarities in morphology, *Yucca* and *Agave* are now included in the same family i.e. Agavaceae.

5.2.3 Satellite Chromosomes : In some chromosomes we find secondary constrictions (eg. maize). Due to this, chromosomal parts are divided as circular components above the secondary constriction. Such parts are called the satellite chromosomes. Interestingly, the distribution of satellite chromosomes are observed only in few groups. Hence, presence or absence of satellite chromosomes can be used in delimiting taxa.

5.2.4 Behaviour of chromosomes during meiosis:- Behaviour of chromosomes at meiosis can also be taken into consideration to identify the relationships among various phylogenetic groups. The kind and degree of pairing in the hybrids, crossing over, translocation lagging, bridge formation, non-disjunction of chromosomes and other cytological events also can be used for classification of groups (Swanson, 1957). Affinities of various groups can also be studied on the basis of giemsa banding patterns.

5.3 CHEMOTAXONOMY :

Chemical characters of plants can also be used in the classification of plants. The application of chemistry to taxonomy is known as chemotaxonomy. This branch of modern taxonomy has solved many taxonomic riddles during last twenty five years. Chemotaxonomy in recent years is also described as chemical plant taxonomy, chemosystematics and plantchemistry. In the chemotaxonomical studies different chemical substances present in various groups of plants are analysed and evidences are collected from these analysis which inturn are used for classification.

The importance of distribution of different chemical substances in taxonomic studies is known since a long time. For example, presence of essential oils made it easy to identify families like Apiaceae and Lamiaceae. In the same way families like Apocynaceae, Asclepiadaceae are distinguished from other by the presence of milky latex.

With the availability of number of modern analytical techniques like chromatography, electrophoresis and X-ray crystallography, chemical basis of taxonomy is gaining momentum. Many chemical compounds like carbohydrates, lipids and also primary and secondary metabolites such as flavonoids, alkaloids, acetylenic compounds, anthocyanins, betacyanins and glycosides occurrence and distribution provide a clue for classification.

5.3.1 Evidences from aminoacid distribution:- Earlier chemotaxonomists largely ignored the distribution of aminoacids, particularly the non-protein aminoacids in different plant groups. However, with the discovery of non-protein aminoacids, such as abrine, canavanine, 3,4—Dihydroxy phenylalanine, mimosine and willandine in seed material, affinities among different plant groups are recognized. Bell (1971) identified seven intrageneric (lower rank than genus) groups in *Lathyrus* (Fabaceae) each with a different amino acid or a group of amino acids. Trchiorech and Hanlet (1967) created four group in the genus *Viccia* (Fabaceae) on the basis of the occurrence of canavanine, b-cyanallanine and also arginine. This grouping agrees with the classification based on cytological and morphological attributes.

5.3.2 Evidences from alkaloids distribution:- The most important chemical compounds used for classification and phylogeny are the alkaloids. These are the characteristics of some families like Amaryllidaceae, Berberidaceae, Fabaceae, Ranunculaceae and Solanaceae. Due to the presence of Isoquinoline alkaloids like prosopine, Fumariaceae and Papavaraceae families are considered as close relatives. Menispermance, Ramunculaceae Fumariaceae and Papavaraceae are also closely related as all these families have barberine type of alkaloids.

In the Leguminosae, the distribution of alkaloids has great taxonomic significance. The dipiperidyllupin alkaloids like hystrin occurs only in *Genista* (Tribe Genisteae) which has been considered as relatively close *Adenocarpus* belonging to tribe sophora. Due to this, the genera *Adenocarpus* and also *Amodendron* deserve to be transferred to tribe genisteae.

5.3.3 Evidence from phenolics distribution:- Phenolics are the largest group of naturally occurring compounds in the form of flavonoids. These can be easily separated by paper chromatography. Phenols can be of immense help in solving taxonomic problems and probably one of the examples in this direction is the enunciation of relationship between Arecaceae and Poaceae. Depsite the fact, that palms and grasses differ greatly in gross morphology, the presence of five leaf flavonoids like glycosyl flavones, tricin, luteolin, bisulphate complex and 5-glucoside link the two families.

The two families Poaceae and Cyperaceae are usually placed near each other in most of classificatory systems because they have same basic flavonoid pattern.

5.3.4 Evidence from the distribution of betalins: Betalins are the red and yellow nitrogenous pigments. Chemotaxonomic studies revealed their presence only in ten families of the order Centrospermae.

The family with xerophytic taxa, Cacataceae usually placed in a monotypic order, the cactales or opuntiales. Infact, this family should be placed under Centrospermae as the members of this family are rich in betalins. On the other hand, Caryophyllaceae and Molluginaceae which lack betalins however, have been still grouped under Centrospermae for other morphological similarities. Based on these characters (absence of betalins), the above two families have been separated from Centrospermae and included under order Caryophyllales.

5.3.5 Other Biochemical Evidences : Based on the chemotaxonomical studies, Shanta Ram (1983) opined that retaining of genus *Coffea* in the family Rubiaceae is valid instead of placing it in a separate monogeneric family called *Coffeaceae*.

The division of Angiosperms into Lignosae (woody taxa) and Herbacae (herbaceous taxa) by Hutchinson (1959) got support from chemotaxonomical studies. It is found that majority of Lignosae contain flavonoids and Hydroxy acids whereas Herbacae posseses flavones and Methoxy acids.

Some times, it is possible to delimit two families based on the distribution of single chemical substance. Capparadaceae and Brassicaceae have been separated into two distinct families because the former contains a special chemical compound called Glucocapparin and the same is absent in the members of Brassicaceae.

Chemistry of DNA and RNA also provide exhaustive information so that relationships between different taxonomic groups can be identified.

5.4. SEROTAXONOMY:

Serology is a branch of biology dealing with the nature and interactions of antigens and antibodies. Serological reactions help us to find out similarities and variations among various taxa. Proteins, which carry useful taxonomic information and they are easy to handle and are the most widely used antigens in sero-taxonomy. The proteins such as storage proteins from seed, stem tubers, algal cells, fern spores, pollen, fruits and leaves can be used as antigens. These proteins are first introduced in the animals such as rabbits or mice in an emulsion form. The amount and nature of precipitate formed by antigen and antibody complex provides inter-relationships amongst various groups of taxa. In general, a high dilution (precipitate density) of the serum signifies close relationships and if no precipitate is formed, releveals distinct or no relationship.

Nuttal (1902) for the first time identified the importance of serology in plant systematics. Mez and Ziengenspect (1926) of Germany on the basis of their serological studies, published a family tree of plant relationships called 'Stommbaum'. Boyden (1954) proved that the two genera *Michelia* and *Magnolia* are closely related by serological analysis.

Tucker (1968) recorded serological relationships amongst various genera of Solanaceae like *Datura, Hyoscyamus, Nicotiana, Salpiglossis* and *Solanum*, Phyto-serological systematic survey of 20 Ranunculacean genera (Jenson, 1968) also revealed their relationships.

Kloz (1971) made important contribution to the serology of Fabaceae by identifying different protein characters. His work also helped in the selection of desirable proteins in plant breeding research.

Desipite significant contribution made in the serotaxonomy, it has so far not gained much importance in the plant systematics. This may be due to constant varations observed in the serological tests. Taxonomists are skeptical about the value of serological evidences in systematics of plant groups. However, in some specific cases, the serological relationship is accurate and thus may provide clue for the phlogeny.

5.5. NUMERICAL TAXONOMY:

The branch of botany concerned with numerical approaches towards taxonomy is called 'The numerical taxonomy'. According to Sokal and Sheath (1973) numerical taxonomy is 'The numerical evaluation of the affinity or similarity between taxonomic units and ordering of these units into taxa on the basis of their affinities'.

Numerical taxonomy is also known as Numerical systematics (Wagner, 1961), Mathematical taxonomy (Jardine and Sibson, 1971), Multivariate morphometrics (Blackith and Reyments, 1971) and Taximetrics (Mayer, 1966).

Adanson, a French botanist was first to moot upon a plan for assigning numerical values to the similarity between organisms. This method of grouping taxonomic units is described as the Adansonian taxonomy.

In recent years data obtained through various branches like palynology, cytology, anatomy, physiology, biochemistry and cytogenetics are analysed more conveniently with the help of computers and also with other mathematical methods. The numerical taxonomy is empirical and aims at developing objective methods. It also provide clues for phylogenetic relations for the evolutionary tendencies more accurately.

Efforts are now being made to identify taxa on the basis of over all similarities by analysing quantitative data of characters, test of significance, standard tests and standard deviation with the help of computers.

The basic procedure in numerical taxonomy starts with the spotting of characters of a taxon and followed by their analysis. Further, the resemblances amongst the individuals are established and are worked out with the help of computers.

In comparision with other methods of taxonomy, the numerical taxonomy has following advantages;

1. Every character will have equal weightage in creating natural taxa.
2. Taxonomy is practiced and viewed as an empirical science.
3. Classifications are based on phenotypic similarity (characters related to the appearance) rather from its supposed phylogeny.

The numerical taxonomic studies involve a series of following repeatable steps.

5.5.1 Selection of Operational Taxonomic Units : The basic unit (lowest ranking taxa) in any particular study are known as Operational Taxonomic Units (OUT). The OUT's could be individuals or other taxonomic units such as species, genus or even a higher ranked taxon. Hence, first these OUT's are selected for numerical taxonomic studies.

5.5.2 Selection of attributes (characters) : In the second phase, a sufficiently large number (n) of suitable characters are selected on OUT's, to obtain a fairly stable and reliable classification. Usually not less than 50 attributes are selected. These characters should be chosen from all parts of the plant and also at all the stages of life cycle. These characters may be grouped into morphological, physiological, chemical, behavioural and ecological units. Further, all these selected characters be given equal weight while creating different taxonomic groups.

5.5.3 Data Processing : To estimate the similarity between pairs of OUT's the basic data are recorded in an Net matrix consisting of 't' (number of OUT's) classified by 'n' characters.

5.5.4 Estimation of Similarity : The resemblance between a pair of OUT's can be expressed in terms of similarity i.e. the percentage of characters in which they agree or disagree.

QUESTIONS

A. Answer the following in about 25 lines.

1. **Write an essay on modern trends in taxonomy.**
2. **Give an account of the application of chemical data to the problems of plant systematics.**
3. **Write an essay on cytotaxonomy.**

B. Answer the following in about 10 lines

1. **Chromosome number - taxonomy**
2. **Satellite chromosomes**
3. **Biochemical evidences**
4. **Serotaxonony**
5. **Taximetrics**
6. **Advantages of numerical taxonomy**

C. Answer the following in 2 or 3 lines

1. **Alpha taxonomy**
2. **Cytotaxonomy**
3. **Karyotype**
4. **Phenols**
5. **Betalins**
6. **Multivariate morphometrics**
7. **OTU**

6

PLANT TAXONOMY

ETHNO BOTANY

6.1 INTRODUCTION

The term Ethno botany was coined by Harsh Berger (1895). According to him, Ethnobotany is 'The systematic study of plants and plant products used by aboriginals for various purposes like food, shelter, medicine, clothing and also for decoration.

Harrington and Fiero Marreo (1916) have defined ethnobotany as 'The study and evaluation of knowledge of all phases of life amongst the primitive societies and of the effect of plant environment upon life, customs, beliefs and history of tribal people. Hence, ethnobotany involves the closer understanding of association of man with surrounding plants and their utilization. Ethnobotanical study is an attempt to find out the mode and scope of influence of plants on man kind.

6.2. HISTORY OF ETHNOBOTANY

Hernandry (1570-75) for the first time conducted ethnobotanical studies on the flora and fauna of Mexico. Faulks (1958) brought out a vivid account of ethnobotany in his book 'Introduction to Ethnobotany'. Janaki Ammal (1954) and Jain (1981) are the first Indian botanists to attempt the systematic studies on Indian ethnobotany. Jain also known as the 'Father of Indian Ethnobotany', and described more than 1,500 different plants used by tribals for food, medicine, fodder, fibres, musical instruments, drinks and also for different religious purposes. Janaki Ammal recorded the important plants of food value like *Dioscorea* used by certain tribal communities of South India.

Melvin R.Gilmore founded the first Ethnobotanical laboratory at the museum of Anthropology, University of Michigan, USA.

6.3 ETHNOBOTANY — ALLIED DISCIPLINES :

Besides fundamental ethnobotanical studies, attempts have also been made by researchers to link various other disciplines like taxonomy, pharmacy, medicine and toxicology with ethnobotany. Studies are also under taken in some specialized fields like ethnomusicology, narcotics, gynecology and fossil ethnobotany. A brief account of various allied disciplines of ethnobotany is given below.

6.3.1 General Ethnobotany : General ethnobotany includes the study of geography, history of various useful plants of a given region. The vernacular names used by the tribals will also be identified in general ethnobotany.

6.3.2 Ethnomycology : It is the study of useful fungi in medicine, and preparation of alcohol by fermentation from yeasts.

6.3.3 Ethnopharmacology : Ethnopharmocology deals with the identification, description and experimental investigations of various ingredients used in recipes prepared by aborigines. It also deals with the effects of drugs on laboratory animals and man.

6.3.4 Ethnomedico-botany : It involves the study of medicines derived from plant. It also gives description of the procedures of medicine preparation, dosage and their efficacy in the control of various ailments.

6.3.5 Ethnobryology and Pteridology : This branch of ethnobotany deals with the use of bryophytes and pteridophytes by man.

6.3.6 Ethnotaxonomy : It deals with identification, classification and naming of various plants and animal species known to aborigines.

6.3.7 Ethnotoxicology : Ethnotoxicology deals with poisonous plants used by tribals to kill animals during hunting. The poisonous substances extracted from the plants are used in arrows to kill animals.

6.3.8 Ethnoagriculture : Ethnoagriculture involves the study of various agricultural practices and harvesting techniques adopted by tribal groups.

6.3.9 Ethnogynaecology : Ethnogynaecology deals with the study of plants used as medicine by tribals during women pregnancy, sterility, abortion and also for family welfare measures.

6.3.10 Ethnomusicology : It deals with the study of music instruments and various plants used for their manufacture by tribals and aborigines. A survey indicated that the wood of mango, *Caryota, Pterocarpus* and *Lagerstromia* are mainly used for making drums. The membranes for the drums are derived from the hides of cow or goat.

6.4 METHODS OF ETHNOBOTANICAL STUDIES :

Ethnobotanists identified following areas of study to unearth unique knowledge about properties and uses of plants from the trbial communities.

6.4.1 Field Studies : By field visits and field studies valuable data can be generated from tribal areas in ethnobotany. It is also possible to gain information on various aspects like tribal's life style, culture and their traditions. Further, by mingling with them one can observe the way they use various plants and plant products for different purposes. This data can be utilized for producing food grains, medicines and textiles for the modern world. Richard Evans Schulter of Harvard University and Rodin of Missori Botanic Gardens spent almost 12 years among tribals conducting field studies.

In India, Jain and his co-workers (1964-79) conducted field studies in Madhya Pradesh. Similarly Gupta (1963) in Bihar, Benerjee (1977), Hemadri (1980), Bhargava (1981) and Rama Rao (1987) in Andaman and Nicobar Islands conducted studies and collected valuable data on ethnobotany.

The field observations at a tribal area comprises two stages 1) Involvement of tribals who are well aquinted with useful plants of the area and recording of field observations along with information provided by them. 2) Collection of useful plants of a particular area and then showing these plants to tribals to get additional information on their importance and uses. In these two cases, vernacular names of the plants which are in use by the tribal communities will also be recorded.

6.4.2 Ethnobotanical Study with help of Herbaria and Musea: The herbaria and musea contain plant collections from various part of the world. These collections provide useful data on the place (locality) of collection along with date, time of flowering and fruiting. The herbarium sheets and field notes preserved in the musea also give us information on vernacular names and their economic importance. These details can be compiled in the form of plant data banks and are useful in ethnobotanical studies. In India, Jain and Dam (1979) thoroghly investigated the Shillong herbarium and were able to identify many useful plants of tribal areas of Shillong division of Assam and Meghalaya.

6.4.3 Ethnobotanical study with the help of Literature : A detailed study of ancient literature undoubtedly provides valuable information as to how some plant species influenced the social and cultural life style of tribes. If proper studies are initiated on ancient literature, vast data on plants and their different uses can be known.

The birth of Indian medicine, the Ayurveda took place during vedic period which dates back to 2000 to 1000 (B.C). Further, an important veda of Hindus, the Atharvna veda, over 114 dhyana slokas provide details on various diseases and their control by using medicines of plant origin. Charaka and Sushruta in their charaka samhita and sushruta samhita respectively gave a vivid description of 600-700 medicinal plants. The works of Kalidasa (4th century AD), Varahamitra (6th Century AD) and Amarasimha (9th century AD) also provide a glimpse into many ethnobotanical aspects.

6.4.4 Ethnobotanical Studies and Sculptures on temples : For the ethnobotanical studies, the sculptures on temples and also on other religious monuments are good source of information. The arts and paintings depict number of plant species known during those times.

Shakti gupta (1986) of Delhi University conducted an extensive survey on the sculptures on temples of India. She unfolded many interesting details on the relationships between the tribals and plants shown in sculptures.

6.4.5 Ethnobotany — Magico religious beliefs : Besides using plants for food, medicine and shelter, they also have been found to be closely associated with customs and mythological rituals of tribal population. Many type of flowers, fruits and even whole plants were in use for religious functions. Even some plants are worshipied still and considered sacred. They are believed to have magical properities.

The well known Basil or Tulsi (*Ocimum sanctum*) similarly Amla (*Emblica officinalis*) are worshipied as an incarnation of Lord Vishnu. The dub grass (*Cynodon dactylon* vern. dharba) is one of the essential items in many religious ceremonies of hindus.

6.5 ETHNOBOTANICAL STUDIES IN ANDHRA PRADESH

Many ethnobotanists and taxonomists have undertook extensive studies on various aspects of ethnobotany in the tribal areas of Andhra Pradesh state. Some important investigations are mentioned below;

a) Jain and his co-workers (1973) studied different medicinal and food yielding plants cultivated by the tribal folk like chenchu, konda reddy, valmiki and gonds.

b) Prakash Rao and Hara Sri Ramulu (1984) conducted an ethnobotanical survey in Srikakulam and identified 52 plant species of economic value.

c) Reddy and Subramanyam (1987) identified different plants used for curing various animal diseases in Chittoor district.

d) Hemadri (1988) published details of over 24 medicinal plants used by the tribal communities living in the districts of Adilabad, Karimnagar and Warangal.

e) Vedavathi (1991) reported about 25 plant species extensively used by the tribals of Rayalaseema for controlling child birth.

f) Nitheswar (1992) reported number of medicinal plants that are employed to cure cough, fever, skin diseases and other ailments in Rampachodavaram tribal area of notth — costal Andhra Pradesh

g) Saiprasad Goud (1995) reported his work on ethnobotany of Kurnool district. He obtained data on plant species used by chenchus, sugalis and erukula tribes to cure human and animal diseases.

At national level following institutes and research centres are involved in the study of various aspects of ethnobotany

1. Central Council for Research in Ayurveda and Sidda (CCRAS)
2. Central Council for Research in Unani Medicine (CCRUM)
3. National Botanical Research Institute (NBRI), Lucknow.
4. National Bureau of Plant Genetic Resources (NBPGR), New Delhi.
5. Birbal Sahni Institue of Paleobotany Lucknow.
6. Central Institute of Medicinal and Aromatic Plants (CIMAP), Lucknow.

QUESTIONS

A. Answer the following in about 25 lines

1. What is Ethnobotany ? Briefly describe the various aspects of ethnobotany.
2. Write an essay on ethnobotany, with special reference to ethnobotanical studies in Andhra Pradesh.

B. Answer the following in about 10 lines.

1. General ethnobotany
2. Methods of ethnobotanical studies
3. Research centres on ethnobotany

C. Answer the following in 2 or 3 lines.

1. Ethnotaxonamy
2. Ethnopharmacology
3. Ethnomedicobotany
4. Ethnoecology
5. Magico relegious beliefs - plants

PART - B

PLANT EMBRYOLOGY

By

Dr. A. Srinivas Reddy

Lecturer Dept. of Botany

S.K.E. Degree College,

Sirpur Khagaz Nagar, Adilabad Dist.

1 PLANT EMBRYOLOGY	INTRODUCTION

1 INTRODUCTION

1.1 HISTORICAL

The branch of botany that deals with development and structure of embryo is known as plant embryology. This definition is given in strict sense of the word. But botanists are looking in broader way and defining the embryology as "the development and structure of microsporangium, megasporongium, male gametophyte, female gametophyte, fertilization, endosperm and embryo".

The work on embryological aspect was initiated only after the invention of microscope by Leeuwenhock (1677). For the first time Grew (1982) stated that the stamens are the male organs of the flower and pollen plays an important role in production of seeds. After conducting several experiments on *Ricinus* and *Zea mays* Comerarius (1694) concluded that the anthers are the male sex organs and gynoecium is the female sex organ of the flower. Amci (1824) explained the role of pollen in fertilization. Hofmeister (1849) observed the embryo formation from the egg cell of embryosac. For the first time syngamy (fusion of male gamete with egg nucleus) and triple fusion (fusion of second male gamete with secondary nucleus) was discovered by Strasburger (1884) in *Monotrapa* and Nawaschin (1898) in *Fritillaria* respectively. Hanstein described the embryogeny in *Capsella* and *Alisma* for the first time. The modern techiniques like histochemistry, autoradiography; electron microscopy, florescence microscopy; microtome and others played an important role in understanding various embryological aspects. Several valuable text books like "Morphology of Angiosperms" (Coulter and Chamberlain, 1903) "Embryologieder Angiospermen" and "Vergleichende Embryologieder Angiospermen (Karl Schnarf, 1929 & 1931); "An introduction to the embryology of Angiosperms" (P.Maheswari, 1950) and systematic Embryology of Angiosperms (G.L.Davis, 1966) have come out with sufficient information on the workers of embryology along with day-to-day scientific journals and Monographs. The work of various workers have thrown light on the developmental sequences in the life history of Angiospermic plant body.

1.2 LIFE HISTORY OF ANGIOSPERMIC PLANT BODY

1.2.1 Sporophytic Phase : (fig-1.1) The hidden tiny plant (embryo of the seed) grows in to a complex sporophyte after seed germination. The angiosperms are highly evolved, more complicated than the lower vascular plants and dominant members of the modern world flora. The plant body is differentiated into root system and shoot system. The

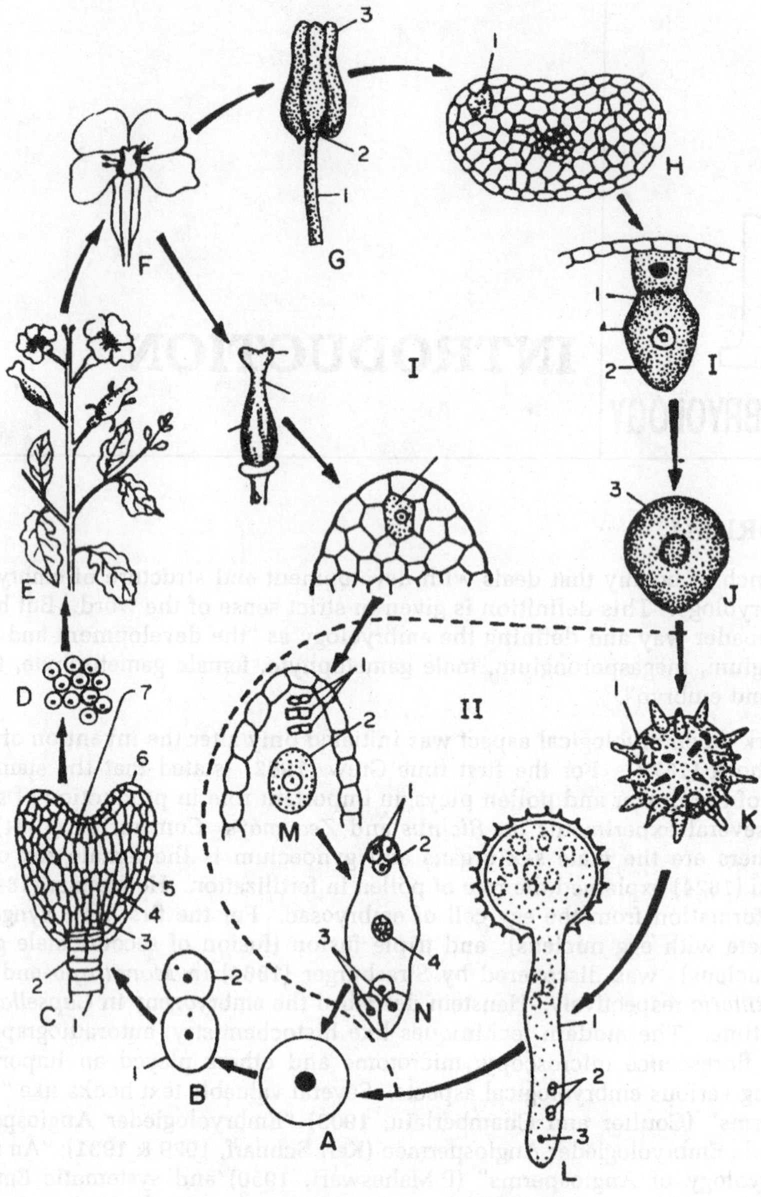

Fig 1.1. : Life history of a typical angiosperm

I - Sporophyte (2X) : A: Zygote; B. Two celled procmbryo : 1. Suspensor Cell, 2. Embryo cell

C: Heart shaped embryo 1. suspensor, 2. radicle, 3. pleurome, 4. periblem, 5. dermotogen, 6. cotyleoon, 7. growing point of plumule

D: Seeds, E: Plant, F; Flower, G: Stamen 1. Filament, 2. Connective, 3. Anther

H: Young anther T.S. - 1 - 5 Porangial initial ;

I-J: Formation of Pollen mother cell : 1. Pariatal cell, 2. Archesporium, 3. Pollen mother cell.

II - Gametophyte (X) : Male ganretophyte (K & L) - K - Pollen grain - 1. Germpore

L: Germinating pollengrain - 1. Pollentube, 2. Male - gametes, 3. Tube nuileus

Female gameto phyte (M & N) : M - Megaspoio genesis, 1. degenerating megaspores ; 2 - Functional megaspore.; N - Embryosac : 1. Synergids,. 2. egg cell, 3. Antipodals, 4. Secondary nucleus

sporophyte is continues to grow vigorously for some period and bear only foliar leaves and vegetative buds. This phase is known as vegetative phase. The vegetative phase is dominant in its life history. Sporophyte posses 2x number of chromosomes. After certain period of vegetative growth it enters into reproductive phase at certain point in its life cycle. During reproductive phase the vegetative buds are transformed into floral buds. In this phase the plant bears flowers, the reproductive organs.

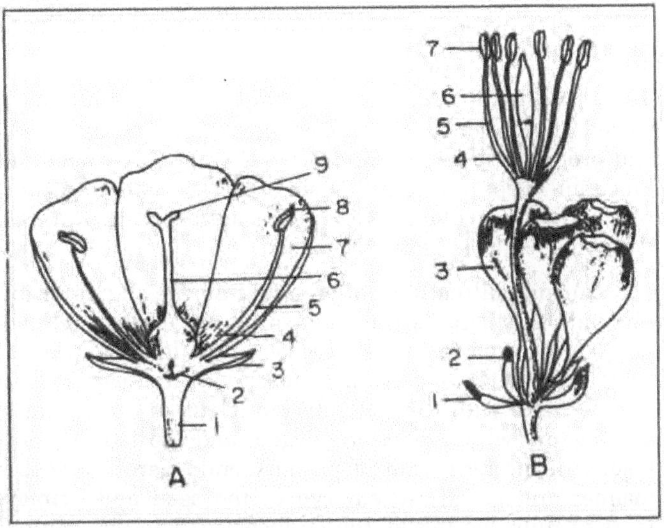

Fig 1.2 : The floral organs - A : L.S. of flower showing floral parts 1. Pedicel, 2. receptacle, 3. Sepal, 4. Ovary, 5. Filament, 6. Style, 7. Petal, 8. Anther, 9. Stigma
B: Flower showing Andhra phore and Gynophore : 1. Sepal, 2. Androphore, 3. Petal, 4. Gynophore, 5. Filament, 6. Ovary, 7. Anther

The flower consists of non-essential organs (calyx and corolla) and essential organs (androecium and gynoecium). (fig-1.2) Androecium (stamens) is considered as male reproductive structures, where as gynoecium is considered as female reproductive organ of the flower. On the basis of presence of kinds of reproductive structures, the flowers are classified into bisexual or hermaphrodite flowers (both reproductive structures are present) and unisexual flowers (only one kind of reproductive structure is present). The unisexual flower may be staminate (stamens present) or pistillate (gynoeciun present). If the plant produces either bisexual flowers or both unisexual flowers the member is described as monoecious. The monoecius member may be monoclinous (bear bisexual flowers) or diclinous (bear staminate and pistillate flowers). On the other hand if the plant bears either staminate flowers (male plant) or pistillate flowers (female plant), they are described as dioecious members. Sometimes few members produce bisexual, unisexual as well as neutral flowers. Such a condition is descried as polygamous (eg. Mango)

A) STRUCTURE OF A FLOWER :- A typical flower has pedicel (stalk) and terminal swollen portion, the thalamus (receptacle) on which borne floral parts in four distinct whorls. From outside, the outer whorl composed of calyx lobes, second whorl composed of corolla, third whorl androecium and inner whorl gynoecium. If the flower has all four floral whorls, the flower is described as complete. If any one of them is missing the flower is said to be incomplete. The complete flower may be perfect (both sexes are present) or imperfect (only one sex is present). The calyx composed of 5-sepals which are greenish colour and protective in function. The corolla composed of five coloured petals which give beauty and fragrance to the flower and attract insects. The androecium consits of five stamens. Each stamen (microsporophyll) is differentiated in to sterile anther filament and fertile bilobed anther. The anther lobes produce four microsporangia. The microsporangium is concerned with the production of haploid microspores (pollen grains) and eventually male gametophyte. The gynoecium is considered as female reproductive sturcutre of the flower that is composed of one to many carpels. The carpel (megasporophyll) produces ovules. Generally ovules (megasporangium with its protective coat) borne on the fertile tissue, the placenta, of the carpel. The arrangement of ovules in the ovary is known as placentation. The megasporangium is concerned with the production of haploid megaspores eventually female gemetophyte. As the sporophyte produces two kinds of spores it is described as heterosporous member.

1.2.2 Gametophytic Phase : The haploid microspores and megaspores represent the gametophytic generation. The microspore germinates in situ (in the microsporangium) and produces male gametophyte. It is much reduced and highly evolved among the plant kingdom. It produces two male gametes. The megaspore on germination produces female gemetophyte or embryosac in the megasporangium. The embryosac has egg apparatus, antipodals and secondary nucleus.

During pollination pollen alight on the stigma of the pistil. Pollengrain germinates in presence of stigmatic fluid and produces pollentube, which penetrates through the stylar tissue and reach to the embryosac of ovule. It carries two male gametes and releases into embryosac. One male gamete fuses with the egg nucleus of the egg cell and second male gamete fuses with the secondary nucleus of the central cell. The former fusion is known as syngamy which gives the zygote and the latter fusion is known as triple fusion which results in the formation of primary endosperm nucleus. The zygote divides in sequential manner and gives rise embryo of the seed and primary endosperm nucleus produces a mass of nutritive tissue, the endosperm. The ovules develop into seeds and gynoecium in to fruit. As the seeds are enclosed inside the fruit wall the members are described angiosperms.

In its life history diploid sporophytic phase and haploid gametophytic phase occur in a cyclical manner to complete its story. The sporophytic phase is independent where as gametophytic phase is dependent on sporophyte and incospicuous. Both the phases follow one after the other in a regular fashion in the life cycle. Hence it is known as alterntion of generations.

2

PLANT EMBRYOLOGY

MICROSPORANGIUM

2.1 INTRODUCTION

"The sporangium that is concerned with the production of microspores (pollengrains) eventually development of male gametophyte is known as microsporangium". In angiosperms the microsporangia are produced on stamens (microsporophylls). Each stamen has a bilobed anther,(each lobe has two microsorangia), connective and anther filament (Fig — 2.1). As the typical anther produces four elongated microsporangia, the anther is said to be tetrasporangiate. Both the anther lobes are connected by column of sterile tissue known as connective tissue (fig — 2.2B). When the anther is young, in each lobe two microsporangia are also seperated by strip of sterile tissue. But at maturity two microsporangia of an anther lobe get joined to form a single locule (compartment) due to the disintegration of sterile tissue. Hence the mature anther has two locules (dithecous). In Malvaceae, *Moringa* and *Wolffia* each anther produces two microsporangia (bisporangiate anther). But at maturity they joined to form single locule (monothecous). In *Arcethobium* the anther has only single horse shoe shaped microsporangium (monosporangiate anther). After maturation the anther dehisces along the stomium and liberates pollen grains. The dehiscence may be longitudinal, transverse, porous (apical pores are formed at the tip of anther lobes) or valvular (window shutter like valves are formed) (fig — 2.3).

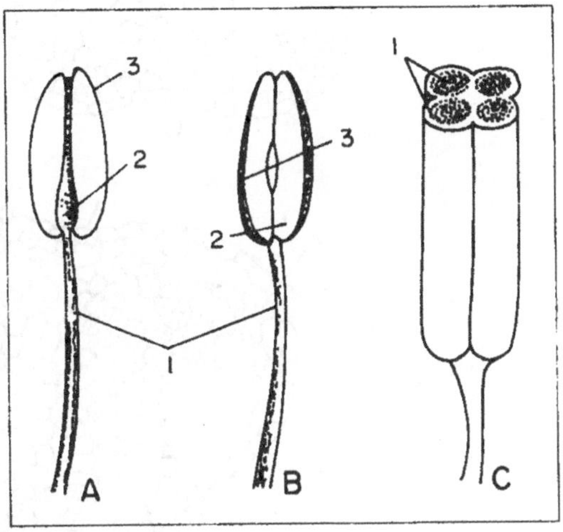

Fig 2.1 : Anther - A: Dorsal view : 1 - Filament, 2 - Connective, 3 - Anther
B : Ventral view : 1 - Filament, 2 - Anther, 3 - Sufures
C : T.S. of anther : 1 - Pollen chambers

2.2 DEVELOPMENT OF MICROSPORANGIUM:- (Fig — 2.2A)

The micro sporangium develops in eusporangiate manner. The very young anther shows homogenous meristematic cells, bounded by epidermal layer. It soon becomes slightly four lobed structure. In each lobe beneath the epidermal layer few cells become more

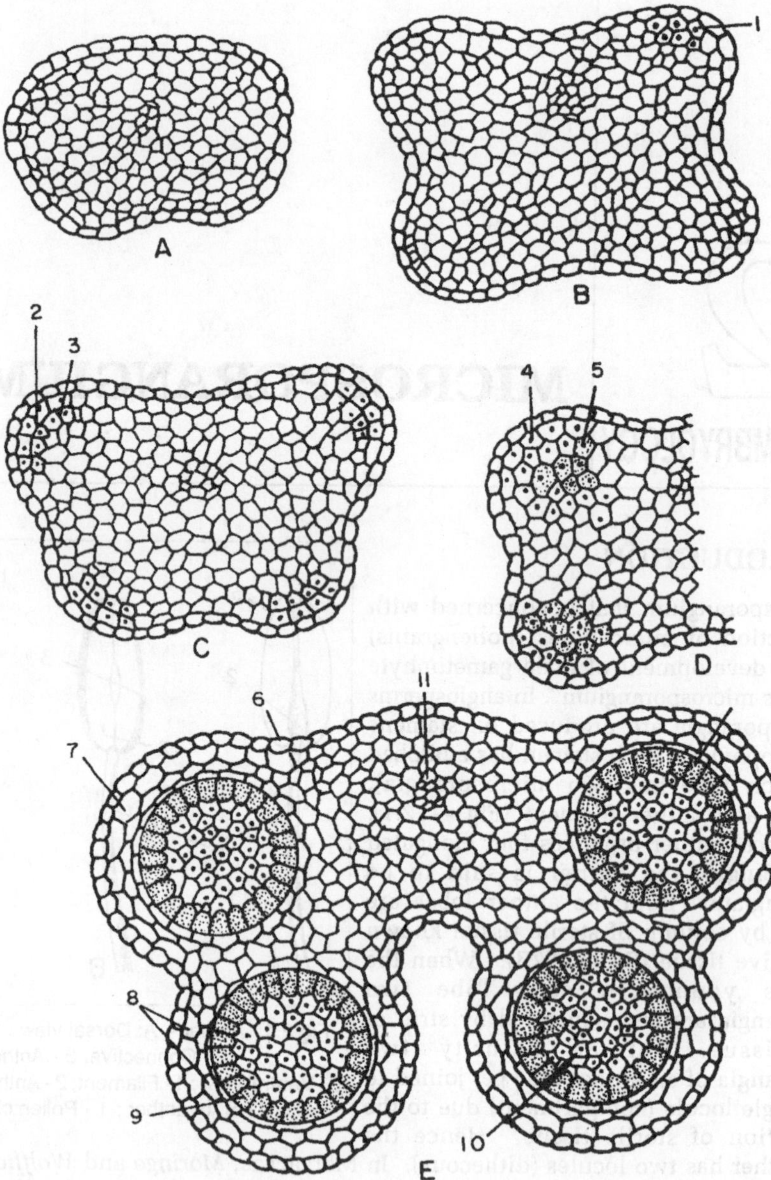

Fig 2.2.A : A - Various stages of the development of microsporangium (A - E).
1. Microsporomgial initials, 2. Primary pariatal cell, 3. Primary sporogenous cell, 4. Parietal tissue, 5. sporogenous tissue, 6. epidermis, 7. endothecium, 8. middle layers, 9. tapetum, 10. Pollen mother cells, 11. Vascular supply.

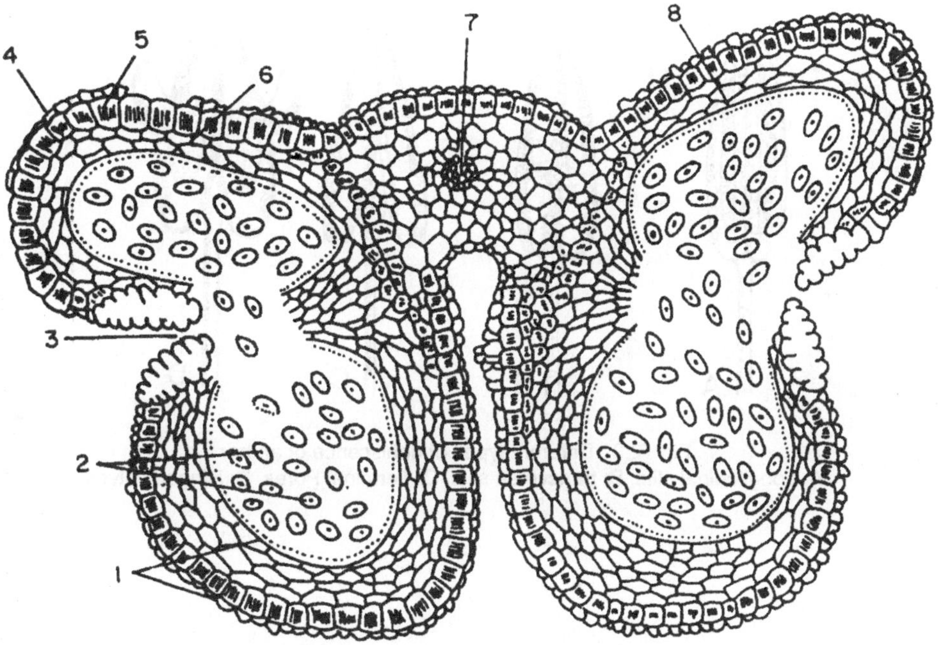

Fig 2.2.B : Transverse section of mature anther
1. Wall of microsporangium, 2. Pollengrains, 3. Stomium, 4. Epidermis, 5. Endothecium, 6. Middle layer, 7. Connective, 8. Tapetum

prominent from the surrounding cells. These cells are larger and slightly radially elongated, contain denser cytoplasm and active larger nuclei. Such well differentiated cells are known as archesporial cells (microsporangial initials). The archesporium comprises either single (rare) or many vertical layers (common) of cells. But in transverse section we can see either only one or plate of cells. The archesporial cells divide periclinally and give rise to outer primary parietal cells and inner primary sporogenous cells. The primary parietal cells divide in anticlinal and periclinal manner and give rise to three to five concentric layers. These layers are called as parietal layers or wall layers and they constitute anther wall. The primary sporogenous cells divide and give rise to sporogenous tissue. The outer most perietal layer differentiates into a distinct layer known as endothecium. The inner most parietal layer that surrounds sporogenous tissue differentiates into a nutritive layer known as 'tapetum'. The layers that are present in between endothecium and tapetum act as middle layers without much differentiation. During anther development the epidermal cells undergo anticlinal divisions in order to cope up with the enlargement of anther.

2.3. STRUCTURE OF MICROSPORANGIUM

The mature microsporangium has outer sterile thick and multi layered protective wall and inner fertile sporogenous tissue.

2.3.1 Wall : The wall of the microsporangium is also known as anther wall. It consist of epidermis, endothecium, middle layers and tapetum.

EPIDERMIS : It is an outer most limiting and protective layer. It is composed of tangentially flattened cells. The cells are closely fitted and have thick walls. Some times the cells may greatly lignified or cutinized.

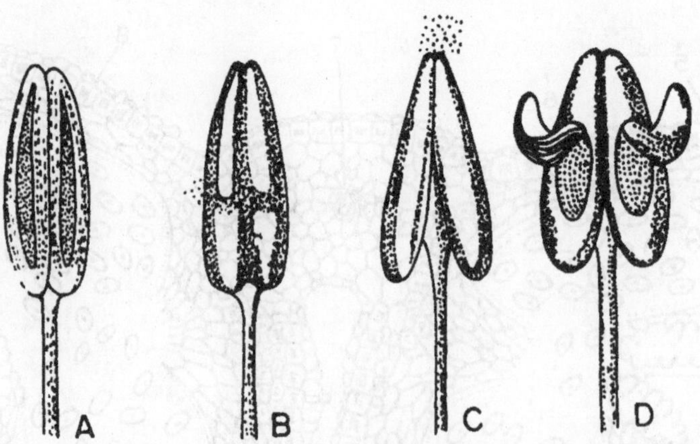

Fig : 2.3 : Different types of dehiscence of anther
A: Longitudinal slits, B: transverse slits, C: Terminal Pores, D: Hinged valves

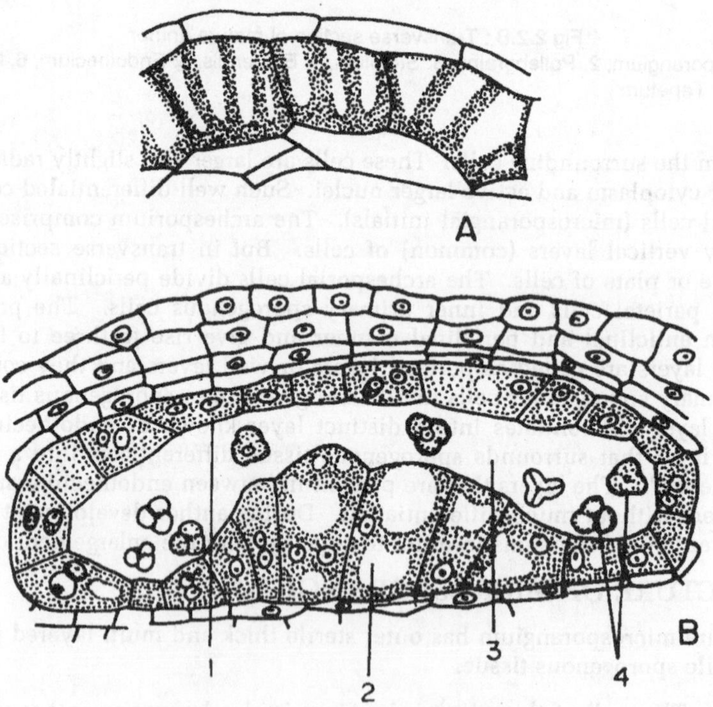

Fig : 2.4 A: Endothecial cells showing fibrous bands; B: Dimorphic tapetum showing, C: Tapetum (derived from connective tissue) and P-tapelum (derived from parietal layer),
1. P-tapetum, 2. C-tapetum, 3. locule, 4. Pollengrains

ENDOTHECIUM (Fibrous Layer) :- It is differentiated from outer most parietal layer and situates beneath the epidermis. In general it is single layered but in some members it may become multilayered. It may be homogenous or heterogenous. It is composed of radially elongated larger cells. Presence of fibrous bands is the characteristic feature of endothecium (fig-2.4A). These bands appear on inner tangential walls and radial walls then run upward, ending near the outer tangential walls. Hence the outer tangential walls remain thin. These bands are composed of α cellulose, pectin and traces of lignin. The hygroscopic nature of endothecial cells play an important role in dehiscence of anther at maturity. Along the line of dehiscence the endothecial cells remain thin walled. Such demarcated thin walled dehiscing region is known as "stomium". The fibrous thickenings may be absent in members of Hydrocharitaceae and cleistogamous (flowers never open) forms.

MIDDLE LAYERS : As name indicates they situate in between well marked endothecium and tapetum. The number of layers varied (1-3) with species to species. They are composed of flattened thin walled cells. These cells store the food material in the form of starch. Generally at the time of formation of pollen, their cells disintegrate liberating their contents into the locule of micro-sporangium. But in *Ranunculus* and *Gloriosa* the middle layers are persistant. In the former member the middle layer cells become densely protoplasmic and looks like tapetum, where as in the latter case due to the appearance of fibrous bands they look like endothecium.

TAPETUM :-

STRUCTURE :- The glandular natured nutritive layer that surrounds the sporogenous tissue is known as tapetum. It is the inner most layer of wall of microsporangium. Most commonly it is single layered. But rarely it is multilayered as in *Vinca rosea* (Boke-1939). It supplies food material to the sporogenous tissue and developing pollen grains. It is composed of larger cells, which have denser cytoplasm, reserve food usual cell organells and prominent nuclei. The number of nuclei and DNA content of a cell vary with cell to cell. Usually the number ranges from 2-16. This situation arises due to nuclear division without cytokinesis. The DNA content of the nucleus either increases due to nuclear fusion or endo mitosis (chromosome duplication and chromatid separation takes place without disintegration of nuclear membrance). Formation of restitution nuclei (mitosis normal upto late anaphase and then two sets of chromosomes are surrounded by common nuclear membrane to form dumbbell shaped nucleus) and polytene (chromosomes duplicate without chromatid separation) are also lead to the increase of DNA content.

In general the tapetum is homogenous and derived from primary parietal layer. But in some members such as *Alectra thomsoni* (fig. 2.4B) the tapetum is dimorphic and composed of two kinds of cells. The portion of tapetum that is composed of smaller cells is known as p-tapetum (situates towards periphery), where as the portion of tapetum that is composed of larger cells is known as C- tapetum (situates towards connective tissue). P-tapetum is derived from parietal layer where as C-tapetum is derived from connective tissue of the anther.

TYPES OF TAPETUM :- On the basis of its behaviour the tapetum of angiosperms is classified in to two types (Maheswari, 1950)

AMOEBOID TAPETUM (Peri Plasmodiattapetum): This type of tapetum is characterised by early breakdown of their cells. When the spore mother cells are at meiotic stage, the dictyosomes of tapetal cells secrete some hydrolytic enzymes. These enzymes disintegrate inner tangential walls of tapetum. The protoplast masses move into the sporangium cavity and fuse to form some sort of sticky substance around the pollen mother cells. This sticky substance is known as periplasmodium and shows plasmodial / amoeboid movement in anther locule. The pollen mother cells or young spore tetrads float in the periplasmodium and absorb food from it. The hydrolytic enzymes (eg. Callase) which is present in periplasmodium is also disintegrate the callose wall of spore tetrads. The liberated young pollen grains are bathed in the tapetal fluid. This type of tapetum is found in *Sagittaria, Tradescantia, Typha, Potamogeton* etc.

SECRETORY (GLANDULAR) TAPETUM : As the name indicates the tapetal cells show secretory nature in supplying of food material continuously to the developing pollen grains without early breakage. Their cells remain intact for longer period upto the formation of pollen grains.

A series of peculiar events occur in tapetal cells during pollen formation which are not found in amoeboid tapetum. At the sporogenous stage the tapetal cells possess several proubish (spherical minute bodies) granules in the cytoplasum along with normal cell organells. At the time of onset of meiosis the proubish bodies and ribosomes increase in number and accumulate near to the inner tangential walls. The plasma membrane is highly convoluted in this region. Thus the space between plasma membrane and inner tangential wall increases. At the time of separation of pollen grains from pollen tetrads the proubish bodies extrude through the plasma membrane and enter into the space between plasma membrane and cell wall. There the proubish bodies are heavily coated with sporopollenin material. Now they are called as ubish bodies. Formation of proubish bodies in the cytoplasm and migration and conversion into ubish granules is a continuous proces.. At this stage the tapetal cells loose their contact with each other. Large vacuoles appear in the cytoplasm and nuclei become irregular at first and finally disorganise. By the action of hydrolytic enzymes the inner tangential walls disappear. The protoplasm of all the tapetal cells along with ubish bodies reach in to the locule of the sporangium. The ubish bodies are involved in the thickening of the pollen exine.

In some members (*Nigella damascena)* at the time of disintegration of inner tangential walls of the tapetum, a new membrane appears around tapetal protoplasm. It is known as tapetal membrane (Bhandari & Ramkishori — 1973).

FUNCTIONS OF TAPETUM :

1. Tapetum plays a significant role in the development of pollen grains
2. It supplies required food materials in different forms to the developing microspores.
3. Tapetum is involved in the production of callase enzyme. It is responsible for the degradation of callose wall of spore tetrad and release of spores from tetrad.
4. Formation of non-functional pollen grains is linked with non-functional tapetum.
5. It involves in the formation of pollen wall.
6. Tapetal plastids synthesize pollen kit substance (complex mixture of hydrophobic lipids and carotenoids) and tryphin (complex mixture of hydrophilic substance and proteins). Pollen kit substance helps in attracting insects and adhering of pollen to the insect body. It also protects the pollen from U.V.Rays.

2.3.2: SPOROGENOUS TISSUE : The fertile tissue that is present in the Central portion of microsporangium and surrounded by nutritive tapetal layer is known as spoigenous tissue. The sporogenous cells either directly function as pollen mother cells (microspore mother cells) or undergo a few mitotic divisions. The cells of last generation act as pollen mother cells (PMC). Each PMC is surrounded by thin cellulose wall. The cells contain dense cytoplasm, all kinds of cell organells and plastids without starch. All pollen mother cells in a sporangium are inter connected by plasmodesmatal connections. Plasmodesmatal connections can also be seen in between PMC's and tapetal cells (fig — 2.5)

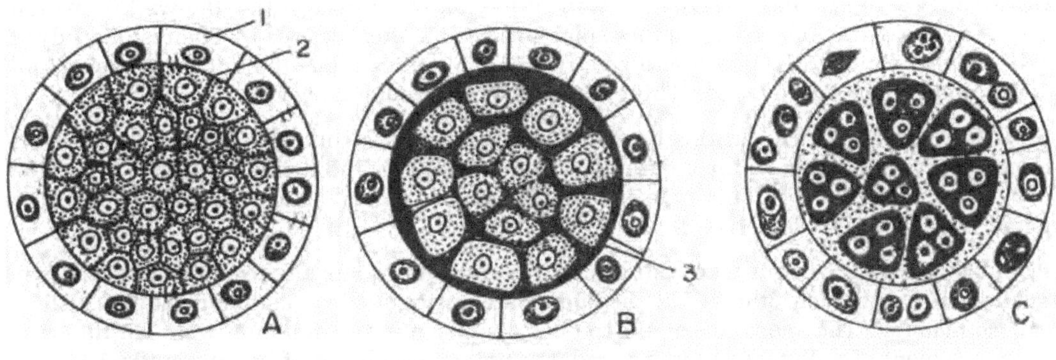

Fig 2.5 : Sporogenous tissue
A : - Plasmodesmatal connections present in between pollen mother cells and topetal cells
1. Tapetum, 2. Pollen mother cells
B : Cytoplasmic channels appeared in between pollen mother cells; 3. Cyto plasmic channels
C : Tetrad stage (cellular connections disappeared)

MICROSPOROGENESIS : The formation of haploid microspores from diploid microspore mother cells (PMC) by meiotic division is known as microsporogenesis. At the time of onset of meiosis the plasmodesmatal connections disappear in between tapetal cells and PMC. The pollen mother cell walls become very thick due to the deposition of additional callose wall material. During this process the plasm odesmatal connections are replaced by cytoplasmic channels. These channels help in the movement of cytoplasmic contents from one PMC to the other. Due to this, a close synchrony is maintained among the PMC during meiosis and finally the channels disappear. Each pollen mother cell after reduction division produces four pollen grains. The process often described as tetrad division. The four pollen grains remain inside the common wall and are known as pollen tetrad. After disintegration of tetrad wall the spores become free from each other.

CYTOKINESIS : (Fig-2.6) The division of cytoplasm after karyokinesis is known as cytokinesis. The cytokinesis in pollen meiosis is described as successive type and simultaneous type.

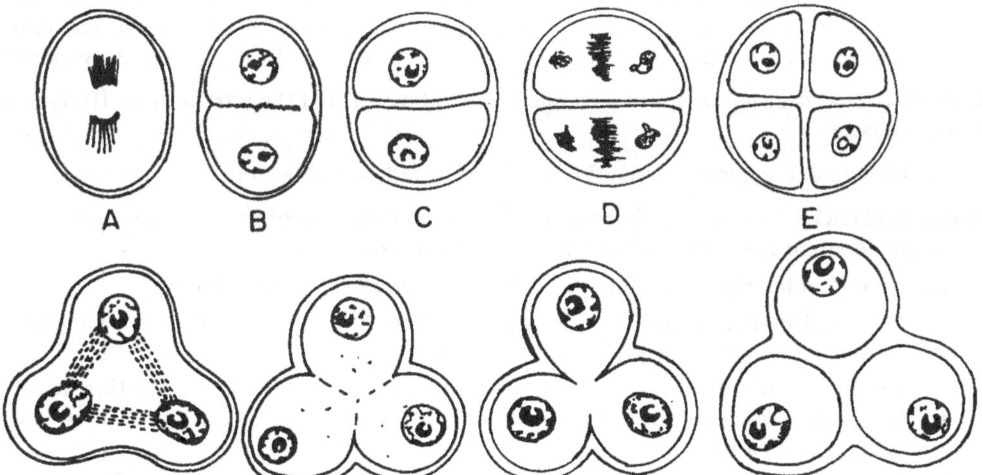

Fig : 2.6 : Wall formation in pollen motehr cells A—E: Successive type, F—I: Simultaneous type

SUCCESSIVE TYPE : In this, type the wall formation is centrifugal (i.e. the wall formation starts in the centre and extends either side) and occurs after meiosis-I. The resultant dyad cell undergo meiotic division II and give rise pollen tetrad. Thus the nuclear division and wall formation go side by side. This type is common among monocots.

SIMULTANEOUS TYPE : In this type the wall formation is contripetal (deposition begins at periphery and meets in the centre) and occurs after meiotic division II (after formation of four haploid nuclei). This type is common among dicots. All the young spores in a tetrad have their own callose walls.

POLLEN TETRADS : In a tetrad all the four spores are independent and not having any ralationship with each other. But the spores are surrounded by common mother wall of callose material. The spores become free after disintegration of common wall. On the basis of arrangement of pollen grains in a tetrad, the pollen tetrads are of five types (fig-2.7)

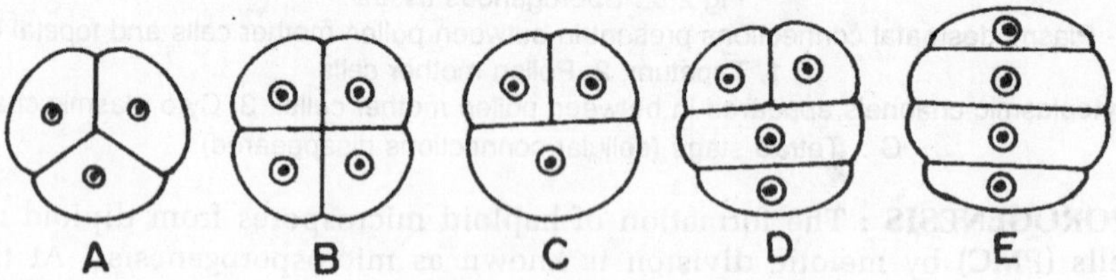

Fig : 2.7 : Arrangement of microspores or pollengrains
A - Tetrahedral, B - Isobilatenal, C - Decussate, D - T.Shaped, E - Linear

TETRAHEDRAL : This type of tetrad formation is very common among dicots. The four micropsores are arranged like the quadrants of a sphere or at four corners of a tetrahedran.

ISOBILATERAL : This type of tetrad formation is very common in monocots. In this type of tetrad the four microspores are arranged at four corners of a square or in one plane (they are diagonally placed) .

DECUSSATE : the microspores are arranged in two pairs. One pair situates at right angles to the other (eg: *Magnolia*)

T-SHAPED : In this type also the microspores are arranged in two pairs. In upper pair the 2 spores situate side by side (Juxtaposed) forming the bar of the 'T'. In lower pair the spores situate one below the other (superposed) and form tail of the 'T', (eg: *Aristolochia).*

LINEAR : The four spores are arranged one above the other in a row (Polyanthes, Asclepiadaccae)

In *Aristolochia elegans* all the five kinds of tetrads are reported.

RARE FEATURES : Generally there should be only four microspores in a tetrad and they should get separate from each other shortly after meiosis.

1. But in *scheuchzeria* the pollen grains occur in pairs called the diads.
2. In *Annona*, *Drosera*, *Typha* etc the four members of a tetrad tend to remain together, such grains are called as *compound* pollen grains.
3. In *Cascuta* a tetrad with eleven microspores is reported. Such a condition is known as *Polyspory*, which might have developed as an abnormality due to the division of four members of a tetrad.
4. In Mimosae members four to sixteen tetrads of a sporngium stick to form a complex compound pollen grain with 16-64 spores. In Asclepiadaceae and Orchidaceae members all the microspores of a sporangium agglutinate to form a single mass known as pollinia.

QUESTIONS

A. Answer the following in about 30 lines.

1. Briefly explain the development and structure of microsporangium upto microspore formation.

B. Answer the following in about 10 lines.

1. Structure of microsporangium.
2. Development of microsporangium
3. Structure of tapetum
4. Functions of tapetum
5. Microsporogenesis
6. Microspore tetrads

C. Answer the following in about 3-5 lines

1. Tetrasporangiate anther
2. Microsporangium
3. Archesporium
4. Endothecium
5. Tapetum
6. Stomium
7. Glandular tapetum
8. Amoeboid tapetum
9. Ubish bodies
10. Periplasmodium
11. Hydrolytic enzymes
12. Dimorphic tapetum
13. Sporo pollenin
14. Pollinium
15. Diad pollen
16. Polyspory
17. Compound pollen grain
18. Microsporangenesis

3

PLANT EMBRYOLOGY

MALE GAMETOPHYTE

3.1 INTRODUCTION

Haploid microspore (pollen grain) represents starting cell of male gametophytic generation. Male gametophyte is concerned with the production of male gametes which are involved in the process of fertilization. Fertilization is an important stage in sexual reproduction that results in the formation of seeds (future generation) in seed bearing plants.

3.2 STRUCTURE OF MICROSPORE

The microspore may be round, oval or fusiform in shape and contains denser cytoplasm. It has haploid nucleus in the centre. All kinds of cell organelles such as dictyosomes, ribosomes mitochondria, endoplasmic reticulum, plastids along with microtubules are present. The food is stored in the form of carbohydrates, proteins and lipid granules. The mature pollen grain is surrounded by thick and highly stratified wall, the sporoderm.

The sporoderm shows various sclupturings (may be smooth, spinose, reticulate, pitted and so on). It comprises two principle layers, the outer thick exine and inner thin intine. The exine is made up of sporopollenin (special cutin) material. It is resistant to physical and biological degradation. Because of this property the pollen grains are well protected from harmful U.V.light and preserved for longer periods in fossil deposits. Exine comprises two sub layers, the outer thick sexine and inner thin nexine. The sexine further differentiated into inner foot layer, central bacula and outer tectum.

The intine has uniform thickness. But at pores region it has thick callose material. Intine is differentiated into outer pectic and inner cellulosic layers.

The sporoderm is not uniform in its thickness. Here and there it has weak (thin) areas known as apertures. These areas are directly or indirectly associated with the germination of pollengrains. The apertures may be short and rounded or elongated. Short apertures are called as pores where as elongated apertures are called as colpii. The apertures may be simple or compound. Pollen grains with pores known as porate and with colpi known as colpate pollen grains. If two types of apertures are present the grain is described as colporate. In cucurbitaceae the germ pores are covered with thick walled "caps".

In insect pollinated members, the pollen grains are covered with sticky oily substance known as pollenkitt substance. It is made up of carotinoids, lipids and proteins. This material is contributed by tapetal cells. This substance gives yellow or orange colour to pollen grains. The biological advantages of pollenkitt are

a) it attracts insects
b) it protects pollin from U.V. light
c) It helps in adhering the pollen to insect body
d) Take part in diseprsal of pollen to some extent due to hydrophobic nature.

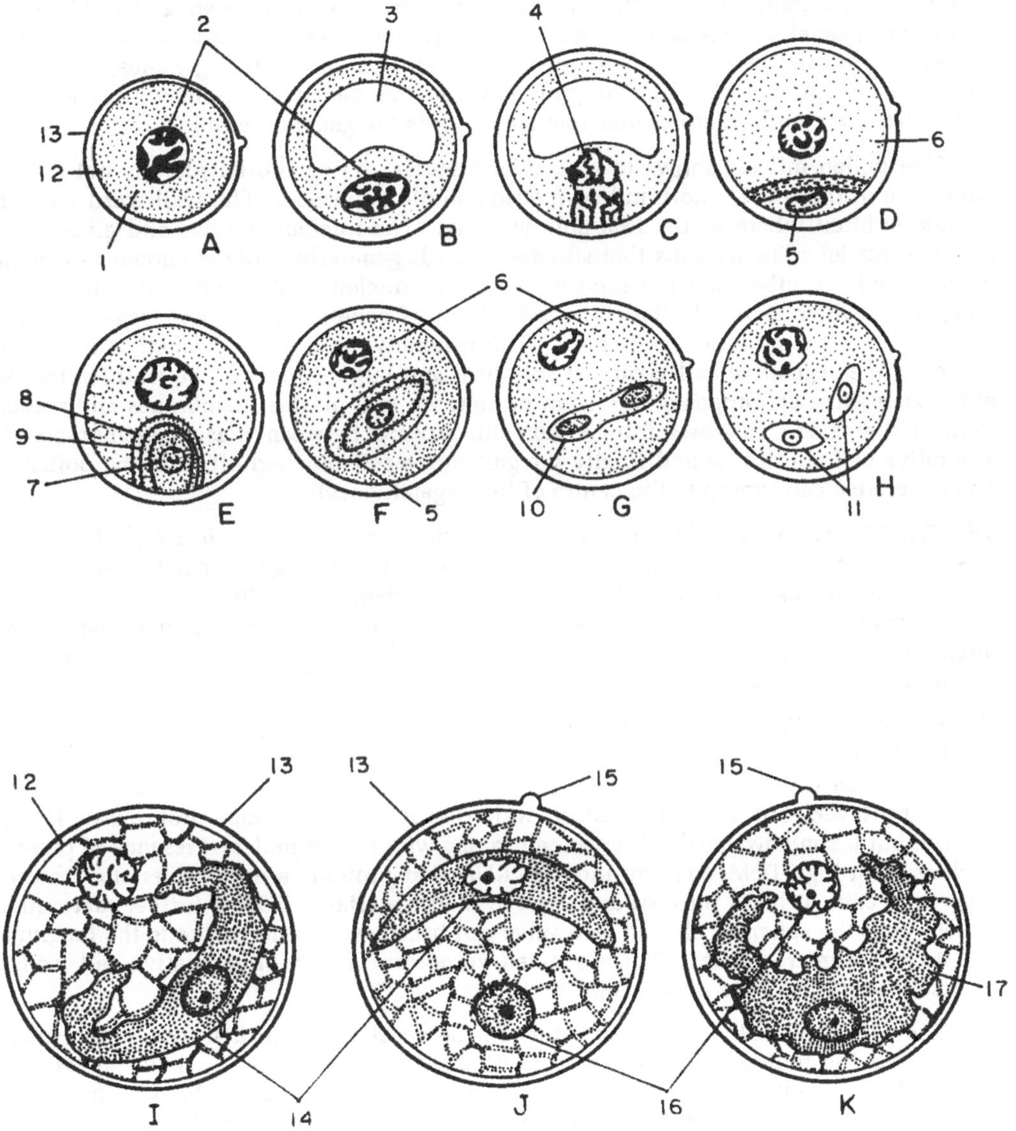

Fig 3.1 : Development of male gametophyte (A - K)

1. Cytoplams, 2. Microspore nucleus, 3. Vacuolre, 4. Division of microspore nucleus, 5. Generative cell, 6. Vegetative cell, 7. Plasma lemma of generative cell, 8. Plasmalemma of vegetative cell, 9. Callose wall, 10. Dividing generative cell, 11. Male gametes, 12. Intine, 13. Exine, 14. Vermiform generative cell, 15. Germpore, 16. Vegetative nucleus, 17. Amoeboid generative cell.

3.3. DEVELOPMENT OF GAMETOPHYTE : (fig — 3.1)

The gametophytic development starts in microsporangium. At the time of development some stricking changes occur in microspore. It increases in its size by absorbing locular fluid of microsporangium. Vacuoles appear in the centre. Protoplasm is pushed towards periphery. The nucleus is displaced from the centre towards one side (pole) of the cell. The displacement is always in a particular direction and marks the position of the generative cell (designated as a generative pole). The position of the generative cell is constant for a species. It is controlled by the genes. The other side of the protoplasm becomes highly vacuolated and marks the position of the vegetative cell (designated as vegetative pole). The chromosomal condensation begins. The cytoplasm shows marked polarisation with respect to the distribution of cell organelles. Most of the cell organelles, reserve food materials except microtubules migrate towards the vegetative pole while fewer cell organelles along with microtubules migrate towards the generative pole.

The pollen mitosis shows tow peculiar features. 1) equatorial plate is not formed at the centre of the cell 2) short and asymetric spindle is formed. The wall ward pole of the spindle is blunt where as the free pole is acute. The nucleus divides and gives rise two daughter nuclei. The nucleus that situates towards generative pole is known as generative nucleus and the other one is called as vegetative nucleus. The cell is divided into two unequal daughter cells. Initially the cytoplasm of generative cell and vegetative cell is separated by two plasma membranes. The generative cell is very small and lenticular. Where as the vegetative cell is almost spherical and larger in size. The wall (callose wall) of the generative cell appears very soon in between two plasma membranes and meets the intine either side. It grows from either side inward between plasma membrane of the generative cell and intine and finally the generative cell pinched off from the pollen wall. The generative cell moves to the centre of the vegetative cell.

VEGETATIVE CELL : It is larger and continous to grow in size. Food is stored in the form of starch, proteins and oil globules. Dramatically some changes occur in vegetative cell. The cell organelles increase in their size and number. Vacuoles disappear. RNA and protein contents increase. The vegetative nucleus becomes irregular in shape and often nucleolus may disappear. At maturity the cell has lesser DNA content and starch accumulates in the plastids.

GENERATIVE CELL : Immediately after pollen mitosis, the generative cell is very small and hemispherical. It attaches to the pollen wall. After detaching from the pollen wall it becomes spherical. At first it is surrounded by its own callose wall and two plasma membrane layers but soon the callose wall disappears. The cell protoplasm is highly reduced but it contains usual cell organnells along with microtubules. Presence of prominent nucleus with high DNA content and less RNA and proteins are characteristic features of generative cell. It lacks plastids and reserve food materials. During further maturation of gemetophyte the spherical shaped generative cell becomes vermiform due to the longitudinal orientation of microtubules. The elongated vermiform nature of generative cell facilitates its easy movement in the pollen tube.

The generative cell may divide either in the vegetative cell or in the pollen tube. In former case the pollen grains are shed at 3 celled stage where as in latter at two celled stage and carried on to the stigma and germinates. The exine ruptures and intine protrudes in the form of a slender tube called pollen tube. The vegetative nucleus moves in to the pollen tube and followed by vermiform generative cell in case of two celled pollen grain. The generative cell divides mitotically and gives rise two male gametes (sperms). Male gametes are spindle shaped, and lie very close to each other. They are definite cells and have their own persistant cytoplasmic sheaths. The pollen tube penetrates through the stylar canal and reach to the ovule in the ovary. The pollen tube contains different enzymes (amylases, invertases, cellulases, pectinases; lipases etc.). The pollen tube enter in to the ovule either through the micropyle (porogamy) or chalaza (chalazogamy) or funiculus/ integuments (mesogamy). The pollen tube discharges male gametes in the embryosac. One male gamete fuses with the egg nucleus (syngamy) and the other fuses with secondary nucleus (triple fusion). Syngamy and triple fusion collectively called as double fertilization which is unique feature of angiosperms.

In gametophytic generation only vegetative and generative cells are present. Prothallial cells are absent as in lower plants. Hence it is regarded as reduced gemetoplytic generation. From the phylogenetic point of view it is highly advanced in the plant kingdom.

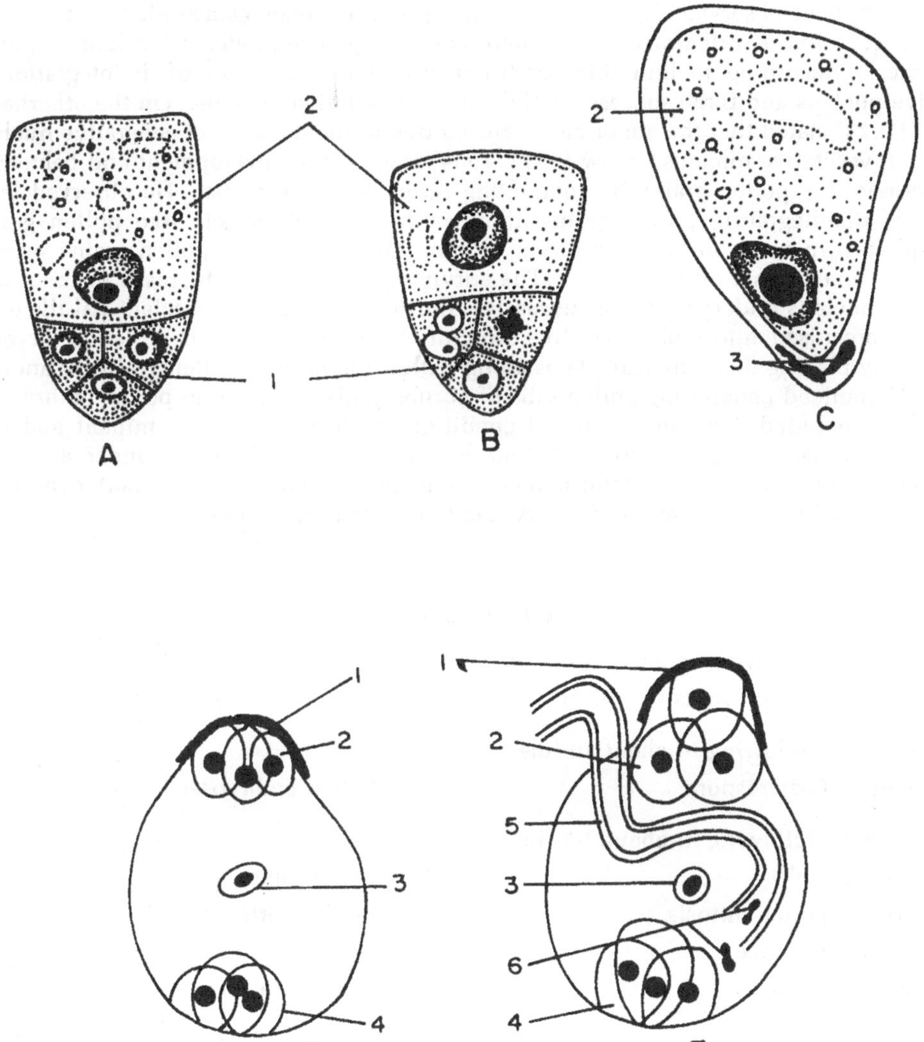

Fig 3.2 : A–C : Pollen development in Cyperus. 1. Non functional Microspores, 2. Functional Microspore, 3. Degenerated Microspores, D – Pollen embryo sac structure, E – Pollen embryo sac receving male cells. 1, Exine, 2. eggapparatus, 3. Secondary nucleus, 4. Antipodals, 5. Pollen tube, 6. Male cells

3.4 : ABNORMAL FEATURES

3.4.1 Pollen Formation in Cyperaceae : In general the pollen mother cell undergoes meiotic division and gives rise to four haploid functional pollen grains. Each pollen grain develops into male gametophyte. But in cyperaceae members, out of four nuclei formed through meiosis, only one is functional. Functional nucleus remains in the centre and non-functional nuclei are pushed to one corner and followed by wall formation (fig — 3.2) and finally degenerate. The functional nucleus divides and gives rise to vegetative cell and generative cell.

3.4.2 Pollen Embryosacs (Nemec Phenomenon) : (fig — 3.2) Generally the developing male gemetophytes are either 2 or 3 nucleated and spherical in form. But Nemec (1898) have observed eight nucleated embryosac like male gametophytes in petaloid anthers of *Hyacinthus orientalis.* Such abnormal male gametophytes are called as pollen embryosacs. According to him this condition arosed as the result of disintegration of generative nucleus and three successive divisions of vegetative nucleus. On the otherhand, de Mole (1923) stated that the abnormality arosed due to the division of generative nucleus rather than vegetative nucleus. Stow (1930, 1934) studied the development and behaviour of embryosac like pollen grains by conducting series of experiments. He observed large number of disintegrating pollen grains along with few embryo sac like pollen grains (gametophytes) in some anthers. According to him the disintegrated pollen grains secrete. "nechrohormones" into anther locule. The surviving pollen grains increase in their size. The pollen nucleus undergoes three successive division to form eight nuclei. All the eight nuclei are organised into embryosac like structure. He also observed pollen embryosac behaviour by placing them on agar plates along with some normal pollen grains of another variety. He noticed penetrating pollen tube of normal pollen grain in to pollen embryosac. Finally he concluded that, under normal conditions male potency is dominant and that lead to the formation of vegetative cell and generative cell. Where as under abnormal conditions in presence of nechrohormones the female potency is dominant over male potency and lead to the formation of embryosac like male gemetophytes.

QUESTIONS

A. Answer the following in about 30 lines.

1. Describe different events that occur in male gametophyte development.

B. Answer the following in about 10 lines.

1. Structure of Microspore
2. Pollen embryosac

C. Answer the following in about 3-5 lines

1. Sporoderm
2. Spindle of pollen mitosis
3. Pollenkit substance
4. Pores and colpii
5. Pollen tube
6. Double fertilization
7. Nectro hormones
8. *Hyacinthus orientalis*

4

PLANT EMBRYOLOGY

MEGASPORANGIUM

4.1 INTRODUCTION

"The sporangium that is concerned with the production of megaspores and eventually the female gametophytes is known as megasporangium". In angiosperms the megasporangium is also known as nucellus. Generally it is surrounded by one or two protective coverings, the integuments. The megaspor-angium with its protective coats is known as the ovule. The ovules born on the placenta (fertile tissue) of the carpel (megasporophyll) the ovary.

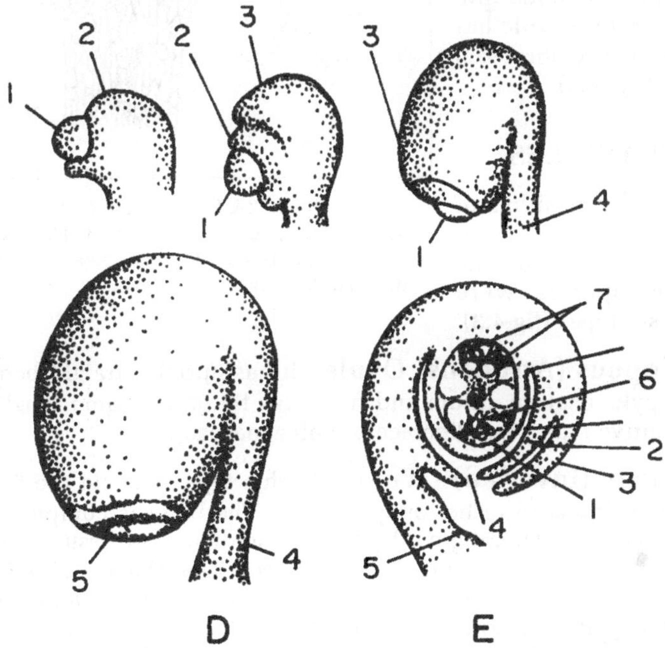

Fig 4.1 : A – D : Stages in the ovule 1 Nucellus, 2. Inne integument, 3. Outer integument, 4. Funiculus, 5. Micropyle
E : L.S. of an anatrapous ovule : 1. nucellus, 2. Inner integument, 3. Outer integument, 4 micropyle, 5. funiculus, 6. Eggapparatus, 7. Antipodals

Group of hypodermal cells of placenta acts as sporangial initials. They divide and give rise small hump of cells known as ovule primordium. The ovule primordium increases in its size and gives rise megasporangium. Simultaniously one or two protective coverings origin from its base and develop around nucellus leaving a narrow slit at its anterior side, the micropyle (fig-4.1). The ovule attaches to the placenta with its stalk called funiculus. The point of attachment of the funiculus to the body of the ovule is known as hilum. In orthotropous ovule the apical region of the nucellus that lies beneath micropyle is known as micropylar region, where as the basal region of the nucellus that lies above the funiculus is known as chalazal region. The ovule receives vascular trace from ovary wall. Beneath the chalaza the vascular trace forks and enter into integuments. A mature ovule has female gametophyte (embryosac) in the nucellus (fig- 4.2)

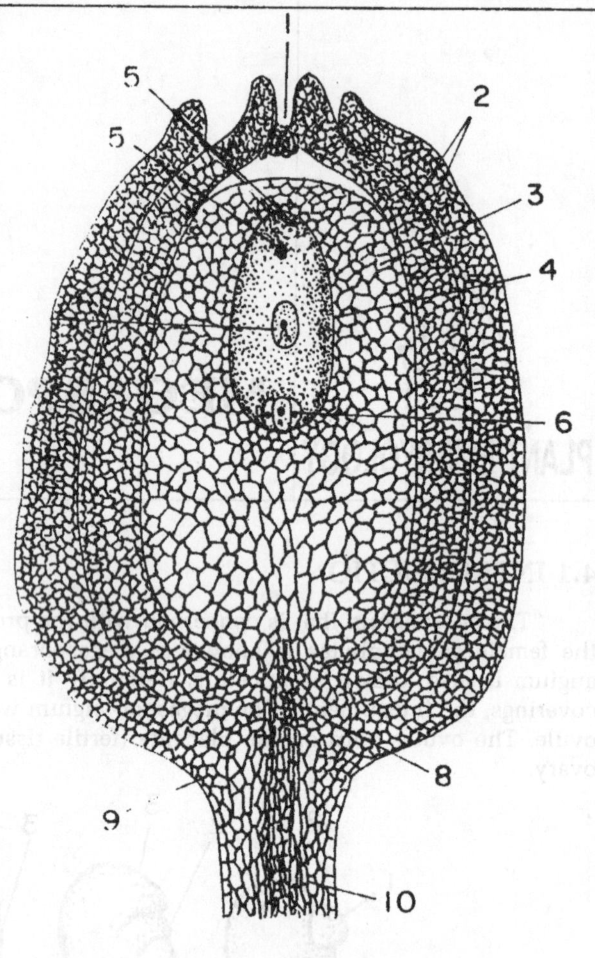

Fig 4.2 : L.S. of mature orthotropous ovule :
1. micropyle, 2. integuments, 3. Nucellus, 4. Embryosac, 5. egg apparatus, 6. antipodals, 7. secondary nuclues, 8. chalaza 9, hilum, 10. funiculus.

4.2 TYPES OF OVULES

On the basis of position of micropyle with respect to the chalaza, hilum and funiculus, the ovules of angiosperms were recognised into six types (fig-4.3)

4.2.1 Orthotropous (Atropous) Ovule: It has upright straight body. In this type of ovule the micropyle, chalaza, hilum and funiculus lie on the same longitudinal axis. It is regarded as primitive type. (eg.: Piperaceae, Polygonaceae).

4.2.2 Antropous (Inverted) Ovule : In this type of ovule the body is straight but not upright in posture as in orthotropous ovule. The body is completely inverted. This situation arises due to unilateral growth of the funiculus. As a result of this the micropyle comes very close to the hilum and the funiculus. The chalaza situates away from the hilum at the opposite end. The funiculus is fused throughout the length of the body of the ovule. This fused portion is known as raphae. The funiculus and hilum remains on one axis where as chalaza and micropyle lie on other axis. Both axes are parallel to each other. This type of ovule is very common among angiosperms, and present in about 80% of members (Davis, 1966) (eg: Gamopetalae and Polypetalae members).

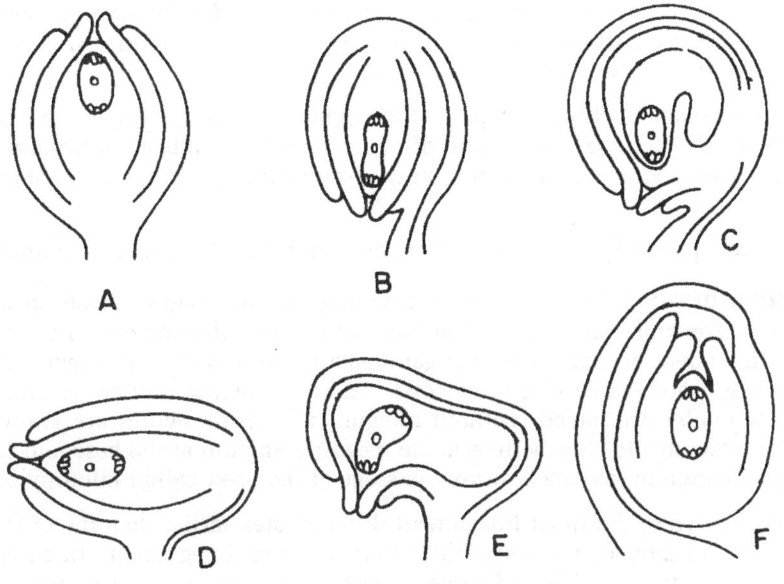

Fig 4.3 : Types of ovules : A: Orthotropous (atrapous) ; B: anatropous ovule; C: amphitropous ovule; D: hemianatrapous ovule; E: campylotropous ovule; F: circinotropous ovule

4.2.3 Hemianatropous (Hemitropous) Ovule : The body is half inverted due to unilateral growth of the funiculus. The body is placed transversely on the funiculus. The micropyle and chalaza lie on one axis where as hilum and funiculus lie on other axis. Both the axes situate at right angles to each other. (eg:- Ranunculus)

4.2.4 Campylotropous : In this type of ovule the curvature is more than that of the hemitropous ovule. The body roughly becomes bean shaped. The curvature effects the nucellar mass, which shows small notch. The embryosac is slightly curved. The micropyle and chalaza do not lie on the same axis. The micropyle, hilum and chalaza are adjacent to one anther. The micropyle faces towards the placenta. The micropyle and funiculus situate on two parallel axes, but chalaza on separate axis at right angles to them (eg. Fabaceae).

4.2.5 Amphitropous Ovule : The curvature of the body is more pronounced than the campylotropous, but less than that of anatropous ovule. Presence of horse-shoe shaped embryosac is characteristic feature of this ovule. The micropyle and funiculus situate side by side but not very close as in anatropous ovule (eg:- Butomaceae, Centrospermales).

4.2.6 Circinotropous Ovule : In this type of ovule the body is straight and upright. The funiculus is very long and surrounds the ovule. (eg: Cactaceae, Plumbaginaceae).

4.3. STRUCTURE OF THE OVULE

The structure of the ovule is very complicated in angiosperms. The typical ovule comprises central nucellar tissue, which is found surrounded by one or two integuments. The ovule attaches to the placenta by funiculus. The point of attachment of the funiculus to the body is known as hilum. In anatropous ovules the funiculus fuses laterally to the body of the ovule. It is known as raphae. The ovule is a bipolar structure. It has micropylar pole (region) and chalazal pole (region).

As a rule appearance of archesporium; development of embryosac; organization of egg apparatus; orientation of radicle during embryogenesis are some of peculiar features associated with the micropylar region (pole) of the ovule.

Chalaza is metabolically active pole of the ovule and has some specific features. Origin of integuments; direction of the growth of embryo sac and embryo; nutrients translocation to the embryosac; endosperm initiation and vascular trace endings are associaed with the chalazal region (pole)

The various parts of the ovule are described in detailed manner here under.

4.3.1 Integuments : "The protective coverings of the megasporangium are known as integuments". Generally an ovule either has one or two integuments. The ovule with one integument is known as unitegmic (eg: gamopetalae) and with two integuments is known as bitegmic (eg: polypetalae and monocots). In some members such as loranthaceae and olacaceae the ovules are naked (without integument). Such ovules are known as ategmic ovules. The integuments closely invests the megasporangium at the base and lateral portion of the megasporangium leaving a narrow passage at the apex called micropyle.

In bitegmic ovules the inner integument differentiates earliar than the outer integument. Generally the outer integument is very thick than the inner integument. In outer integuments stomata, chloroplasts, sclereids and some times latex cells may also present.

4.3.2 Micropyle : "The narrow passage that is present at the apical portion of the ovule is known as micropyle". It is the main path for penetrating pollentube. In bitegmic ovules the micropyle may be formed either by inner integument alone or by both the integuments. The passage that is formed by outer integument is known as exostome and inner integument is known as endostome. If both are present on straight line the micropyle is known straight micropyle. In some members the micropyle becomes zig-zag due to slight curvature of outer integument. The micropyle is filled with mucilaginous substance that is produced by nucellar apex and inner integument. The chemotropic substance that is secreated by synergids of embryosac is also deposited in micropyle. The substance provide nutrition and guide to the developing pollen tube.

4.3.3 Nucellus : "The central massive tissue (megaspormgium) that is surrounded by integuments is known as Nucellus". It is the main nutritive tissue for the developing embryosac as well as developing endosperm. The quantity of the nucellus may be varied with species to species. On the basis of quantity of the nucellus the ovules are classified into crassinucellate ovules and tenuinucellate ovules.

A) CRASSINUCELLATE OVULE : In this type of ovules the nucellus is massive and generally the ovules are bitegmic. Phylogenetically these are primitive type. The archesporial cell appears beneath the nucellar apidermis. It divides transversely giving outer parietal cell and inner sporogenous cell (fig-4.4). The parietal cell divides anticlinally and periclinally forming mass of tissue that pushes the sporogenous cell (megaspore mother cell) deeper into the nucellus. In some members the nucellar epidermal cells beneath the micropyle divide and give rise a cap like structure known as nucellar cap. The nucellar cap is composed of columnar cells with denser protoplasm and abundant RNA (eg: *Ornithogalum*). In members of Caryophyllaceae, Euphorbiaceae and Cucurbitaceae the nucellus projects into the micropyle forming a nucellar beak. The cells of nucellar beak contain strach and proteins.

B) TENUINUCELLATE OVULE : In this type of ovule the nucellus quantity is very less and the ovules are unitegmic. Tenuinucellate condition is advanced and found only in advanced members. The archesporial cell differentiates immediately below the nucellar

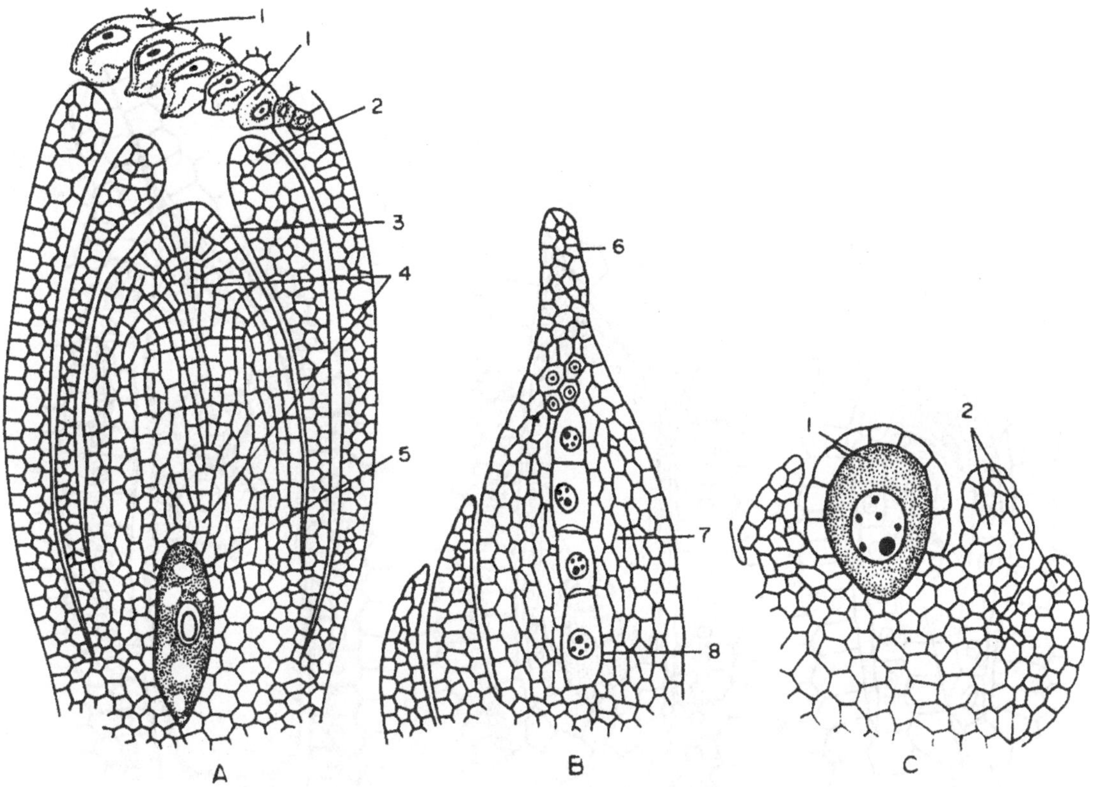

Fig 4.4 : A & B : Longitudinal section of crassinucellate ovule showing well developed nucellus. 1. enlarged cells of obturator, 2. integument, 3. nucellar epidermis, 4. parietal tissue, 5. megaspore mother cell, 6. nucellar beak, 7. well developed nucellus 8. megaspore tetrad
C: Tenuinucellate ovule showing poor nucellus : 1. megaspore mother cells, 2. integuments at the base of megaspore

epidermis. (fig-4.4). It directly functions as megaspore mother cell. It is surrounded by single layered nucellus.

Generally the nucellus is consumed by developing embryosac or endosperm. In some members such as Piperaceae and Polygonaceae the nucellus remains persist in the mature seed. Such a nutritive tissue is known as perisperm (fig-7.5(c)). In members of Podostomaceae, beneath the developing embryosac the nucellar cells degenerate to form a cavity known as psuedoembryosac (fig-7.5B).

4.3.4 Endothelium : (fig-4.5)

"The specialised nutritive layer that is differentiated from the inner most layer of integument to supply the food material to the developing embryosac is known as endothelium".

In unitegmic tenuinucellate ovules the scanty nucellus degenerates at an early stage of embryosac development. So that the embryosac comes in contact with the integument. In this circumstances the inner layer of integument becomes nutritive. As it is derived from integument and functions like tapetum of anther, it is described as integumentary tapetum. Generally the endothelium is single layered. But in Compositae members it is multilayered. In sunflower it is 10-12 layered. The endothelium is composed of radially elongated, glandular natured cells. The cells are rich in cytoplasm along with carbohydrates, proteins

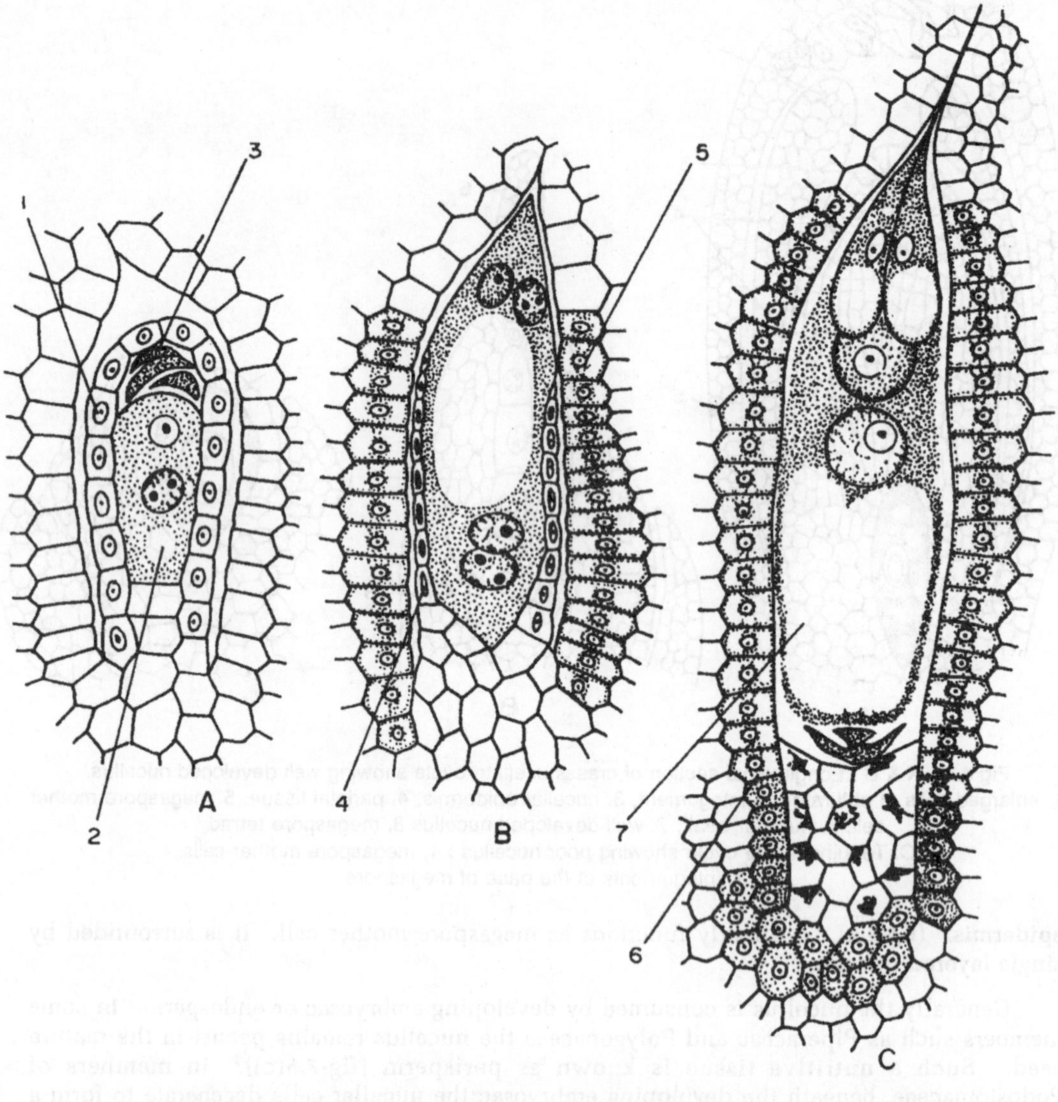

Fig 4.5 : Various stages in the formation of endothelium in *Lobelia trigona* (A – C)
1. nucellar epidermis, 2. developing embryosac, 3. degenerating meyasproes, 4. degenerating nucellar epidermis, 5. formation of integumentary tapetum from inner layer of integument, 6. integumentary tapetum, 7. Mature embryosac.

fats, enzymes, ascorbic acid and RNA. Distinct plasmodesmatal connections exist in between endothelial cells and embryosac as well as among endothelial cells. The cells may be uninucleated or multinucleated and diploid or polyploid in condition.

4.3.5 Obturator : (fig-4.6) : "The uniseriate, multi-cellular, glandular natured hairy out growths, that are associated with directing the pollen tube towards micropyle is known as obturator. The structure may be formed from placenta (Euphorbiaceae); funiculus (eg:

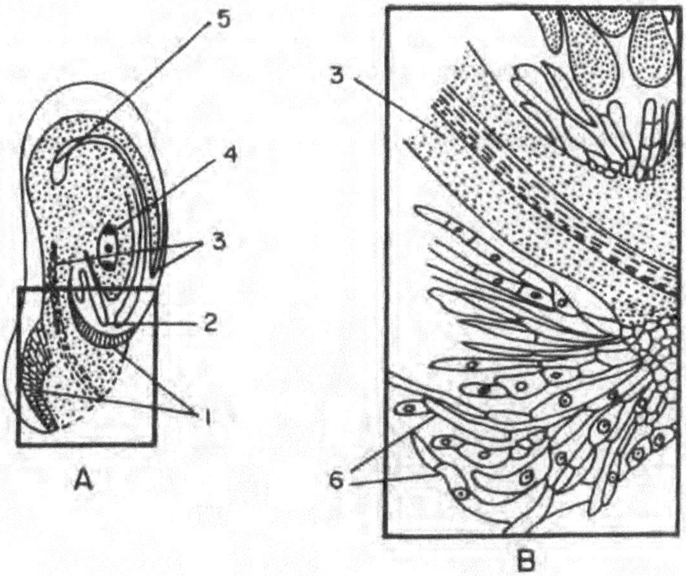

Fig 4.6 : Structure of obturator (A & B) 1. Obturator, 2. Micropyle, 3. Funiculus, 4. Embryosac, 5. Airspace, 6. Obturator showing glandular hairs in magnificaiton

Acanthaceae); integuments (eg: *Myriocarpa*) or stylar canal (Thymeliaceae). The cells are rich in active protoplasm, E.R; dictyosomes, vesicles and other cell organelles. The obturator produces some sort of sticky substance and provide nutrition, mechanical and chemical guidance to the growing pollen tube. It degenerates after fertilization.

4.3.6 Epistase and Hypostase : Group of cells that are differentiated from nucellar epidermis and present above the embryosac, like a cap is known as epistase. The cells of epistase are radially elongated and have either suberized or cutinised cell walls (eg: *Costus, Castalia*).

Group of cells that are differentiated from nucellar cells in the chalazal region below the embryosac and above the vascular supply is known as hypostase (fig-4.7). The cells contain little protoplasm and have lignified walls. Starch, proteins, lipid like reserve materials are reported in hypostase cells.

Different functions are attributed to hypostase by different workers are as follows.

1) According to van Tieghem (1901) hypostase acts as boundary for the growing embryosac and prevents the embryosac elongation beyond chalaza.
2) According to Johanson (1928), it stabilises water balance of resting seeds during dry season.

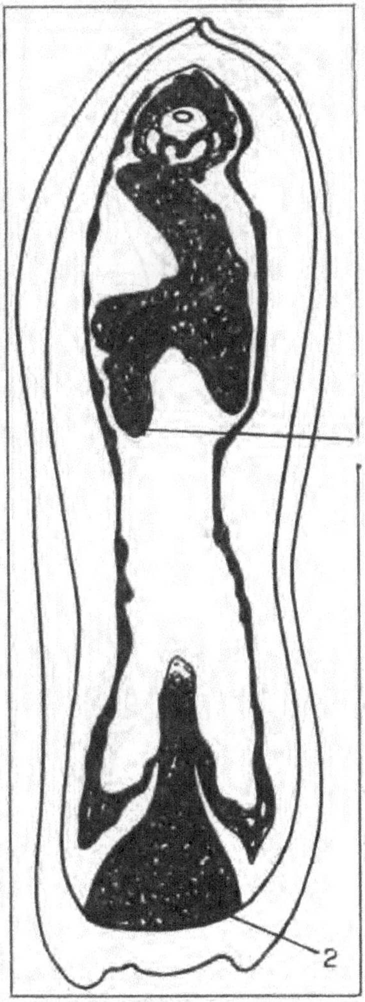

Fig 4.7 : L.S. of Young seed of *Zostera marina*
1. embryo, 2. hypostase

3) Goebal (1933) feels that the hypostase is related with the water economy of the embryosac.

4) According to venkat Rao (1953) it is the tissue that connects vascular tissue to embryosac and plays active role in rapid transportation of food materials to the embryosac.

5) It not only translocates nutrients to embryosac, but also acts as reservoir of food material (Tilton, 1980).

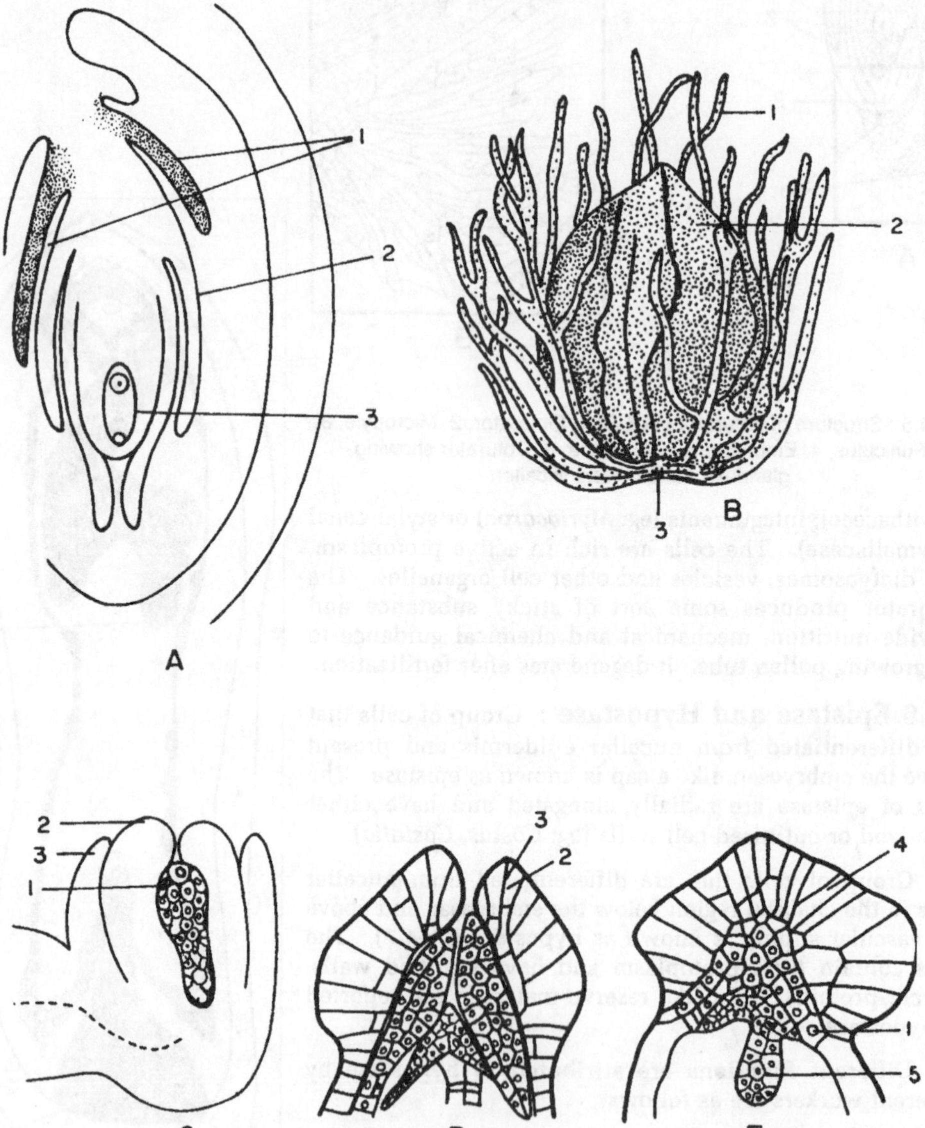

Fig 4.8 : A – Development of aril around the ovule 1. aril; 2. integuments, 3. embryosac
B : Structure of arilloid, 1. arilloid, 2. seed, 3. hilum, C – E : Development of operculum :
1 Nucellus, 2. inner integument, 3. outer integument, 4. operaculum, 5. embryo

6) According to Haberlandt hypostase secretes various enzymes and hormones essential for the growth of embryosac.

4.3.7 Aril and Arilloid : The annular out growth that is formed from funiculus and surrounds the ovule like an integument is called aril (fig-4.8-A). According to Maheswari aril is a new structure that is formed only in few taxa and some times represents a third integument. In *Pithecolobium* the aril is fleshy and edible. In *Myristica* it is red coloured, dry, leathery and branched. It forms the 'mace' a valuable spice.

The aril like out growth that is formed from the micropylar region of the integument and surrounds the ovule is known as arillode or arilloid (false aril) (fig-4.8-B). It helps in seed dispersal.

4.3.8 Operculum : In some members such as *Lemna* the integumentary cells at the tip region divide and give rise a caplike structure with compact cells around micropyle (fig-4.8C). It is known as operculum.

4.3.9 Caruncle : Caruncle is the spongy pad of tissue that is produced by outer integument in micropylar region (eg. *Ricinus*). It involves in absorption of water during seed germination as well as seed dispersal to some extent.

4.3.10 Coma : The tuft of hairs present around the seeds of Asclepiadaceae members is known as coma. Actually these are the integumentary hairs of the ovule. It helps in seed dispersal.

4.4 MEGASPOROGENESIS

In young ovule at the micropylar region a single hypodermal (rarely several) cell becomes more prominent. This cell is larger than the surrounding cells and contain larger active nucleus and denser cytoplasm. It is known as archesporial cell. In crassinucellate ovules the archesporial cell divides transversely and produces outer primary parietal cell and inner primary sporogenous cell. The parietal cell divides and produces mass of parietal tissue. It pushes the sporogenous cell into deeper portion of the nucellus. The sporogenous cell functions as megaspore mother cell (MMC). In tenuinucellate ovules the archesporial cell directly acts as megaspore mother cell. Plasmodesmatal connections are present between megaspore mother cell and surrounding cells. At first MMC has thin cellulose wall, but before onset of meiosis callose material is deposited. The MMC undergoes I meiotic division and give rise a dyad. The dyad cell undergoes II meiotic division and give rise four haploid magaspores. The megaspores are arranged in linear tetrad. Out of four megaspores generally the micropylar three are smaller in size and non-functional. The Chalazal megaspore is larger in size and functional. The functional megaspore germinates and produces female gemetophyte where as the non functional megaspores degenerate eventually.

QUESTIONS

A. Answer the following in about 30 lines.

1. Describe different structures of ovule.

B. Answer the following in about 10 lines.

1. Types of ovules
2. Integuments
3. Epistage and hypostage
4. Nucellus
5. Endothelium

C. Answer the following in about 3-5 lines

1. **Integumentory tapetum**
2. **Nucellus**
3. **Micropyle**
4. **Tenuinucleate ovule**
5. **Crassinucleate ovule**
6. **Exostome and endostome**
7. **Fumiculus**
8. **Nucellar beak**
9. **Ategmic ovule**
10. **Obturator**
11. **Chalaza**
12. **Raphae**
13. **Hilum**
14. **Aril and arilloid**
15. **Megasporogenesis**
16. **Coma**

5

PLANT EMBRYOLOGY

FEMALE GAMETOPHYTE

5.1. INTRODUCTION

Functional haploid megaspore represents gametophytic generation. The megaspore produces female gametophyte or megagametophyte or embryosac. The development of female gemetophyte in angiosperms is endosporic i.e. formed within the megaspore wall. Mostly the embryosac is 8 nucleated and 7 celled structure (fig-5.1). The micropylar region of the embryosac is occupied by 3 celled egg apparatus, where as the chalazal region is occupied by three antipodals. The remaining portion of the embryosac is known as central cell, which has central larger vacuole and two polar nuclei.

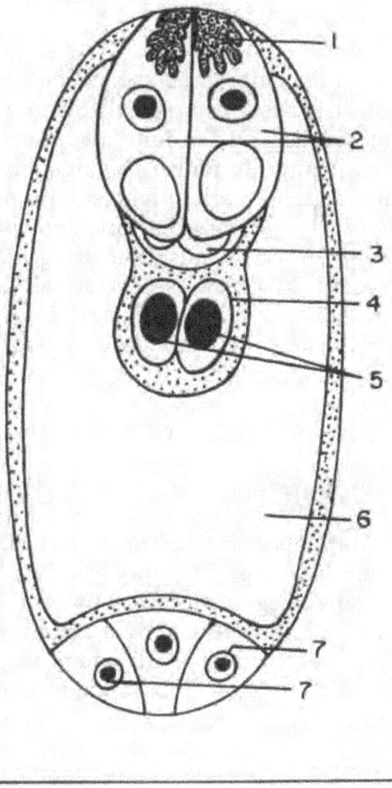

Fig 5.1 : Structure of typical embryosac. 1. filiform apparatus, 2. synergids, 3. egg cell, 4. central cell, 5. polar nuclei, 6. vacuole, 7. antipodals

5.2. DEVELOPMENT OF EMBRYOSAC: (fig-5.2)

During mega gametogenesis, the functional megaspore increases in its size into a sac like structure. At first the nucleus situates in the centre and the cytoplasm is non vacuolated. During further development small vacuoles appear either side of the nucleus. The nucleus divides and gives rise two daughter nuclei. The two nuclei migrate to either poles of embryosac and all the vacuoles fuse to form a big vacuole in the centre. Each nucleus divides twice and give rise four nuclei at each pole. The four nuclei at the micropylar region constitute micropylar quartet and four nuclei at the chalazal region constitute chalazal quartet. All the eight

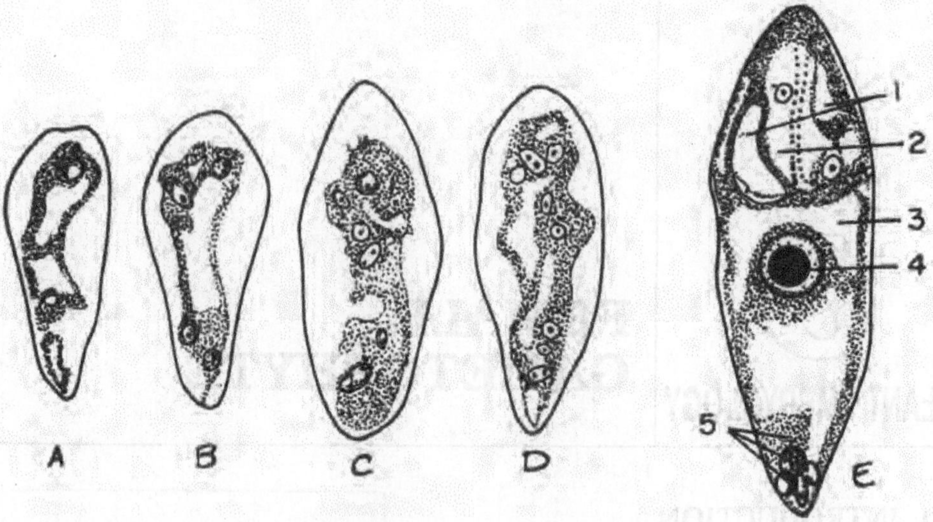

Fig 5.2 : Development of typical embryosac (A – E)
1. synergids, 2. egg cell, 3. central cell, 4. secondary nucleus, 5. antipodals

nuclei of the embryosac are present in common cytoplasm. As the embryosac posses eight nuclei, it is described as eight nucleated embryosac. Now the cellular organization occurs in embryosac. Of the four nuclei at the micropylar region three nuclei organise into egg apparatus and the fourth free nucleus acts as upper polar nucleus. Similarly three nuclei of the chalazal quartet organise into antipodals, and the fourth nucleus acts as lower polar nucleus. The lower polar nucleus migrates towards upper polar nucleus and situates side by side. After organisation of egg apparatus and antipodals, the central portion of the embryosac is known as central cell. It is occupied by big vacuole and two polar nuclei situate beneath the egg apparatus. The two polar nuclei fuse to form a diploid nucleus which is known as secondary nucleus. As this mode of embryosac development occurs in majority angiosperms, it is designated as common type. For the first time this type of embryosac was described by Strasburger in *Polygonum divaricatum.* Hence, this mode of development is described as 'polygonum type'.

5.3. STRUCTURE OF MATURE EMBRYOSAC

The embryosac has its own wall. It is composed of cellulose and pectin materials. The common type of embryosac is a 7-celled structure (fig-5.1). At the micropylar region three celled egg apparatus is present. At the chalazal portion three antipodals are present. The central big cell is known as central cell. The cells of egg apparatus and antipodals are uninucleated and haploid, where as the central cell is either binucleated or diploid. All the 7-cells of the embryosac are inter connected by each other by delicated plasmodesmatal connections.

5.3.1 Egg Apparatus :

It situates at the micropylar region of embryosac. It consists of central egg cell and lateral two synergids. All the three cells are inter connected by plasmodesmata.

A) EGG CELL :- The size and structure of egg cell vary with species to species. It shows common walls with synergids and central cell. It has thick wall at the micropylar region

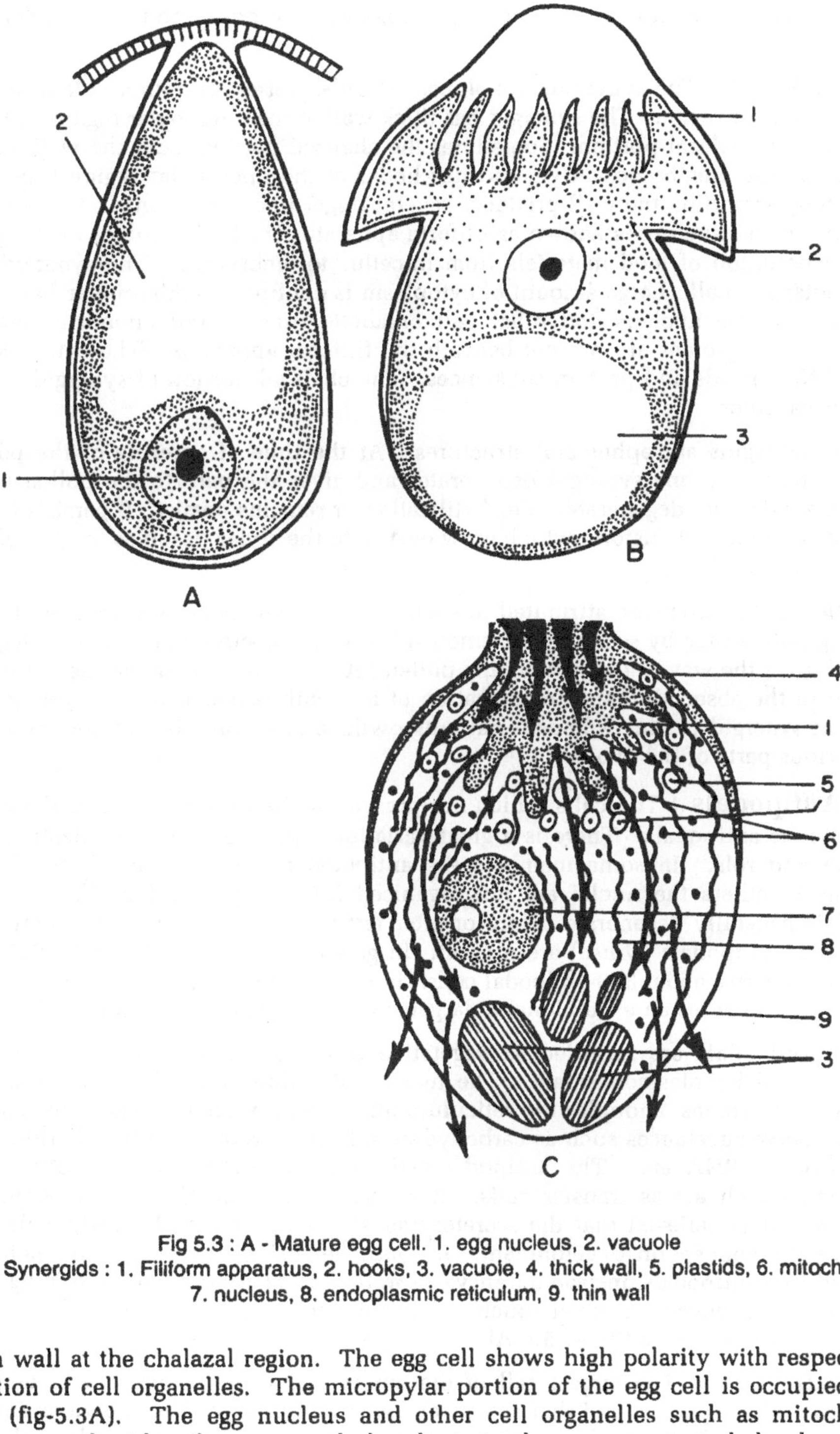

Fig 5.3 : A - Mature egg cell. 1. egg nucleus, 2. vacuole
B & C : Synergids : 1. Filiform apparatus, 2. hooks, 3. vacuole, 4. thick wall, 5. plastids, 6. mitochondria, 7. nucleus, 8. endoplasmic reticulum, 9. thin wall

and thin wall at the chalazal region. The egg cell shows high polarity with respect to the distribution of cell organelles. The micropylar portion of the egg cell is occupied by big vacuole (fig-5.3A). The egg nucleus and other cell organelles such as mitochondria, dictyosomes, plastids, ribosomes and abundant starch are present at chalazal portion of the egg cell. In general the egg cell lacks hook like structure which we find in synergids. But if synergids are absent in egg apparatus, the egg cell has hook like structures and wall ingrowths as in synergids at the micropylar region. The hooks and wall ingrowths show haustorial behavior.

B) SYNERGIDS : The synergids are elongated cells presei.. on either side of the egg cell at the micropylar region. The synergid has thick wall at the micropylar region and gradually thins towards chalazal portion. Sometimes the chalazal portion lacks the wall, but instead has two plasma membranes. Synergid has hooks at the micropylar region (fig-5.3 B). At the micropylar portion the wall produces several finger like projections in to the cytoplasm. This prominent structure is known as filiform apparatus (FA). It involves in the absorption and transportation of food materials from nucellus to embryosac. The synergid is also a highly polarised cell. Large amount of cytoplasm is confined to micropylar half of the cell (fig- 5.3 C). The nucleus and other cell organells such as mitochondria, endoplasmic reticulum, dictyosomes are present beneath the filiform apparatus. The cytoplasm is also rich in RNA, lipids and protein substances. The chalazal portion of synergid is occupied by larger vacuoles.

The synergids are ephemeral structures. At the time of entering of the pollen tube into the embryosac one synergid degenerates and gives the path to the pollen tube. The other synergid either degenerates after fertilization or remains persist. If persists it produces an elongated tubular haustoria which grow even into the stylar tissue as in *Quinchemalium chilense*(fig. 5.4 A).

Different functions are attributed to synergids. 1. Synergids play an important role in attracting pollen tube by secreting chaemotropically active substance 2. The degenerating synergid gives the way to the penetrating pollen tube 3. Filiform apparatus of the synergid involves in the absorption and translocation of food substances from surronding tissue 4. Persistant synergid produces haustrorial outgrowths and involve in sucking food materials from various parts of the ovule.

5.3.2 Antipodals : The cells which are present at the chalazal pole of the embryosac are known as antipodals. There is a great variation regarding their organization, number, structure and role. In some members the antipodal nuclei are absent (eg. *Oenothera*), where as in others the nuclei are not organised into antipodals (eg. Sapotaceae). The antipodals generally degenerate before or after fertilization or remain persistant. In some members 11-antipodals (*Drusa*) are present. In grasses the three antipodals divide several times and give rise about 300 antipodal cells. In some instances the cells may be polyploid. In *Acotinum napellus* an exceptional enlargement of antipodals are reported (fig — 5.4 C).

The antipodals are regarded as vegetative tissue of the embryosac. The cells are interconnected by plasmodesmata. The maize antipodals have abundant mitochondria, Plastids, dictyosomes, endoplasmic reticulum and several vesicles. The cells also contain various reserve substances such as carbohydrates, lipids, proteins, polysacharides, ascorbic acid, oxidases, RNA etc. The antipodal cells have papillate wall projections into the cytoplasm, which act as transfer cells. It suggests their nutritive role in nourishing to embryosac. It is believed that the secretory substances produced by antipodals may also influence the endosperm development. In *Quinchemalium* the antipodal nuclei are not organised into antipodals, instead a transverse wall appears around nuclei forming antipodal chamber. It produces branched much elongated houstoria, which penetrate into ovular tissue to absorb nutrients (fig — 5.4 A)

5.3.3 Central Cell : The largest cell of the embryosac which is present in between egg apparatus and antipodals, and that acts as endosperm mother cell is known as central cell. It is occupied by big vacuole which is a reservoir of sugars, aminoacids and inorganic salts. The vacuole is surrounded by thin film of cytoplasm. The central cell either has two polar nuclei or diploid secondary nucleus beneath egg apparatus. The cytoplasm of

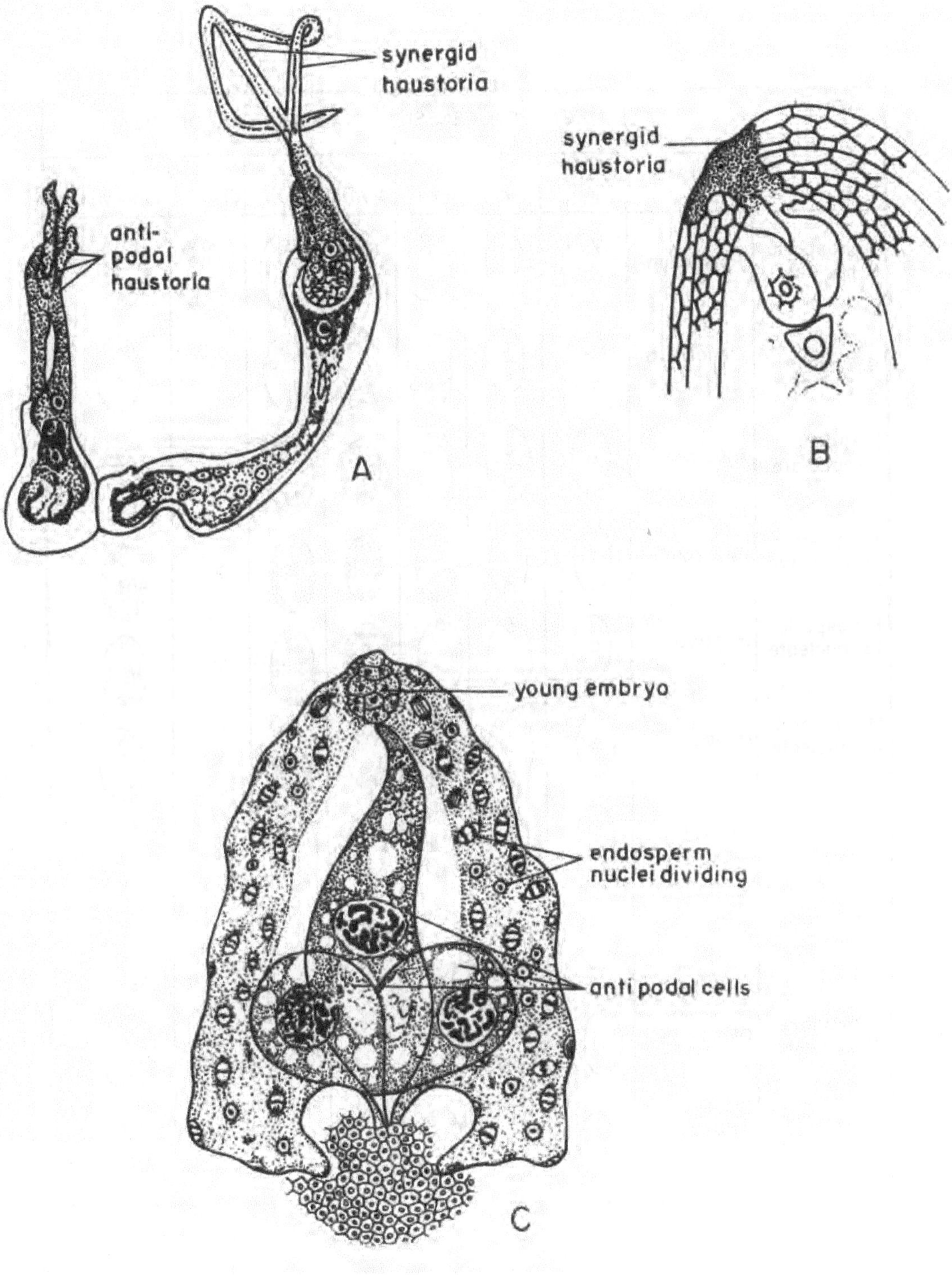

Fig 5.4 A : Mature embryosac of *Quinchematium chilense* showing synergid and antipodal haustoria
B : Synergid hausboria projecting through the micropyle in *cortaderia selloana*
C : Embryosac of *Aconitum napellus*

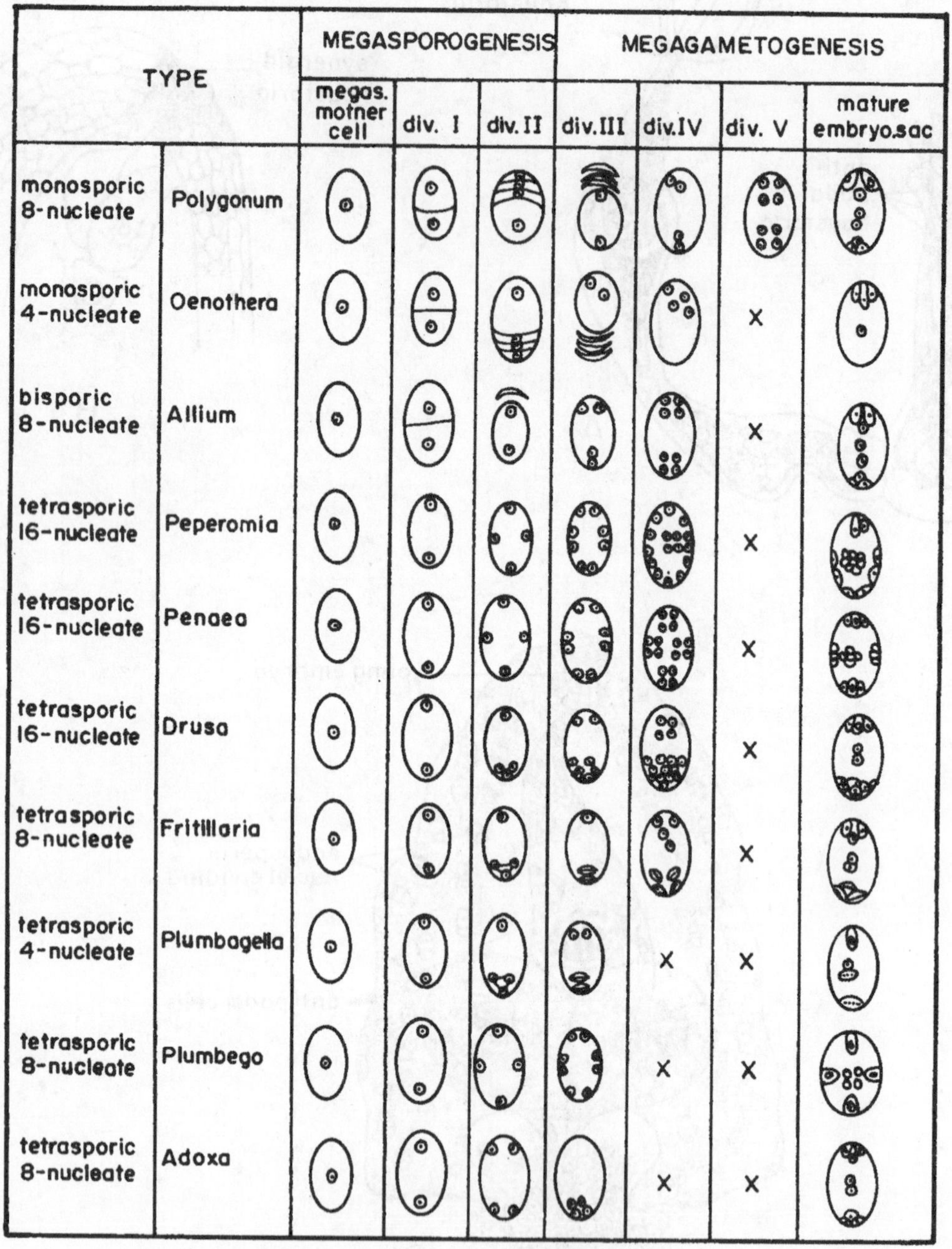

Fig 5.5 Important types of embryosac development within the angiosperm

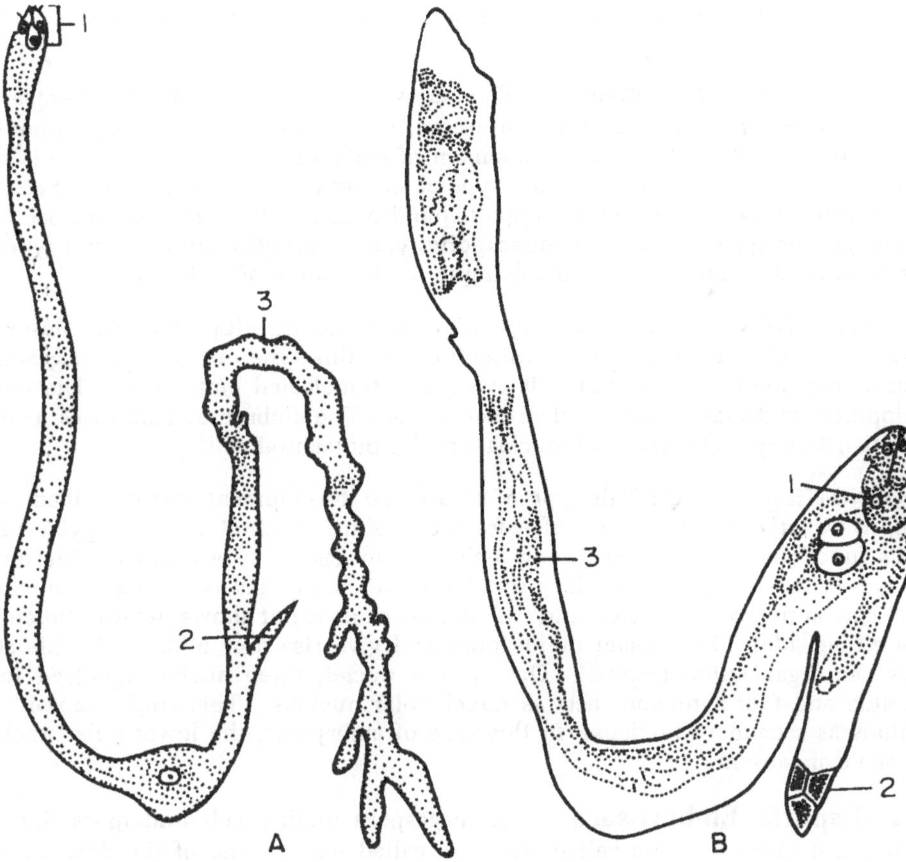

Fig 5.6 : A : Amature embryosac of *Santalum album* showing chalazal caecum - 1. egg apparatus, 2. antipodals, 3. caecum
B : A mature embryosac of *Comandra Umbellata* showing micropylar caecum - 1. egg apparatus, 2. anti podals, 3. caecum

central cell is rich in mitochondria, endoplasmic reticulum, dictyosomes, ribosomes, spherosomes and glyoxysomes. The cell has reserve food in different forms. Central cell shows cell wall projections at micropylar and chalazal regions. These projections absorb the food materials from the surrounding nucellar tissue. In *Aquilegia vulgaris* the central cell produces skirt like evaginations at the chalazal region. They surround antipodals, grow into nucellus and digest it by enzymatic action (Rifot, 1973). In some members (eg. *Santalum*; *Comandra*) the central cell produces tubular haustorial structures or caeca to absorb food materials (fig-5.6)

5.4 TYPES OF EMBRYOSAC :

(fig-5.5) Depending on the number of megaspores / megaspore nuclei taking part in the development of embryosac (mega gametogenesis), the embryosacs are classified in to monosporic, bisporic and tetrasporic types. Again these three main types were classified into different sub types on the basis of 1. Which side (wether chalazal or micropylar) megaspore is taking part in the development of embryosac, 2. Into how many mitotic divisions the megaspore nucleus is undergoing and giving rise total number of nuclei, 3. The way of embryosac organisation and subsequent arrangement of cells in embryosacs and their ploidy etc.

5.4.1 Monosporic Embryosacs : The megaspore mother cell (MMC) undergoes meiotic division and give rise linear tetrad. Out of four megaspores one megaspore either chalazal or micropylar, takes part in the development of embryosac. Hence, this type is called as monosporic type. All the nuclei in this type of embryosac are genetically identical, because they are formed by the division of single haploid nucleus. Hence the embryosac is said to be homogenous type. Again this monosporic type is classified into two sub types on the basis of which side megaspore is involving in the formation of embryosac.

A) POLYGONUM TYPE : This type of embryosac development was described by Strasburger (1879) in *Polygonum divaricatum*. In this type out of four megaspores, the chalazal megaspore is functional. It gives rise 8-nucleated embryosac. The embryosac development and organisation is described earlier. The embryosac comprises 3-celled egg apparatus, 3-antipodal cells and binucleate or diploid central cell.

B) OENOTHERA TYPE : This type of embryosac development was described by Geerts (1908) in *Oenothera lamarkiana*. After formation of linear tetrad, the micropylar megaspore acts as functional one and remaining 3-chalazal megaspores degenerate. The megaspore increases in its size and the nucleus divides mitotically giving two daughter nuclei. Both nuclei stay at micropylar region and the embryosac does not shows polarity (nuclei do not move either side). The nuclei divide once and give rise four nuclei. At this stage the embryosac organisation begins. Out of four nuclei, three nuclei organised in to egg apparatus and fourth nucleus acts as upper polar nucleus. This single haploid nucleus functions as secondary nucleus. In this type of embryosac, the lower polar nucleus and antipodals are absent.

5.4.2 Bisporic Embryosacs : The megaspore mother cell undergoes first meiotic division and gives rise two celled structure called dyad. One of the dyad cells act as functional megaspore and undergoes II meiotic division. Other dyad cell degenerates. The cell with two nuclei is equivalent to two megaspores and the nuclei are genetically different. This cell gives rise embryosac and hence called bisporic type. The embryosac nuclei belong to two different genetic constitutions. Hence the embryosac is heterogenous. The two nuclei stay at opposite poles of the cell. Bisporic embryosacs are 2-sub types.

A) ALLIUM TYPE : This type has been described by Strasburger (1879) in *Allium fistulosum*. In this type the chalazal dyad cell with two nuclei is functional. The two nuclei undergo two mitotic divisions and give rise 8-nucleated embryosac with 4-nuclei at each pole. Further organisation and structure of embryosac is similar to polygonum type.

B) ENDYMION TYPE : This type of embryosac is reported in *Endymion hispanicus* by Battaglia (1958). In this type the micropylar dyad cell with two nuclei is functional. Both nuclei divide twice and give rise 4-nuclei at micropylar region and 4-nuclei at chalazal region. The organisation and structure is similar to polygonum type.

5.4.3 Tetrasporic Embryosacs: The diploid nucleus of MMC undergoes reduction division and gives rise four haploid nuclei. The divisions are not followed by wall formation as in monosporic and bisporic types. So that the mother cell posses four haploid nuclei in common cytoplasm. This four nucleated cell directly functions as coenomegaspore and take part in the development of embryosac. As the coenomegaspore is equivalent to four megaspores, the type of embryosac is described as tetrasporic type. It is more heterogenous than the bisporic type of embryosac. They are classified into 7-types on the basis of presence of number of nuclei in embryosac before organisation. The 7-types are grouped into three groups such as 8-nucleated; 16-nucleated and 4-nucleated embryosacs.

A) 8-NUCLEATED EMBRYOSACS :

a) Adoxa type : It was reported in *Adoxa moschatella* by Johnson in 1879. In coenomegaspore two nuclei situate at micropylar region and other two at chalazal region. They divide once and give rise eight nuclei, four at each pole. The organization and sturucture of embryosac is similar to polygonum type.

b) Plumbago type : This type of embryosac development was described in *Plumbago Capensis* by Haupt (1934). In coenomegaspore, out of four nuclei one situates at micropylar region one at chalazal and two at lateral portion of the cell. The four nuclei divide once and give rise 8-nuclei, two in four groups. One nucleus from each group migrate to central portion and fused to form tetraploid secondary nucleus. The micropylar nucleus organised into single egg cell and chalazal nucleus into single antipodal cell. The lateral two nuclei are also cut off by walls.

c) Fritillaria Type : This type of embryosae was described in *Lilium bulbiferum.* In coenomegaspore out of four nuclei, one situates at micropylar region and three at chalazal region. The chalazal nuclei fused to form a triploid nucleus. The haploid and triploid nuclei undergo two mitotic divisions giving four nuclei at each pole. Out of four haploid nuclei at micropylar region three are organised into egg apparatus. The fourth nucleus acts as upper polar nucleus. Out of four triploid nuclei at chalazal portion three are organised into antipodals and fourth nucleus acts as lower polar nucleus. Both polar nuclei fused to form tetraploid secondary nucleus.

B) 16-NUCLEATED EMBRYOSACS :

a) Drusa Type: This mode of embryosac development was described by Hakansson (1923) in *Drusa oppositifolia.* In coenomegaspore one nucleus situates at the micropylar region and three at the chalazal region. All the nuclei divide twice and give rise four at micropylar and twelve at chalazal portion. The micropylar three nuclei are organised into egg aparatus. The chalazal eleven nuclei are organised into antipodals and remaining two haploid nuclei fused to form diploid secondary nucleus.

b) Penaea Type : Stephens (1909) described this type of embryosac in *Penaea* sps. The four nuclei of coenomegaspore are arranged as in plumbago type and undergo two mitotic divisions to form 16-nuclei, which are arranged in four groups. From each group one nucleus migrates to the centre and fused to form tetraploid secondary nucleus. The remaining three nuclei of each group are organised into triads. Usually the micropylar triand acts as egg apparatus and chalazal triad acts as antipodals.

c) Peparomia Type : *Peparomia pellucida* exhibits this mode of embryosac development (Campbell, 1899; Johnson, 1900). Upto the formation of 16-nuclei, four in each group, is similar to penaea type. The organised embryosac has an egg apparatus with one egg cell and one synergid; six peripheral cells and a central cell with 8-polar nuclei.

C) 4-NUCLEATED EMBRYOSAC :

a) Plumbagella Type: This type of megagametogenesis was reported in *Plumbagelle micrantha.* The developmental sequences upto the formation of one haploid nucleus at the micropylar region one triploid nucleus at the chalazal region is similar to that of fritillaria type. The nuclei divide once and give rise four nuclei, two at each pole. After organisation the embryosac has single haploid egg cell, one triploid antipodal cell and tetroploid secondary nucleus in the central cell.

5.5 HAUSTORIAL BEHAVIOR OF EMBRYOSAC

The embryosac surface has food absorbing capacity from surrounding nucellar tissue. In some members the persistant synergids and antipodals as well as central cell show haustorial properties and extensions.

The persistant synergid either swells or elongates into tubular structure and penetrates through the nucellar tissue and reach into micropylar region in several compositae members. In *Quinchamalium chilense* the synergid produces much elongated tubular haustorial out growths (fig — 5.4 A) which penetrate into the stylar tissue to suck the food materials. In *Cortaderia* the filiform apparatus produces several finger like projections outside, which reach to the micropylar portion of the ovule.

In *Quinchamalium* the antipodal chamber produces tubular structure (fig — 5.4 A) that penetrate through the funiculus and reach to the placental tissue of the ovary. In maize and Poppy the wall ingrowths of antipodals look like transfer cells which have absorptive nature.

In several taxa the central cell produces an elongated tubular extensions or caeca from micropylar region or chalazal regions. They grow in to ovular and carpellary tissues and function as haustoria. In *santalum album* and *Comandra umbellata* the central cell produces an elongated curved caecum from chalazal region and micropylar region respectively (fig.- 5.6). In same acanthaceae members (eg. *Nelsonia*) intrestingly the developing embryosac protrudes out of the micropyle and buried itself in the placental tissue. The embryosac organisation occurs in the placental nutritive tissue.

5.6 NUTRITION OF EMBRYOSAC

Nucellus is the chief nutritive tissue of the embryosac. The nutrients enter into embryosac from the chalazal portion. The hypostase involves in the supply of nutrients, enzymes, and hormones to embryosac. Synergids, antipodals and caeca also play an important role in supply of food materials to embryosac. In tenuinucellate ovules the integumentary tepetum involves in nourishing the embryosac.

QUESTIONS

A. Answer the following in about 30 lines.

1. Discuss the development and structure of normal type of embryosac.
2. Describe different kinds of female gametophytes you have studied.

B. Answer the following in about 10 lines.

1. Development of polygonum embryosac.
2. Structure of typical type of embryosac
3. Synergids
4. Monosporic embryosacs
5. Haustorial behaviour of embryosac

C. Answer the following in about 3-5 lines

1. Coenomegaspore
2. Monosporic embryosac
3. Bisporic embryosac
4. Tetrasporic embryosac
5. Egg apparatus
6. Synergids
7. Filiform apparilus
8. Caecum
9. Epistage and hypostage
10. Nutrition of embryosac

6 PLANT EMBRYOLOGY	ENDOSPERM

6.1. INTRODUCTION

"The chief nutritive tissue that is formed after fertilization and supplies required food material to the developing embryo is known as endosperm". At the time of fertilization one male gemete fuses with the egg nucleus (Syngamy) and the other with secondary nucleus (triple fusion) and gives rise to zygote and primary endosperm nucleus respectively. The zygote develops into well organized embryo of the seed (future hidden plantlet). Primary endosperm nucleus divides and gives rise undifferentiated food laden parenchymatous tissue around developing embryo. It stores food material in the form of carbohydrates, proteins, fats etc., Functionally it is comparable to the female gametophyte of gymnosperms. In gymnosperms it is formed before fertilization (pre fertilization product) and is haploid, where as in angiosperms it is formed after fertilization (post fertilization product) and is in general triploid. In some angiospermic members it may be diploid (*Oenothera*) or polyploid (*Plumbago, Pepormia* etc). The ploidy of the endosperm is determined by the number of polar nuclei take part in the formation of secondary nucleus and is constant for a species. The developing embryo either consumes entire endosperm or only part of it. In former condition the seeds do not posses endosperm, so they are called as non endospermic or exalbuminous (pea, bean, gram, tamarind etc.,). In latter condition the seeds are described as endospermic / albuminous (coconut, castor, maize, rice, wheat etc.,). In non-endospermic seeds the cotyledons store the food material. In Orchidaceae, Podostomaceae and Trapaceae the endosperm is not formed.

6.2. TYPES OF ENDOSPERM

Depending upon the mode of development, three types of endosperms (Nuclear, cellular and helobial) have been described in angiosperms (fig-6.1). Davis (1966) has recorded nuclear endosperm in 161 families; cellular endosperm in 72 families and helobial endosperm in 17 families.

6.2.1 Nuclear Endosperm : "The endosperm in which the nuclear divisions are not followed by wall formation is known as nuclear endosperm". In this case the primary endosperm nucleus divides and gives rise two daughter nuclei which is not followed by wall formation. The two daughter nuclei divide successively several times without wall formation. So that thousands of free nuclei are formed in endosperm mother cell (Central

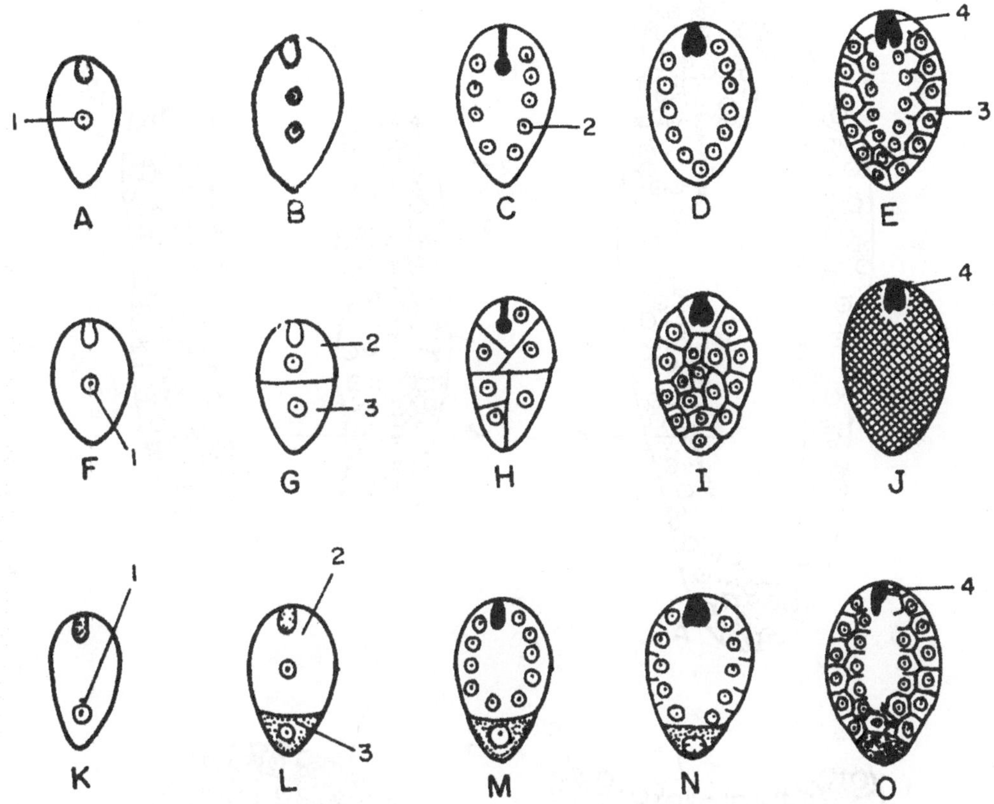

Fig 6.1 : Main types of endorsperm development
A – E : Nuclear type. 1. Primary endosperm nucleus, 2. Free nuclei, 3. Wall fermation, 4. embryo
F – J : Cellular type : 1. Primary endosperm nucleus, 2. micropylarchamber, 3. Chalazal chamber, 4. embryo
K – O : Helobial type : 1. Priamry endosperm nucleus, 2. Micro pylar chamber, 3. Chalazel chamber, 4. embryo

cell of embryosac). The cytoplasm increases in its quantity and small vacuole appears in the center. The vacuole increases in its size and the nuclei are pushed towards periphery. This free nuclear state either persists until it is consumed by the embryo or become cellular. The process of cellularization generally begins from the periphery and proceeds towards center of the cell (centripetal manner). But in some members it proceeds from apex to the base.

In general the entire endosperm becomes cellular (fig.-6.2). But in several taxa the degree of cellularization varies with species to species and cellularization confined to only micropylar part of embryosac (Fig.- 6.3). In *Phaseolus* the cellularization is very little and is confined to around embryo. The remaining portion remains free nuclear. In *Crotalaria*, *Lomatia* and *Grevillea* the micropylar half of the embryosac becomes cellular and chalazal half remains free nuclear state. The chalazal portion elongates to form an elongated tubular, vermiform, coenocytic appendage. It acts as haustoria in absorbing food material from ovular and extra ovular tissues. In *Crotalaria* the terminal portion of haustorium swells and acts as sucking vescicle. In *Lomatia* additional finger like projections (Secondary

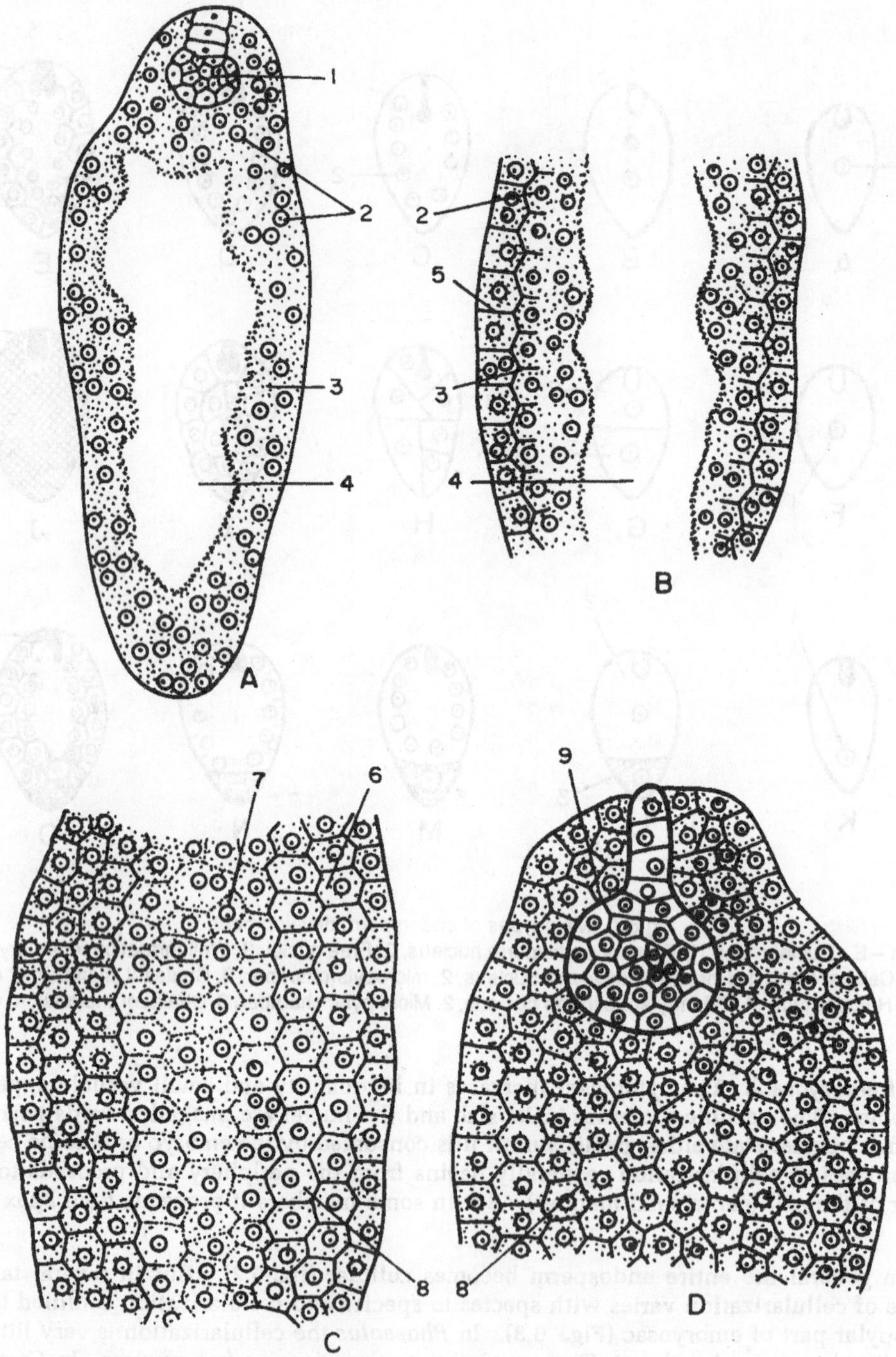

Fig 6.2 : Formation of nuclear endosperm
(A – D) 1. embryo, 2. free nuclei, 3. cytoplasm, 4. vacuole, 5. cell wall formation, 6. cell walls laid down, 7. cell walls developing, 8. endosperm, 9. embryo

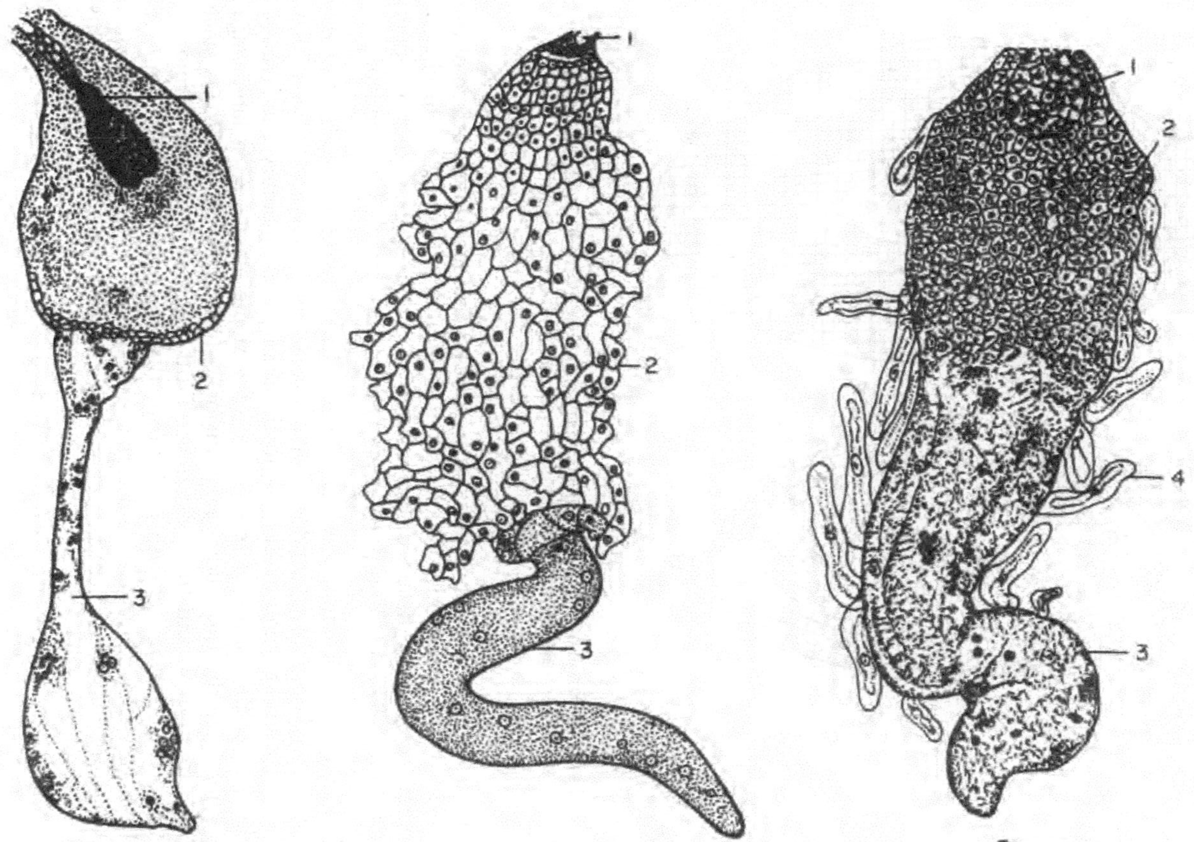

A B C

Fig 6.3 : Nuclear endosperm

A : In Crotalaria : the upper half of the endosperm becomes cellular where as the lower half remains free nuclear

B : In Grevillea lower chalazel free nuclear portion became vermiform haustorium

C : In Lomatia finger like unicellular structures develop on cellular endosperm and free nuclear chalazal haustorial portion.

1. embryo, 2. cellular endosperm, 3. free nulcear haustorium, 4. unicellular out growths.

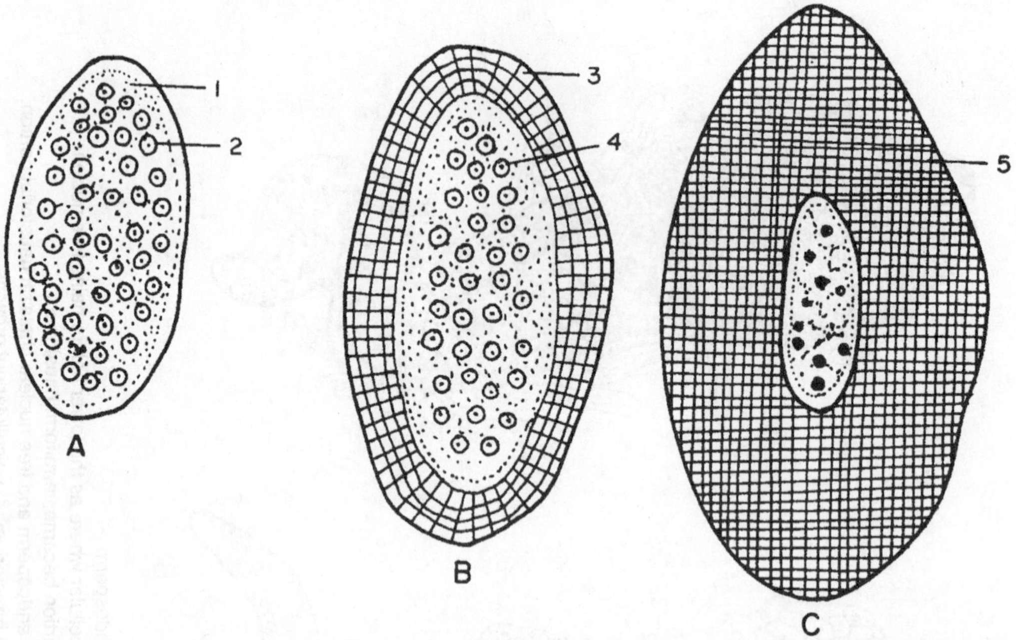

Fig 6.4 : Development of endosperm in coconut (A – C) 1. cyloplasm, 2. free nuclei, 3. peripheral cellular endosperm, 4. central free nuclear endosperm, 5. matured endosperm

haustoria) which are present all over the endosperm increase the absorbing surface of the endosperm. In *Grevillea* the haustorium is aggressive type and penetrates in to chalzal portion by disintegrating chalazal tissue of the ovule. At later stages the coenocytic appendage become cellular. This cellular portion may be called as secondary endosperm.

The mode of development of endosperm is very peculiar and interesting in coconut (*Cocas nucifera* Fig.-6.4). The primary endosperm nucleus undergoes a number of free nuclear divisions. In early stages when the fruit is about 50mm long the embryosac contains denser fluid in which float thousands of free nuclei of variable sizes. At a later stage, when the fruit is 100mm size the free nuclei are transformed into free cells. At this stage we can see suspended free nuclei and free cells in the fluid along with growth hormones, amino acids and sugars. During the growth of the fruit the free cells gradually settle on the embryosac wall and gives rise few layered cellular endosperm. Free nuclear divisions, formation of free cells and their settlement is a continuous process for some period. This multilayered spongy tissue is known as coconut meat. Its thickness increases due to cell divisions in cellular portion. In mature fruit milky fluid contains sugars and amino acids etc.

6.2.2 Cellular Endosperm (Fig.-6.5) : "The endosperm in which the nuclear divisions are followed by wall formation is known as cellular endosperm". Generally the first division is transverse and it may be equal or unequal. But very rarely the division may be vertical or oblique. The two nuclei divide several times followed by wall formation. Thus the embryosac becomes cellular.

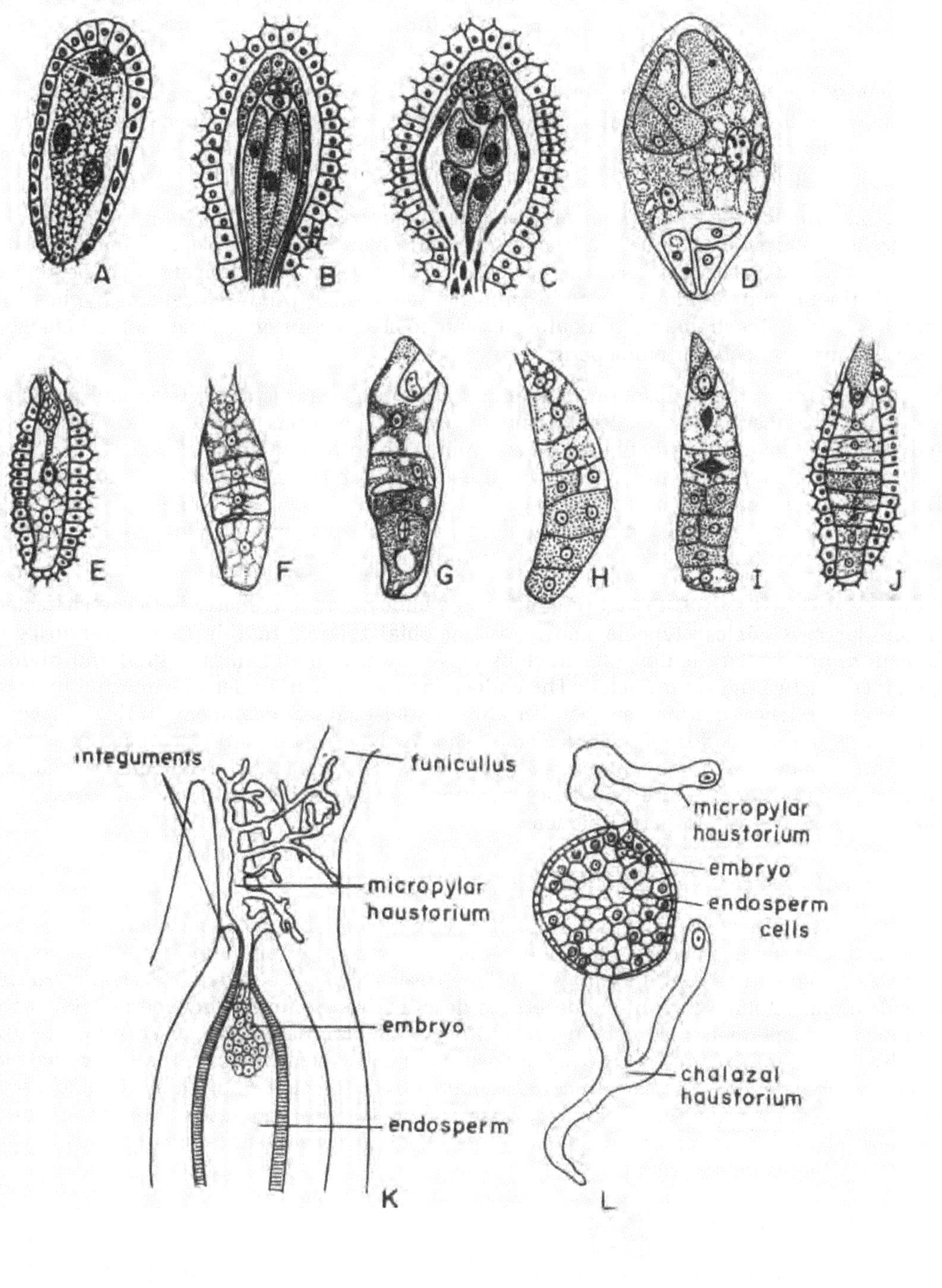

Fig 6.5 A – D : Formation of Cellular endosperm in *Adoxa*
E – J : Formation of cellular endosperm in *Villarsia* K – Well branched micropylar haushorium in *Impatience*
L – Micropylar and chalazal haustoria of *Nemophila*

In *Adoxa* the endosperm mother cell undergoes two vertical divisions and gives rise to four similar cylindrical cells. These cells undergo a transverse division and give rise to eight cells which arrange in two tiers. The further divisions are irregular in cellular endosperm formation.

In *Villarsia* the endosperm mother cell undergoes three successive transverse divisions and gives rise to row of eight cells situated one above the other, in a linear fashion. These cells undergo rapid and repeated divisions in different planes and gives rise to massive cellular endosperm.

In *Impatience* the endosperm mother cell undergoes unequal division and gives rise to smaller micropylar chamber and larger chalazal chamber. Micropylar chamber divides transversely and produces three smaller cells. The upper cell elongates and penetrates through the micropyle and produces a gaint haustorium. It produces several hyphae like branches, which penetrate into funiculus of ovule to absorb nutrients. The chalazal chamber gives rise massive cellular endosperm.

In *Nemophila* the endosperm mother cell undergoes transverse division and produces two equal chambers. The chalazal chamber directly functions as haustorial mother cell and produces an elongated tubular haustorium with lateral tubular branch. The lateral branch penetrates through the funiculus and even reach to the placental tissue of the ovary. The micropylar chamber divides and gives rise a central cell and an upper cell. The upper cell directly gives rise curved tubular micropylar haustorium. Only the central cell is responsible for the formation of massive cellular endosperm.

6.2.3 Helobial Endosperm : (Fig.-6.6) : The endosperm that shares both the characters of cellular and nuclear type is known as "helobial type". In this type of endosperm formation, the primary endosperm nucleus move towards the chalazal region and divides and gives rise two daughter nuclei. The endosperm mother cell undergoes unequal division and produces larger micropylar chamber and smaller chalazal chamber. In this respect it shares the cellular endosperm character. The nucleus of microphylar chamber is very active and undergoes large number of free nuclear divisions. The nucleus of chalazal chamber is very inactive and undergoes only few free nuclear divisions. In this respect it shares with nuclear endosperm character.

6.3. RUMINATE ENDOSPERM : (Fig.- 6.7A,B)

In general the endosperm has smooth outer surface. But in some members such as Annonaceae, Aristolochiaceae, Palmae etc. the surface counter part shows unevenness and irregular along with several infoldings. In these cases we can see several ridges and grooves in endosperm. Such a type of endosperm is described as ruminate endosperm. So far the ruminate endosperm is recorded in 32 families. The rumination is caused either by the activity of the seed coat or by growing endosperm it self. In former case the inner seedcoat layer produces several peg- like structures, which grow and cause rumination in endosperm. In latter case either the endosperm increases in its volume and reach into the irregular seed coat plane and become ruminate or the peripheral portion of endosperm shows irregular growth during its development.

6.4 MOSAIC ENDOSPERM :

"The endosperm that is composed of two or three different coloured patches that are arranged in irregular pattern (mosaic pattern) is known as mosaic endosperm". These patches also differ in genetic makeup. In *Zea mays* the endosperm is made up of red and white patches. It may be formed either of the following ways.

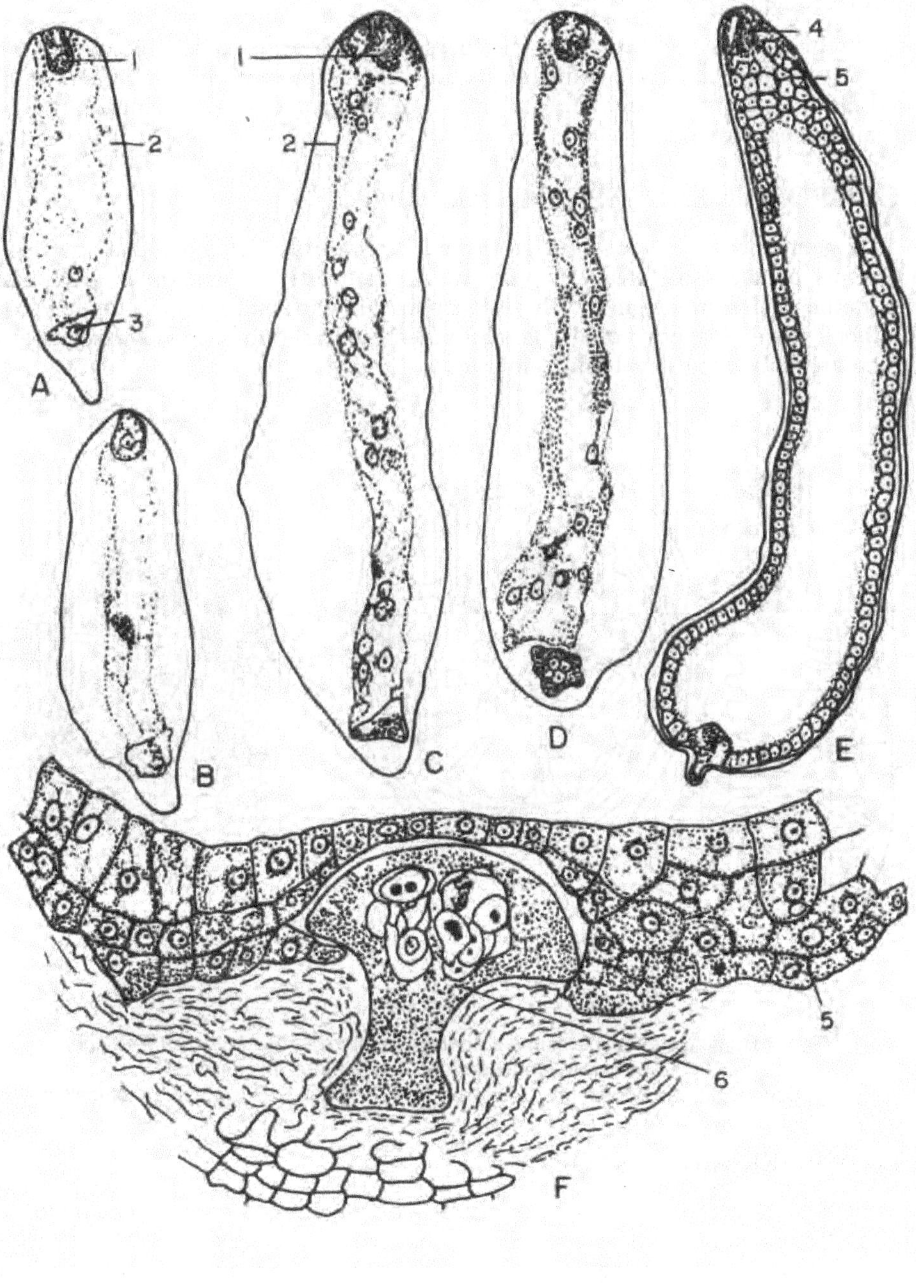

Fig 6.6 Formation of helobial endosperm (A – F)
1. Zygote, 2. Micropylarchamber, 3. Chalazal chamber, 4. Embryo, 5. Endosperm, 6. Chalazal chamber

a. After syngamy the second male gamete does not fuses with secondary nucleus. The secondary nucleus and second male gamete lie side by side and they directly act as primary endosperm nuclei. They divide independently and produce endosperm cells of different in colour and genotype.

b. Sometimes the second male gamete fuses with one of the polar nuclei and the second polar nucleus does not involve in the fusion. The primary endosperm nucleus (fusioned diploid product) and the other haploid polar nuclus involve independently in the formation of mosaic endosperm.

6.5 COMPOSITE ENDOSPERM : (Fig.- 6.7C)

"If a gaint mass of endosperm is formed by the union of more than two developing endosperms of the same ovary is known as composite endosperm". In Loranthaceae members the ovales are ategmic. The megasporangia with embryosacs situate very close to each other. The endosperms of all the embryosacs are fused to form a composite endosperm in which are embedded developing embryos.

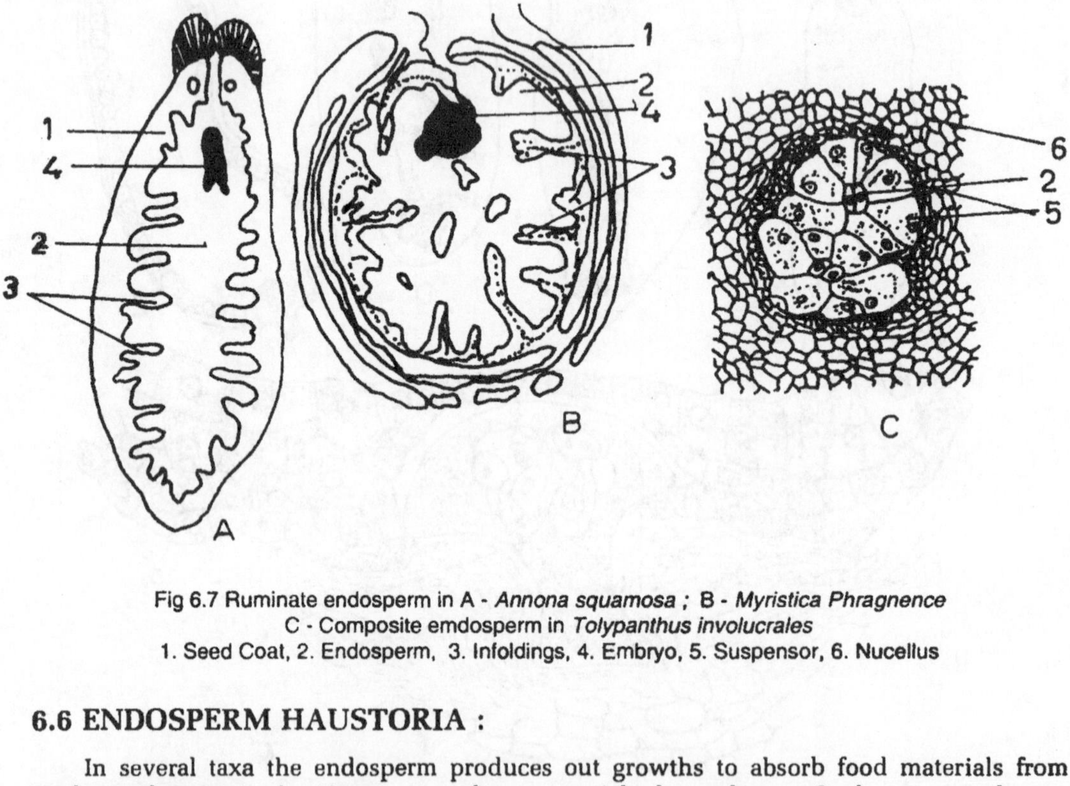

Fig 6.7 Ruminate endosperm in A - *Annona squamosa* ; B - *Myristica Phragnence*
C - Composite emdosperm in *Tolypanthus involucrales*
1. Seed Coat, 2. Endosperm, 3. Infoldings, 4. Embryo, 5. Suspensor, 6. Nucellus

6.6 ENDOSPERM HAUSTORIA :

In several taxa the endosperm produces out growths to absorb food materials from ovular and extra ovular tissues in order to nourish the embryo. Such out growths are called as endosperm haustoria. The haustoria may be unbranched and tubular or well branched hyphae like. It may secrete different cell wall degrading enzymes while penetrating in to the surrounding tissues. Their length varies with species to species. The longest endosperm haustorium (16 mm) is reported in *Echinocystis lobata* of cucurbitaceae family.

Torosian (1971) studied the fine structure of haustoria of *Lobelia dunnii*. He observed several wall ingrowths and plasma membrane invaginations in haustorial cells. These invaginations act as transfer cells in sucking and transporting food materials. It gives us strong evidence about absorptive nature of endosperm haustoria.

On the basis of origin the haustoria may be primary or secondary. If the haustoria are formed from the daughter cells of endosperm mother cell, they are described as primary endosperm haustoria. Sometimes after formation of endosperm, the endosperm cells produce additional absorptive appendages. They are described as secondary endosperm haustoria. The haustoria are produced by all kinds of endosperms. Based on their postion the haustoria are classfied into micropylar, chalazal and lateral types.

6.6.1 Micropylar Haustoria : This type of haustoria are very common in cellular type of endosperm. It may be single tubular structure as in *Nemophila* (Fig.- 6.5L) or well branched and hyphae like as in *Impatience* (Fig. — 6.5K). The aggressive natured haustoria of *Globularia* penetrates into micropyle, integuments and funiculus regions of the ovule. Sometimes they may penetrate in to ovary wall through placental tissue by sucking food

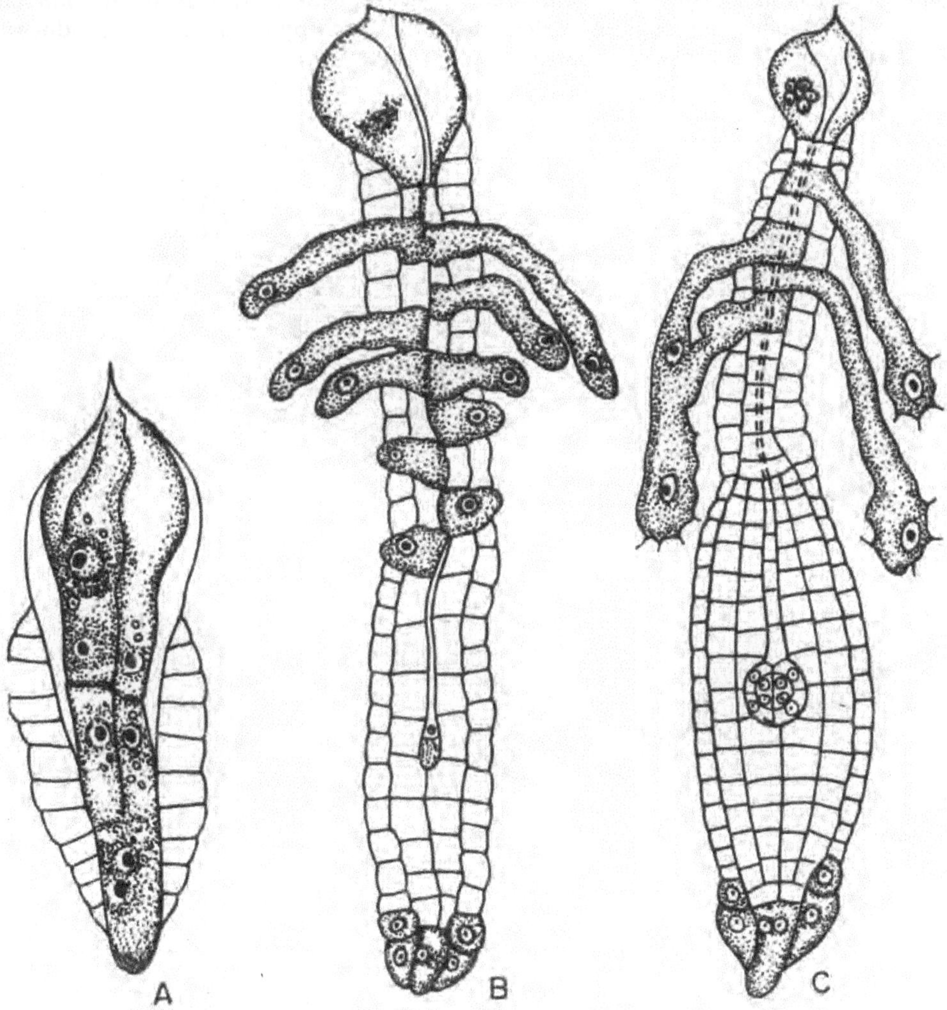

Fig : 6.8 Origin and development of secondary haustoria in *Centranthena hispida*

material. In *Centranthera* the micropylar endosperm cells produce active much elongated tubular secondary haustoria beneath inactive micropylar primary haustorium. (Fig.-6.8).

6.6.2 Chalazal Haustoria : Chalazal endosperm haustorium is wide spread in angosperms. This type is very common in nuclear and cellular type of endosperms. They may be single tubular as in *Grevillea* (fig.- 6.3) or may posses finger like structures (*Lomatia*). In *Iodina* (fig.-6.9) aggressive chalazal haustorium tip is much branched and looks like a coralloid structure. The chalazal haustoria also penetrate into integuments, chalazal portion and fertile placenatal region. In *Globularia* the aggressive chalazal haustorium is well branched and destroy the integumentary cells by creating large lacunae in integuments.

6.6.3 Lateral Haustoria (fig.—6.10) Occurrence of laterel haustoria ("diverticulae") is very rare and only scantly reports are available in literature. In *Agrostemma* in which cellular endosperm is described, slender tubular lateral appendage appears from the endosperm mother cell. After first division of primary endosperm nucleus one of the daugheter nucleus passes into the tube. The tube without separating from the mother cell elongates and acts as haustorium, which persists even in the mature seed. In *Monochoria* where helobial type of endosperm is described, the micropylar larger cell produces two tubular haustoria towards chalazal portion.

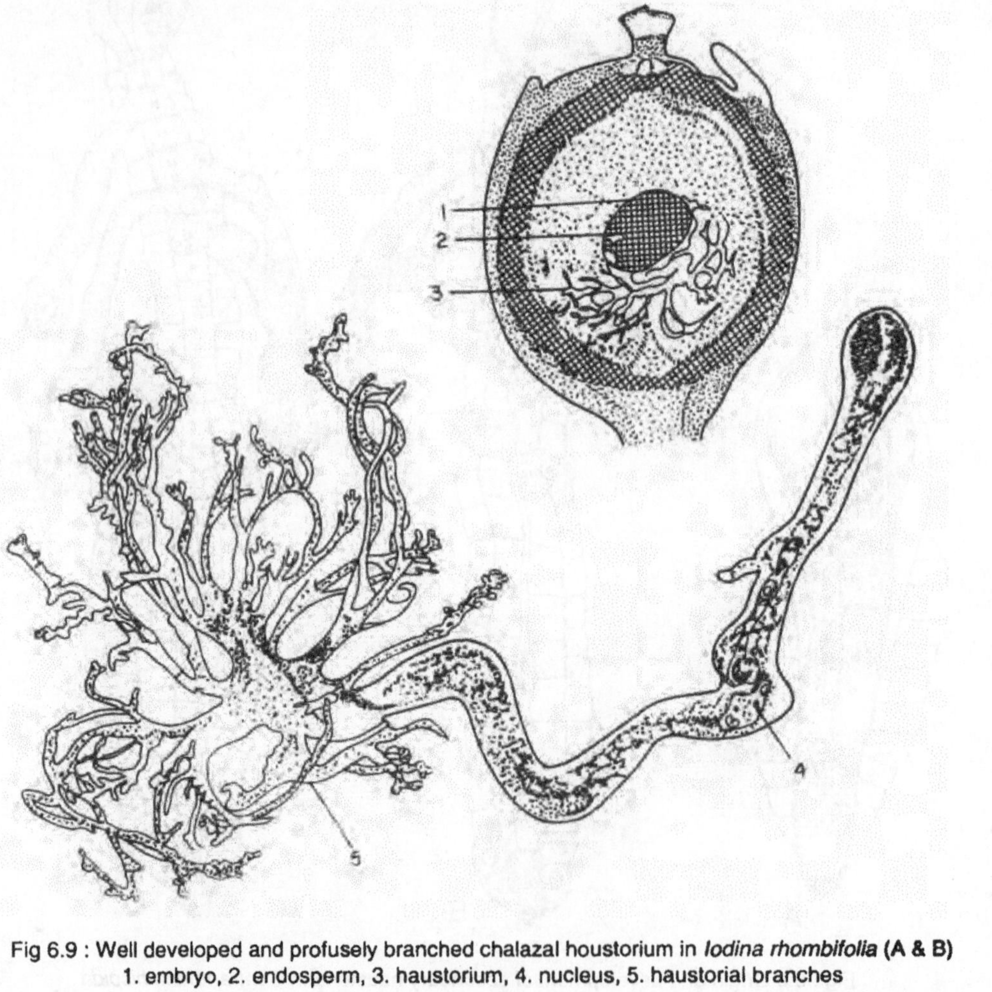

Fig 6.9 : Well developed and profusely branched chalazal houstorium in *Iodina rhombifolia* (A & B)
1. embryo, 2. endosperm, 3. haustorium, 4. nucleus, 5. haustorial branches

6.7 CYTOLOGY

Most commonly the endosperm is triploid. But in some members it may be diploid (*Oenothera*) or polyploid (*Penea*; *Peparomia*). In latter case more than two female gametophyte nuclei act as polar nuclei and take part in the fusion. The ploidy of endosperm nuclei may also increased due to endomitosis or nuclear fusion during its development. The highest ploidy (24576 'n') is reported in *Arum maculatum* endosperm. Normally the endosperm is colourless, but in *Raphanus*, *Crinum* and *Viscum* it is chlorophyllous. In general the endosperm is composed of morphologically, physiologically and genetically similar kind of cells. But in cereals the outer most 2-5 layers of cells become highly specialized morphologically and physiologically. Such layers are known as aleuron layers. Their cells contain aleuron grains (protein storage organelles). In Maize mosaic (morphologically and genetically) endosperm is described.

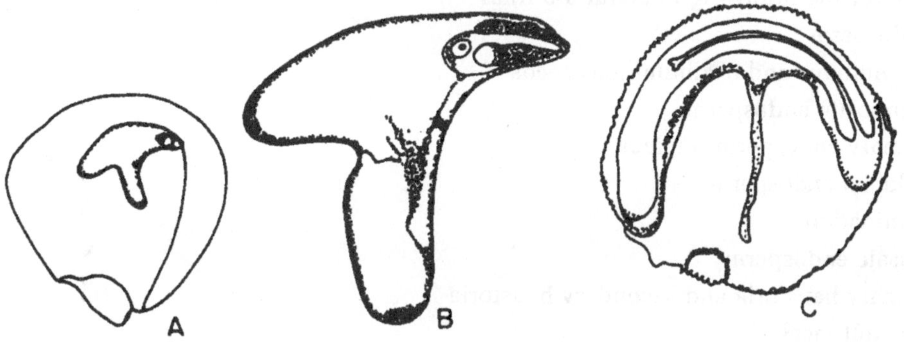

Fig 6.10 : Formation of lateral haustorium or diverticulum in embryosac of *Agrostemma githago*
A : L.S. ovule showing early stage in formation of diverticulum
B : enlarged view of embryosac from 'A'
C : L.S. nearly mature seed, showing large embryo and persisting diverticulum

6.8. FUNCTIONS

1. Endosperm stores food material in the form of carbohydrates, proteins and fats.
2. It supplies required food to the developing embryo.
3. In case of endorspermous seeds it nourishes the seedling.
4. If the endosperm is not formed properly, it adversely affect the development of embryo.
5. Endosperm regulates the precise mode of embryo development.
6. Young coconut endosperm is rich in various growth regulating substances such as auxins, cytokinins and gibberellins along with different amino acids; sugars, inorganic ions etc.
7. Immature endorsperm is potent for inducing divisions. Hence coconut milk is used in tissue culture experiments. The coconut milk induces formation of callus and differentiation of embryoids.
8. Large quantities of carbohydrates, aminoacids, protein complexes, Indoleacetic acid, Zeatin Cytokinin and several purine compounds are traced in corn endosperm at milkystage.
9. endosperm is the main source of food to the human beings.

QUESTIONS

A. Answer the following in about 30 lines.

1. What is endosperm ? Discuss different types of endosperm in Angiosperms
2. Write an essay on endosperm haustoria

B. Answer the following in about 10 lines.

1. Cellular endosperm
2. Nuclear endosperm
3. Ruminate endosperm
4. Functions of endosperm
5. Endosperm development in coconut
6. Endosperm haustorium

C. Answer the following in about 3-5 lines

1. Endosperm
2. Albuminous and exalbuminous seeds
3. Composite endosperm
4. Primary endosperm nucleus
5. Helobial endosperm
6. Rumination
7. Mosaic endosperm
8. Primary haustoria and secondary haustoria
9. Coconut meal
10. Cellular endosperm
11. Nuclear endosperm
12. Endosperm haustorium
13. Diverticulum
14. *Echinocystis lobata*

7

PLANT EMBRYOLOGY

EMBRYO

7.1. INTRODUCTION :

The rudimentary plant or hidden future plantlet that is present in seed is known as embryo. It is produced by the zygote after systematic and predetermined divisions. The embryo is a bipolar structure and it has potentiality to form a complete complex sporophyte.

After fertilization the egg is called as the zygote. The zygote is the totipotent cell to form mature embryo. In general the zygote enters into resting stage (dormant period). The resting stage is varied with species to species. The resting period is shorter where the endosperm is cellular than the nuclear. During dormancy several striking changes occur in zygote. 1. The vacuoles disappear. 2. The size of the zygote decreases. 3. Large amount of cytoplasm accumulates at the chalazal region of the cell. 4. Dictyosomes become active and increase in their number. 5. The endoplasmic reticulum gets closely associated with the plasma membrane. 6. The cell organelles cluster around the nucleus. 7. New cell wall is formed around the zygote. 8. The ribosomes aggregate to form polysomes. 9. Plasmodesmatal connections disappear from surrounding cells. 10. After the end of dormancy metabolic activities begin.

When the zygote is ready to divide it shows polarity with respect to the distribution of cell organelles. The vacuoles reappear, so that the cell gradually increases in its size. The vacuoles are confined to the micropylar region of the cell. In this region the cytoplasm is scanty and posess only fewer cell organells. This pole is designated as basal pole. Denser cytoplasm, active nucleus is present at the chalazal part of the zygote. The nucleus is surrounded by active mitochondria and large number of plastids. This pole is designated as apical pole. The polar determination in zygote is gene controlled.

Generally the zygote divides transversely and gives rise two unequal cells. Larger cell that is formed at the basal pole (micropylar side) is known as basal cell (cb) and smaller cell that is formed at the apical pole (chalazal side) is known as apical cell (ca). Rarely the first division may be vertical or oblique as in Loranthaceae and *Triticum* respectively. From the two celled stage unitl the initiation of organs the embryo is commonly called as proembryo. The early stages of embryogeny in dicots and monocots are similar but variations begin at a later stages. Hence, they are discussed separately in this text.

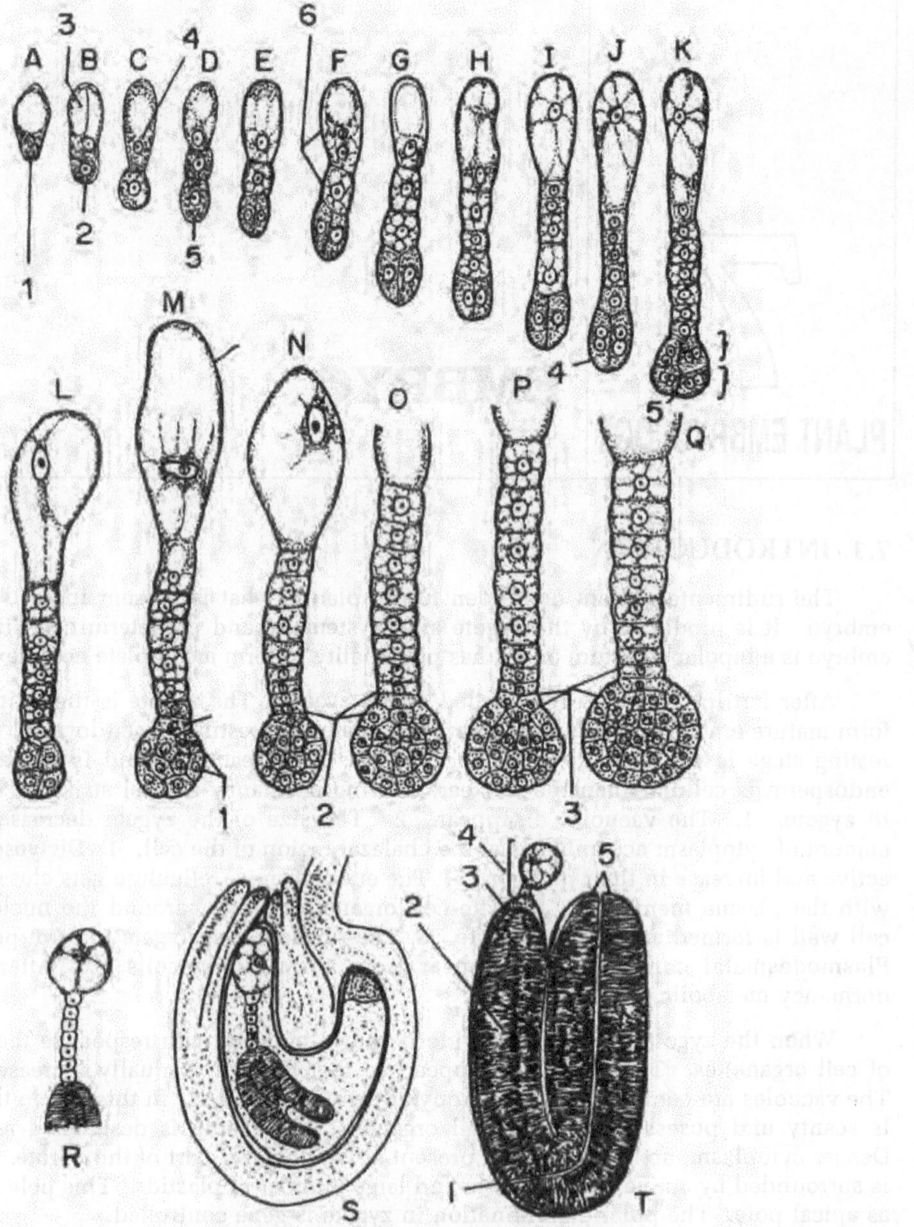

Fig 7.1 : (A - Q) Development of the embryo of *Capsella bursa-pastoris* (Cruciferae type) - 1. zygote, 2. apical cell (ca), 3. basal cell (cb), 4. basal dividing cell, 5. 3 celled proembryo, 6. suspensor cells, 7. quadrant, 8. octant, 9. dermatogen, 10. hypophysis, 11. hypophysis cells, 12. initials of periblem and pleunome.

S : Developing embryo in immature seed. ;

T: Mature Dicot embryo - 1. shjoot apex; 2. precambium, 3. tip of root, 4. suspensor, 5. cotyled ons

7.2 EMBRYOGENY IN DICOTS :

Johnson (1944) and Maheswari (1950) have classified the dicotyledonous embryos into six main types. They are 1. Crucifer or onagrad type (eg. cruciferae), 2. Asterad type (eg. Asteraceae), 3. Solanad type (eg. Solanaceae), 4. Caryophyllad type (eg. caryophyllaceae) 5. Chenopodial type (eg. chenopodiaceae) and 6. Piperad type (eg. piperaceae). The classification is mainly based on the plane of cell division in the apical cell at the 2-celled proembryo and the contribution of the basal cell (cb) and apical cell (ca) in the formation of embryo proper. Among 6-types crucifer type of embryogeny is described by Hanstain in *Capsella bursa-pastoris* (Cruciferae member) in detailed manner for the first time. Hence, in this text this model is discussed.

The zygote divides transversely and gives rise two unequal cells. The larger cell that situates towards micropylar side is known as basal cell (cb) and smaller cell that situates towards chalazal side is known as apical cell (ca) (fig-7.1).

The basal cell divides transversely giving two superposed cells (lower 'm' and upper 'ci' cells), where as the apical cell divides vertically giving two juxtopposed cells. As a result inverted T-shaped ('^') 4-celled proembryo is formed.

The two cells of apical cell undergo another vertical division and give rise a quadrant. The second vertical division is at right angles to the first. The cells of quadrant divide transversely giving 8-cells, the octant. In octant the 8-cells are arranged in two tiers of 4-each. The lower tier is designated as 'l' and upper tier is designated as l_1. The 'l' tier is destined to give rise the stem tip and cotyledons where as the upper 'l' is hypocotyle and radicle.

During octant stage the two cells, 'm' and 'ci' divide transversely and give rise a row of 6-10 suspensor cells. This elongated suspensor anchors the embryo to embryosac and pushes it into the endosperm. The upper most cell of suspensor becomes swollen and vesicular (v). It acts as haustorium. The lowest cell of the suspensor is designated as hypophysis (h). The hypophysis cell divides transversely and vertically and give rise 8-cells arranged in 2-tiers of four cells each. These give rise root cortex and root cap initials.

The 8-cells of an octant divide periclinally forming a peripheral dermatogen layer (it gives future epidermis). Again the 8-central cells divide another periclinal division producing another peripheral layer beneath dermatogen layer. The inner layer is composed of periblem and pleurome initials (gives future cortex and stele). The octant cells undergo few vertical and transverse divisions and give rise globose proembryo. In the central top region of the globose embryo group of cells act as shoot apex initials and either side two groups act as cotyledonary initials. The shoot apex initials have slow rate of cell division where as cotyledonary initials have faster rate of cell divisions. As a result of cell divisions in these groups, the globose embryo becomes heart shaped. The hypocotyle as well as the cotyledons soon elongate in size. Due to lack of sufficient place in developing seed the embryo becomes curved like horse — shoe shape.

The mature dicot embryo has an embryonal axis and two cotyledons. The cotyledons attach to the embryonal axis laterally. The portion of the embryonal axis above the level of cotyledons is known as epicotyle. The tip of epicotyle is occupied by meristematic tissue, known as plumule (embryonic shoot). The portion of the embryonal axis below the level of cotyledons is known as hypocotyle and terminal portion known as radicle (embryonic root). During seed germination the plumule along with epicotyle gives shoot system of the plant. Redicle gives root system. Hypocotyle portion that situates between epicotyle and radicle is a transitional zone of stem like structure. The cotyledons (food storage organs) digest and absorb the food from the endosperm and function as first photosynthetic leaves.

7.3. EMBRYOGENY IN MONOCOTS

In monocots the development of embryo varies with species to species. Several workers have described the embryogeny in various taxa. Among the monocots, the graminae family is highly evolved. The development and structure of grass embryo is differ from that of other monocots. As the embryo shows some unique features in development and structure, embryogeny in graminae members with reference to *Triticum* (Graminad type) is taken as example and discussed in this text.

The mode of development of embryo in *Triticum* has been studied and described in great detail by Batygina (1969). She has observed fixed pattern of embryogeny in four species of *Triticum*. Regular occurrence of oblique divisions in early embryogeny is characteristic feature of *Triticum* embryo.

In polarized zygote the first nuclear division is followed by oblique wall formation. The first division is unequal and gives rise smaller apical cell (ca), towards chalazal side and larger basal cell (cb), towards micropylar side (fig. 7.2). The cell 'cb' divides obliquely giving lower 'ci' and upper 'm' cells. The 'Ca' cell divides and gives rise two cells situate side by side in slanting position. The wall formation in 'Ca' cell is perpendicular to the

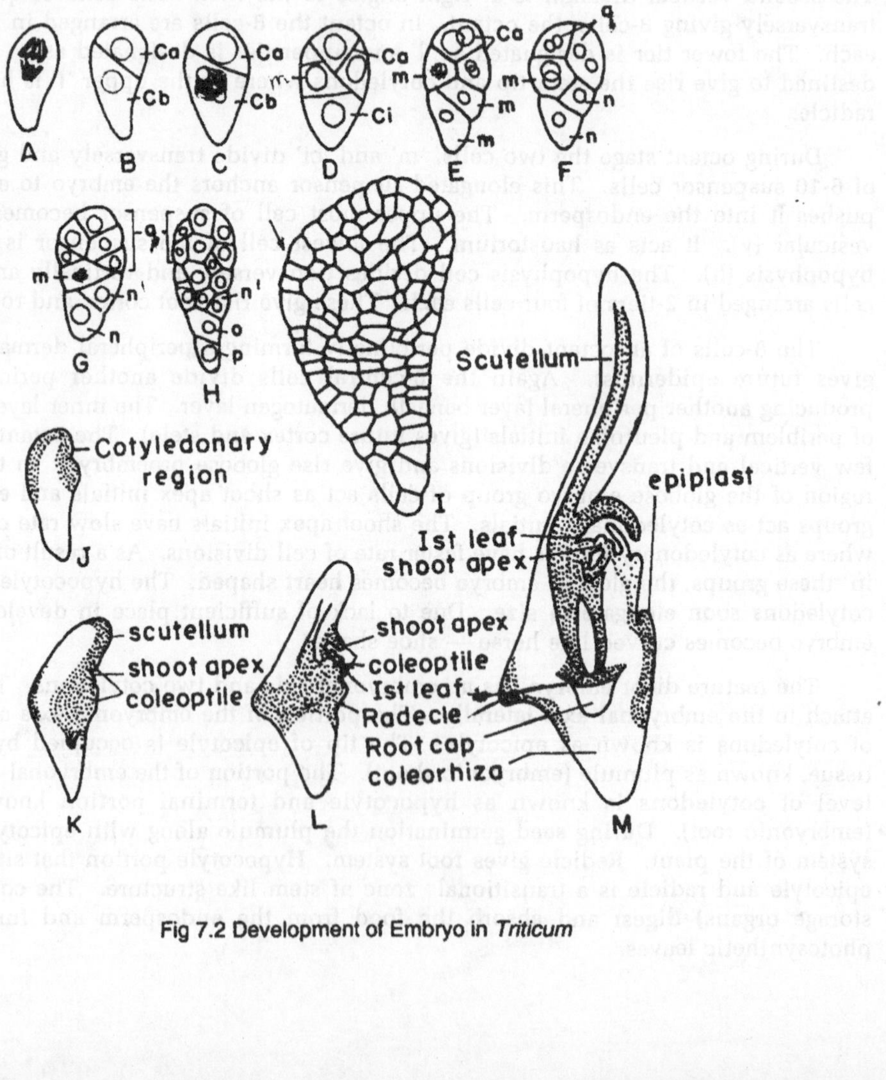

Fig 7.2 Development of Embryo in *Triticum*

first wall formed in zygote. As a result the four celled proembryo looks like roughly inverted T-shaped ('^'). The orientation of walls in 'T' shaped proembryo is quite different from that of 'T' shaped proembryo of dicots and other monocots. The orientation characterizes the *Triticum* (Wheat) embryogeny.

The 'Ci' cell divides and gives rise 'n' and 'n[1]' cells. In this division the cell wall orientation is similar to first zygotic division. The two cells of 'Ca' divide and give rise a typical quadrant 'q'. The plane of cell division is as first division in 'Ca' but at right angles to it. The cell 'm' divides vertically forming two cells. Up to this stage the cell wall orientation is typical. Further divisions occur in various planes.

After formation of 32-celled proembryo, the organogenesis begin. Single cotyledon or scutellum is the first organ that is initiated at first. Its differentiation begins with the growth in the apical-lateral region (arrow marked) of the proembryo (fig-7.2 I). During this growth one group of cells belonging to 'q', 'ni' segments may act as cotyledonary initials. As these initials divide actively in various planes, single cotyledon is formed. Appearance of other cotyledonary group is suppressed at an early embryonic stages, and it is gene controlled. As the cotyledon elongates a notch or constriction appears opposite to it in a 'Ca' sector. The constriction demarcated the scutellum from rest of the embryonic tissue. The shoot apex primordial and coleoptile primordial appear very close to the constriction. The radicle differention begins endogenously in the central portion of the embryo. Slowly root cap initials appear around the meristematic radicular portion. The terminal cells of the 'q' divide and give rise epicotyle. The leaf primordia originate beneath the shoot apex.

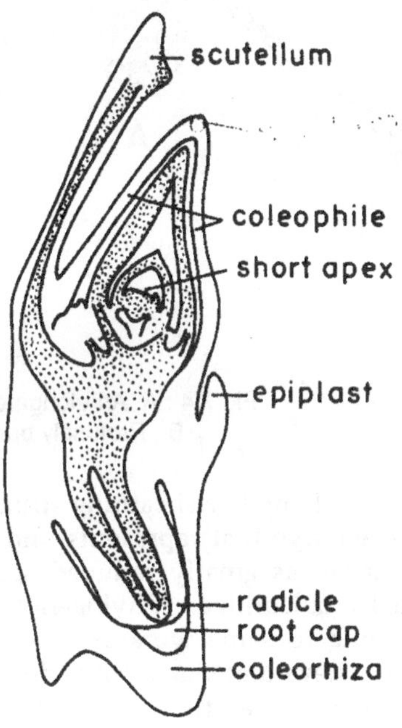

Fig 7.3 L.S. of *Triticum mature* embryo

In median longitudinal section the embryo shows embryonal axis and laterally attached single cotyledon (scutellum) (fig-7.3). The axis that situates at the level of scutellum is known as epicotyle. It comprises of shoot apex, leaf primorida and hollow foliar structure, the coleoptile. The coleoptile has an apical pore. The portion of the embryonal axis below the level of scutellum is called as hypocotyle, which ends with radicle. The radicle is surrounded by root cap. The radicle and root cap regions are enclosed by coleorhiza. Coleorhiza is the sheath like structure which represents undifferentiated lower part of the proembryo. Opposite to the scutellum the coleorhiza has small outgrowth called epiblast.

There were different opinions among scientists regarding unique structures, such as scutellum, coleoptile, epiblast and coleorhiza which are found only in grass embryo but not in other monocots. According to Arber (1934) the scutellum and coleoptile jointly constitutet the cotyledon. It is believed that the cotyledon is modified in such a way to perform two different functions. The scutellum absorbs the food materials from the surrounding tissue, the endosperm. The finger like projections of the cells of scutellum involves in absorption. The coleoptile, the cotyledonary sheath protects the shoot apex.

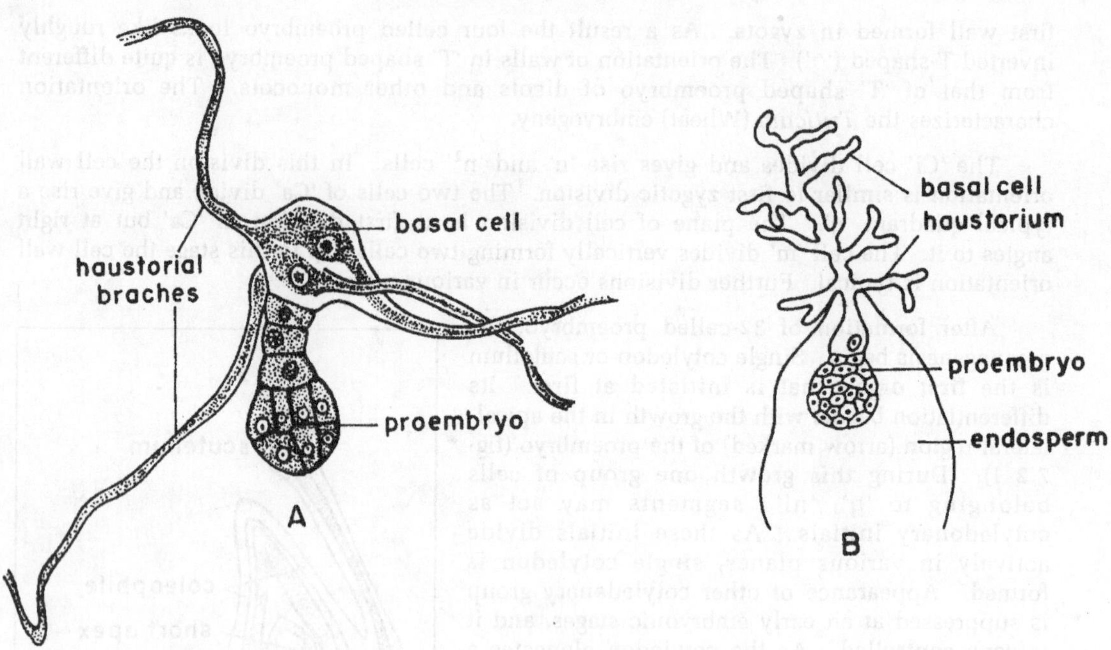

Fig 7.4 : A: An elongated tubular haustorial branches in *Dicraea stylosa*
B : Profusely branched haustorium in *sedum ochroleucum*

Presence of another sheathing structure the coleorhiza is anther peculiar feature of *Triticum* (grass) embryo that represents undifferentiated lower part of the proembryo. The epiblast is regarded as greatly reduced second cotyledon (coulter, 1955) but not the out growth given by coleorhiza. Phylogenetically monocot embryo is advanced over dicot embryo. According to some workers one cotyledon is suppressed during development, where as others believe that single cotyledon is formed by the union of two cotyledons. In conclusion one can say that the grass embryo shows phylogenetic significant features over other monocots.

7.4. SUSPENSOR

The suspensor is present at the radicular end of the proembryo. It attaches the embryo to the embryosac and pushes it into the endosperm during development. It is regarded as temporary embryonic root. Because it involves in sucking and transportation of food materials to the developing embryo. The suspensor is an ephemeral structure and in mature seed it is found as remnants.

The suspensor shows great variation with respect to its size, shape and number of its component cells. It may be single celled, uniseriate filament of 5-10 cells; biseriate and much elongated with approximately twenty pairs of cells; multi seriate and massive structure or may look like bunch of grapes. In *Pisum* the suspensor is composed of multinucleated (coenocytic) much elongated tubular cells placed in two pairs.

In several taxa the suspensor cells show pronounced increase in their size. In *Dicraea* during early embryogenesis, the basal cell swells and produces four elongated tubular haustorial branches. They penetrate in between two integuments. (fig-7.4). In some members the suspensor haustorial branches may reach into the endosperm, micropyle and other ovular tissues. One good example can be cited here to explain their extent of development and the haustorial nature. In *Sedum ochroleucum*, a member of Crasullaceae, at two celled proembryo stage the basal cell swells to form vesicle like structure. This vesicle penetrates through the nucellus and produces tubular brnaches on either side of the embryosac. The branches penetrate into inner integument and digest its cells by enzymatic action. The branches grow further in between the integument and raphae. In this region the branches produce several branched hyphal filaments which look like fungal mycelium. They spread inter as well as intra cellularly like fungal hyphae.

Various workers have reported different substances such as amino acids, proteins, carbohydrates, polysaccharides, RNA, Large number of mitochondria, endoplasmic reticulum, plastids along with maximum accumlation of gibberellic acid in suspensor cells of different plants. Alpi *et al* (1975) have observed 30 times more Gibberellic acid in suspensor cells than in heart shaped embryo of *Phaseolus coccineus*. After complete formation of cotyledons the suspensor cells show the symptom of degeneration. In this stage the gibberellin quantity is very poor in suspensor and very high in embryo proper, this suggests the translocation of gibberellic acid from suspensor in to the embryo.

7.5. NUTRITION OF EMBRYO

7.5.1 Endosperm : It is the chief nutritive tissue of the embryo. The endosperm stores food material in different forms such as carbohydrates, proteins and fats. The tissue is also rich in inorganic ions, amino acids, sugars, growth regulating substances such as IAA and cytokinins etc. The endosperm has the capacity to develop micropylar as well as chalazal haustorial structures, which involve in the absorption of nutrients from various parts of the ovule.

7.5.2 Perisperm : The food laden nucellar tissue that remains persists in the seed around the endosperm is known as perisperm. (fig-7.5C). Generally the nucellus is consumed

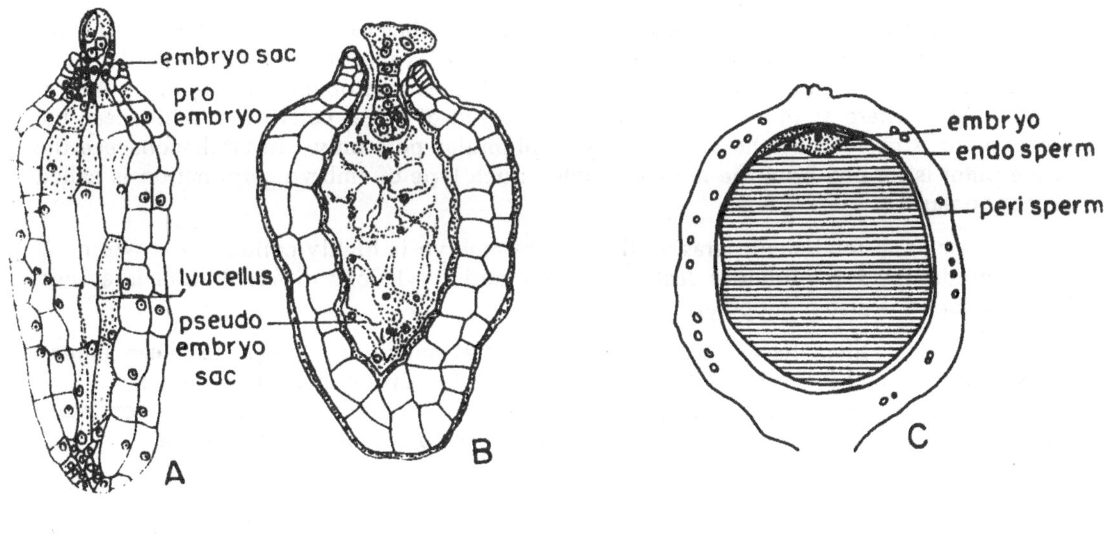

Fig : 7.5 A : Formation of psendoembryosac, B : Proembryo at quadrant stage, C: L.S. of piper nigrun fruit (Pre-globular stage)

by developing embryosac as well as embryo. But in some members such as Piperaceae, Amaranthaceae; Capparidaceae the nucellus remains in the seed. In P*eper nigrum* the nucellar cells below the embryosac divide faster, increase in their size. They contain denser protoplasm and store food materials in the form of oil globules, starch etc. The endosperm is present between embryo and perisperm. The nutrients present in perisperm pass through the endosperm and reach to the embryo. In piper 95% of the seed is occupied by perisperm.

7.5.3 Chalazosperm : The food storing prominent tissue that appears at the chalazal portion at the time of development of embryo and supplies nutrients to it is known as chalazosperm. In *cyanastrum* endosperm and major portion of nucellus degenerate during seed development. In such conditions the nucellar cells at the chalazal region above the vascular trace divide actively and give rise loosely arranged cells. These cells store the food materials in the form of starch and fat.

7.5.4 Pseudoembryosac : The gaint embryosac like structure that is formed by the disintegration of nucellar cells below the actual embryosac, and take part in the nourishment of develping embryo is known as pseudoembryosac. At the time of formation of embryosac the nucellar cells below it elongate. Their cell walls become very thin and disappear to form a cavity with multinucleated protoplasm and reserve food materials (fig. 7.5 A, B). In Podostomaceae members endosperm is absent. Its role is taken up by pseudoembryosac.

7.5.5 Suspensor Haustoria : In some members the endosperm is either poorly developed or absent. In such cases the suspensor produces active haustorial structures. They absorb food materials from various parts of the ovule and supply to the developing embryo.

7.7.6 Under Developed (Reduced) Embryos : The developed embryo has an embryonal axis with plumule and radicle along with laterally attached one or two cotyledons. But in some members such as Orchidaceae, Orobanchaceae, Balanophoraceae, Convolvulaceae etc, the organogenesis is failed during embryogeny. As a result, underdeveloped and reduced embryos are formed. Such embryos are mainly found in saprophytic and parasitic genera.

In Orchidaceae and Orobanchaceae the mature seed has ovoid or sub globose embryo. The pole of the embryo that situates towards micropyle is morphologically radicular pole. The chalazal pole is morphologically plumular pole. The radicular pole is composed of smaller cells, where as the chalazal pole is composed of larger cells. In Orchidaceae the entire plant is derived by the sole activity of plumular pole only. In Orobanchaceae the entire plant is formed from the radicular pole. Such type of embryo germination is called as monopolar.

In Cascuta, a total stem parasite, the mature embryo is highly coiled like a worm. It is surrounded by few plumular scales. Radicle and cotyledons differentiation is absent. Its germination is also monopolar.

In Eranthis hiemalis (Ranunculaceae) a seed at shedding stage has pear shaped undifferentiated embryo without organ initiation. But it has an elongated suspensor. Few weeks after falling the seed on the ground, the embryo undergoes complete differentiation and shows dicot embryo strucuture. Then only the seed germination takes place.

QUESTIONS

A. Answer the following in about 30 lines.

1. Describe the development of Dicot embryo.
2. Describe the development of monocot embryo.

B. Answer the following in about 10 lines.

1. Structure of Triticum embryo
2. Suspensor
3. Reduced embryoes

C. Answer the following in about 3-5 lines

1. Proembryo
2. Hypophysis
3. Embryonic root
4. Scutellium
5. Epicotyle and hypocotyle
6. Coleoptile and coleorhiza
7. Suspensor haustorium
8. Pseudo embryosac
9. Perisperm
10. Chalazosperm
11. Dermatogen
12. Periblem and pleurome

8

PLANT EMBRYOLOGY

POLYEMBRYONY

8.1. INTRODUCTION

The phenomenon of occurrence of more than one embryo in a developing seed is known as polyembryony (poly = many; embryony = embryogeny). It is quite common in Gymnosperms, but rare in Angiosperms. Its occurrence was reported for the first time in orange seeds by Antony Van Leeuwenhock (1719). However, the term was coined by Brown (1859). Even through more than one embryo is formed in the developing seed, the additional embryos do not reach to the maturation stage. They may remain arrested at an early stage of development or may degenerate during further maturation of seed. As a result, in general the mature seed posses single well developed embryo. However, there are some reports in the literature regarding germinable biembryonate seeds (fig-8.1). Except in few members (*Citrus*, *Mangifera*) poly-embryony occurs as an abnormal feature. Different types of twins such as diploid-diploid (Randolph, 1936; Hagerup, 1947); diploid- haploid (Kappert, 1933; Ramaiah *etal* . 1935), diploid-triploid (Yamamoto) were reported in various taxa.

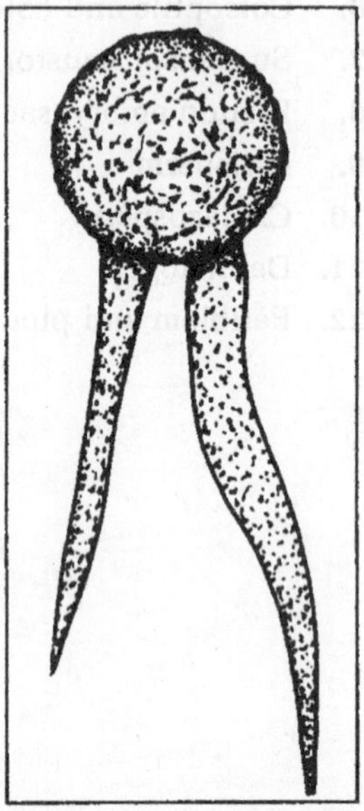

Fig 8.1 Biembryonate seed

8.2 TYPES OF POLYEMBRYONY

On the basis of their origin the polyembryony is recognized into two types. They are spontaneous polyembryony and induced polyembryony (table 8.1).

8.2.1 Spontaneous Polyembryony : The naturally occurring polyembryonic state is known as spontaneous polyembryony. Ernest (11918), and Schnarf (1929) have recognized two types of spontaneous polyembryony. They are true polyembryony and false polyembryony.

TRUE POLYEMBRYONY : In this type of polyembryony the additional embryos either may be derived from cells of embryosac or sporophytic cells (nucellus / integuments). The true polyembryony is further classified into cleavage polyembryony and adventive polyembryony.

a. Cleavage Polyembryony : Formation of additional embryos, due to the cleavage or proliferation of proembryo is known as cleavage polyembryony. It is very common in Gymnosperms and in few Angiosperms (orchids). In broader sense the development of additional embryos from the cells of embryo sac such as synergids and antipodals along with normal zygotic embryo is described as cleavage polyembryony.

TABLE — 8.1
POLYEMBRYONY

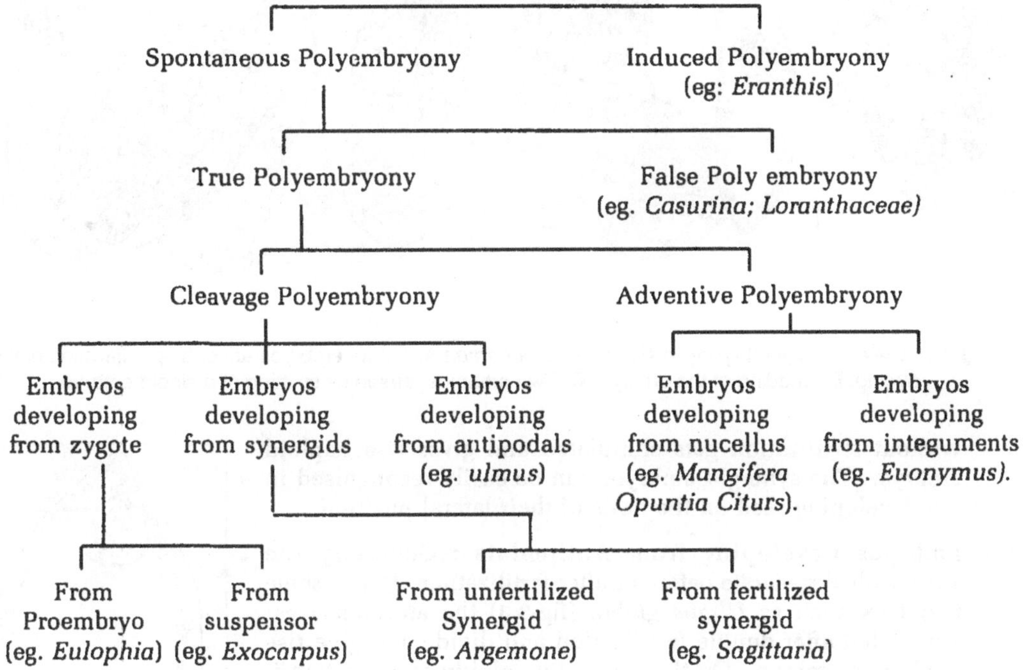

i) **Embryos developing from Proembryo** : Swamy (1943) worked extensively on *Eulophia epidendrae* a member of orchidaceac, and recorded three different modes of supernumerary embryo formation as described below (fig. 8.2. A,B,C)

1. The zygote divides irregularly to form a mass of cells. The cells lying towards chalazal end divide independently and give rise two to three embryos.

2. The proembryo may produce small bud like structures from its tip or periphery which grow into embryos.

3. The tissue of proembryo cleaves into few filamentous structures, which ultimately develop into independent embryos.

Suspensor polyembryony is reported in *Exocarpus*. During embryogeny the suspensor cells proliferate and act egg like which divide and give rise to additional embryos.

ii) **Embryos developing from Synergids** : Apart from haustorial behavior, the synergids behave abnormally and take part in the formation of additional embryos. In *Sagittaria* the synergid gets fertilized during normal process of fertilization. This is brought about by the entry of more than one pollen tube into the embryosac. In such instances both developing embryos are diploid. In some members such as *Argemone* the synergid

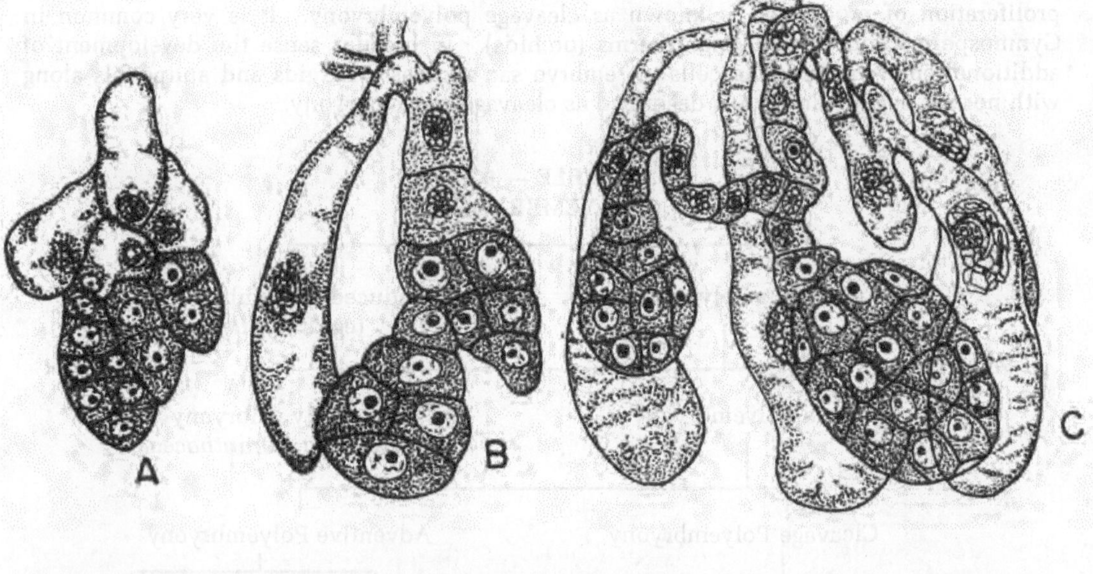

Fig 8.2 : Clevage polyembryony in *Eulophia epidendraea* A. Three embryos arisenby proliferation of the gygote. B. Budding of the embryo, C. Two embryos arisen by splitting of a single embryo

without fertilization gets stimulated and gives rise haploid embryo. The synergid embryos can be easily recongnised in the developing seed on the basis of their lateral position.

iii) Embryos Developing from Antipodals : Generally the antipodals degenerate before or after fertilization. But in some members such as *Ulmus glabra* (fig-8.3) the antipodals get stimulated after double fertilization and divide and give rise to globose embryo. On the basis of its ploidy and position it can be easily distinguishable from actual zygotic embryo.

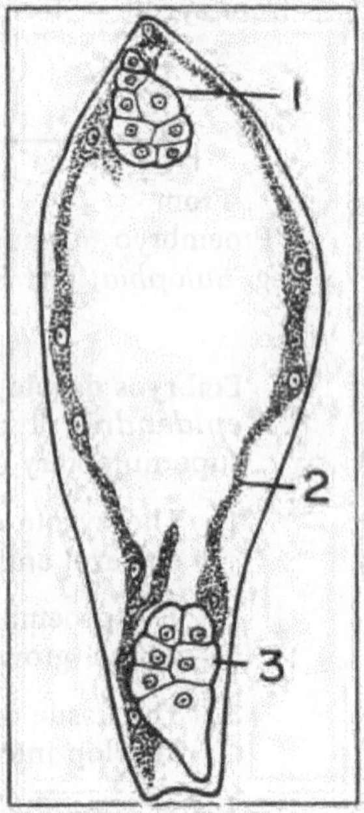

Fig 8.3 Embryosac of *Ulmus glabra*
1. Zygotic embryo
2. Endosperm
3. Antipodal embryo

8.2 ADVENTIVE POLYEMBRYONY

The development of additional embryos from the maternal sporophytic tissue such as nucellus and integuments is described as adventive polyembryony.

i) Embryos developing from nucellus : Nucellar polyembryony is very common in *Mangifera Opuntia* and *Citrus* in which crassi nucellate ovules (nucellus well developed) are present. In *Mangifera* at the time of fertilization the peripheral cells of the nucellus around the embryosac become more prominent with respect to their size, denser protoplasm with starchy contents. After fertilization such differentiated nucellar cells behave like egg and produce heart shaped embryos around embryosac. They gradually pushed into the endosperm during further seed maturation (fig-8.4). The adventive embryos do not show synchronous development.

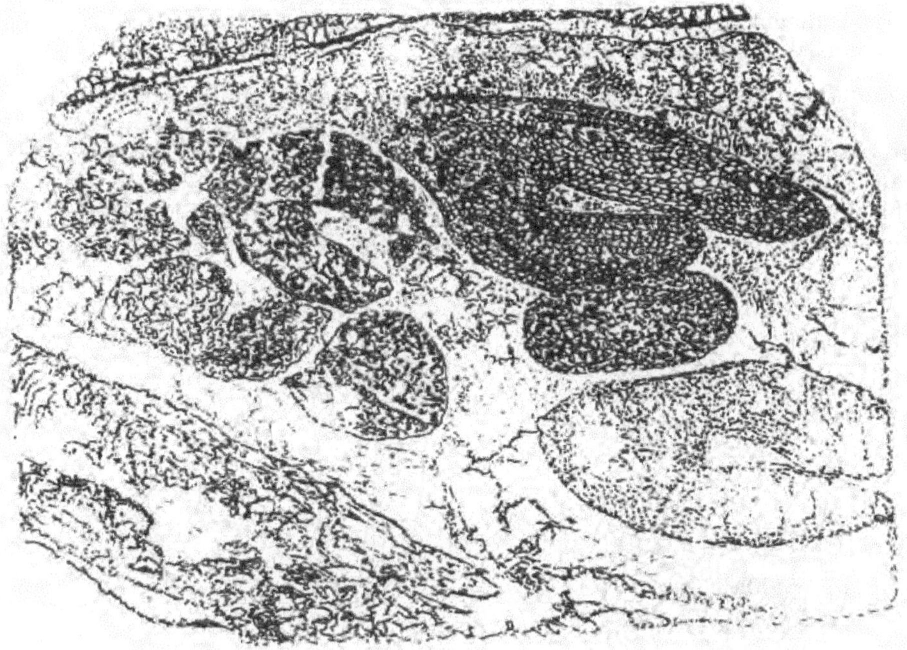

Fig 8.4 L.S. of *Citrus microcarpa* seed showing nucellar embryos in various stages of development

It is interesting to note that in *Opuntia dillenii* egg apparatus, antipodals and polar nuclei degenerate by the time of fertilization. There is no chance in the formation of zygotic embryo. The entry of pollen tube stimulates some of the nucellar cells present around embryosac. They divide and give rise large number of adventitious embryos, but ultimately only one enters into maturation stage (Bhatnagar and Bhojwani, 1974).

In *citrus unshie* as many as 40 additional embryos have been reported.

ii) **Embryos developing from Integuments :** Integumentary embryos are reported in tenuinucellate ovules in which the quantum of nucellus is very less. In *Spiranthes australis* a member of orchidaceae, both male and female gametophytes are non functional and actual process of fertilization is lacking. (swamy, 1948). In this genus the integumentary cells of epidermal sub epidermal layers become egg like, proliferate and give rise to 2-6 integumentary embryos.

B) FALSE POLYEMBRYONY : The development of more than one embryo in a seed due to occurrence of many embryosacs is known as false poly embryony. This situation arises due to the formation of more embryosacs in the ovule and entry of separate pollen tubes in to separate embryosacs (eg. *Casurina, Citrus*). Multiple embryoscs in a single ovule may arise from same megaspore or two different megaspores or from nucellar cells as an abnormality.

8.2.2 Induced Polyembryony : Experimentally created poly embryony in the laboratory conditions by using different chemical substances is known as induced poly embryony. Different chemical substances such as 2, 4-D; 2,4,5-T; N.A.A. and Acidic buffers (PH-4.0) are used in induction of polyembryony.

In *Eranthis hiemalis* the seed at shedding stage contain undifferentiated pearshaped embryo. After shedding the differentiation begins while the seed is in soil. Haccius (1955) obtained twin embryos from freshly harvested *Eranthis* seeds by treating them with chemicals (fig-8.5). Along with the chemical substances the other methods employed for the artificial production of polyembryony are a) delayed pollination b) subjection of flower to low or high temperatures just after fertilization and c) application of x-rays or foreign pollen.

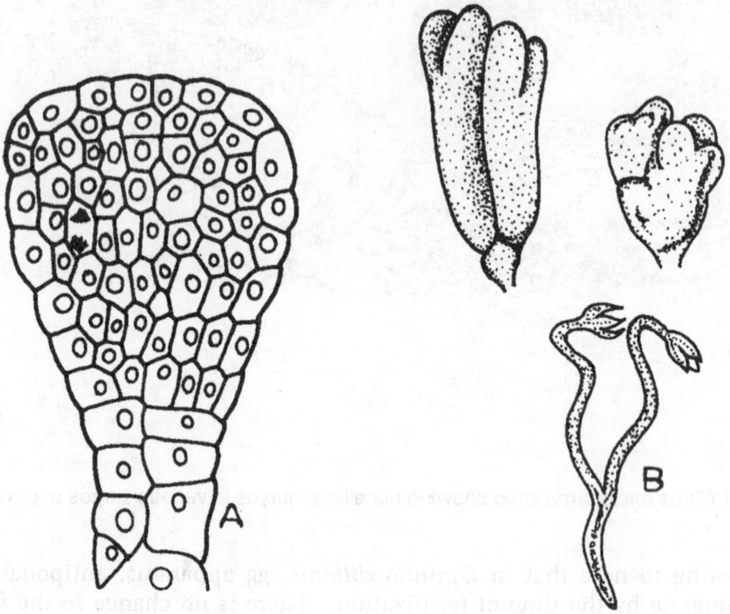

Fig 8.5 A : pear shaped embryo of *Eranthis* seed, B: Induction of twin embryos by chemicals

In nature the formation of embryo is restricted to the ovular tissues. It was assumed that embryos require special physical and chemical environment which is present only inside the "magic bath" of embryosac. By providing suitable environmental and nutritional conditions the cells of the plant body can be stimulated to grow into viable embryos. Such embryos formed in tissue culture method are variously interpreted as embryoids or somatic embryos or accessory embryos or supernumerory embryos or adventive embryos. Embryoids have been obtained from the cells of zygote (*Solanum*), nucellus (*Citrus*), anther (*Datura*) stem (*tobacco*), root (*carrot*), fruit (*Cucurbita*) and leaf segments.

In vitro development of embryoids in carrot is a two step process. First step requires proliferation medium (pf. Medium) which contains 1 mg L^{-1} of 2,4,-D, an inducing agent.

Hence pf-medium is also called as induction medium. Second step requires embryo development medium which contains 0.1 mg L^{-1} of Auxin, a growth stimulating substance. The callus is initiated on proliferation medium. After some period localized groups of meristematic cells called embryonic clumps (EC) appear on the callus. In repeated sub culturing the embryonic clumps continue to multiply without maturation into embryoids. If these embryonic clumps are transferred on to the embryo development medium, each clump develops in to an embryoid.

8.3. CAUSES OF POLYEMBRYONY

Different opinions were put farward by embryologists to explain polyembryony in various ways.

Haberlandt (1922) proposed the "necrohormone theory". According to him the degenerating nucellar cells may produce stimulating substances called necrohormones. These substances may induce the adjacent cells to divide and form adventive embryos during normal zygotic embryogenesis.

Leroy (1947) and Baker (1960) explained recessive gene theory. According to them set of recessive genes are responsible for polyembryonic condition in few taxa. In *Mango* the mono embryonatae forms possess dominant genes whereas polyembryonatae forms possess recessive genes. The set of recessive genes might be resulted and expressed their character after recombination during hybridization.

According to Frusato *et al* (1957) different factors such as 1) Age of the tree (more embryos are formed in older trees) 2) nutritional status (more embryos are formed with increased food supply) and 3) Fruit set (higher in years of higher fruit set) are responsible for the occurrence of polyembryony in citrus.

8.4. SIGNIFICANCE OF POLYEMBRYONY

Polyembryony has significant role in horticulture, cytology, genetics and plant breeding.

1. Nucellar adventive embryos provide uniform seedlings of parental type.
2. In vegetative propagative members the vigour may lost after repeated propagations by cutting. In such members the nucellar adventive embryos provide vigorous generation.
3. The nucellar embryos are free from diseases. In monoembryonatae citrus cultivars virus free clones can be raised by culturing nucellus.
4. Haploids are of great use in cytological studies.
5. Homozygous diploids can be raised from haploids by colchicine treatment.

QUESTIONS

A. Answer the following in about 30 lines.

1. What is polyembryony ? Explain different kinds of polyembryony with suitable example.

B. Answer the following in about 10 lines.

1. Causes of polyembryony
2. Adventive polyembryony
3. Experimental induction of polyembryony
4. Significance of polyembryony.

C. Answer the following in about 3-5 lines

1. **Spontanious and induced polyembryony.**
2. **False polyembryony**
3. **Callus**
4. **Embryoids**
5. **Why the embryosac is called as "Magic bath"**
6. **Nectrohormone theory**
7. **Adventive embryoes**
8. **Proliferation medium**
9. **Cleavage polyembryony**
10. **Recessive genetheory**
11. **Embryonic clumps**

9

PLANT EMBRYOLOGY

POLYNOLOGY

9.1 INTRODUCTION : "The branch of science that deals with the study of pollen grains is known as polynology". The word polynology was coined by Hyde and Williams (1845). The pollen grain represents first cell of male gametophytic generation. The pollen grains are formed in microsporangium of anther lobes. Well matured pollen grain contains denser cytoplasm with haploid nucleus. The spore is covered by thick spore coat called sporoderm. The sporoderm shows some weak areas known as apertures. The sporoderm shows various ornamentations. The pollen grains exhibit great deal of variation with respect to their size, shape, sporoderm features etc. (fig. 9.1).

9.2 POLLEN MORPHOLOGY

To study the detailed morphological features of pollen grains one has to concentrate on 1) Polarity 2) symmetry 3) size 4) shape 5) Apertures 6) Pollen wall stratification and 7) Exine ornamentations.

9.2.1 Polarity : Usually the pollen grains are formed in tetrads. In a tetrad the surface of the pollen grain that faces towards the center (equatorium) is known as proximal pole (fig. 9.2 A). The pole apposite to the proximal pole (p.p) is known as distal pole (d.p.) A hypothetical line that joins proximal and distal poles is known as polar axis (p.a.). Another hypothetical line that can be drawn at right angles to the plar axis within the pollen grain is known as equatorial axis (e.a.). If the equatorial axis divides the pollen grain in to two equal parts, such pollen grain is termed as isopolar pollen grain. If the pollen grain is not divided into equal parts such grain is called as heteropolar pollen grain. The schematic illustration of a pollen grain is called "Polynogram" (fig. 9.2).

9.2.2 Symmetry : The pollen grains of equal size and shape are called symmetric pollen grains. The pollen grains of unequal size and shape are called asymmetric pollen grains. The symmetric pollen grains may be radial or bilateral.

9.2.3 Size : The size of the pollen grains varies from 10 ìm (*Myosotis alpestris*) to as much as 200 ìm (*Cucurbita pepo*). To measure the size of the pollen grain, polar diameter (p) and equatorial diameter (E) are taken in to consideration. On the basis of diameter pollen grains are classified in to six groups (Erotman, 1940) namely perminute (less than 10 ìm), minute (10-25 ìm), media (25-50 ìm) Magna (50-100 ìm), Permagna (100-200 ìm) and giganta (more than 200 ìm).

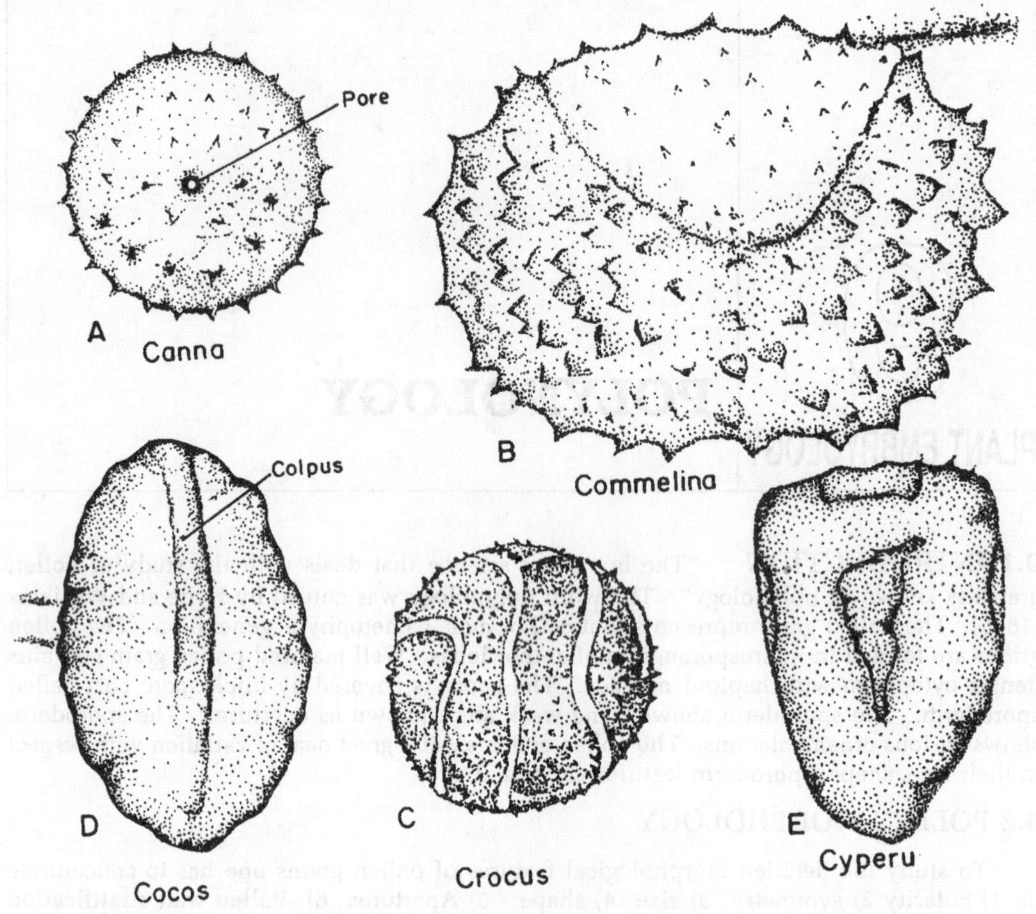

Fig 9.1 Various kinds of Pollengrains showing size, shape and ornamentation

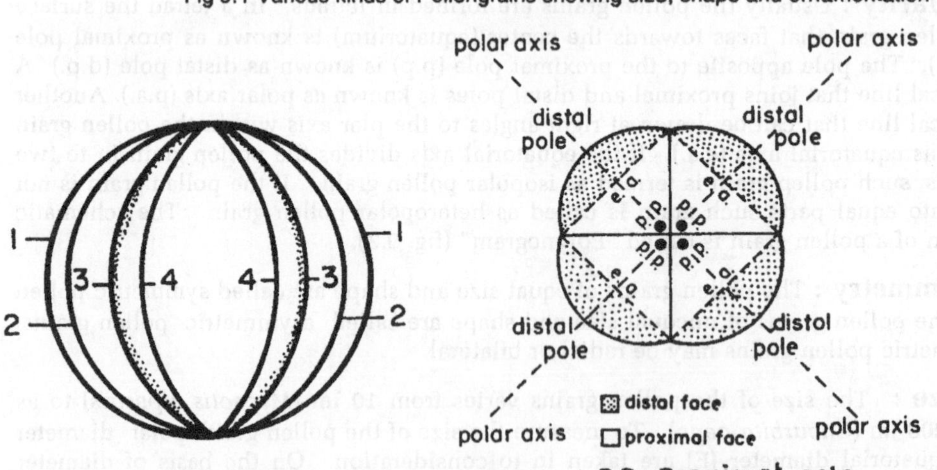

Fig 9.2A Polarity of the pollen tetrad; P.P. - Proximal pole; B: Pollen shapes 1. Shperical 2. Sub-prolate, 3. Prolate 4. Perporolate

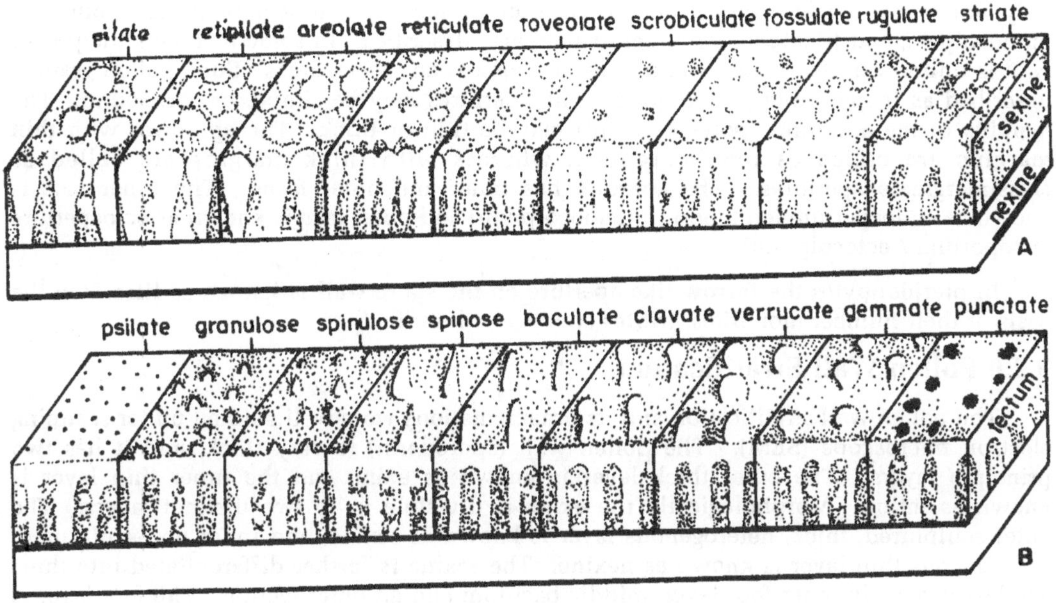

Fig 9.3 : Exine strata and surface patterns of pollen grains

9.2.4 Shape :

The shape of the pollen grains varies from spherical, (fig: 9.2-B) oval, oblong, lenticular, triangular, flat and so on. The shape of the pollen grain is determined by multiplying the ratio of polar diameter (P) and equatorial diameter (E) by 100. (i.e. P/E X 100 = shape of Pollengrain). On the basis of above equation 8 kinds of shapes are determined as mentioned in table 9.1.

TABLE : 9.1

S.No.	Shape	P/E X 100
1.	Peroblate	Less than 50
2.	Oblate	Between 70-75
3.	Sub oblate	Between 76-88
4.	Oblate-Spheroidal	Between 89-100
5.	Prolate-Spheroidal	Between 101-114
6.	Sub Prolate	Between 115-133
7.	Prolate	Between 134-200
8.	Per prolate	Above 200

9.2.5 Apertures :

The weak areas which are present on pollen surface and are directly or indirectly associated with pollen germination are called as apertures. The apertures play an important role in polynology and taxonomy. They may be small and circular or elongated and furrow like. Small opertures are called as pores whereas elongated furrows are known as colpi (fig: 9.1). The pollen grains with pores are termed as porate, where as with colpi are termed as colpate pollen grains. The pollen with pores and colpi are described as colporate. The pollen grains are described as monoporate (one pore), triporate (with 3 pores) and poly porate (with many pores) on the basis of presence of number of pores. Similarly monocolpate, and tricolpate pollen grains are also described. In some members the pollen wall is uniform in its thickness and does not possess any weak areas. Such pollen grain is described as inaperturate pollen grain. The apertures may be simple or compound. The margins of the apertures (pores or colpi) may be thin or thick. The apertures with thin margins are called as tenuimarginate, whereas with thick margins are called as crassimarginate apertures. The aperture has inner and outer faces. The inner face is known as endoaperture (endoporium/endocolpium) and outer face is known as ectoaperture (ectoporium / ectocolpium).

In pteridophytes the furrow like aperture on the spore wall is known as lites. On the basis of their number monolites, trilites spores are described.

9.2.6 Pollen Wall Stratification :

The pollen wall stratification and its ornamentations were well studied under scanning electron microscope (SEM). The pollen wall (sporoderm) is very thick, and it has two principle layers. The outer thick layer is known as exine and the inner thin layer is known as intine. Morphologically the exine is divisible in to two distinct layers. The outer sculptured, thick, heterogeuous layer is known as sexine; and inner non-sculptured, homogenous thin layer is known as nexine. The sexine is further differentiated into three sub layers namely inner foot layer, middle baculum and an outer tectum. The foot layer is some times described as an outer layer of nexine. But its chemical composition is similar to sexine. The baculum (columellae) is composed of rod like elements, and it is over roofed by tectum or tegellum. Actually the tectum is the layer that is formed by the union of terminal heads of rod like (columellae) elements. The tectum may be smooth or shows various types of processes.

9.2.7 Exine Ornamentation :

Exine surface shows various types of ornamentations. The sexine (ectine) of the exine is composed of radial rods (columellae). When the distal surfaces of the rods are free and bright, and the intervening walls are dark, the pattern is called as pilate (fig. 9.3. A). The rods in most of the grains are fused to form different types of patterns having depressed areas (called lumina) and the inter vening areas between lumina (called muri). When a network is produced by lumina and muri it is called as reticulate. In the reticulate pattern, if the rods are incompletely fused, the pattern is called as retipilate. If the lumina are circular and closely placed it is called as foveolate; if they are distantly placed it is called as scrobiculate. Some times the lumina may be elongated (fossulate), parallel (striate), anastamosing (regulate) and so on

The exine posses various kinds of outgrowths (excrescences). The out growths may be in the form of minute granules (granulose) (fig-9.3 B) or in the form of small spines (spinulose). When the outgrowths on the exine are in the form of tubercles, it is called as tuberculate. If the tubercles are long with pointed ends, the pattern is called as spinose, if they are rod shaped it is called as baculate and when they are club shaped or drum stick like the ornamentation is described as clavate. In some cases the out growths may be rounded and wart like without constriction at the base (verrucate) or wart like with constriction (gemmate).

The exine ornamentation pattern is much complicated and attractive in insect pollinated members.

9.3 NPC SYSTEM

The apertures play an important role in the classification of pollen grains. For the

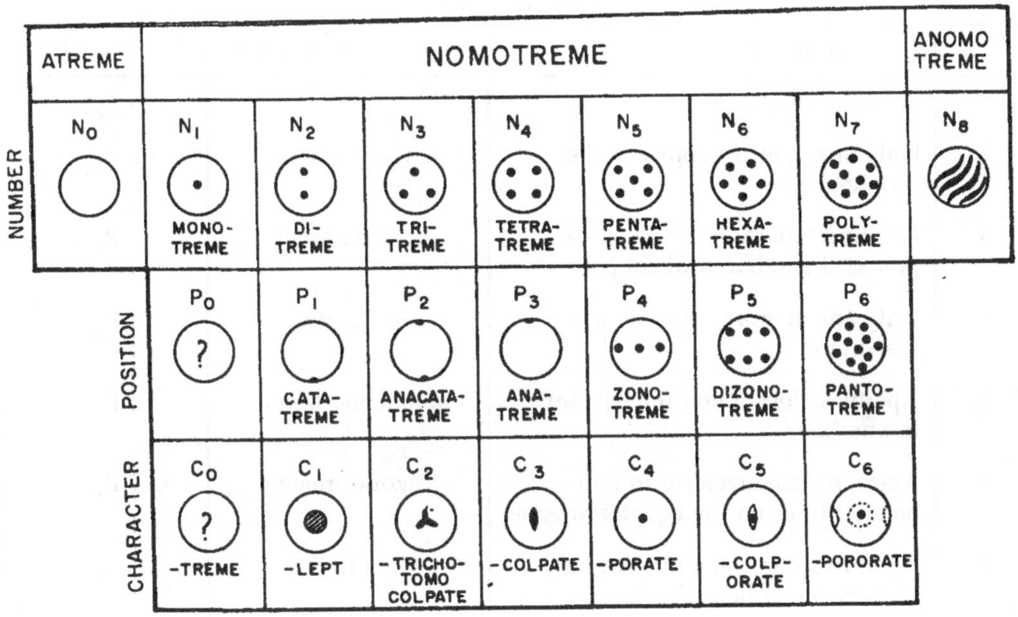

Fig 9.4 : NPC System

first time Erdtman (1969) has given importance to apertures and proposed NPC system of classification of pollen grains. NPC system is based on the number (N), position (P) and characters (C) of the apertures. (fig- 9.4). Under this system the term "treme" has been recomended in place of "aperture" while preparing keys for the classification.

9.3.1 Number of Apertures : On the basis of number of apertures 9 types of pollen grains are recognized as mentioned in table 9.2.

S.No.	Number of Apertures	Type of Pollen Grain	Designated as
1.	Nil	Atreme	N_0
2.	One	Monotreme	N_1
3.	Two	Ditreme	N_2
4.	Three	Tritreme	N_3
5.	Four	Tetratreme	N_4
6.	Five	Penta treme	N_5
7.	Six	Hexatreme	N_6
8.	Seven	Poly treme	N_7
9.	Irregular	Anomo treme	N_8

9.3.2 Position of Apertures : On the basis of position of the apertures the pollen grains are classified into 7 types as mentioned in table 9.3.

TABLE — 9.3

S.No.	Position of aperture	Type of Pollen grain	Designated as
1.	Position unknown	?	P_0
2.	Only one aperture found at the proximal pole	Catatreme	P_1
3.	One aperture found at proximal pole and another at distal pole.	Anacatatreme	P_2
4.	Only one aperture present at distal pole	Anatreme	P_3
5.	Apertures located on the equator in one line	Monozono treme	P_4
6.	Apertures are arranged in two parallel lines on the equatorial zone	Dizono treme	P_5
7.	Apartures distributed uniformely all over the pollen surface.	Pantotreme	P_6

9.3.3 Characters of Apertures

On the basis of characters of apertures 7 types of pollen grains are recognized. (Table — 9.4)

TABLE — 9.4

S.No.	Character of aparture	Type of Pollen Grain	Designatred as
1.	Character not known	?	C_0
2.	One thin split aperture (Leptoma) is present	Mono lept Catalept (Leptoma on proximal face) / Analept (Leptoma on distal face)	C_1
3.	Pollen wall has 3 slit colpus	Trichotomocolpate	C_2
4.	One to many unbranched colpi	Colpate	C_3
5.	One to many circular single pores	Porate	C_4
6.	Pollen wall with single pore and colpi	Colporate	C_5
7.	The pore outer face is (ectoporium) narrow and inner face (endoporium) is broader	Pororate	C_6

9.3 POLLEN MORPHOLOGY OF HIBISCUS :

The pollen grains of *Hibiscus* are spherical with a diameter of 73.8 x 76.6 ìm. The grains are isopolar, radially symmetrical and polyporate. As many as 30-40 pores are distributed throughout the surface of the pollen grain. The pores are circular in outline with 6-7 micro meters in diameter. The pores are arranged in sigmoid (S-configuration) pattern. The pores have crassimarginate aperture (pore margin has thick wall).

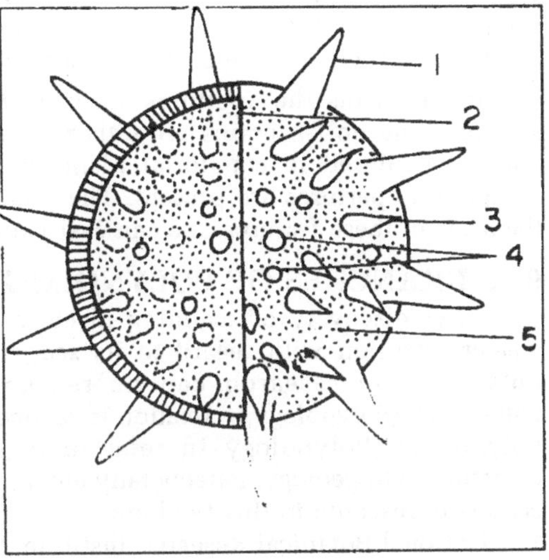

Fig 9.5 Pollen morphology of *Hibiscus*
1. spines, 2. columellae, 3. apertures, 4. inter spinal regions, 5 eruine

The pollen grain has very thick stratified wall. Sporoderm is differentiated into outer thick exine, and inner thin intine. Exine is about 7.5 micro meters, and it is further differentiated into outer thick, sculptured, heterogenous layer called sexine and inner thin non sculptured, homogenous layer called nexine. Sexine composed of radially arranged rod like structures called bacula or columellae. The apical swollen regions of the columellae fuse to form a layer called tectum. The pollen grains are spinose (fig-9.5). Each spine is about 19 micro meters in length and 8 micro meters in diameter. The base of the spine is bulbous and apex is pointed. In general the spines are unbranched rarely branched. The inter spinal region of the pollen wall is granular.

9.5. POLLEN MORPHOLOGY OF ACACIA

The pollen grains of *Acacia* are polyads. Polyads are radially symmetrical. In a polyad 16 pollen grains are aggregated in such a way to form a compound pollen grain (fig — 9.6). Out of 16 spores, 8 spores are arranged in the center of the polyad in the form of 4+4 in two tiers. These eight spores are peripherally surrounded by remaining 8 microspores. The polyad is bilaterally flattened with a diameter of 40-50 micro meters. Spores are inaperturate. Pollen wall is very thick and it is differentiated into outer thick exine and inner thin intine. The exine is 2-3 microns in thickness. Exine is differentiated into outer sculptured sexine and inner non sculptured nexine. Sexine is thicker than nexine. The exine ornamentation may be granular or reticulate.

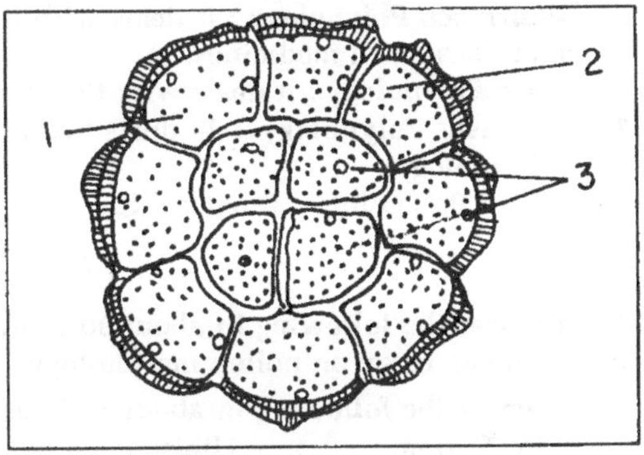

Fig 9.6 Polyad of Acacia.
1. peripheral grains, 2. central grains

9.6. POLLEN MORPHOLOGY OF GRASSES

The grass pollen grains are spheroidal to ovate in out line. The diameter ranges from 50-105 ìm. The grains are heteropolar (not divisible into two equal parts), and mono porate. The pore is confined to distal region of the pollen grain. The pore is circular in its out line with a diameter of 3 micro meters. The pore is slightly protrudes with thickned margin which is characteristic feature of grass pollen grain (fig-9.7).

The pollen wall stratification and ornamentation is very simple. It has two principle layers called outer thick exine and inner thin intine. The exine is about 1-5 microns in its thickness. It is further divisible into outer sexine and inner nexine. Sexine and nexine are equal in their thickness. Sexine shows ornamentations where as nexine is smooth.

In monocots the apertures are generally confined to distal pole. The apertures are simple. The pollen grains are either monoporate or monocolpate. In dicots the aperture may be confined to equatorial plane or distributed throughout the surface. In general compound apertures are present. The pollen grains may be tricolporate or polyporate. On the basis of these features one can easily distinguish monocots from Dicots.

9.7. BRANCHES OF POLYNOLOGY

After discovery of scanning Electorn microscope (SEM) the polynology is receiving greater attention with regard to its scope in science. The polynological research can be recognized into basic polynology (Polynology in relation to taxonomy) and Applied Polynology (Polynology in relation to medicine, crime detection. Oil geology, Paleobotany etc.,). In India the chief centres of research in this field are

1. National Botanical Research Institute (Lucknow)
2. Birbal Sahani Institute of Palaeobotany (Lucknow)
3. Bose Institue (Calcutta) and
4. Osmania University (Hyderabad)

Polynology developed into several branches.

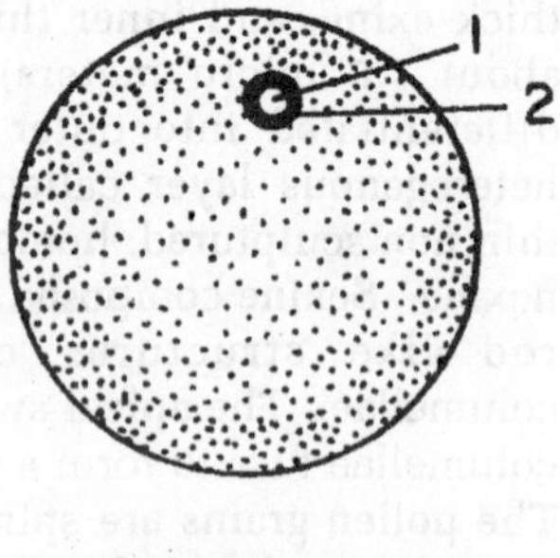

Fig 9.7 Grass Pollengrams
1. Pore 2. Annuleus

1. **Polynotaxonomy** : It deals with the study of pollen morphology in relation to taxonomy.
2. **Geo-Paleo Polynology :** It deals with the study of fossil pollen grains.
3. **Aero Polynology:** It deals with the study of pollen grains found in atmosphere.
4. **Melitto Polynology :** It deals with the study of pollen grains found in honey.
5. **Pharmaco Polynology** : It deals with the study of pollen grains which are used in the manufacture of medicines.
6. **Iatro Polynology :** It deals with the study of pollen grains which cause allergy.
7. **Forensic Polynology :** It deals with the study of pollen grains with reference to criminology.
8. **Capro Polynology** : It deals with the study of pollen grains found in animal excreta.

QUESTIONS

A. Answer the following in about 30 lines.

1. Write an essay on pollen morphology.

B. Answer the following in about 10 lines.

1. NPC System 2. Hibiscus pollen morphology 3. Acacia pollen morphology
4. Grass pollen morphology 5. Branches of polynology

C. Answer the following in about 3-5 lines

1. Polynolergy
2. Proximal pole and distal pole
3. Polar axis and equatorial axis
4. Isopolar and heteropolar pollen
5. Tenuimarginate and crassimarginate aperture.
6. Colporate pollen grain
7. Endoporium and ectoporium
8. Monozonotreme and dizonotreme
9. Atreme and polytreme
10. Sexine and nexine
11. Basic polynology and applied polynology
12. Polynotaxonomy
13. Pharmaco polonology
14. Forensic polynology

PART - C

PLANT PHYSIOLOGY

By
Dr.S.Janardhana Chary
Principal (Retd.)
S.R.A.S. Govt. College
Kothagudem, Khammam A.P.

1 INTRODUCTION

PLANT PHYSIOLOGY

Plant Physiology deals with various life-activities occurring in plants, which include absorption of water and mineral nutrients from the soil, assimilation of carbon, respiration, growth, development etc. These processes are to be understood in full to improve our plant resources, so as to meet the rising demands of human population on the planet.

In recent years Physics, Chemistry and Biochemistry have helped greatly in studying and understanding the life-activities in greater details, even at the molecular level. So it is inevitable to have the basic knowledge of Physics, Chemistry and Biochemistry for comprehensive and complete understanding of the life-processes. The knowledge of Physics helps in understanding the physical processes like Diffussion, Osmosis, imbibition etc. The knowledge of Chemistry helps in isolation and identification of Chemical compounds present in plants and that of Biochemistry in identifying chemical transformations occurring in the plants. Biology, therefore, is a combination and manifestations of physico-chemical processes. As a result of which branches like Biophysics and Biochemistry are becoming increasingly important in the elucidation of the life-processes.

The other branches of botany like Ecology, Taxonomy, Genetics and plant breeding also help us in understanding the complexities of the processes. The knowledge obtained from these branches are being applied in agricultural practices.

The application of Plant Physiology Principles led to the green revolution and self-sufficiency in agriculture in food production in India. The Knowledge of mineral nutrition of plants helped the agriculturists to detect the deficiency of mineral elements in the soil, on the basis of deficiency symptoms produced in the plant. Supply of deficient elements to the soil, besides the application of appropriate combination of mineral nutrients at the appropriate time for their efficient use have helped in increasing the crop-yield. Several growth promoting substance like Auxins, Gibberellins, cytokinins are employed for achieving various benefits like increasing crop-yield, weed control, prevention of pre-mature fruit drop and for the production of seed less varieties. Auxins are used in vegetative propagation of many ornamental plants. Auxins are also used in inducing roots to the cuttings of fruit-yielding plants like Mango etc. Gibberellins are used in several vegetable and fruit-yielding crops for increasing the size of fruits like grapes. Gibberellins are also used in delaying in ripening of citrus fruits, cytokinins are efficiently used in inducing roots. In Groundnut the increased crop-yield and increased oil production has been recorded by the application of cytokinins. Ethylene is being efficiently used in artificial ripening of the fruits. Thus

the knowledge and application of plant physiology principles helped in increasing the quality and quantity of crop-productivity.

The chapters in this book are written in detail in accordance with the common-core syllabus adopted by all the universities of Andhra Pradesh, incorporating the recent developments in the concerned fields.

2

PLANT PHYSIOLOGY

ISOLATION OF CELL ORGANELLES

2.1 INTRODUTION

The study of ultrastructure of different cell organelles and their functions in different metabolic activities is an important component of cell biology. For this, a method known as cell fractionation is widely employed. With the help of cell fractionation coupled with other techniques one can study the detailed structure of nucleus, mitochondria, plastids, ribosomes, lysosomes, plasmamembrane, endoplasmic reticulum and golgicomplex. The cell fractions thus obtained are also used *in vitro* studies.

Cell fractination can be achieved by three step procedure.

1. Homogenization
2. Centrifugation
3. Analysis

2. 2 HOMOGENIZATION

In the tissue homogenization process, first the cell wall or plasmamembrane are removed by applying enzymes like cellulase or lyase. This results in the formation of a suspension of cell fragments and sub cellular fractions. In general, the destruction of animal cell membrane is easier than the plant cell membrane as latter possess cellulose cell wall. Once the suspension of cell fragment is obtained, through tissue homogenization in pestle and motor waring blendor or teflan plungen and also by employing ultrasonic integrators. In some cases a device called French Pressure cell (pressure of 10^8 Pa) is also used. However, the use of homogenization apparatus depends on the plant material selected for cell fractionation.

The solution used for homogenization should contain a buffer so that the fluctuations in pH caused by organic acids present in the cell vacuoles can be balanced. Further, an osmotic solution in the form of 0.25 M sucrose is used as homogenization medium to prevent the damage to organelle membranes due to changes in osmotic pressure. It is also observed that during homogenization, some phenolic compounds present in the cells may be converted to quinines in the presence of phenolic oxidase enzyme which may cause damage to cell organelles. Hence, to absorb phenolic substances, another compound called polyvinyl pyrrolidone is mixed to homogenization medium. This technique has some the major use of homegenization is a preliminary step for the separation of cellular components disadvantages as it cause morphological alterations in some tissues and enzymes may be inactivated by organic solvents. Chelating agents such as EDTA or EGTA may be added to remove divalent cations such as mg^{+2} or ca^{+2} for successful fractionation.

2.3 CENTRIFUGATION

Homogenization of cell components is followed by the separation of organelles and sub-cellular structures for which centrifugation process is required. This can be achieved by an instrument known as ultra centrifuge. Based on centrifugal force and on different attributes like structure and density, the cell organelles sediment in the centrifuge tubes. This process is also described as differential centrifugation. The sedimentation rate or sediment coefficient is different for different cell organelles. Usually 1×10^3 sedimentation coefficient is taken as basic unit and the multiples of this are shown as Swedberg unit (S) for different organelles.

Basically a centrifuge machine consists of following components 1. Rotor (where centrifuge tubes are arranged) and 2. Motor. The cell suspension poured into tubes which are run by motor. At present different types of centrifuges are available for cell fractination

2.4 HIGH SPEED CENTRIFUGE

The high speed centrifuges are available with maximum rotor speed in the region of 25,000 rev/min, for generating a relative centrifugal field of above 60,000g and have a total capacity of upto 1.5 dm^3. This instrument is often used to collect microorganisms, cellular debris, larger cellular organelles. However, they cannot generate sufficient centrifugal force to sediment smaller organelles such as ribosomes.

2.4.2 Analytical Ultracentrifuge : In this centrifuge maximum rotor speed is about 75000 rpm generating a centrifugal force of about 5,00,000 g.

Besides above centrifuge types small bench centrifuges (4000 — 600 rpm), small microfuges (8000-13,000 rev/ min^{-}) continuous flow centrifuges, preparative ultracentrifuges and analytical ultracentrifuges are also available for cell fractionation.

2.5 METHOD OF CENTRIFUGATION

The homogenate suspension first poured into centrifuge tubes of rotor. Then centrifuged for 10 minutes at 700g. Based on sedimentation coefficient, the nucleus, cell wall sedinents settle at the base of the tubes (sedimentation — I). It is also described as nuclear fraction. In the next stage the supernatant — I of the first tube is transferred to another tube and then re-centrifuged at 20,000g for 15 mins. In this process cell organeles such as mitochondria, lysosmes and peroxisomes sediment (sedimentation — II) and it is called as mitochondrial fraction. In the third stage, the supernatant (supernatant- II) is taken into another tube and results in the sedimentation of ribosome, golgi comlpex (intracellular micro segments).

In the last stage, the supernatant obtained in the third stage (supernatant — II or microsomal supernatent) contains soluble nucleic acids, polysaccharides and lipid droplets. These are separated by applying chromatographic and dialysis techniques.

2.6 ANALYSIS

The cell organellels thus separated by centrification are analysed for their structure, chemical composition by following spectrophotometry technique and also by using radioisotopes.

3

PLANT PHYSIOLOGY

CELL MEMBRANES & PERMEABILITY

3.1 INTRODUCTION

All the cells are bounded by a thin membrane called plasmalemma or plasma membrane. Membranes are chemically lipoproteins with polysaccharides. Proteins are of three types; structural proteins, enzymes, and carrier proteins, of these structural proteins form the backbone of membrane. Enzymes are catalytic proteins. The common carrier proteins are; lecithin and cholesterol. In addition to these certain ions like Ca, Zn, Mg are also present in the membrane.

The structure and functions of cell membrane have been studied in the previous year.

3.2 PERMEABILITY

Permeability may be defined as the ability of membrane to allow the diffusion of substances through it. Cell wall is a permeable membrane. It permits the diffusion of all substances through it. Plasma membrane is differentially permeable and permits the diffusion of water, some solutes where as impermeable to some other solutes. Alcohol (C_2H_5OH), ether (CH_3COCH_3) and chloroform ($CHCl_3$) are diffused through the membrane rapidly. An increase in —OH groups in a compound decreases its permeability. Glycerol ($C_3H_5(OH)_3$) with 3(OH) groups penetrate less rapidly than alcohol with 1 (OH) group. Further, cell membranes are impermeable to polysachlarides, phospholipids and proteins. The rate of diffusion of inorganic salts depends on the valancy of ions. For example monovalent cations and anions penetrate more rapidly than divalent and trivalent cations and anions.

3.3 SALT ABSORPTION

Mineral salts are available in the soil either in the form of free ions in the soil solution or as adsorbed as ions on the surface of clay micelle. It is believed that the adsorbed ions on clay particles are absorbed by the roots of plants possibly by ion exchange method. Two theories have been put forward to explain the ionic exchange.

3.3.1 Carbon dioxide hypothesis: According to this theory CO_2 produced by roots during respiration reacts with water to form carbonic acid. Carbonic acid in turn disassociate into hydrogen ions (H^+) and bicarbonate ions (HCO^-_3). The hydrogen ions are exchanged with cations (K^+, Na^+ etc) adsorbed on the soil particles and bicarbonate ions with anions adsorbed on the soil particles. Thus both cations and anions are exchanged and absorbed by the roots. But the theory fails to explain movement of ions between the clay particles and roots of the plant.

3.3.2 Cation Exchange hypothesis:

3.3.2 Cation Exchange hypothesis: This theory explains the movement of ions between the clay particles and roots. It postulates that the ions are not static but always in a state of oscillation around their adsorption site. When the oscillation volumes of ions both on the roots and colloidal particles overlap each other, ion exchange occurs (see fig. 3.1)

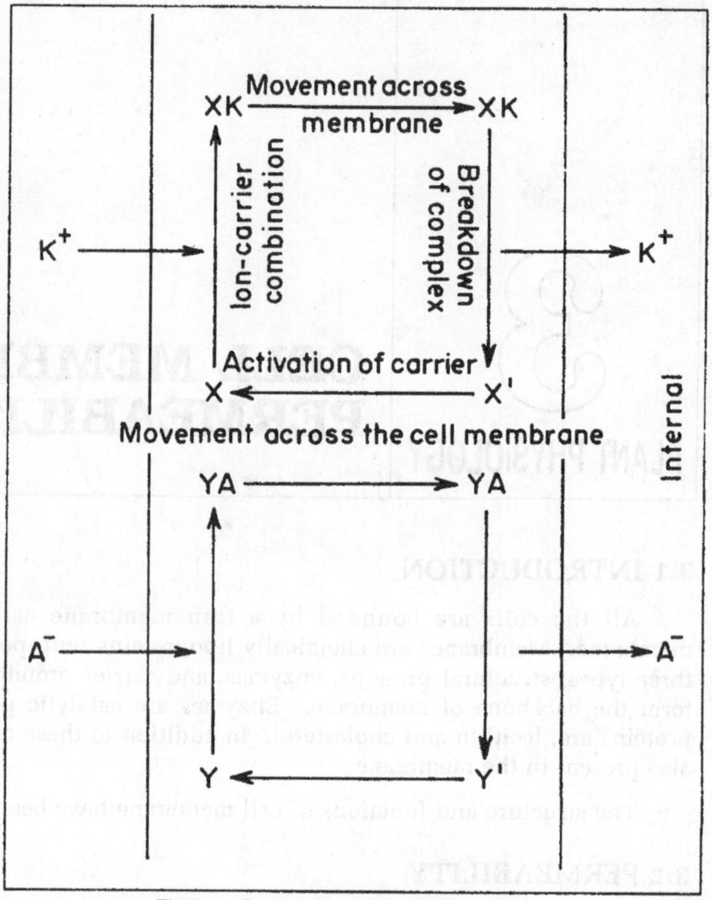

Fig 3.1 : Carrier concept (Honert, 1937)
A. Anion, K = Cation X, Y = Carriors

3.4 MECHANISM OF SALT ABSORPTION

Several theories have been proposed to explain the mechanism of salt absorption. They can be grouped into two groups

1. Passive absorption 2. Active absorption.

3.4.1 Passive absorption: Passive absorption is by physical forces. It may be called as physical absorption. The process is not affected by any metabolic inhibitors and doesn't require any expenditure of energy. It includes the theories of diffusion, Mass flow, Ion exchange, Donnan's equilibrium etc.

3.4.1.1 Mass (Bulk) flow theory: According to this theory the ions are taken up by the roots along with the flow of water, under the influence of transpiration. It is presumed that 'Transpiration pull' which is developed inside the plant due to the active transpiration is responsible and acts as driving force for ion uptake. The effect is direct. The increase in transpiration rate increases the ion uptake. However, this theory fails to explain salt accumulation against osmotic gradient.

3.4.1.2. Ion Exchange theory: According to this theory the hydrogen ions (H^+) and hydroxyl ions (OH^-) are readily available adsorbed on the surface of the cell membrane. The cations and anions in the soil solution are exchanged freely. The cations are exchanged with hydrogen ions and anions are exchanged with hydroxyl ions. This process does not involve the participation of metabolic energy.

3.4.1.3. Donnan's Equilibrium: It presumes that there are fixed or indiffusable anions present inside the membrane. In order to maintain equilibrium more number of cations enters into the cell, besides normal uptake. The equilibrium can be represented as follows.

$$\frac{\text{Cl (out side)}}{\text{Cl (inside)}} = \frac{K^+ \text{(inside)}}{K^+ \text{(outside)}}$$

When fixed anion are present inside the membrane, cations in addition to normal exchange would be absorbed to maintain the equilibrium. Thus cation concentration increases in the internal solution compared to the outside soil solution. The same type of mechanism operates in case of fixed cations on the inner side of membrane.

3.4.1.4 Objections to passive absorption : Several strong objections are raised against the passive absorption concept. Some of the objections are as follows:

1. Passive absorption fails to explain the accumulation of salts or ions against osmotic concentration gradient. For example in *Nitella translucens* the conc of K^+ ions in the plant is 1000 times more than the conc K^+ ions in the outside water. Even then the uptake of K^+ ions continue against the conc. gradient.
2. The theory fails to explain the relationship between salt accumulation and respiration. Increase in the rate of respiration also increases the rate of salt accumulation. The salt uptake is inhibited by metabolic inhibitors like chloroform etc., which inhibits the respiration.

These objections led scientists like Lundegardh, Epistein etc., to support active absorption.

3.4.2 Active absorption: It envisages the involvement of metabolic energy in the uptake of mineral salts. It includes the theories like carrier concept, cytochrome pump hypothesis, ATP theories etc.

3.4.2.1. Carrier Concept: Honert (1937) proposed this theory. According to this theory there are certain carriers in the cell membrane, which may be organic molecules or vesicles. Carrier carry the ions from the outer space to the inner space by forming ion carrier complex. The complex releases the ion on the inner space. The ions, which are released on the inner space cannot move out and thus are accumulated. Separate carriers are involved for transportation of cations and anions. The scheme is explained in the fig. 3.1.

The carrier concept is supported by the fact that when the concentration of salts in the outside medium is increased, the rate of uptake is also increased. But after some time, the rate of uptake becomes constant, even though the concentration of the salts in the outside medium is increased. The reason can be that all the carrier sites remain occupied with ions, no matter even if outside cone is increased and the rate of uptake becomes constant.

Selective absorption of ions by roots also supports the carrier concept. Different types of ions are absorbed by different rates, suggesting the presence of specific carriers.

3.4.2.2. Cytochrome — Pump hypothesis: Lundegardh and Burstom (1933) found correlation between anion absorption and respiration. When a plant is transferred from water to salt solution, the rate of respiration increased. The increased respiration is termed as "Salt Respiration".

Later Lundegrdh (1950) expressed the opinion that anion absorption is independent of cation absorption and oxygen concentration gradient exists from the outer surface to the inner surface of membrane indicating oxidation at the outer surface and reduction at inner surface of the membrane. The transport of anions takes place through the cytochrome

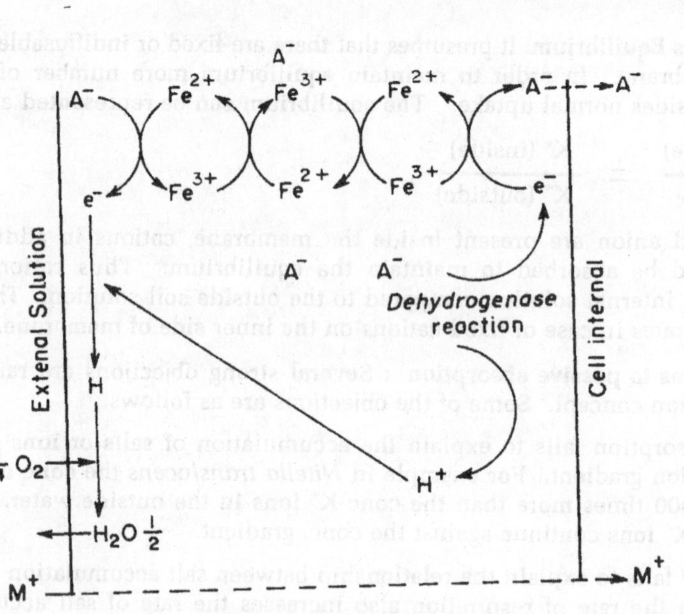

Fig 3.2 Lundegardh Cytochrome - pump Hypothaeis

system. Cytochromes carry the anions from outer surface to the inner surface of the membrane (fig. 3.2).

According to Lundegardh (1950) protons (H^+) and electrons (e^-) are produced at the inner surface of the membrane due to hydrogenase reactions. The electrons move outwards via. Cytochrome chain. The electron is released by taking anion in its place. The released electron unites with proton and oxygen to form water. The anion picked up by the cytochrome on the outer membrane is transported to the inner surface and released at the inner surface of the membrane. Wheres as cations are absorbed passively.

The cytochrome pump theory was widely accepted by scientists because of many experimental evidences in support of the theory. Yet there is some criticism against it. 2-4-dinitrophenol (DNP) which inhibits oxidative phosphorylation, increases respiration. Robertson *etal* (1951) found that on application of 2-4 D, respiration rate increased but salt absorption decreased. Robertson *etal.* on the basis of above finding, expressed the opinion that phosphorylation should have been included in the theory.

The objections raised against the theory are as follows.

i. The theory explains the mechanism for anion absorption only. No proper explanation for cation absorption.

ii. It explains common mechanism for all kinds of anions. It does not explain the selectivity and differential rates of absorption based on selectivity.

iii. It operates only in aerobic forms. But in anaerobically respiring forms such an elaborated electron transport system is lacking.

3.4.2.3. Protein — Lecithin theory : Bennet Clark (1956) proposed this theory to explain the absorption mechanism for both cations and anions. According to this theory protein lecithin is synthesized and hydrolysed in a cyclic manner. Lecithin picks up ions (both cations and anions) at the outer surface of the membrane forming lecithin ion complex which moves to the inner space, releases the ions on hydrolysis of the complex — ATP

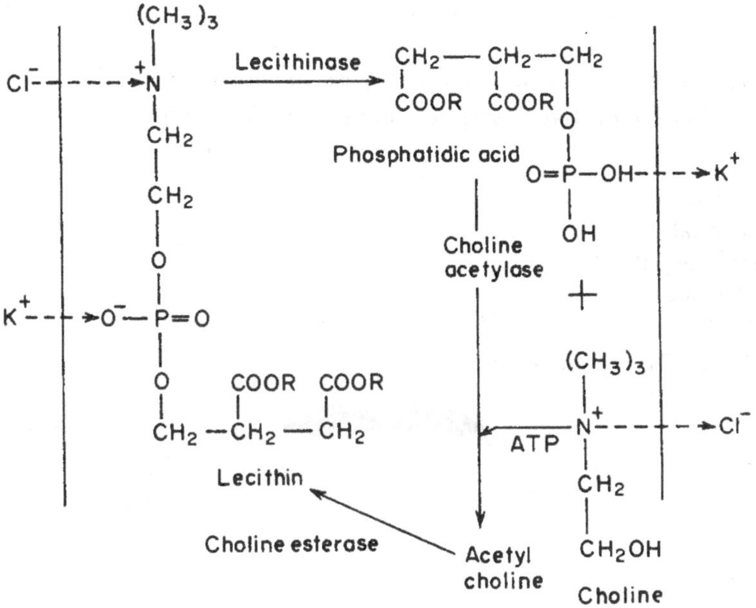

Fig 3.3 Protein - Lecithin theiry (Bennet Clark, 1956)

molecules are required during the synthesis of lecithin. Thus the process occurs in cyclic manner (see fig. 3.3).

3.4.2.4. ATP theories : Laties (1957) and sutchiffe (1959) proposed ATP theories to explain cation uptake. According to them ATP molecules are hydrolysed to make available the energy for energising ion pumps. It is made through the enzymes, which are found existing in cell membranes. The role of ATP in cation uptake is of two types.

In first case the organic compound is first phosphorylated, which on dephosphorylation combines with cation. The cation is released when phosphorylation occurs again.

In second case phospharylated organic compund combines with cation and cation is released on hydrolysis of the ion compound complex.

It is believed that both passive and active absorption mechanisms are involved in salt uptake though, there are several criticisms against both the mechanisms. So much more work is needed to fill up the gaps.

QUESTIONS

I. Essay type :

1. Describe Permeability.
2. Describe the mechanism of Salt absorption.
3. Explain the carrier concept in detail.
4. Give an account of Lundegardh's cytochrome hypothesis.
5. Describe Protein-Lecithin theory. How it has fulfilled the objections raised against the previous theories.

6. **What is salt respiration? Explain.**
7. **What is passive absorption. Describe the various theories proposed to explain passive absorption of salts.**
8. **How active absorption of salts differ from passive absorption.**
9. **Differentiate between active absorption and passive absorption.**

II. Short Notes :

1. **Ion-Exchange**
2. **Salt respiration**
3. **Carrier concept**
4. **Cytochromes**
5. **Lecithin**
6. **Hoagland's Experiments**

4

PLANT PHYSIOLOGY

MINERAL NUTRITION

4.1. INTRODUCTION

Plants require supply of large number of substances from outside. This supply is called nutrition. If both organic and inorganic substances are required for the growth of the organism, the organism is called *heterotrophic*, but if only inorganic elements are required, the organism is called *Autotrophic*. All the green plants are autotrophic, since they synthesize their own organic requirements and depend on outside supply of inorganic elements only. Animals, Non-green plants are heterotrophic.

All green plants, obtain their inorganic requirements directly or indirectly from soil, excluding carbon, hydrogen and oxygen. Since all these elements are obtained from minerals, the elements are known as mineral nutrients and the nutrition is called mineral nutrition.

Analysis of plant ash reveals the presence of large number of elements. As many as 90 elements have been recorded. But all the elements found in the plant are not essential for the growth of the plant. A large number of them are non-essential. Mere presence of element in the plant cannot be taken as criterion for their essentiality.

Carbon, hydrogen and oxygen are known as framework elements. Bulk of the plant body is composed of these elements, as they are present in cell wall and protoplasm. Carbohydrates, fats, proteins, cellulose are composed of these elements. Nitrogen, sulphur and phosphorous are known as protoplasmic elements, since proteins and nucleic acids are composed of these three elements. Cell wall requires calcium, chlorophyll requires magnesium for their synthesis. Iron, copper, zinc and manganese etc are present in enzymes. Apart from this many other elements are required for the synthesis of plant metabolites.

To find out whether an element is essential or non-essential for the growth of the plant, the plant is grown in controlled cultural conditions, in the absence of particular element. If the plant grows normally the element is non-essential and if it does not grow normally, the element is considered to be essential.

Arnon and Stout (1939) proposed three criteria to decide the essentially of an element. The criteria are as follows.

An element is not considered as essential unless.

a) A deficiency of it makes it impossible for the plant to complete the vegetative or reproductive stage of its life cycle.

b) Such deficiency is specific to the element in question and can be prevented or corrected only by supplying this element and

c) The element is directly involved in the nutrition of the plant quite apart from its possible effects in correcting some unfavourable microbiological or chemical conditions of the soil or other culture medium.

The experimental procedure is growing plants from seedling stage in a nutrient medium, which contains all elements regarded as essential, except one element under consideration. If normal growth is not possible in the absence of the element under consideration and deficiency symptoms are developed in the plant, the element under consideration is termed as essential element. Further, the addition of the element under consideration alone, and no other element can correct or prevent the deficiency symptoms.

In recent years, there has been growing awareness that Arnon and Stout criteria are inadequate and have inherent defects. For example chlorine is regarded as essential element, but it has been shown that Bromine can replace it in some plants, although at much higher concentration. There can be no doubt regarding the essentiality of chlorine but if Arnon and Stout's criteria are followed strictly, it is not fulfilling the second condition of complete specificity of an element. So it is the consensus of many plant physiologists that only the first condition of Arnon and Stout criteria is to be met to decide the essentiality of an element.

Another suggestion is certain elements can be designated as functional or metabolic elements rather than as essential element. For example chlorine and bromine may be designated as functional element, rather than an essential element.

4.2 HYDROPONICS OR SOLUTION CULTURE EXPERIMENTS

The system of growing plants in soil less cultures or solution cultures is known as hydroponics (fig. 4.1). It has been developed on large scale to promote higher yield of crop plants.

In solution cultures plants are grown in pure nutrient medium containing desired elements in proper proportions. Pyrex glass jars are used as containers. The containers are covered on outer surface by black paper to protect the roots from direct sun light and to check algal growth in the medium. Aeration for the roots is provided. pH of the nutrient medium is maintained and nutrient solutions are changed frequently. Knop's, Sach's and Hoaglands culture media are commonly employed. Hydroponics has become quite popular during recent past for large scale production of vegetables like tomatoes, potatoes, carrots, lettuce and ornamental plants like roses.

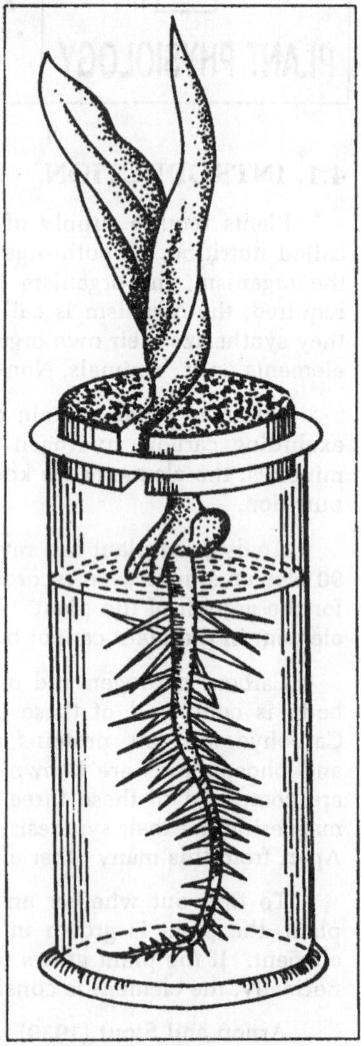

Fig 4.1 : Solution Culture Experiment

Hydroponic has certain advantages over the sand culture or agriculture practices. The advantages are.

1. Controlled nutrients solution can be maintained. The nutrient solution contains the desired elements in proper proportions.
2. The growth of bacteria or other contaminants is checked
3. The culture solution is frequently replaced to avoid accumulation of toxic substances.
4. The culture solution is aerated
5. Weed growth is eliminated
6. Crop production is high compared to the normal cultivation. Tomatoes yielded an average of 200 tons per acre as against 15.20 tons per acre under normal cultivation. Paddy grown by this method yields 11,400 lbs of rice per acre. Hydroponics has vast scope for further development in the near future in view of its advantages and high yield.

4.3 CHEMICAL FERTILIZERS

One of the important contributions of nineteenth century to agriculture was the discovery that soil fertility and crop yield can be increased by adding inorganic nutrients to soil. Before nineteenth century the common method for increasing crop production was to apply plant and animal debris, (Manure, Compost etc). Another common method was crop rotation. Crop plants are rotated with leguminous plants periodically to increase the fertility of the soil. But the scientific explanation for the increased fertility of the soil due to the cultivation of leguminous plants was known only in the late nineteenth century.

It was due to the work of Liebig in Germany, Boussingault in France, Lawes and Gilbert in England, it became clear that losses of nutrients in the soil could be replaced by addition of inorganic salts to the soil. Today addition of mineral nutrients to the soil has become common practice for agriculturists. Salts of nitrogen, phosphorus and potassium — N.P.K. — are used in large amounts because these three elements are most likely deficient in many soils. In addition to macro-nutrients, the importance of micro-nutrients in crop production has been recently realised.

4.3.1. Nitrogen Fertilizers

Nitrogen element is most important element for the growth and development of the plant and for good crop production. Nitrogen is essential constituent of proteins, nucleic acids, chlorophyll, enzymes etc, Nitrogen is supplied to the plants mainly in three forms.

1. Urea
2. Ammonium sulphate
3. Calcium ammonium nitrate

4.3.1.1. Urea (NH_2 Co NH_2) : It is white, crystalline hygroscopic solid substance. Nitrogen is present in amide form. Easily soluble in water, hence readily available to the plants.

Manufacturing Urea : First ammonium carbonate is synthesized by the reaction of liquid ammonia with liquid Co_2 in silver coated vessels.

$Co_2 + 2NH_3$ ------------> NH_2 $COONH_4$ (Ammonium Carbonate)

The ammonium carbonate is heated at high atmosphere pressure at 130^{0}C to get urea. It is now in liquid form. Liquid urea is condensed in vaccum evaporators to get the crystalline urea.

4.3.1.2. Ammonium Sulphate ($(NH_4)_2 SO_4$) : It is in a white crystalline form. It contains 21% nitrogen in ammonia form, 24% sulphur in the form of sulphate. Ammonium sulphate causes acidity to the soil, so to reduce the acidity calcium carbonate is to be used. In our country 22 lakh tons of ammonium sulphate is being produced in 14 factories. Sindhri fertilizer factory, where ammonium sulphate is produced, is the largest fertilizer factory in Asia.

Ammonium sulphate is synthesised by a gypsum method. The main raw materials are gypsum ($CaSO_4$, $2H_2O$), ammonia, Co_2, water.

The process takes place in three steps. First Ammonia is produced by combining nitrogen with hydrogen. Ammonia is now reacts with Co_2 and water to produce ammonium carbonate. Gypsum is now reacts with ammonium carbonate to get ammonium sulphate and calcium carbonate. The equations are as follows:

1. $N_2 + 3H_2$ ----------> $2\ NH_3$
2. $2NH_3 + Co_2 + 2H_2O$ -------------> $(NH_4)_2 Co_3 + H_2O$
3. $(NH_4)_2 Co_3 + CaSo_4$ ----------------> $(NH_4)_2 So_4 + CaCo_3$

4.3.1.3 Calcium-ammonium nitrate : It contains 26% nitrogen in the form of ammonia and nitrate in equal proportions. Nitrate form is readily available to the plants. Ammonia form is first reduced to nitrate before it is being, used by the plants.

4.3.2 Potasium Fertilizers

Potassium is a constituent of important metabolites like proteins, fats, chlorophyll, carbohydrates etc. potassium is required for high crop yields. Indian soils are not deficient of potassium. Murate of potash, potassium sulphate are two important phosphorus fertilizers. Murate of potash contains 60% of potassium, where as potasium sulphate contains 48-50% of potassium. In addition to it sulphur is also present to the extent of 17%. Murate of potash is being extensively used for all types of crops. Potassium sulphate is costly. It is used for tobacco, sugar cane, grapes etc.

4.3.3 Phosphate Fertilizers

4.3.3.1. Super phosphate : It is often used phosphorus fertilizer in India. It contains 16% water-soluble phosphorus. It also contains 2/3 Gypsum. It supplies calcium, sulphur required for the crops. Mostly useful for the ground nut crop. Rock Phosphate is made to react with sulphuric acid. Super phosphate and calcium sulphate are formed.

$Ca_3 (PO_4)_2 + 2H_2 SO_4 + 2H_2O$ ----------------> $Ca(H_2PO_4)_2 + 2CaSO_4, 2H_2O.$

4.3.3.2. DAP (Diammonium Phosphate) : It contains 46% phosphorus, 18% nitrogen. Phosphorus is easily soluble in water. The DAP is suitable for all types of soils. It is being extensively used in India.

4.3.3.3. Urea Phosphate : It is a complex fertilizer. Nitrogen, phosphorus, potassium is present in the ratio of 28-28-0 or 24-24-9 or 20.20.0. Most useful for paddy, sugar cane etc.

QUESTIONS

I. Essay type Questions :

1. What are the essential elements, non-essential elements. How to identify.
2. What are the criteria of arnon and stout for determining the essentiality of an element.
3. Describe the Hydroponics and their benefits.
4. What are trace elements. Write an account of their role in the nutrition of plant.
5. Differentiate between Macro-nutrients and Micro-nutrients.
6. Describe the role played by elements K, Na, Ca, Fe in the life of the green plant.
7. Write an account of role played by any two macro-nutrients and two micro-nutrients.

II. Short Notes :

1. Arnon and stout criteria
2. Soil-less culture experiments
3. Hydroponics
4. Micro-nutrients
5. Trace elements.

5

PLANT PHYSIOLOGY

PLANTS AND WATER RELATIONS

5.1 INTRODUCTION

The major constituent of the plant body is water. The protoplasm of the cell contains about 70 — 90% of water by weight. It is more so in the case of hydrophytes, the plants living in water, where the percentage of water is upto 97% by weight. Water is present throughout the plant body starting from the underground parts of the plants to the aerial parts of plant including leaves, buds, flowers and fruits. Water is essential for carrying out various vital processes of life in the plant.

5.2 SIGNIFICANCE OF WATER

The significance of water in the life of the plant can be understood by studying its functions.

1. Water forms the major constituent of protoplasm.
2. Water is good solvent, in which mineral nutrients enter into the plant body from the soil solution. The transportation of nutrients from one part of the plant to another part; from one organ to another organ and from one tissue to another tissue and from cell to cell takes place through the water medium only.
3. Water is one of the raw materials required for the photosynthesis.
4. Water is a reactant in many of the metabolic processes for ex. TCA cycle.
5. Water maintains the turgidity of the cells there by their form and structure is maintained.
6. Water is necessary for germination of seeds. Germination of seed is initiated only after the absorption of water by the seed.
7. Water is responsible for variety of movements in plants. These include the movement of guard cells in opening and closing of stomata, the nocturnal foldings of leaflets in certain plants like sensitive plant (*Mimosa pudica*) etc.
8. The elongation of cells in growth and development after the cell division depends upon absorption of water.

9. In tropical plants water plays an important role in regulating temperature of the plant. It is believed that plants loose about 90-95% of water only for maintaining the optimum temperature of the plant.

5.3 CHARACTERSTIC FEATURES OF WATER MOLECULE

Water is a substance having unique physical and chemical properties.

1. The molecular weight of water is very less compared to other substances. It is only 18.
2. The formation of water molecule from oxygen and hydrogen is shown below. The hydrogen atom has one electron, the oxygen has six electrons in the valence orbitals. There is a sharing of electrons in pairs between the oxygen and hydrogen atoms, forming the covalent bond. Covalent bonds are very strong, therefore, the water molecules are very stable.

    H

    H° + H° + O ---------------------> H O

    Fig. The distribution of electrons in the formation of water molecule.

3. Water acts as universal solvent because of its dipolarity and hydrogen bonding. Because of dipole character the positive side is attracted to negative charges and its negative side to positive charges. The Hydrogen bonds are developed between the hydrogen atom of one water molecule to oxygen atom to another molecule. Hydrogen bonds can also be formed between solids and water molecules.
4. The boiling and freezing temperatures of water are 100°C and O°C respectively, where as the boiling temperatures of hydrogen and oxygen, which are structural components of water are 253°C and 180°C respectively.
5. The specific heat of water is high i.e. 37°C. The high specific heat of water is due to strong attraction of water molecule to one another. Certain amount of heat is required to break its hydrogen bonds. The quantity of heat required to raise the temperature of 1g of substance of 1°C at 15°C is called specific heat. The high specific heat of water helps plants maintain stable internal temperatures. When large fluctuations in temperature in the external atmosphere of the plant takes place, it will have little effect on its internal temperature of the plant.
6. The water is having high surface tension. It helps in ascent of sap in plants to rise to greater heights.
7. Water is less affected by solar radiation. Water absorbs ultraviolet and visible radiation to only a limited extent.

5.4 PLANTS AND WATER RELATIONS

5.4.1 Diffusion, Osmosis, Imbibition and water potential

To understand the mechanism of absorption of water by plants, certain physical phenomena like diffusion, Osmosis, Imbibitition and water potential etc are to be understood, because the movement of substances between the plant and environment and from cell to cell in plants, takes place by physical forces.

Diffusion: Diffusion is defined as the movement of molecules or ions of a given substance from region of higher concentration to a region of lower concetration till their concentrations become equal and a state of dynamic equilibrium is achieved. Even after the equilibrium is achieved the diffusion continues but it is not detectable, because the movement of molecules or ions continue in both directions at equal rate. For example, when copper sulphate crystal is placed in a beaker containing water the molecules or ions of copper sulphate gradually move from the region of higher concentration to the region of lower concentration, till the concentration become uniform throughout the water. As a result of diffusion of copper sulphate molecules the water becomes blue coloured. Similarly when the sugar is added to the milk, the milk becomes sweet due to the uniform distribution of sugar molecules in the milk.

The diffusion of particles of one substance does not depend upon the diffusion of particles of another substance. In a system if two and more substances are present, they diffuse independently and depend mainly on their concentration gradient. Neither the direction nor the rate of diffusion is affected by rate of diffusion of other substance present in the system. Each substance diffuse according to its own concentration gradient.

The rate of diffusion of a substance depends on factors such relative density, temperature, concentration gradient and concentration of the medium. Hydrogen diffuses four times as fast as oxygen and five times as fast as carbondioxide. Rise in temperature increases the rate of diffusion. The steeper the concentration gradient, higher the rate of diffusion. The concentration of medium also affect the rate of diffusion of a substance in it. Higher the concentration of medium the lesser the rate of diffusion. Substances tend to diffuse more freely and at faster rate in air than in the liquid medium.

The potential ability of substances to diffuse from the region of higher concentration to region of lower concentration is called as "Diffusion Pressure"

5.4.2 Osmosis : Osmosis can be defined as diffusion through semi-permeable membrane; it is defined as, when two solutions of different concentrations are separated by a semi-permeable membrane, the movement of water (solvent) from a weaker solution into the stronger solution through the semi-permeable membrane is called Osmosis. Thus osmosis is a special type of diffusion. It differs from diffusion in two ways;

1. In osmosis semi-permeable membrane is involved, while in diffusion no such membrane is required

2. In osmosis only solvent (water) moves but solute does not move. Where as in diffusion both solvent and solute move in both ways.

Osmotic pressure can be defined as pressure required to prevent the entry of water (solvent) into osmotic system through the semi-permeable membrane. Osmotic pressure of solution is proportional to molecular concentration. If the solute is non-electrolyte the osmotic pressure is directly proportional to the molar concentration. In case of electrolytes the osmotic pressure depends on degree of ionization. Electrolytes exerts a greater osmotic pressure because grater part of electrolyte is in an ionized condition in the solution and increased number of ions cause higher osmotic pressure.

Osmotic pressure is now referred as 'osmotic potential'. It is expressed with negative sign and with the addition of the solute it becomes more negative.

5.4.3 Turgor Pressure : The pressure that develops inside the cell due to the osmotic entry of water and exerts against the cell wall is called Turger Pressure (TP). The cell wall also exerts an equal and opposite pressure against the TP is called wall pressure (WP). As the turgor pressure and wall pressure are equal and opposite the cell becomes turgid. It does not burst.

5.4.4. Water Potential (y) : The diffusion pressure of a solution is always less than its pure solvent. The difference in pressure between the solution and its solvent is known as Diffusion Pressure Deficit (DPD) or suction pressure (SP). It is indicative of water deficit of a solution. It is now called as "water potential". It is indicated with a Greek letter Psi (y) and is measured in bars. The difference between the free energy of water molecules in pure water and free energy of water in solution is termed as water potential. Since water potential (y) of pure water is zero (0), the addition of solute particles to the water reduces the free energy of water, and the negative value increases. Thus y of solution is always less than zero. So the highest value is zero.

The water potential in most of the plants, rooted in well watered soils, ranges from 2 to 8 bars. Most of the plants cease growth when water potential falls to 15 bars. Leaves of herbaceous plants are not likely to recover if the water potential drops to 20 to 30 bars. In conditions of severe drought the leaves of desert plants have greater ability to survive, when subjected to lower water potentials such as 30 to 60 bars. Viable air-dried seeds have very low water potential of 60 to 100.

Water potential is determined by three factors.

$\psi w = \psi m + \psi s + \psi p$

Where, ym is matric potential, ys is solute potential or Osmotic potential, ψp is pressure potential or turgor pressure

Since in plant system ym is disregaraled, the equation can be represented as follows

$\psi w = \psi s + \psi p$

The form solute potential (ψs) is used in place of Osmotic pressure (OP). But (OP) is expressed in bars with + sign, where as solute potential (ψs) is expressed in bars with the negative sign.

Solute potential depends on total number of solute particles (Molecules or ions) in solution rather than their charge. For example one molecule of potassium chloride in solution disassociates into 2mol of ions, Hence 0.5 molal solution of potassium chloride has solute potential approximately equal to that of 1mol of sucrose solution. Another example, calcium chloride in solution disassociates into 3mol of ions. Hence 0.33 molal solution of calcium chloride has the same solute potential as 1.0 molal of sucrose solution. The solute potential of solution, having several solutes, is the sum total of individual solute potentials contributed by each of the solute.

5.4.5 Pressure Potential : The pressure developed inside the cell due to the entry of water into the cell is called Turgor Pressure. The turgor Pressure is equal and opposite to wall pressure.

The pressure potential, unlike that of osmotic potential, is positive. In herbaceous plants, the leaf cells have a turgor pressure ranging from +3 to +5 bars during the day time and +15 bars or more during the night time. Some times, during rapid transpiration the turgor pressure may be zero or negative.

In fully turgid cell, the osmotic potential (ψs) and pressure potential (ψp) are equal and oppsite, so the water potential (ψw) becomes zero.

For example

ψs = -10 bars — ψp is +10 bars

$\psi = \psi s + \psi p$

$\psi = -10 + 10$

ψ = zero

And in case of flaccid cell

ψp = zero

$\psi = \psi s + \psi p$

ψ = -10 bars + 0 bars

ψ = -10 bars.

Another example

A and B are two different Osmotic systems. In 'A' osmotic system ψs = -16 and ψp = 8 and in 'B' osmotic sytem ψs = -12 bars ψp = 2 bars. So ψ in 'A' system will be

$\psi = \psi s + \psi p$

$\psi = -16 + 8$

ψ = - 8 bars

In 'B' osmotic system

$\psi = \psi s + \psi p$

$= -12 + 2$

ψ = -10 bars

So the water moves from 'A' to 'B' osmotic system.

5.5 ABSORPTION OF WATER BY PLANTS

The plants absorb water mainly through the roots. The roots are the water absorbing organs of the plant. Hydrophytes, Plants living in water, absorb the water through their entire surface. A few plants absorb the water from the atmosphere through the leaves eg. Lycopodium; Orchids absorb water from the atmosphere through modified roots having special type of tissue called 'Velamen'.

5.6 ROOTS

The entire root system does not take part in the absorption of water. Most of absorption of the water takes place through the tender parts of the root system, mainly through the regions bearing root hairs. Root hairs are unicellular structures arise as an out growths from the cells of outer wall of the root, known as piliferous layer, in the maturation region of the young root.

A root hair is a modified cell of the epiblema or piliferous layer meant for the absorption of water. The wall of the root hair consists of cellulose and pectic substances. Both the substances are hydrophilic i.e, water loving, help in attraction of water and transfer to plasma membrane enclosing cytoplasm. The cytoplasm contains nucleus and a central large vacuole. Vacuole contains cell sap.

Each root-hair cell acts as an osmotic system. The cell sap has an osmotic pressure about 2-5 atm or even more. The cytoplasmic membrane acts as Semi permeable membrane.

The root hairs penetrate between the particles of soil, reaches the soil solution. The soil solution contains small quantities of salts such as nitrates, carbonates, phosphates, sulphates of sodium, potassium and calcium etc. The soil solution has osmotic pressure less than cell sap of the root-hair. So the water enters the vacuole of the root hairs from the region of less concentration to the region of more concentration through the plasmamembrane, which acts as semipermeable membrane.

The absorption of water by root-hairs depend not only on osmotic concentration but also on DPD (Diffusion Pressure Deficit or suction pressure). The water from the root hair cell move across the cortex of the root, through the endodermis and pericycle and into the xylem vessels. The xylem vessels are having zero turgor pressure. So its diffusion pressure deficit is more compared to pericycle cells. So the water is drawn into the xylem. Thus the movement of water from root-hair cell to the xylem vessels is due to the increasing gradient of diffusion pressure deficit from root hair to xylem vessels.

5.7 MECHANISM OF WATER ABSORPTION

Two different theories have been proposed to explain the precise mechanism of water absorption.

1. Active absorption
2. Passive absorption

5.7.1 Active absorption : In this process the water is absorbed as a result of activities in the root itself and shoot system has no role to play in this process. It is again of two types. i) Osmotic mechanism ii) Non-osmotic mechanism

5.7.1.1. Osmotic mechanisms : Atkins (1916) and priestely (1920) proposed this theory to explain the mechanism of absorption of water. According to this theory the water is absorbed due to differences in osmotic pressure between soil water and cell sap of root hairs. It doesn't require any expenditure of energy. The osmotic pressure of the soil solution is less than 1 atm pressure but that of cell sap is generally 2 atm, which may go up to even 8 atm. Because of this difference in osmotic pressure the water moves into the root. The movement of water across the cortex of the root, endodermis, pericycle is also due to osmotic concentration gradiency.. The existence of root pressure is the best evidence in support of this theory.

However, there are some objections to this theory. The major objection is how the water moves into the xylem vessels from cortical cells? The xylem sap is metabolically inactive and has low osmotic value 1.e less than 2atm, as against the osmotic pressure of cortical cell. So the water has to move against the concentration gradient, which requires some other mechanism. Further tracheids are without semi-permeable membranes.

5.7.1.2 Non-Osmotic Mechanism : Thimann (1951) and kramer (1959) proposed that water absorption is an active process and occurs due to non-osmotic process. This process requires expenditure of energy. The energy is obtained from respiration.

The non-osmotic mechanism of water absorption is supported by the following facts.

1. The non-osmotic mechanism of water absorption requires expenditure of energy. This energy is supplied from respiration. Thus there is link between the water absorption and respiration. It is evident from the fact that when respiratory inhibitors like KCN, Chloroform are applied, there is decrease in the rate of respiration and also rate of absorption of water.

2. The rate of root pressure is more in day time than in the night time. Similarly the rate of respiration is more during the day time than in the night.

3. The application of auxins enhances the rate of respiration and also the absorption of water.

The above facts indicate the positive involvement of energy (respiration) in the process of absorption of water.

5.7.1.3 Passive absorpiton : According to this theory the forces responsible for absorption of water originates in the transpiring shoots rather than in the root itself. These forces develop due to transpiration. In rapidly transpiring plant, the transpiration causes negative tension in the xylem vessels and tracheids. During transpiration the DPD in the mesophyll cells in leaf increases, which result in movements of water from the xylem cells to mesophyll cells. As the water is present as continuous column from leaves to the roots through the xylem vessels, the deficit is transmitted up to the root hairs through endodermic and cortical cells. The negative force developed in the xylem vessels draws the water into the plant from soil solution. Thus the water absorption is passive and the forces responsible for water absorption generates in the transpiring shoot system and roots have no role to play in the absorption process except that the roots are in contact with soil solution.

There are several evidences in support of this theory.

1. The rate of absorption of water is approxinately equal to the rate of transpiration.
2. It explains the movement of water against the concentration gradient.
3. It is purely physical process. It requires no expenditure of energy.
4. The movement of water by osmosis, is very slow and it cannot cope up with requirements of water by the transpiring cells of leaves. But the movement of water by passive forces is greater than simple osmotic forces and can meet any demand of water by the transpiring cells of leaves.
5. In transpiring plant, the cut ends of stems are capable of absorbing water due to development of negative tension in the shoot system due to transpiration.

5.8 FACTORS AFFECTING THE ABSORPTION OF WATER

The factors affecting the rate of absorption of water can be broadly classified into two categories.

1. External factors 2. Internal factors.

5.8.1 External factors : Most important external factors are availability of soil water, concentration of the soil solution, soil temperature and soil aeration.

5.8.1.1 Availability of soil water : The water in soil present in different forms i.e. capillary water, hygroscopic water and gravitational water etc. of all these forms of water, capillary water is useful to the plants. The water present in spaces between the soil particles and thin film of water around the soil particles is known as capillary water. Hygroscopic water is held firmly around soil particles, hence it is of no use to plants. Gravitational water is beyond the reach of the plants, as it percolates deep into the soil due to gravitation force. Field capacity may be defined as the amount of water held under the field conditions, when the soil is wet with rain or irrigation, after the excess water is drained off. Field capacity differs from soil to soil depending upon the size of soil particles. Permanent wilting percentage is the percentage of soil water left in the soil, when the leaves first exhibit the signs of permanent wilting. The amount of water between the field capacity and the permanent wilting point (PWP) is called available water. The increase in water beyond the field capacity reduces the rate of absorption because it affects the aeration of the soil and decrease in soil water diminishes the absorption of water.

5.8.1.2. Concentration of the soil solution: The soil water contains many elements in dissolved state. On account of these elements the concentration of the soil solution changes. If the soil solution is highly concentrated than the cell sap of the root hairs, water is drawn out of the plant rather than absorption of a water into the plant. In such soils plants fail to grow because of physiological dryness of the soil.

5.8.1.3. Soil temperature : The optimum temperature for the absorption of water by plants is 20° to 30°C. Low temperature reduces the rate of absorption and high temperature increases the rate of absorption . A very high temperature kills the cells.

5.8.1.4. Soil aeration:_ The absorption of water is facilitated in well aerated soils than in poorly aerated soils. The probable reason may be reduction in respiration rate by roots in poorly aerated soils. Deficiency of oxygen inhibits the growth and metabolism of the roots. Further, accumulation of Co_2 increases the viscosity of protoplasm and decrease the permeability of cell membranes. Deficiency of oxygen and accumulation of Co_2 affect severely the process of absorption of water by plants

5.8.1.5. Internal factors : The root system play major role in the absorption of water by the plants. Number of root hairs account for the rate of absorption. However, the formation of root hairs depends upon environmental factors. In Gymnosperms root hairs are absent instead, mycorrhiza hyphae absorb the water from the soil.

Some correlation has been observed in the metabolism of the roots and absorption of water. Application of anaesthetics like chloroform and KCN reduced the rate of respiration and also the rate of absorption of water. Thus metabolism of the root play an important part in the absorption of water.

QUESTIONS

I. Essay type Questions :

1. Explain the mechanism of absorption of water by plants.
2. Waater is a good solvent? Discuss.
3. Give an account of importance of water in the life of a green plant.
4. What is diffusion. How it differs from Osmosis.
5. Define Osmosis and Osmotic pressure.
6. Explain the role played by Osmosis in the absorption of water by plants.
7. Explain water potential and its relationship with Osmotic potential and pressure potential.
8. Cell is an Osmotic system - Discuss.
9. What is active absorption of water. How it differs from passive absorption of water.

II. Short Notes :

1. Osmotic potential
2. DPD
3. Water potential
4. Active absorption
5. Passive absorption
6. Imbibition
7. Role played by root-hairs in the absorption of water.

6

TRANSPIRATION

PLANT PHYSIOLOGY

6.1. INTRODUCTION

Plants absorb large quantities of water from the soil irrespective of their requirements. Only 1% of the water is utilized by the plants, 99% of water is lost from the aerial parts of the plant, either in the form of water vapour or rarely in liquid form. The loss of water from living tissue from aerial parts of the plant in the form of water vapour is termed as *transpiration* and in the form of liquid is called as *guttation*.

Table 6.1 : Differences between Transpiration and Guttation

Transpiration	Guttation
1. The water is lost from aerial parts of the plant in the form of water vapour	The water is lost from aerial parts of the plant in liquid form.
2. Transpiration occurs during the daytime	Guttation occurs mainly during the night time.
3. Transpired water is pure	Guttated water contains salts and sugars
4. Transpiration takes place mainly through the stomata, occasionally through the lenticels and cuticle	Guttation occurs through the special structures called hydathodes or water pores found on leaf tips or margins
5. Transpiration has cooling effect on the plant.	No such effect is noticed
6. It is a controlled process	It is an uncontrolled process
7. It does not depend on root pressure	It depends on root pressure.

6.2 KINDS OF TRANSPIRATION

Transpiration is of three types 1. Stomatal transpiration 2. Cuticular Transpiration 3) Lenticular Transpiration.

6.2.1 Cuticular Transpiration : Certain amount of transpiration takes place through the cuticle of leaves and herbaceous stems. Cuticle is a wax layer on the outer wall of epidermis, to prevent the loss of water. Cuticle is impermeable to water. The thickness of cuticle varies in various types of plants. Generally cuticles thin in plants growing in normal conditions, thick in desert plants and completely absent in plants growing in water. The rate of cuticular transpiration depends on the thickness of the cuticle.

6.2.2 Lenticular Transpiration : Leuticular transpiration amounts to 0.1 percent of the total transpiration. Lenticels are special structures developed in the bark of stem, containing loosely arranged mass of cells, through which the loss of water takes place. Leticular transpiration is neglible compared to the total weight of the plant.

6.2.3. Stomatal Transpiration : Most of the transpiration takes place through the stomata. Stomatal transpiration amounts to 80 to 90% of the total transpiration.

6.3 STRUCTURE OF STOMATA

Stomata are minute pores mostly situated in the epidermis of leaves and also on herbaceous stems. Each stomatal pore is surrounded by two guard cells. The guard cells are kidney shaped in dicots and dumble shaped in monocots. Each guard cell has a cytoplasmic lining and a central vacuole. The cytoplasm contains nucleus and many chloroplasts. The chloroplasts are functionless. They are incapable of participating in photosynthesis. The epidermal cell surrounding the guard cells are modified cells called 'subsidiary cells'.

The size and shape of stoma and guard cells vary from plant to plant. The number of stomata in a specified area also varies and depends on plant and environment. Xerophytes possess more number of stomata per unit area than the mesophytes. The sub merged plants lack stomata.

6.3.1 Stomatal Types : Five categories of stomatal types have been recognized on the basis of their distribution.

6.3.1.1 Apple or Mulberry type : Stomata confined to only lower surface of the leaves are called hypostomatic plants. Eg. Apple, Peach, Mulberry.

2. Potato Type : Stomata are distributed on both the surfaces of leaves but more on the lower side and less on the upper side. Such leaves are called amphistomatic leaves Eg. Potato.

3. Oat Type : Stomata are distributed equally on both the surfaces of leaves. Such leaves are called amphistomatic and isostomatic Eg. Maize, Oats.

4. Wild Lily type : Stomata are found distributed only on the upper side. Such leaves are described as epistomatic Eg. Water lily, Nymphea etc.

5. Potamogeton type : Stomata are altogether absent or even if present they are vestigial. Such leaves are called Astomatic plants Eg. Potamogeton and submerged plants.

6.1.3.2 Mechanism of stomatal opening and closing :

The opening and closing of stomata mainly depend upon the turgidity of guard cells. The stomata open when the guard cells are turgid and stomata are closed when the guard cells are flaccid. The size of the pore depends on the degree of turgidity of guard cells (fig. 6.1).

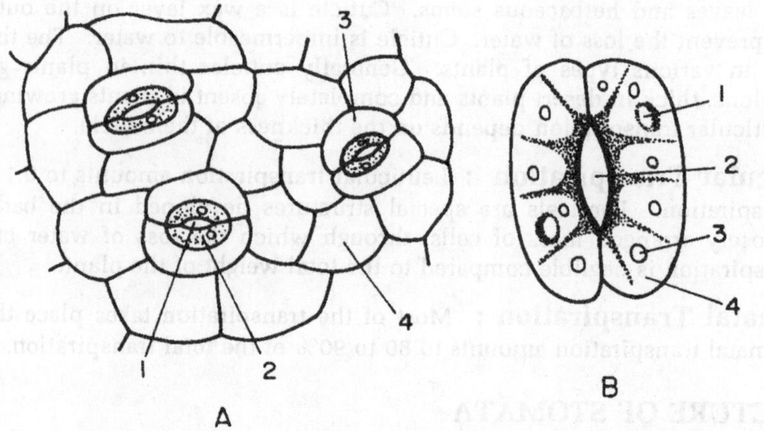

Fig 6.1 Structure of stomatal apparatus B. Single Stomata
A = 1 Stomatalpore, 2. Guard Cell, 3. Subsidiary Cell, 4. Epidermal Cell

The inner wall of the guard cells surrounding the pore is thick and inelastic and outer wall is thin and elastic. When the turgidity of guard cells increases due to the entry of water into the guard cells from adjacent cells, the outer thin wall stretches and inner wall is being thick and inelastic becomes concave. As a result of which the stomatal pore widens and opens, when the turgidity decreases the guard cells come back to their original position and the stomata close. Thus the opening and closing of the stomata depend upon the changes in turgidity of the guard cells. But how the changes in turgidity of the guard cells occur is a matter of controversy. Several theories have been proposed to explain the changes in turgidity of guard cells. Some of the theories have been considered under.

6.3.1.2.1 Von Mohl theory of Photosynthesis in guard cells : Von Mohl (1956) observed the presence of chloroplasts in guard cells. He also observed that stomata open during the day time and close during the night.

On the basis of his observations, he proposed that the chloroplasts present in guard cells participate in photosynthesis during the day time synthesize sugars, which dissolve in the cell sap of guard cells, increasing the osmatic concentration of the guard cells resulting in the increase of turgidity of guard cells, as a result of which stomata open. The sequence of events leading to the opening of stomata can be represented as follows.

Light ---------> Synthesis of carbohydrates -------> Dissolution of carbohydrates ------>
In guard cells in the cell sap of guard cells

1 2 3

Increase in the osmotic concentration of cell sap of guard cells ---------> Increase in turgidity of guard cells due to endosmosis (entry of water from neighbouring cells -------------> Stomata open

4 5 6

However this theory was not accepted because of major objection raised against the theory. The chloroplasts present in guard cells are incapable of synthesizing the carbohydrates.

6.3.1.2.2 Theory of starch Sugar Hypothesis : Lloyd (1908) proposed this hypothesis. According to this hypothesis the opening and closing of stomata is due to changes in turgidity of guard cells, which is associated with the conversion of starch to sugar and vice versa in guard cells. According to Lloyd, the amount of starch in guard cells increases during the night times (stomata closed) and decreases during the day time (stomata open).

Lotfield (1921) also observed similar changes. Sayre (1926) made further studies on pH changes. He observed that stomata remain open in alkaline pH even in the night times and closed in acidic pH even in the day time. He proposed that pH determines inter conversions of starch to sugar and vice versa.

The various steps involved in opening and closing of stomata may be summaried as follows.

Table 6.2 steps involved in opening and closing of stomata

Day Time	Night Time
1. Co_2 released due to respiration consumed during the daytime in photosynthesis.	1. Co_2 released from the respiration is accumulated in the inter cellular spaces due to non-occurance of photo-synthesis
2. pH of the guard cells increases	2. pH of the guard cells decreases
3. An increase in pH favours hydrolysis of starch into sugars.	3. A Decrease in pH favours formation of starch from soluble sugars.
4. Due dissolution of sugars in the cell sap of guard cells the osmotic concentration increases.	4. Due to conversion of sugars to starch the Osmotic concentration decreases.
5. Due to endosmosis water enters into the guard cells from the neighbouring cells.	5. Water moves out of the guard cells to subsidiary cells.
6. The turgor pressure of the guard cells increases.	6. The turgor pressure of guard cells decreases. The guard cells become flaccid
7. As a result the stomata open	7. As a result the stomata close.

6.3.1.2.3 Theory of starch Glucose Inter conversion : Sin and Tung (1984) discovered the presence of phosphorylase enzyme in the guard cells. The enzyme phosphorylase catalyzes the reaction in the presence of inorganic phosphate as follows.

$$\text{Starch} + 1\text{P} \underset{\text{pH 5}}{\overset{\text{pH 7 Phosphorylase}}{\rightleftharpoons}} \text{Glucose} - 1 - \text{Phosphate.}$$

Steward (1964) criticised above scheme and pointed out that unless glucose —1-phosphate is further converted to glucose and inorganic phosphate, no appreciable change in osmotic concentration can be obtained. Hence Steward proposed a modified scheme, which can be represented as follows (fig. 6.2).

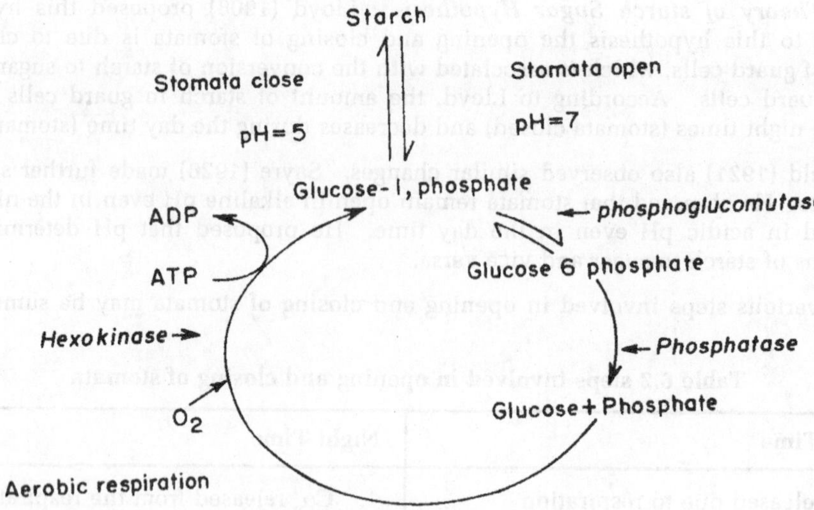

Fig 6.2 Steward's (1964) Scheme

Light ---------> Photosynthesis in mesophyll -------------> consumption of Co_2 from the intercellular spaces ---------> Increase of pH in the guard cells ---------> Enzymatic conversion of starch into glucose ---------> Increase in osmotic concentration of call sap cells due to dissolution fo Glucose in cell Endosmosis ---------> Entry of water in to guard cells ---------> guard cells become turgid ---------> stomata open. Reverse changes lead to closure of stomata.

However, the steward's modified scheme has been subjected to criticism on the following points.

1. In some plants guard cells lack starch, yet the opening and closing of stomata occurs as usual
2. The Co_2 concentration to which the guard cells are exposed in the atmosphere of intercellular spaces, is insufficient to induce any major change in pH of guard cells, required for the conversion of starch to glucose.
3. Starch $\rightleftharpoons$ sugar interconversion process is too slow to account for opening and closing of stomata.
4. Stomata can open independent of Co_2 concentration by light and temperature.

6.3.1.2.4 Theory of Glycolate metabolism : Taking into account the fact that Co_2 concentration in the environment of stomata is insufficient to bring out opening and closing of stomata, Zelitch (1963) proposed Glycolate theory. According to him the production of glycolate in the guard cells play an important role in the opening of stomata. Glycolate is produced under low concentration of Co_2. Glycolate give rise to carbohydrates and also it participates in the production of ATP, which provide energy for the opening of stomata.

The process can be summarized as follows.

1. NADP + ADP + P_3 + H_2O ---------> NADPH + H+ ATP $+1/2O_2$

2. Glyoxylate + NADPH + H^+ ---------------------------> (Glyoxlate reductase) Glycolate + NADP +

3. Glycolate + 1/2 O_2 ---------------------------> Glyoxylate + H_2O_2

4. H_2O_2 ------------------> H_2O + 1/2 O_2

The theory is supported by the following facts.

1. £- hydroxy sulphonate, which is inhibitor of glycolic acid oxidase, prevent the opening of stomata.
2. A correlation exists between the conconcentration of £-hydroxy sulphonates, the concentration of glycolic acid and the degree of stomatal opening.
3. Low conconcentration of oxygen inhibits both glycolic acid metabolism and stomatal opening.
4. Addition of glycolic acid opens the stomata.

Though this theory explains the opening and closing of stomata under conditions of limited supply of Co_2, yet it failed to explain the following points.

1. It fails to explain the opening of the stomata in the light even in the absence of Co_2, when there is no Co_2 assimilation or glycolate metabolism.
2. It fails to explain how stomata widely open in the darkness in response to Co_2 removal.

6.3.1.2.5 Theory of Proton Transport : (Active K^+ - transport) : Levitt (1974) proposed the proton transport concept to explain the mechanism of opening and closing of stomata. According to this theory K^+ ions play a prominent role in the opening and closing of stomata. The guard cells take the potassium ions from the neighbouring cells and open, release the K^+ ions out and close. The adjacent epidermal cells act as storage cells.

The transport of potassium ions into guard cells is an active mechanism and involves a hydrogen — potassium ion exchange pump in the plasma membrane. The potassium ions in the guard cells are neutralised by organic acids especially malic acid

$$R(COOH)_2 \rightleftharpoons R(COO) + 2\,H^+$$
$$H^+ \rightleftharpoons K^+$$

Malic acid provide H+ ions for exchange for K+ ions and anion to R(COO) to balance the charges of K^+ ions.

The synthesis of malic acid in illuminated guard cells accompanies the influx of K^+ ions, but the exact mechanism is not understood fully, It may be due to convertion of starch to malic acid in guard cells or the excretion of malic acid from chloroplasts into the cytoplasm of guard cells due to the effect of light.

Noggle and Fritz (1976) summarized the sequence of events as follows:

Light -------> Malic acid Production -------> Disassociation of malic acid into Hydrogen (H+) and Malate ions -------> Influx of K^+ in exchange of H^+

Transport of -------> Potasim malate Into the cell sap | Osmotic ------> entrance of water into guard cells | Increase of turgor -------> pressure | Stomata open

The closure of stomata involves participation of inhibitor hormone, Abscissic acid (ABA). ABA inhibits K^+ uptake in dark by changing the diffusion and permeability of guard cells. K^+ ions move out of the guard cells. ABA induces acidification in guard cells which result in lowering of pH in guard cells. Lower pH favours the conversion of sugar to starch, resulting in the falling of osmotic pressure. Water moves out of the guard cells. Guard cells become flaccid and stomatal pore is closed.

6.4 STOMATAL OPENING IN SUCCULENT PLANTS

The stomata in succulent plants, which carriey on crassulean acid metabolism (CAM), open in the night times and remain closed during the day time. Nishid (1963) explained the mechanism. According to him in succulent plants during the night times, there is incomplete oxidation of carbohydrate in respiration and accumulation of organic acids therefore, Co_2 is not released and as a result stomata remain open during the night time. Whereas, during the daytime the accumulated organic acids undergo oxidation, releasing the sufficient quantity of Co_2, which causes the stomata closed

$$\underset{\text{(glucose)}}{2\,C_6H_{12}O_6} + 3\,O_2 \xrightarrow[\text{time}]{\text{During night}} \underset{\text{(Malic acid)}}{3\,C_4H_6O_5} + 3\,H_2O$$

$$C_4H_6O_5 + 3\,O_2 \xrightarrow[\text{Time}]{\text{During day}} 4\,Co_2 + 3\,H_2O$$

6.5 PLANT ANTITRANSPIRANTS

Any substance or material used to reduce the transpiration rate termed as anti-transpirants.It is evident that plants absorb large quantities of water from the soil, but only a fraction of it is utilized by the plants for its various metabolic activities. Some times the rate of transpiration becomes equivalent to rate of absorption of water from the soil. Under drought conditions transpiration becomes injurious to plant and some times it become fatal to the plant. Hence, it is desirable to reduce the loss of water from the plants. It will be beneficial to agriculturists. In recent years efforts are made to find out anti-transpirants which reduce the loss of water from the plants without affecting the gaseous exchange in respiration and photosynthesis and plant growth.

Some of the examples of anti-transpirants are colourless plastics, silicone oils, low viscosity waxes, phenyl mercuric acetate, abscissic acid and Co_2 etc. Of these colourless plastics, silicon oils, low viscosity waxes are sprayed on leaves with the object of forming a thin film over the leaves, which is permeable to Co_2 and oxygen but not to water. Phenyl mercuric acetate, when applied in low concentration (10^{-4} M), gave good results, resulted in partial closure of stomata. Similarly abscissic acid also useful in inducing partial closure of stomata. Co_2 also acts as good antitranspirant. A little rise in Co_2 concentration from the normal concentration of 0.03% to 0.05% induced the partial closure of stomata. Co_2 also inhibits photorespiration, which is advantageous to the plant.

6.4 FACTORS AFFECTING THE TRANSPIRATION

The factors affecting the rate of transpiration can be broadly classfied into two categories viz. External factors and Internal factors.

6.4.1 External factors : The factors which originate in the environment external to the plant are called external factors such as humidity of the air, temperature, wind velocity, light, atmospheric pressure etc.

6.4.2 Humidity of the air : The relative Humidity of the air affect the rate of transpiration to a great extent because the rate of transpiration greatly depends on the DPD gradient between the internal atmosphere of the leaf and outside atmosphere. Generally the internal atmosphere of the leaf is nearly saturated, while outside atmosphere is relatively unsaturated. Hence DPD gradient exists. Therefore, the water vapour diffuses from the leaf to the outside atmosphere. The transpiration is slow when the air is humid. The rate of transpiration increases with the decrease in the humidity of the air.

6.4.3 Temperature : Temperature increases the rate of transpiration. Temperature effect is two fold. 1) by increasing the rate of evoporation of water from cell surface 2) by decreasing the humidity of the air, thereby steepening the DPD gradient between the external atmosphere and internal atmosphere of leaf.

The increase in rate of transpiration follows vant Hoff's rule to a certain extent that is for every 10°C rise in temperature the rate of transpiration becomes double.

6.4.1.3 Wind velocity : The increase in wind velocity increases the rate of transpiration by removing the humidity air from the immediate vicinity of plant. Transpiration rate is high in mild wind but winds of high velocity causes the closure of stomata, transpiration is retarded.

6.4.1.4 Light : Light has no direct effect on the rate of transpiration but it affects the transpiration indirectly in two ways. Firstly by controlling the opening and closing of stomata and secondly by increasing atmospheric temperature. The rate of transpiration increases with the increase in intensity of light and decreases in dark.

6.4.1.5 Atmospheric Pressure : The reduction of atmospheric pressure reduces density of the external atmosphere there by permitting diffusion of water vapour from the leaf. The plants growing on hills exhibit higher rates of transpiration because of low atmosphere pressure on hills.

6.5.2 Internal factors

6.5.2.1 Stomatal frequency : The number of stomata per unit area of leaf surface is called stomatal frequency. Stomatal frequency varies from plant to plant and also depends on environment of the plant. The increase in stomatal frequency increases the rate of transpiration. It also depends on the degree of opening of stomata.

6.5.2.2 Structure of leaves : Certain plants reduce their leaf size to avoid excessive transpiration. Xerophytic plants have needle like leaves. Eg. *Pinus*. Certain plants have thick cuticle and wax like substances on the leaf surface. In several other plants stomata are sunken in pits Eg. Cycas.

Several other factors such as intercellular spaces, amount of spongy tissue, orientation of leaves also affect the rate of transpiration.

6.5.2.3 Significance of transpiration : Transpiration is often decribed as "Necessary or unavoidable evil". Some others described it as beneficial to plants. Let us examine the above statement in detail.

Transpiration is Advantageous :

Transpiration is considered to be advantageous to the plants for the following reasons.

1. It helps in the ascent of Sap.

2. It is supposed to play an important role in the absorption of water and mineral salts from the soil.
3. It is supposed to play an important role in the translocation of solutes in the plant.
4. It helps in maintaining the optimum temperature of plants.
5. It has a cooling effect on the plant
6. It brings about the opening and closing of stomata, which indirectly facilitate the process of photosynthesis and respiration

Transpiration is unavoidable and Dangerous to the plant.

It is very well known fact that plants transpire large quantities of water and only a small quantity of it is utilized by the plants for its metabolic activities. Loss of water by the plants does not serve any useful purpose to the plants. On the otherhand, transpiration becomes lethal to the plants because of excessive loss of water. in drought conditions.

Transpiration is unavoidable because as long as stomata open for gases exchange during the respiration and photosynthesis, the loss of water through the stomatal pores is unavoidable.

Transpiration is necessary because plants absorb enormous amount of radiant energy from the sun. Only less than one percent is being utilized by the plants. Transpiration is the main process for dissipating sun's excess energy falling on leaf. Thereby keeping the temperature of plant under control.

QUESTIONS

I. Essay type Questions :

1. What is transpiration. How it differs from evoporation.
2. Describe the different kinds of transpiration.
3. Explain the mechanism of stomatal openingn and closing and its related theories briefly.
4. Explain starch $\rightleftharpoons$ sugar hypothesis.
5. Describe the role played by K^+ ions in the mechanism of stomatal opening and closing.
6. What is Guttation? How it differs from transpiration.
7. Transpiration is a "necessary evil". Discuss
8. Write an essay on the role played by environmental factors on the rate of transpiration.
9. Explain steward's hypothesis and its validity.
10. Explain the role played by pH onn the stomatal mechanism of opening and closing.
11. Describe zelitech's hypothesis.

II. Short Notes :

1. Cuticular transpiration
2. Guttation
3. Hydathodes
4. Astomatons plants
5. Lenticels
6. Guard cells
7. Stomata Structure

7

PLANT PHYSIOLOGY

ENZYMES

7.1 INTRODUCTION

Many biochemical reactions taking place in the cell are catalysed by certain specialized organic substances called enzymes. They accelerate the biochemical reactions without being used up and without affecting the equilibrium of the reactions. Hence enzymes are called as 'Biological Catalysts'.

The use of enzymes by man can be traced back to pre-historic times, when ancient Greeks used enzymes for the production of wine by fermentation process. Other practices in which the activity of enzyme was used were the making of cheese and vinegar etc.

Much of the credit for working on enzymes goes to Louis pasteur, a great French Scientist, who named the active principle in the fermentation process as 'ferment' (a name used for enzyme at that time). He held the view that 'ferments' were active only in the living cells. Buchner (1897) demonstrated that living cells were not necessary for the fermentation process, but the extract obtained from yeast cells was responsible for fermentation process. He named the active principle as 'zymase' which literally means 'in the yeast'. The next significant discovery in the study of enzymes was the isolation and purification of an enzyme called 'urease' by James. B.Sumner (1926). Later several other enzymes were isolated and characterised.

7.2 CHEMICAL COMPOSITION OF ENZYMES

All enzymes are proteins but not all the proteins are enzymes. Many enzymes contain only protein. Some enzymes contain non-protein part in addition to protein. If the enzyme contains only protein, it is called 'simple enzyme'. If the enzyme contains non-protein part besides protein, it is called 'conjugated or Holo enzyme'. The protein part of the conjugated enzyme is called "apo enzyme" and the non-protein part is known as "prosthetic group". The prosthetic groups are of two types:

1. **Co-factors or Activators** : If the prosthetic group is of metallic or inorganic ions. Eg. Fe^{++}, Zn^{++}, Mg^{++} etc.

2. **Co-enzymes** : If the prosthetic group is an organic compound Eg. NAD, NADP, FMN, FAD etc.

Enzymes are active only when both apoenzyme and prosthetic groups are firmly associated. If they are separated they become inactive. Many co-enzymes act as donor or acceptor of atoms that are added to or removed from the substrate. They play an important part in oxidation- reduction reactions.

7.3 PROPERTIES OF ENZYMES

1. Enzymes are giant molecules. The molecular weight of enzymes ranges from 35,000 to 4,83,000.

2. Enzymes are chemically proteinaceous and physically colloids. Because of their colloidal nature enzymes provide large surface area for the reactions to take place.

3. Enzymes are catalytic in their action. They speed up the rate of reaction without affecting their equilibrium.

4. Enzymes a.e active in extremely small quantities. But they are capable of bringing about chemical change in the large amount of substrate. It is like a scissor, which can be used to cut tons of cloth. The number of moles of substrate converted per minute by 1 mole of enzyme is called turn-over number. This number is different for different enzymes and vary from 100 to over 3,00,000.

5. Enzymes activity can be inhibited by certain inorganic poisons such as salts of heavy metals, which partially or completely inhibits the enzymes activity. In some cases the poisons can kill the enzymes completely.

6. Enzymes are thermolabile i.e. heat sensitive because of their protein nature.

7. Enzymes are pH sensitive. The enzyme activity depends upon the pH of the medium. At some particular P^H of the medium the enzymes show the maximum activity.

8. Enzymes are specific in their action. They act only on a particular substrate. Enzyme specificity can be classified as follows.

i) **Absolute Specificity** : A specific enzyme acts on the specific type of substance. Eg. urease acts on urea alone.

ii) **Absolute group specificity** : These enzymes act on a specific chemical bond or group, though such group may occur in different types of substances Eg. Alcohol dehydrogenase enzyme acts on all substance having alcoholic groups.

iii) **Related group specificity** : These enzymes catalyse the changes of all those chemical groups, which are totally or partially similar in structure. Eg. Trypsin can act on peptide bonds between the amino acids; arginine and lysine and also on their ester bonds.

iv) **Optical rotation specificity** : Some enzymes catalyse the change of one type of optical isomer but not its opposite optical isomer. Eg. a - amino acid oxidase can bring about chemical changes in laevo rotatory amino acid only but not their dextro-rotatory forms.

7.4 ENZYME ACTION

Michaelis and Menton (1913) proposed a theory on mode of enzyme action. According to them every enzyme has got certain active sites for the attachment of the substrate molecules. Accordingly enzymes and substrate molecules form a weakly bound enzyme substrate complex, which on hydrolysis give rise to end products. This theory can be represented by the following equation.

Enzyme + Substrate ---------> Enzyme — Substrate complex ------> End products + Enzyme

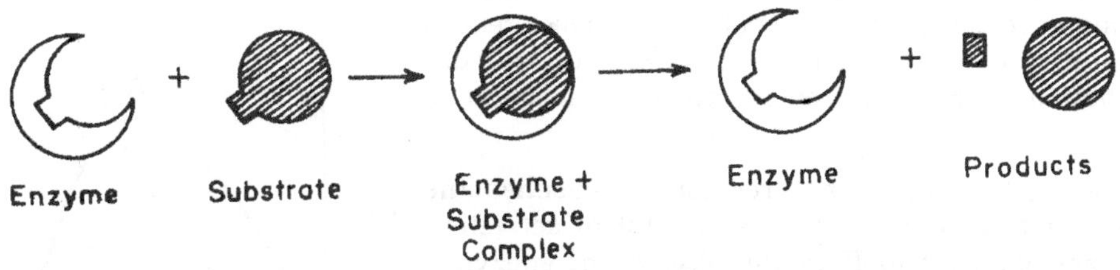

Fig. 7.1 Luck and key theory to explain enzyms action

7.4.1 Lock and key theory (fig 7.1): Fischer (1898) proposed 'Lock and key theory' to explain the mode of enzyme action. This theory was supported by Paul Fildes & D.D.Wood. They proposed that a particular key can open a particular lock. Similarly a particular enzyme can act only on a particular substrate. Every enzyme has specific active site, where a particular substratum can be attached. This theory is based on physical contact between substratum and enzyme molecule. Enzyme remains unchanged but the substratum is broken into products.

The best evidence in support of Lock and Key theory is provided by 'Competitive inhibiton'. The substances which inhibit the activity of enzymes are called 'inhibitors'. When the medium contains real substrate molecules and also other molecules which resemble in structure with that of real substrate molecules, then there will be competition between the real and false molecules for the occupation of the active site of enzymes, resulting in the reduction in the rate of enzyme activity. This phenomenon is called Competitive inhibition. For example the enzyme succinic dehydrogenase catalyses the dehydrogenation of succinic acid. Malonic acid resembles succinic acid in structure (fig. 7.2). If succinic acid and molonic acid molecules are present in the medium, then there is a competition between the succinic acid molecules and malonic acid molecules. Some of the malonic acid molecules adsorb on the surface of the enzyme thereby making the enzyme inactive. As a result of which the rate of reaction is inhibited. It is like a wrong key struck up in the lock.

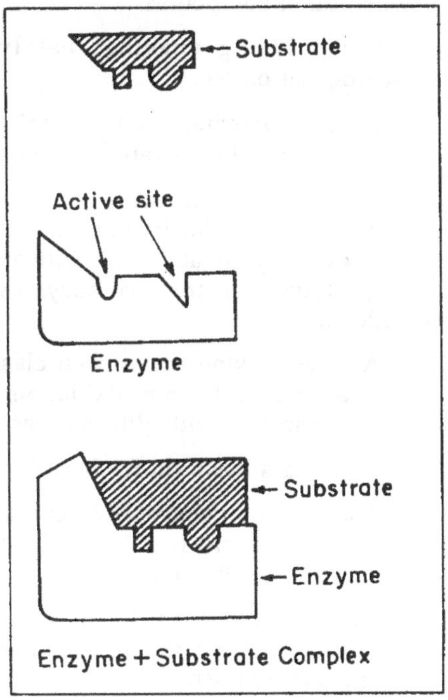

Fig. 7.2 Induced - fit theory

7.4.2 Induced-fit theory : This theory was proposed by Koshland (1959). According to him the substrate molecules induces certain structural changes in the enzyme molecule. As a result the enzyme molecule changes its shape according to the requirement of substrate molecule (fig 7.2).

7.5 ENZYME KINETICS

Enzymes act as catalysts and bring about chemical changes in the cell at normal temperature and pressure by lowering the activation energy and speed up the chemical reaction without undergoing any change either in quality or in quantity.

The above facts can be explained in this way. Every reactant molecule has to cross over a hill-like barrier to reach the active state. Unless the reactant molecules get the activation energy, it cannot reach the active state. It cannot cross the barrier. Enzymes reduce the quantity of energy barrier i.e. it reduces the amount of activation energy required for chemical reaction to occur.

The fig.7.3 shows that reactant A reaches the activated state (AX) by getting X amount of activation energy and changes to B in the absence of enezyme. In the presence of enzyme the reactant A requires only Y amount of activation energy (less than X) and changes to B.

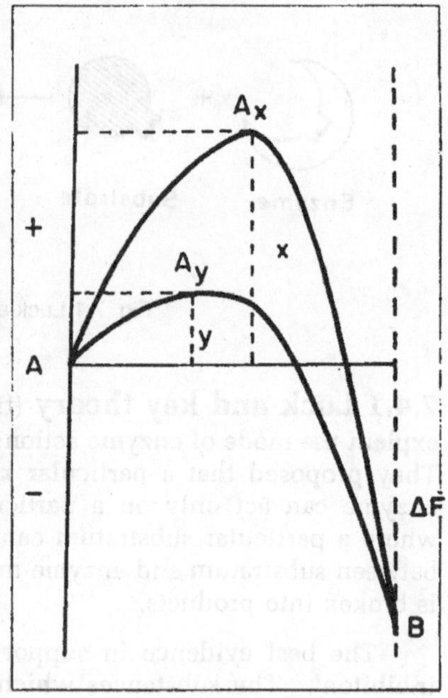

Fig 7.3 Enzyme Kinetics

It is presumed that substrate molecules are adsorbed on the surface of enzyme molecules. The enzyme and substrate form an enzyme — substrate complex. Within this complex some of the bonds in the substrate break and rearrange to form products much more rapidly than in the absence of enzymes. For example the amount of energy required for breaking off H_2O_2 is 18000 cal/mol but in the presence of enzyme it is only 6400 cal/mol.

7.6 CLASSIFICATION OF ENZYMES

Earlier enzymes have been classified into two main categories. 1) Hydrolysing enzymes. 2) Desmolysing enzymes.

Hydrolysing enzymes catalyse the hydrolytic reactions utilizing the water for the dis-association of compound.

Desmolysing enzymes catalyse the reactions like oxidation, reduction and dehydrogenation etc.

Later transferring enzymes, which were originally in hydrolyzing enzymes, are separated as third group called 'Transferring enzymes'.

The above classification doesn't serve any purpose to-day as the number of enzymes increasing rapidly day by day. In 1961, the International union of biochemists had realised the inadequacy of above classification and proposed a modern classification based on reaction types, substrate, co-enzymes etc. The salient features of modern classification are as follows:

i. All the enzymes have been classified into six major classes, based on six main types of reactions. Such as oxidation reaction, transfer of groups, hydrolysis, lytic reactions, isomerisation and addition reactions.

ii. Each class has been further divided into sub-classes, sub-sub classes.

iii. The enzyme name should contain two parts; first part being the name of the substrate and the second part being the type of reaction catalysed. The enzyme name should end with suffix 'ase'.

iv. Each enzyme has got a code number consisting of four digits, in which the first digit is of class, second of sub-class, third of sub-sub class and the fourth digit is the enzyme number.

v. Additional information about the type of reaction may be given in parenthesis.

The six classes of enzymes are as under

7.6.1 Oxido-reductases : This group of enzymes catalyse oxidative and reductive reactions by removal or addition of hydrogen, oxygen or electrons from or to the substrate.

Eg: EC 1.1.1.1 = Alcohol : NAD oxidoreductase (dehydrogenating)

C_2H_5OH + NAD $\rightleftharpoons$ Acetaldehyde + NADH + H^+

7.6.2 Transferases : This group of enzymes catalyse transfer of groups from one compound to other. Transfer of groups include aldehyde, ketone, acyl, alkyl, glycosyl, phosphate, sulphur and amine groups.

Eg: EC 2.7.1.1 = ATP - D-hexose phosphotransferase (hexokinase).

D-hexose + ATP $\rightleftharpoons$ D-hexose phosphate + ADP

7.6.3 Hydrolases : Hydrolysing enzymes catalyse the hydrolytic reactions in which a water molecule is used, resulting in the cleavage of the substrate. Hydrolysing enzymes have long list of enzymes such as peptidases, phosphatases, sucrases, maltase, lipase etc.

$$\underset{\text{sucrose}}{C_{12}H_{22}O_{11}} + H_2O \xrightarrow{\text{Invertase}} \underset{\text{glucose}}{C_6H_{12}O6} + \underset{\text{fructose}}{C_6H_{12}O_6}$$

7.6.4 Lyases : These enzymes remove groups from substrates by a mechanism not involving hydrolysis.

Eg: 4.1.2.7 = Ketose — 1 — phosphate aldehydelase (Aldolase).

Fructose 1- 6 diphosphate $\rightleftharpoons$ Dihydroxyacetone Phosphate + Phospho-glyceraldehyde

7.6.5 Isomerases : These enzymes catalyse different types of isomeric reactions including optical, geometrical and positional isomers.

Eg. a - alanine $\rightleftharpoons$ D-alanine

Glucose — 6 — phosphate $\rightleftharpoons$ Fructose — 6- phosphate

7.6.6 Ligases or Synthetases : These enzymes catalyse the condensation of two or more molecules coupled with cleavage of ATP or some other nucleocide triphosphate.

Eg. 6.2.1.4 = Succinate : Co-enzyme A ligase (GDP) (Succinic thiokinase)

Succinate + Co-enzyme A+GTP $\rightleftharpoons$ GDP + Phosphate + Succinyl COA

7.7 FACTORS AFFECTING THE ENZYME ACTIVITY

Like all other chemical reactions, an enzyme catalysed reaction is also affected by several external factors. The following factors affect the rate of enzyme activity.

7.7.1 Substrate Concentration : When the enzyme concentration is kept constant, changes in substrate concentration affect the rate of enzyme activity (fig 7.4).The enzyme molecules are normally much large than its substrate molecules. When the substrate is in low concentration, the active sites on the enzyme molecule are not fully occupied. So the enzyme is not working at maximum efficiency. If the substrate concentration is gradually increased then there will be gradual increase in the rate of reaction till the maximum efficiency is achieved, beyond which there will not be any further increase in rate even if substrate concentration is increased. This is because at the maximum rate of enzyme activity all the active sites of enzymes are fully used. Further increase of substrate concentration will not increase the rate of enzyme activity.

7.7.2 Enzyme Concentration : If the substrate concentration is sufficient, then the rate of enzyme activity increases with the increase of enzyme concentration. The rate is proportional to the increase of enzyme concentration (fig.7.5).

7.7.3 Temperature : Enzymes, because of their protein nature, are sensitive to temperature. Therefore, the changes in temperature affect the enzyme activity. It is observed that with the rise of 10ºC, the reaction rate is doubled. The maximum temperature for enzyme activity ranges from 50ºC to 55ºC. Many enzymes become inactive at 60ºc. The optimum temperature for enzyme activity ranges from 25ºc to 35ºc. The minimum temperature for enzyme activity is 0ºc. Enzymes freeze at 0ºc temperature but when the temperature is raised they become active again. So this is reversible change. (fig. 7.6).

7.7.4 Effect of pH : The changes in the H-ion concentration of the medium affect the enzyme activity. Every enzyme has got a maximum activity at some optimum pH, increase or decrease of pH of the medium reduces the enzyme activity. Some of the optimum pH values of enzymes are given below.

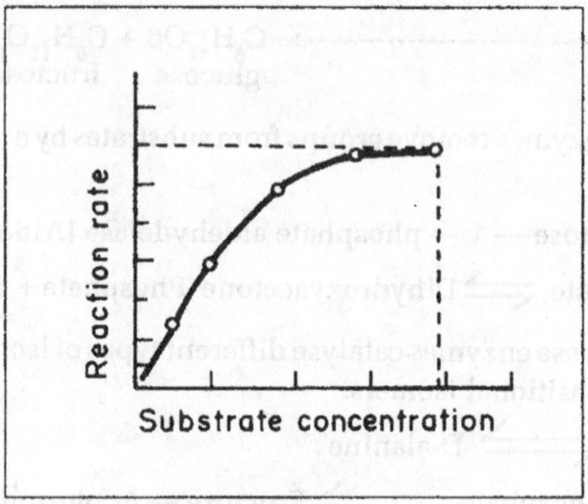

Fig 7.4 Effect of Substrate concentration on enzyme activity

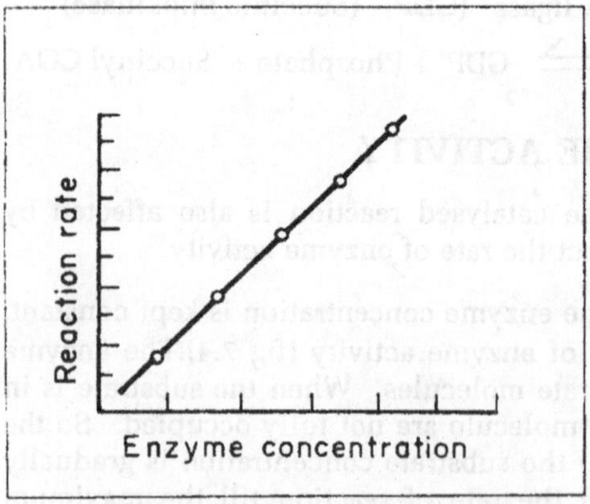

Fig 7.5 Effect of enzyme concentration on enzyme activity

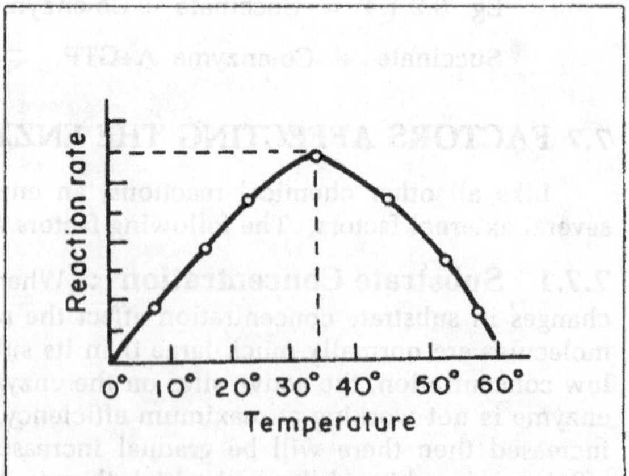

Fig 7.6 Effect of temperature on enzyme activity

Enzyme	Optimum pH of the medium
Pepsin	2.0
Invertase	4.5
Peroxidase	5.0
Cellobioase	5.0
Maltase	7.0
Amylase	7.0
Urease	7.0
Catalase	7.0
Trypsin	8.0

7.7.5 Effect of Inhibitors : The substances which decrease or destroy, the activity of enzymes are called 'Inhibitors'. Inhibition is of two types 1. Competitive inhibition 2) Non-competitive inhibition. Competitive inhibition is already explained. Some substances like poisons destroy the activity of enzymes permanently. Some of the end products also inhibit the enzyme activity and the phenomenon is called 'end product inhibition'.

QUESTIONS

A. Essay type Questions :

1. Write an essay on enzymes
2. Enumerate the main characterstics of enzymes
3. Explain the mode of enzyme action
4. Describe the classification of enzymes with the suitable examples
5. Give a brief account on factors affecting the enzyme activity
6. Describe the various theories proposed to explain the mechanism of enzyme action
7. Explain with suitable examples the specificity of enzyme action
8. Give the examples for co-enzymes and explain their importance in the metabolism of plants.
9. "All enzymes are proteins, but not all the proteins are enzymes" comment.

B. Short Notes :

1. Competitive inhibition
2. Coenzymes
3. Co-factors
4. Enzyme specificity
5. Apo enzyme
6. Active Centre

C. Define the following in one or two sentences.

1. Prosthetic group
2. Turn over number
3. Activation energy
4. Inhibitors

8 PLANT PHYSIOLOGY

PHOTOSYNTHESIS

8.1. INTRODUCTION

Photosynthesis is the most important biological process on earth which not only provides food to heterotrophic organisms but also cleans the environment by releasing large amount of oxygen. It is estimated that about 1.55 X 10^{11} tonnes of dry matter is produced by this physiological event by fixing about 200 billion tonnes of carbon by plants. The carbon fixed by plants of remote past was ultimately transformed into reserves of fossil fuels, namely coal, petroleum and natural gas.

Martin Kamen (1963) defines Photosynthesis as 'A Process in which electromagnetic energy is converted to chemical energy that can be used for various biosynthetic processess'. Thus, photosynthesis is the conversion of light energy into chemical energy, where, plants absorb light energy through the chlorophylls and this energy is utilized for reducing carbon-di-oxide to sugar with the help of water.

The photosynthesis equation can be shown in the following way.

$$6CO_2 + 12\,H_2O \xrightarrow[\text{Chlorophyll}]{\text{Light}} C_6\,H_{12}\,O_6 + 6O_2 + 6H_2\,O$$

8.2. HISTORY

Stephane Hales (1727) for the first time suggested that plants derive their nutrition partly from air and also require light for food preparation. It was Nicholas de Saussure (1804) while performing the first experiment on photosynthesis, demonstrated that CO_2 and O_2 are exchanged and water participates in the process. Mayer (1842) recognised the importance of sun light in the photosynthetic process and observed that solar energy available in the form of electro-magnetic radiation would be converted into chemical energy. Julius Sacchs (1864) reported the presence of starch grains in the chloroplasts exposed to sun light which is a product of photosynthesis.

By the end of 19th century, it was clear that plants absorb CO_2 and convert to organic matter releasing O_2 as by-product and water was found to be an essential ingredient along with light as a source of energy.

Blackman (1950) an English plant physiologist, indicated that the photosynthesis consists of two phases 1. a fast light dependent reaction phase and 2. a slower dark reaction phase independent of light which requires carbondioxide.

Blackman further suggested that while light reaction is a real photochemical reaction, the dark reaction or Blackman reaction is a thermal or biochemical reaction. Robert Hill and Scarisbrick (1937) earlier proposed that the isolated chloroplasts in the presence of light, water and a suitable hydrogen acceptor such as potassium ferric cyanide produce O_2 in the absence of CO_2. This experiment proved that light induces splitting of water molecule which is known as photolysis of water. It also provided sufficient evidence that O_2 thus released comes from the splitting of H_2O but not from CO_2. Now the process of O_2 evolution by chloroplast in the light is popularly known as "Hill reaction" in the honour of Robert Hill. The reaction can be represented in the following way.

$$2H_2O \xrightarrow[\text{Chlorophyll}]{\text{Light}} 4H^+ + 4e^- + O_2$$

American chemist, Samuel Ruben and his co-workers (1941) using ^{18}O in the form of $H_2{}^{18}O$ further confirmed that O_2 evolved in photosynthesis comes from water but not from CO_2 and the same concept has been accepted with the works of Alen Stemler and Richard Radmer in 1975.

The understanding about photosynthesis expanded rapidly after 1950. It was Calvin and Benson used $^{14}CO_2$ and traced the path of carbon in dark phase. This pathway of CO_2 reduction is known as Benson — Calvin cycle or Malvin Calvin cycle for which Calvin won noble prize in 1964.

Daniel Arnon (1954) studied the mechanism of ATP generation during electron transport in light phase. Emerson (1957) identified the participation of two pigment systems in the light reaction namely, PSI and PS II. Hill and Bendall (1960) have shown that the two photosystems operate in a specific 'Z' series in photosynthesis. Later, Mitchell (1961, 1979) proposed a chemiosmotic hypothesis for the production of ATP and won nobel prize in chemistry in 1978.

The green algae like *Chlorella, Scenedesmus* and *Chlamydomonas* have been widely used for the photosynthesis experiments. Some other notable contributions which lead to the present day concept of photosynthesis will be discussed in the following sections.

8.3. THE PHOTOSYNTHETIC APPARATUS :

In algae and higher plants, the photosynthesis is carried out in sub-cellular organelle called chloroplast. The photosynthetic apparatus are located in chloroplasts more specially in the thylakoid membrane. On the other hand, in prokaryotes such as cyanobacteria and phototrophic bacteria, they are found on the cell membrane.

The photosynthetic apparatus located in the thylakoid membrane consists of different components. These include,

i) Type — II — reaction center
ii) Type —I — reaction center
iii) The Cyt $^+$ b6 complex
iv) ATP synthatase complex consisting of two co-factors (CF_1 and CF_0)
v) Light harvesting chlorophyll a/b binding (CAB) protein complex (LHC-II)
vi) Mobile light harverting chlorophyll a/b binding complex (mobile LHC).

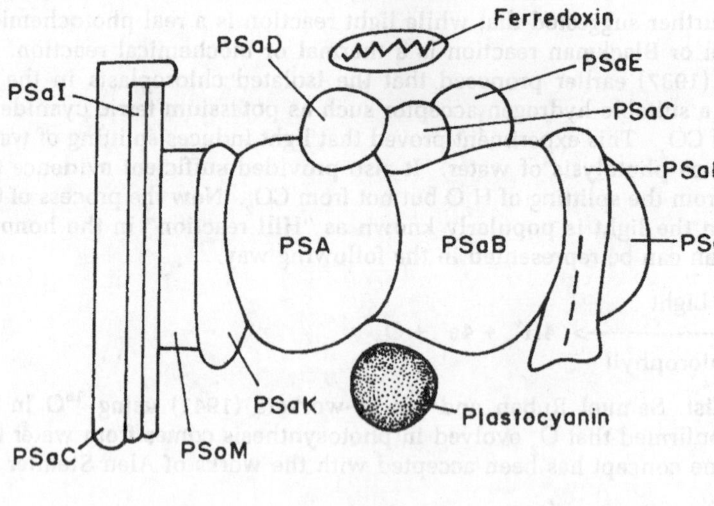

Fig. 8.1 Structural Organization of PSI

First four of these components are involved in the electron and proton transport, while last two are associated with the initial harvesting or absorbing of solar energy (Fig. 8.1A)

In the photosynthetic bacteria, the photosynthetic apparatus consists of a reaction center (RC) and two light harvesting complexes (LH_1 and LH_2). In higher plants and also in cyanobacteria, a photochemical reaction center (PRC) and the light harvesting center (LHC) also described as antenna are associated together to form a photosystem. There are two such photosystems, viz, PS II and PS I present within the thylakoid membrane and are linked with each other through a thermo-chemical bridge with several electron carriers like plasto-quinone and plastocyanin (fig. 8.1b). Now it is known that LHC or the antenna complex performs main function of capturing photons with high efficiency and transfer them to reaction center. The reaction center then converts the solar energy into chemical energy.

The structure of LHC was actually described in 1975, but the detailed information came from the work on a cyanobacterium *Mastigocladus laminosum*. The atomic structure of LHC found in purple non-sulphur photosynthetic bacteria has been described by McDermott and his team (1995). Another notable contribution is the discovery of type II reaction center in *Rhodopseudomonas viridis* and *Rhodobacter sphaeroides* by Deisenhoffer, Huber and Michel of Max Planck institute in Germany.

8.3.1 Some Basic Features of Photosynthetic : The chloroplasts in higher plants are generally lens or disc shaped and range from 3^{-10} x 1^{-5} mm in size. The number of chloroplasts per cell vary. The alga *Micromonas* contains a single chloroplast and many higher plants posses about 100-150 chloroplasts per cell. In mesophyll cells of wheat it was found that approximately 70% of cell area is occupied by chloroplasts. This large number facilitates the diffusion of CO_2 in large amount through the intercellular spaces of mesophyll tissue and increases the light capturing capacity of chloroplasts.

The chloroplast under electron microscope found to have a two layered membrane described as chloroplast envelope. Each layer of the membrane is about 5 nm thick and the space between two membranes is 10-20 nm. The envelope is made up of lipo-proteins and has fluid mosaic structure similar to other cell organelles. The chloroplasts of some plants show the invagination of the inner membrane into granum particularly in *Sorghum*, maize and sugarcane. This invagination results in the increased surface of granum facilitating higher photosynthetic rate.

The colourless hyaline portion of the chloroplast embedded in the envelope is called stroma. It contains low molecular weight metabolites and proteins. The most important proteins are the enzyme proteins, which are necessary for carbon reduction reactions. Sometimes clear starch grains are also observed in the stroma. It also contains DNA and ribosomes.

The other part of the chloroplasts is the green coloured granular part called granum (pl. grana). The grana are made up of flattened sacs called thylakoids. These are packed together in such a way that grana look like a pile of twenty five paise coins. There may be 40 to 60 grana per chloroplast and each granum may be made up of 2 to 100 or more thylakoids. Grana are linked together by inter granal lamellae or stromal thylakoids known as frets. The area between two thylakoids is termed as partition. The grana — fret system is now regarded as an extension of single membrane organization. The thylakoid membranes show distinct inner and outer surfaces. The former is connected with stroma and latter with lumen. The lumen that is described infact is the inner space between two membranes and generally filled with water and nutrients.

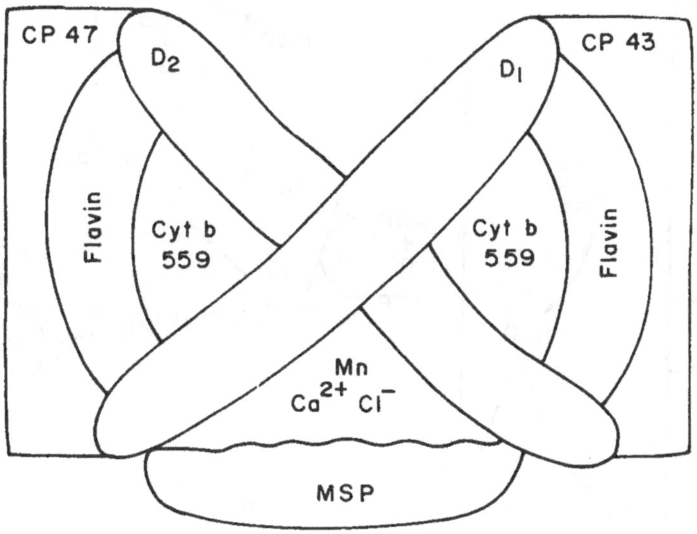

Fig. 8.2 Structural organiztion of PS-II

8.2.2 Ultra - Structure of a Thylakoid : As we know that the first photo-chemical reactions occur in the thylakoid membranes. Hence, the understanding of molecular organization of thylakoids is necessary.

Branton and Part (1967) who studied the detailed structure of thylakoids observed that the thylakoid membranes possess two sub-units. The one sub-unit is 17.5 nm width and 9 nm thickness and another is 11 nm wide and 9 nm thick. The membranes contain lipids and proteins.

The chlorophyll molecules present in the thylakoid membrane are stacked into some discrete units showing dimensions of about 18 nm long, 15.6 nm wide and 10 nm thick. These units have been described as quantasomes and they are thought to be the light harvesting photosynthetic units. One unit is believed to contain about 230 chlorophyll molecules together with other accessory pigments. The atomic weight of these units is about 200 million daltons (fig.8.2B).

It is of interest know that some modern techniques gave an entirely new concept to the molecular structure of thylakoid membranes. As stated earlier like other unit membranes, these also have a bilipid layer and along with protein layer, but the constitution of the substances vary much. The lipids are about 50% and the other 50% constitute galacto — lipids and only a small amount of phospholipids. The lipids further have large amount of fatty acid, linolenic acid and linoleic acid. Some other spherical granules of different units are also observed on either side of the thylakoids when exposed to high resolution microscopy. These granules are made up of proteins. Recent researches also gave a detailed the nature of pigment arrangements in the membranes. It is found that the pigments are

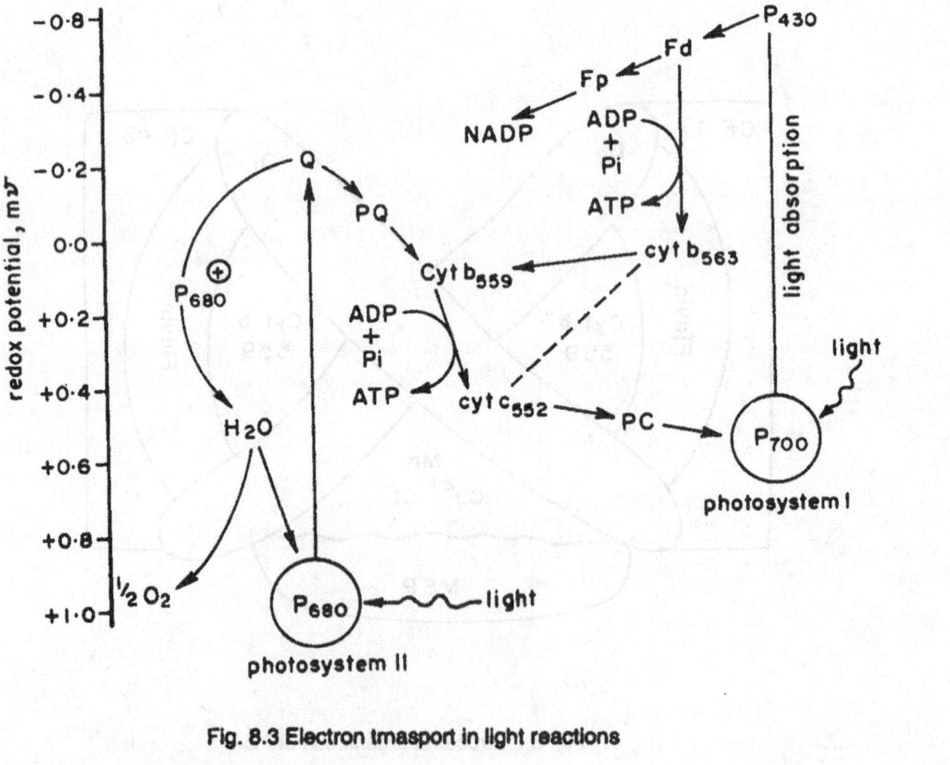

Fig. 8.3 Electron trnasport in light reactions

covalently bonded with proteins forming a pigment-protein complexes. Further, the light capturing capacity of pigments depend on their arrangement in the membrane. The chlorophyll molecules are placed perpendicular to the long axis of thylakoid membranes which invariably increases the light harvesting efficiency (fig. 8.2C).

8.4 PHOTOSYNTHETIC PIGMENTS

Chlorophyll pigments of the chloroplasts are responsible for the efficient capture of solar energy. They absorb visible light and some infra-red radiation. Chlorophyll pigments are about nine types. These are chlorophyll a, b, c, d, e, bacterio chlorophylls (B-Chl) a, b, *Chlorobium* chlorophyll 650 and 660. However, chlorophyll 'a' is the major light harvesting pigment of green plants. In these plants chlorophyll a,b pigments are present in 3:1 ratio. In some algae, besides chlorophyll a and b, Chlorophyll c, d and e are also present. In photosynthetic bacteria, the bacteriochlorophyll is abundant. The chlorobium chlorophyll 650, 660 are the characteristic feature of sulphur bacteria.

All types of chlorophylls show almost similar type of molecular structure. For example, chlorophyll a is constituted of four substituted pyrrole rings and fifth non-pyrrole ring. The characteristic five ring porphyrin derivative is called pheophytin. It also has a long isoprenoid chain consisting of the alcohol phytol (fig. 8.3) esterified to a carboxyl group substituent in ring four. The four N_2 atoms of four pyrrolle rings are linked through coordinate bond to a magnesium atom (Mg^{2+}) forming a complex. The head part is hydrophilic. Attached to the head is a long hydrophobic chain of terpene alcohol called phytol, which forms the tail part of chlorophyll. The empirical formula of chlorophyll a is $C_{55} H_{72} O_5 N_4$ Mg. In higher plants besides chlorophyll-a chlorophyll-b is also present. Chlorophyll —b differs from chlorophyll-a by having a formyl group (-CHO) in place of methyl group ($-CH_3$) on ring II. The empirical formula of chlorophyll b is $C_{55}H_{70}O_6N_4Mg$.

8.5. MECHANISM OF PHOTOSYNTHESIS

8.5.1 Photo chemistry : Light absorption and its utilization is the primary requirement of phytosynthesis. This process is carried over in the chlorophyll molecule i.e. in the protein complexes of the thylakoid.

Chlorophyll pigments absorb violet, blue, orange and red light while carotenoids absorb only the blue and ultra-violet light. When pigment extract is dissolved in an organic solvent like ether and light rays are allowed to fall on this solution, some specific wave lenths of light are absorbed by the extract solution, whereas other wave lengths are transmitted through it. The relative absorbance of various wavelengths in a purified pigment can be measured with spectrophotometer. The absorbed rays are graphically represented and such a graph is known as 'Absorption spectrum'.

Chlorophyll 'a' shows maximum absorption at 660 nm and the other at 430 nm. Chlorophyll — 'b' has absorption peaks at 640 nm and 453 nm (Table — 1). The other pigments like carotene and xanthophylls have absorption bands lying in between 430 nm and 490 nm.

It has been suggested that chlorophyll 'a' is the most efficient pigment of photosynthesis and chlorophyll 'b' or phycobilin take second place in efficiency. The carotenoids are poor in capturing light.

When, a visible light photon (physical unit of light) strikes a chlorophyll molecule, it releases an electron of chlorophyll (on porphyrin ring) into an outer molecular orbital. This state of chlorophyll molecule is called excited state. There are atleast four states of excited chlorophyll molecules.

a) first singlet state
b) first triplet state
c) second singlet state
d) second triplet state.

A red light photon excites an electron from a ground level molecular orbital to the first excited singlet state (S_1) and the higher energy photon excites electron to the second singlet state (S_2). The S_2 state electron drops down to S_1 state with the loss of energy mostly in the form of heat. S_1 state level electrons also drop down to ground state level. This return is accompanied with the emission of light in the region. This process is called fluorescence, and some energy is lost in this way is used in photosynthetic reactions. Another process is also observed resulting in the emission of light called phosphorescence, when triplet state electrons return to ground state. But most of the energy of triplet state is used for photochemical reactions.

The energy of the absorbed photons appear as heat and fluorescence and phosphorescence and only a part of it as chemical energy. It is the chemical energy that is utilized in dark reactions of photosynthesis for the ultimate assimilation of CO_2 into carbohydrates. This conversion of light energy into chemical energy takes place in specialized photosynthetic units or clusters called photosystems. In each such unit, there would be about 200 to 300 chlorophyll molecules and about 50 carotenoid molecules, but they are ultimately transfer their excitation energy to a site where chemical reactions occur. This site is described earlier is called the reaction centre. All other pigment molecules in such a unit or cluster are called LHC or antenna molecules.

TABLE : 8.1. Comparison between PS - I and PS — II

S.No	Photosystem — I	Photosystem - II
1.	PS-I consists of pigment molecules absorbing both longer and shorter wave lengths	PS-II consists of pigment molecules, which absorb only shorter wavelengths.
2.	The reaction center is P700	The reaction center is P680
3.	PS-I lie on the outer surface of the thylakoids	PS-II occur on the inner surface of the thylakoids
4.	Molecular O_2 is not released	Molecular O_2 is released by photolysis of water
5.	PS-I participates in cyclic as well non-cyclic flow of electrons	PS-II is involved only in non-cyclic flow of electrons.

Table : 8.2. Photosynthetic Pigments : Their distribution and light absorption (in nm)

Pigment	Distribution	Light absorption (in nm)
I. *Chlorophylls*		
Chl-a	All Photosynthetic organisms except Photosynthetic bacteria	429, 410, 660
Chl-b	All green plants except diatoms, red, brown and cyanobacteria	453, 430, 642

Pigment	Distribution	Light absorption (in nm)
Chl-c	Diatoms and brown algae	645
Bacterioviridin	Green-Sulphur bacteria	750 or 760
Bacterio - chlorophyll	Purple Sulphur bacteria	800, 850, 890
II. *Carotenoids*		
Carotene	Many higher plants and red and a few brown algae	420, 440, 470
Carotene	All photosynthetic organisms except purple sulphur bacteria	425, 450, 480
III. *Xanthophylls*		
Lutene	Principal xanthophylls of green leaves and green and red algae	425, 445, 475
Violaxanthol	Xanthophylls of higher plants	425, 450, 475
IV *Phycobilins*		
Phycocyanin	Principal phycobilin of cyanobacteria	618
Phycoerythrin	Main phycobilin pigment of red algae	490, 546, 570

8.5.2 Emerson enhancement Effect : In higher plants, two photosystems, PS II and PS I are involved in converting light energy into chemical energy. Involvement of two photosystems was initially suggested by Robert Emerson and his colleagues (1957). Earlier to this, Emerson *etal* (1943) measured the efficiency of photosynthesis of *Chlorella* in a monochromatic light. They observed that the quantum yield (it is the number of O_2

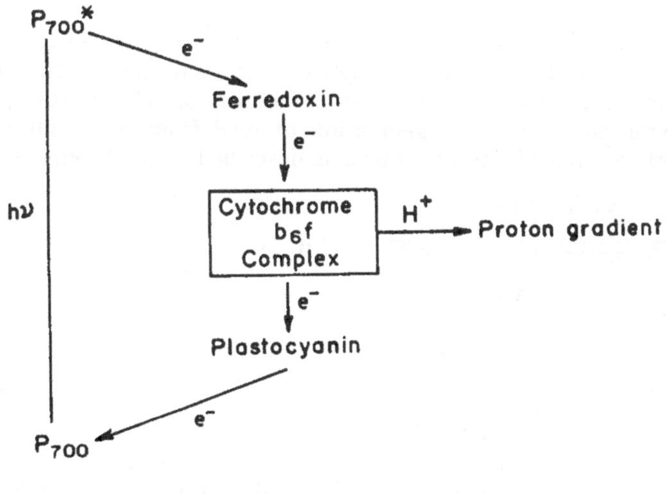

Fig. 8.4 The cyclic photophosphorylation

molecules evolved per light quanta) and two light quanta are required to move one electron. Such four of electrons are involved in the evolution of one molecule of oxygen. Therefore, eight quanta are needed for the evolution of one O_2 molecule (also described as quantum requirement). Emerson found that the quantum yield decreased very sharply towards the far red end of the spectrum (680 nm and longer) (fig.8.4A). Since, this decrease in quantum yield is observed at the far — red region of the spectrum it is described as 'red drop' (fig. 8.4B).

Emerson further observed that if the red light of shorter wavelength (< 680 nm) was superimposed with light of far red light (>680 nm) (or both supplied simultaneously), the rate of photosynthesis or the quantum yield was greater than the total yield when both these wave lengths were given separately (not superimposed). This increase in photosynthetic efficiency or conversely the quantum yield is called the Emerson effect or Emerson enhancement effect. Many findings later also have supported the observations on increased quantum yield. Peavey and Gibbs (1975) identified that the photosynthetic yield was increased by about 25% when two lights were supplied simultaneously to the isolated chloroplasts of spinach thus confirming the synergistic effect of lights on photosynthesis. Where, one helped the other in carrying out the process. This lead to the conclusion that two groups of pigments, one absorbing in lower range and other absorbing in higher range are involved in the photosynthesis (PS II and PSI).

8.5.3 Z — Scheme (involvement of two photosystems) : The discovery of Emerson enhancement effect clearly indicated the existence of two photosystems. In this case, PSI includes the pigments which absorb light of wave length longer than 680 nm and designated as P_{700}. On the otherhand, PS II pigments absorb light at 680 nm and also less than this wavelengths and is known as P 680. The two systems correspond to two reaction centres of energy trap where energy of one form (light energy) is converted to other (chemical energy).

Both the photosystems together from morphologically distinct physical entities and differ in some characters (see table 8.1). The two photosystems act in a series to produce NADPH and ATP which are the eventual products of light reactions or the assimilatory power. In this process the electrons flow from water to NADP to NADPH. The pathway of electron transport from water to NADP involving two photosystems is called 'Z scheme' as graphic representation of the transport pathway based on redox potential (in mv) of various molecules involved in the process gives an appearance of alphabet 'Z' (fig 8.5).

8.5.4 The Two Photochemical Processes : The two photosystems as described earlier absorb light of different wave lengths. The light energy absorbed by the pigments of these systems is used to generate ATP. Synthesis of ATP by this process is known as Photophosphyrlation. Thus, the generation of ATP from ADP and inorganic phosphate (Pi) in the light reaction of photosynthesis is described as photophoshorylation.

$$ADP + Pi \underset{\text{ATpase}}{\overset{\text{ATP synthetase}}{\rightleftharpoons}} ATP + H_2O$$

It was observed that during the transfer of electrons from cytochrome b_{559} (an electron carrier) to cytochrome — f in PS II to PS I chain, a large drop in the electron potential takes place, this is coupled with the generation of high energy ATP molecule. The chemi-osmotic hypothesis proposed by Peter Mitchell (1961, 1968) explains the mechanism of ATP generation. The fundamental principle of Mitchell concept is the generation of a proton gradient across the thylakoid membrane during electron transfer between PS II and PS I. The protons (H^+) move from outside to the inner side of the thylakoid (Fig. 8.5).

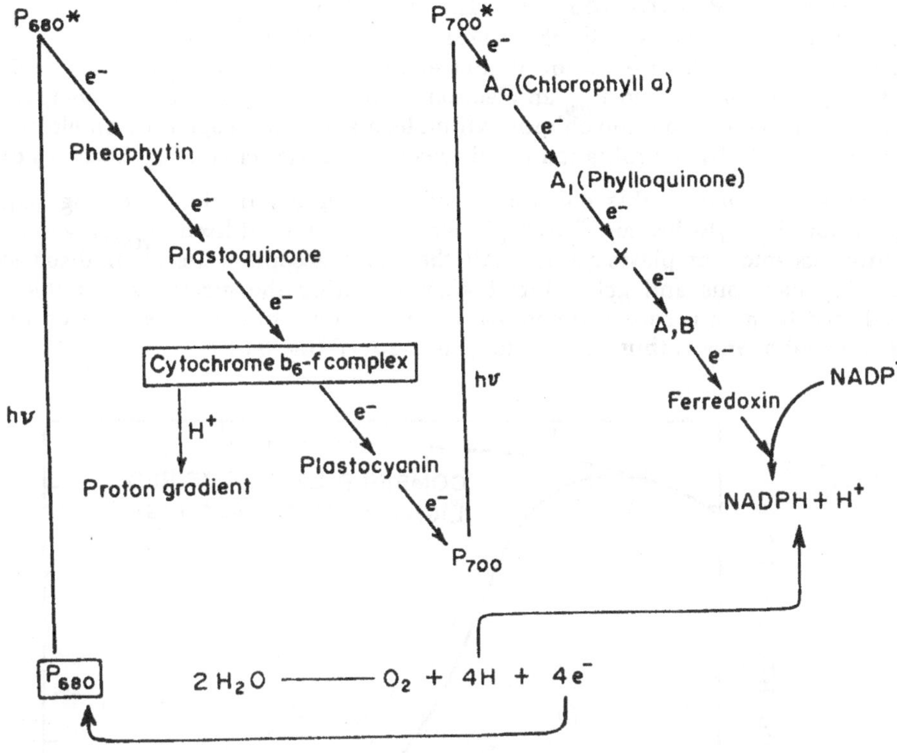

Fig. 8.5 Non-cyclic Photophosphorylation

Thus, photo-induced electron generates a gradient of H^+ across the thylakoid membrane, so that the acidity inside the thylakoids is higher than the surrounding medium, due to this difference in pH, the energy of electron transfer is used in generation of high energy ATP molecules from ADP+ Pi.

In the photosystem — I, the excited electrons return to the same chlorophyll molecule after releasing the excess of energy. Hence, the generation of energy in the form of ATP by this method of electron transfer is as cyclic photophosphorylation. In PS II, the electrons from excited chlorophyll molecules are trapped by NADP (used in the formation of $NADPH_2$) and do not cycle back to chlorophyll, hence it is called non-cyclic photophorylation. Here, the electron lost by chlorophyll molecule is replaced by electrons from H_2O molecule. This because, a kind of electron 'hole' in P680 component of PS II. Hence, this hole is filledup by an electron (or electrons) from H_2O either directly or through another electron carrier called 'Z'.

It has been observed that photosynthetic cells have evolved a special type of water splitting catalytic system, popularly known as water oxidising clock or oxygen evolving complex. This system splits a pair of water molecule into electrons (e^-) protons (H+) and molecular oxygen.

$$2H_2O \xrightarrow[\text{Mn + (Water splitting enzyme complex)}]{\text{4 Photons}} 4H^+ + 4e^- + O_2$$

Now let us study the detailed aspects of cyclic and non-cyclic photophosphorylation processes of light phase.

8.5.5 Cyclic Photophosphorylation: The process of cyclic photosynthetic phosphorylation involves only photosystem — I. A photon of light excites an electron of chlorophyll molecule and the excitation energy is eventually passed on to the reaction center P_{700}. In the excited P_{700} an electron acquires a large amount of energy. This 'hot' electron is expelled from the chlorophyll molecule and is accepted by an electron acceptor Ferredoxin (Fd). By the initiation of this electron transfer chain, P_{700} becomes oxidised.

Ferredoxin passes this electron to other acceptors in the following sequence, i.e. cytochrome b_6, cytochrome 'f' and plastocyanin. The oxidised P_{700} serves as a natural electron acceptor for plastocyanin. All the above acceptors are in oxidised state before accepting electrons and get reduced after accepting the electrons. As the electron is transferred from an acceptor to another, certain amount of energy is released in each step. The released energy is thus utilized to generate ATP (fig. 8.6).

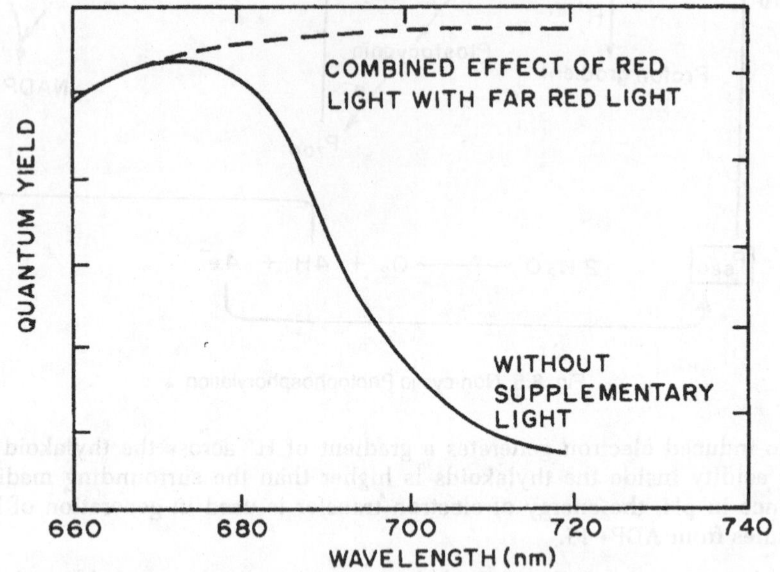

Fig 8.6 Graph showing Emerson enhancement effect

8.5.6 Non-cyclic Photophosphorylation: It involves photosystem — I aswell as photosystem — II. Both of these systems absorb light photons. Non-Cyclic photophosphorylation can be studied under the followng three events (fig 8.7A).

a) Transfer of electrons from P_{700} for the formation of $NADPH_2$

b) Replacement of ejected electon of P_{700} from P_{680}.

c) Photolysis of water and replacement of ejected electron to P_{680}.

8.5.7 a) Transfer of electrons from P_{700} for the formation of $NADPH_2$: The excited electrons given out by activated P_{700} are accepted by Ferredoxin (Fd) as in cyclic photophorylation. In the next step, the reduced 'Fd' transfers the electron to another acceptor NADP (Nicotinamide adenine dinucleotide phosphate) and as a result NADP is reduced to $NADPH_2$. Since the electron from P_{700} has been transferred to NADP and it can not come back to P_{700}. It (P_{700}) remains in oxidised state.

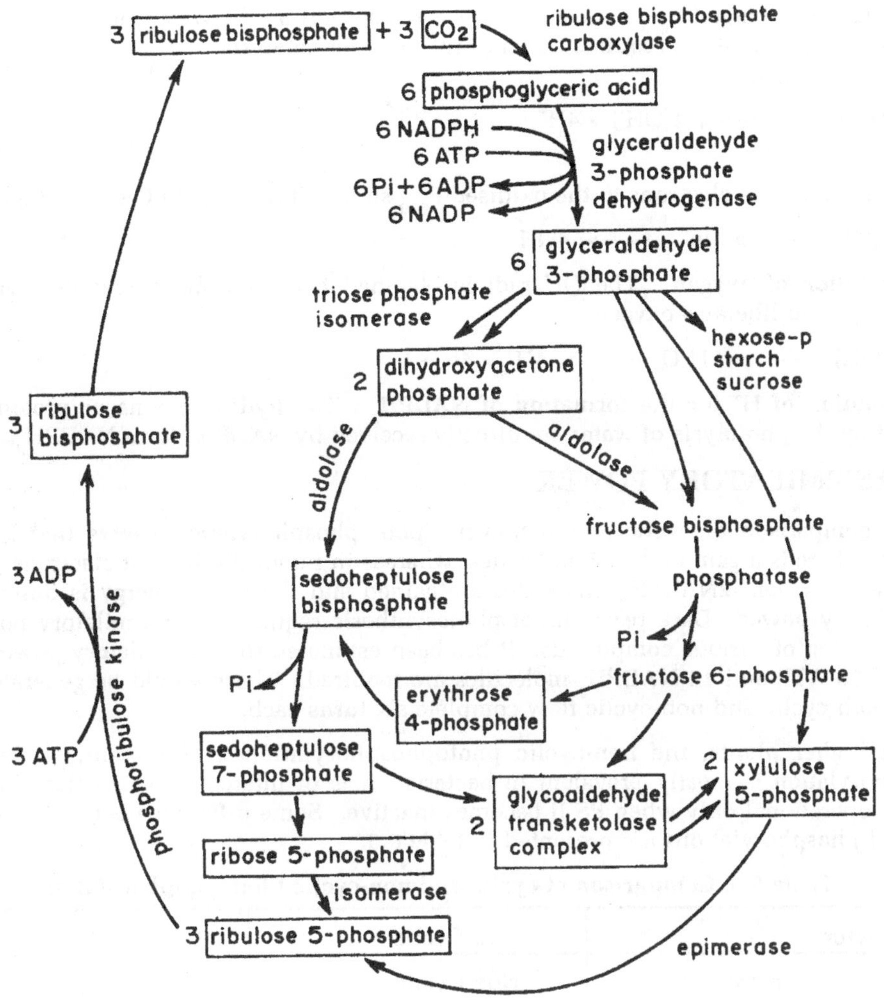

Fig. 8.7 Reactions of Calvin Cycle

8.5.8 b) Replacement of ejected electon of P_{700} by electron from P_{680} :

P_{700} is non-functional in oxidised state and has to become functional again. In order to become functional, it must receive an electron. This deficient electron comes from P_{680}. Accordingly, when P_{680} absorbs a photon of light it gives out an electron and thus becomes positively charged. The ejected electron from P_{680} passes to P_{700} through the following series of electron carriers.

P_{680} ---------> Plastoquinone --------> Cyt b6 --------> Cyt f --------> Plastocyanin --------> P_{700}

In this series of electron transfer, enough energy is released. When electron is transferred from Cyt b6 to cyt f, released energy is utilized to synthesize ATP from ADP + Pi.

8.5.9 c) Photolysis of water : Break down of water molecule under the influence of light is known as photolysis of water. It occurs to fulfil the following three objectives.

1) Replacement of ejected electron of P_{680} :

The positively charged P_{680} molecule exerts a strong pull on the water molecule splitting it into H^+ and OH^- ions. It requires Cl^- and Mn^{2+} ions for catalysis.

$$4H_2O \xrightarrow[Cl^-]{Mn^{2+}} 4(OH^-) + 4H^+$$

The OH^- donate its electrons to the oxidised P_{680} so that it once again becomes functional.

$$4\ (OH)^- \longrightarrow 4e^- \longrightarrow 4OH$$

2) evolution of oxygen . The OH radicle obtained from the above reaction eventually forms H_2O and liberates oxygen

$$4(OH) \longrightarrow 2H_2O \longrightarrow 4H^+ + 4e^- + O_2$$

3) Donation of H^+ for the formation of $NADPH_2$: The hydrogen ions or protons ($4H^+$) formed by the photolysis of water are directly received by NADP to form $NADPH_2$.

8.6. ASSIMILATORY POWER

A comparison of cyclic and non-cyclic photophosphorylation shows that in cyclic pathway there is a gain of 2ATP molecules, whereas in noncyclic flow of electrons, besides 1ATP and oxygen, $2NADPH_2$ molecules are gained and the entire energy is described as assimilatory power. Dark reactions of photosynthesis require this assimilatory power for the reduction of various compounds. It has been estimated that assimilatory power in the form of 18 ATP and 12 $NADPH_2$ molecules are required. These would be generated only when both cyclic and non-cyclic flow complete six turns each.

In higher plants, the non-cyclic photophosphorylation is dominant, while cyclic phosphorylation is mostly prevalent in bacteria. It is of interest to know that the cyclic flow is operational only when PS II becomes inactive. Some difference between the above types of phosphorylations are presented in table 8.3.

Table 8.3: Comparison of cyclic and non-cyclic Photophophorylation

Character	Non-Cyclic	Cyclic
Pathway of electrons	Non-Cyclic	Cyclic
First electron donor	Water	P_{700}
Last electron acceptor	NADP	P_{700}
Products	ATP $NADPH_2$	Only ATP
Photosystems involved	PS-I and PS-II	Only PS-I
Inhibition	Dichlorophenyl Dimethyl Urea DCMU	Insensitive to DCMU

In recent years, the possibility of an alternative to Z-scheme for photosynthesis light reactions was suggested. For instance, it was opined that light phase could be completed by PS II alone independent of PS I or it is possible that PS II should be able to bring about Photolysis of water and also able to reduce NADP (table 8.8). Greenbaum (1996) gave an evidence while working with *Chlamydomonas reinhardtii*. According to him that PS II carries out all reactions in the absence of PS I. It is suggested that PS II has evolved first

Fig 8.8 Kranz anatomy as seen in C4 Plant
1. Epidermis, 2. Mesophyll cells, 3. Bundle sheath cells, 4. Xylem, 5. Phloem

under anaerobic conditions that existed before oxygen started accumulating in the atmosphere due to photosynthesis. It is also speculated that PS I may have appeared to stabilize PS II under changing atmospheric conditions.

Structural organization of the two photosystems also have been studied in recent years. It has been identified that the PS II from plants, algae and bacteria is made up of many proteins, cofactors and ionic elements. More than two polypetides have been identified in higher plants. Pistorious (1993) who extensively worked on a *Cyanobacterium synechoccus* found that two polypeptides CP 47 (Mw 44 — 51 KD) and CP43 (Mw 30 — 34 KD) are involved in binding with chlorophyll a peptides D_1 and D_2 (Mr 30-34 KD) are at the main reaction center where primary photochemistry and charge separation takes place. Two apoproteins of cytochrome-b 559 are also present. Further, a manganese (Mn) stabilising protein (MSP) is also present in the complex. Regarding the structural complexity of photosystem-I, it is made up of heteromeric proteins and functions as light driven PC-Fd oxido reductase. As many as 13 polypeptides have been found to present in PS I with approximately 100 chlorophyll — a molecules along with several beta-carotenes, two phytoquinone molecules and three fe-s clusters. Chitnis (1996) proposed a structural model of PS I. Accordingly two proteins Psa A and Psa B form the photoreactive centre of PSI. Light harvesting and charge separation takes place, at this reaction center. Other proteins like Psa AL, PsaK, PsaF, PsaG, PsaH and PsaN are also present in the photosystem — I complex which mostly useful in the organisation of light harvesting complex (8.7B).

8.7 CARBON-DI-OXIDE ASSIMILATION OR DARK REACTIONS OF PHOTOSYNTHESIS

The light independent or dark reactions of photosynthesis reduces the CO_2 to glucose of starch. The overall dark reaction which takes place in several steps.

These reactions take place in the stroma of chloroplasts and does not require light but it depends upon the products of photochemical reactions of photosynthesis as shown above in the form of ATP and $NADPH_2$. The sequence of reactions involving CO_2 fixation and formation of complex carbohydrates is also known as 'path of carbon' in photosynthesis.

8.7.1 Hostorical aspects of dark reactions : Discovery of radioactive carbon and its use in photosynthetic research helped in understanding the pathway of conversion of CO_2 to carbohydrates. Ruben and his team (1940) by using a short lived radioactive carbon (^{11}C) proved that the fixation of CO_2 proceeded during dark immediately following a period of illumination and that an intermediate compound was formed in the synthesis of carbohydrates. Later, ^{14}C isotope (more stable) was used to trace the pathway of CO_2 assimilation. Another technique called paper chromato graphy was also used to trace the carbon pathway. The first stable product in the form of phosphoglyceric acid (PGA) was identified which is a 3-carbon compound after 2-7 seconds of exposure to light. The dark reactions or the light independent phase of photosynthesis occur through one of the following two pathways under two different conditions.

1. Calvin — Benson or C_3 pathway
2. Hatch and Slack or C_4 pathway

8.7.2 Calvin Cycle or C_3 Pathway : The credit for tracing the pathway of reduction of carbondioxide into one molecule of glucose goes to Melvin Calvin, James Basham and Andrew Benson (1946-53) of United States of America.

Calvin and his team grew two algae, *Chlorella* and *Scenedesmus* in ordinary CO_2. Later, these were grown again a culture flask supplied with radioactive CO_2 ($^{14}CO_2$) for a short period and the algal cells were killed immediately in boiling ethanol (or methanol). The radioactivity in various compounds of algal cells were separated by paper chromatography technique and was estimated. When algal cells were exposed to $^{14}CO_2$ only for five seconds or less, it was observed that the first compound present was phosphoglyceric acid (PGA). This product was isolated and chemically degraded and it was found that most of ^{14}C is in carboxyl position of PGA P — O — CH_2 — CHOH — ^{14}COOH.

Further work in this direction revealed the entire sequence of reactions in CO_2 assimilation, i.e. the sequence of reactions right from the entry of $^{14}CO_2$ in to the cells till the formation of glucose. Since the first stable product in the series of reactions is a three carbon compound (PGA), this pathway is also known as C_3 pathway. It is also known as Calvin or Calvin and Basham Cycle (fig.8.9A). Calvin was awarded Nobel Prize in 1962 for this pioneering research work.The various biochemical reactions of C_3 pathway can be studied under the following three headings.

1. Carboxylation of Ribulose bi-Phosphate (RUBP)
2. Glycolytic reversal or Reduction reactions
3. Regeneration of RUBP.

8.7.2.1 Carboxylation of RUBP : Addition of CO_2 to a compound is called carboxylation. In the first step of C_3 pathway, Ribulose 1,5 biphosphate (also known as ribulose di-

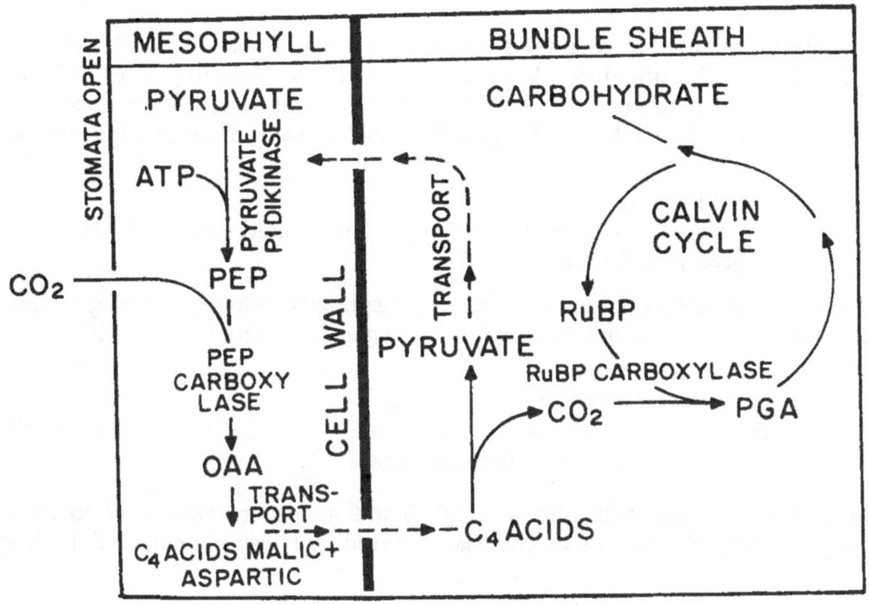

Fig 8.9 Compartmentation of C4 reactions in mesophyle and bundle sheath cells

phosphate), a phosphoglycerated 5-carbon sugar acts as primary CO_2 acceptor and combines with Co_2. The enzyme responsible for the reaction is Ribulose 1,5 biphosphate carboxylase (conversion of inorganic carbon into organic carbon).

The RUBP 1,5 biphosphate carboxylase is also known as carboxydismutase. This enzyme is the most important component of fraction-I protein. The enzyme is located on stromal surfaces of thylakoid membrane and it is the major protein present in the leaf and consists of two kinds of of 16 sub-units (8 large and 8 small) with a molecular weight about 5.5 X 105. The enzyme further has about 80 sulphydryl groups and requires Mg^{2+} ions for its optimum activity. It is considered to the major protein on earth. Besides catalysing carboxylation of RUBP, it can also oxygenate this substrate to produce phosphoglycolate in a specialized process called photorespiration. In view of this additional activity of enzyme, the enzyme is also named as Ribulose biphosphate carboxylase—oxygenase (RUBISCO). The carboxylation reaction catalysed by RUBISCO is as follows:

When a mole of five-carbon sugar ribulose 1,5 — biphosphate accepts a CO_2 a mole of six-corbon compound is formed which is most untable and form immediately two molecules of phosphoglyceric acid is formed.

$$\begin{array}{c} CH_2O\ (P) \\ | \\ C{=}0 \\ | \\ CHOH \\ | \\ CHOH \\ | \\ CH_2O\ (P) \end{array} + CO_2 \underset{Mg^{2+}}{\overset{RUBISCO}{\rightleftharpoons}} \begin{array}{c} CH_2O\ (P) \\ | \\ 2\ CHOH \\ | \\ COOH \end{array}$$

RUBP — Phosphoglyceric acid (PGA)

Hence, PGA (a 3 carbon compound) is the first stable product of Calvin cycle.

8.7.2.2. Glycolytic reversal or reduction reactions : In this phase of dark reaction of C_3 pathway PGA forms phosphoglyceraldehyde (PG Ald) by utilizing ATP and $NADPH_2$ synthesised in the light phase (two step reaction). In the first step, PGA is phosphorylated by ATP in the presence of an enzyme phosphoglycerate kinase to produce Diphosphoglyceric acid (DPGA)

Phospho-
12 PGA + 12 ATP -----------------------> 12 Diphosphoglyeric acid (3C) + 12 ADP
glycerate Kinase

In the second step, DPGA is now reduced to phosphoglyceraldehyde or Glyceraledehyde 3-Phosphate (G3P). In this reduction process hydrogen is donated by $NADPH_2$ formed in light reactions.

NADP Glyceraldehyde
12 DPGA + 12 $NADPH_2$ --> 12 GAP (3C) + 12 NADP
3 Phosphate dehydrogenase

In the next step, out of 12 molecules of GAP, 5 molecules are converted to Dihydroxy acetone Phosphate (DHAP) and this reaction is catalysed by triose phosphate dehydrogenase enzyme.

Triose
5 Phosphoglyceraldehyde ⇌ 5-dihydroxy acetone Phophate (3C)
Phophate dehydrogenase

The rest of Phosphoglyceraldehyde (3GP or PG Ald) combines with 3 molecules of Dihydroxyacetone Phosphate to from 3 molecules of Fructose 1,6 biphosphate.

Aldolase
3GAP + 3DHAP -----------------------> 3 Fructose 1-6, diphosphate (6C)

Although the trioses glyceraldehyde phophate and dihydroxy acetone phosphate are the first carbohydrates to be synthesized, many other sugars are synthesized from these trioses. Hence, Fructose 1,6 biphophate is the first hexose to be synthesized in photosynthesis. This Fructose 1,6 — biphosphate is dephosphorylated to Fructose 6-Phosphate (F6p) in the next reaction which is catalysed by the enzyme phosphateae. In the subsequent reactions, regeneration of RUBP is initiated by series of intermediate reactions, before that one molecule of glucose is produced from 3 fructose 6 phosphate.

3F6P -----------------------> 1 glucose (6C) + 2 F6 P

8.6.2.3 Regeneration of Ribulose 5 Phosphate : In the first step of Calvin cycle CO_2 is accepted by RUBP and it enters the C_3 pathway. Later, once glucose is formed, RUBP must be regenerated so that it once again made available to accept CO_2 for subsequent carboxylation reactions. Following are the various steps resulting in the regeneration of RUBP. In the first step a 6 carbon compound Fructose 6 phosphate reacts with Glyceraldehyde 3 phosphate to give rise to one molecule each of Erythrose 4 phosphate and Xylulose 5-Phosphate. This reaction involves the clevage of F6P and condensation of one of the clevage products (2C) to another molecule of GAP. The reaction is catalysed by the enzyme transketolase.

Transketolase
Fructose 6 Phosphate + Glyceraldehyde Phosphate -->
Xylulose 5 phosphate (XU5P) + Erythrose 4 Phosphate (E$P)

2 molecules of dihydroxy acetone phosphate and 2 molecules of erythrose 4 phosphate condense together in the presence of aldolase to derive 2 molecules of 7 carbon compound, Sedoheptulose 1,7 diphosphate (SDP)

Aldolase
2 Erythrose 4 Phosphate + 2 Dihydroxy acetone Phosphate ----------> 2Sedoheptulose 1, 7 diphosphate

In the next reaction Sedoheptulose 1,7 diphosphate is converted to Sedoheptulose 7-phosphate (S7P) in the presence of Phosphotase enzyme. This is followed by cleavage of Sedoheptulose 7-Phosphate and the condensation with GAP resulting the production of Ribose 5-Phosphate (R5P) and xylulose 5 Phosphate (X 5P).

TPP
2 Sedoheptulose 7 Phosphate + 2 Phosphoglyceraldehyde ---------
transketolase

2 ribos 5 phosphate + 2 xylulose 5 phosphate

At this stage, two xylulose 5 phosphate are epimerised through a specific epimerase enzyme to form Ribuloses 5-phosphate (RU5P) and this is followed by isomerisation of ribose 5-phosphate to ribulose 5-phosphate to form ribulose 5-phosphate.

Epimerase
Xylulose 5 phosphate -----------------------------> Ribulose 5 Phosphate
(Xu 5P) (RU5P)

Isomerase
Ribose 5 Phosphate -----------------------------> Ribulose 5 Phosphate
(R5P) (RU5P)

In the last step of reactions, the Ribulose 1-5 biphosphate is regenerated to become the acceptor of CO_2 (or for carboxylation reaction) in the presence of ATP

Ribulose
RU5P + ATP -------------------------------------> Ribulose 1-5, biphosphate
5 Phosphate Kinase (RUBP)

The regeneration of RUBP signals the completion of C_3 pathway.

If one examines the above series of reactions of C_3 pathway, as described above, three phases of Co_2 assimilation are clearly seen.

Carboxylation Phase

CO_2 + RUBP ----------------------------> PGA

Reductive Phase

PGA + ATP -----------------------------> DPGA

DPGA + $NADPH_2$ -----------------------------> G3P

Regenerative Phase

Regeneration of RUBP from a series of reactions.

Besides above three phases, a fourth products synthesietic phase is also observed. In this phase synthesis of various metabolic end products such as sugars, organic acids, amino acids and fatty acids formation is observed (fig. 8.9b).

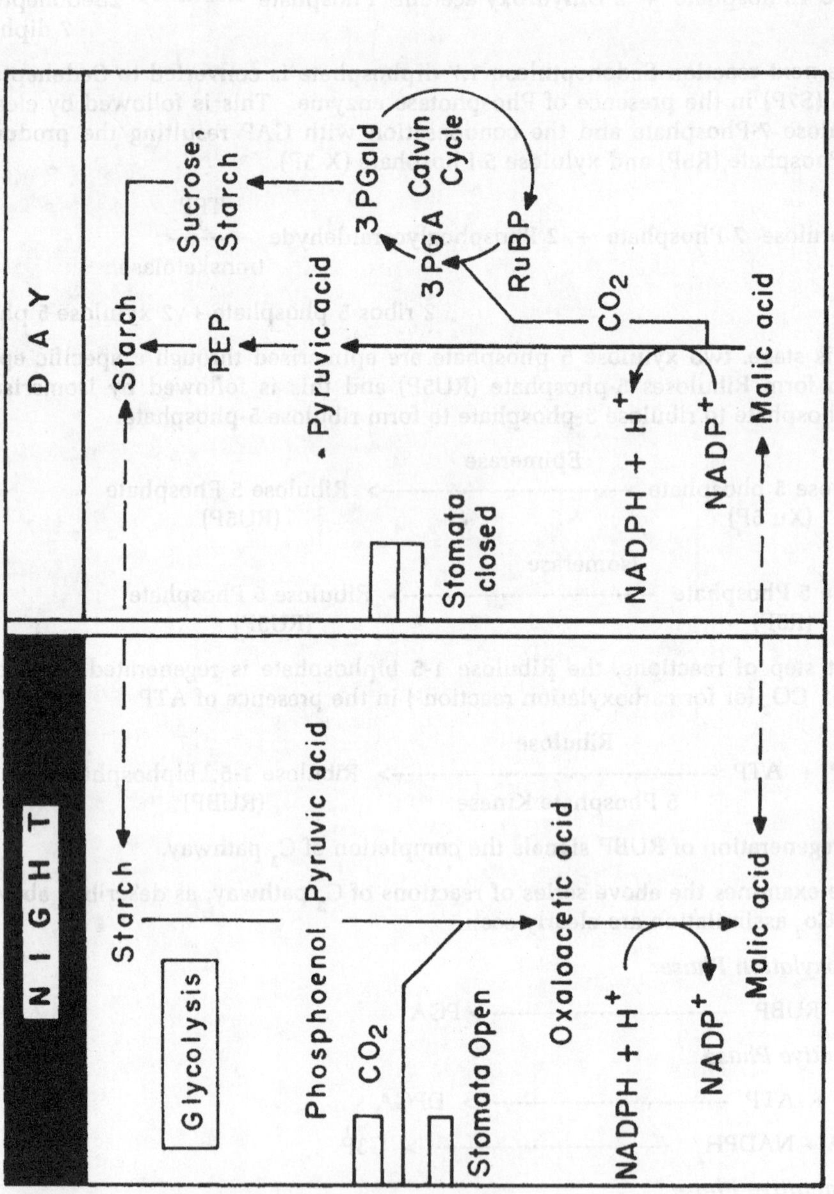

Fig 8.10 Events leading to CO2 fixation in CAM Plants

8.6.3 The Hatch and Slack or C4 Cycle : It was thought till the year 1965 that all green plants fix CO_2 through Calvin cycle. But in 1965 Kortschak, Hart and Burr while studying the mechanism of CO_2 fixation in sugarcane fed with $^{11+}CO_2$ identified that most of the labelled carbon compound appeared in the form of malic acid and asparatic acid. Subsequently, two Australian Plant Physiologists, Hatch and Slack (1970) discovered that radioactive CO_2 is foxed as oxalo acetic acid instead of phosphoglyceric acid. Oxaloacetic acid is the first stable product of carboxylation which is a 4C carbon compound. The primary acceptor of CO_2 in this case is Phospho Enol Pyruvic acid (PEP). Hatch and Slack also outlined the entire series of reactions involved in Photosynthetic CO_2 assimilation in Sugarcane leaves. Later this pathway was also observed in many tropical and sub-tropical grasses including corn (*Zea mays*) and also in some millets. Some dicotyledons belonging to Centrospermae also found to have this carbon pathway. The plants having Hatch and Slack reaction are referred to as C_4 species as the first stable compound in photosynthesis is a 4-carbon compound in the form of oxaloacetic acid (OAA). Downton (1975) listed over 117 genera of plants belonging to 10 families of Angiosperms such as Aizoaceae, Amarantaceae, Asteraceae, Chenopodiaceae, Cyperaceae, Poaceae, Euphorbiaceae, Nyctaginaceae, Portulaceae and Zygophyllaceae are C_4 plants.

The common examples where C_4 pathway is observed are *Digitaria, Eragrostis, Saccharum Chloris, Sorghum, Zea mays, Amaranthus* and *Tribulus* spp. However, interestingly in wheat, oats, rice and bamboo CO_2 fixation follows Calvin Cycle.

8.7.3.1. Characteristic features of C_4 species : The plants having Hatch-Slack pathway have a peculiar type of leaf anatomy known as Kranz anatomy (Kranz = Wreath or border) which is different from that of C_3 species.

8.7.3.2. Kranz anatomy : The vascular bundles of leaves of C_4 plants are surrounded by a layer of paranchymatous cells known as bundle sheath cells (fig. 8.10) that separate them from mesophyll cells. The mesophyll cells and bundle sheath cells are in close contact with one another. They are connected by plasmodesmata.

The bundle sheath cells of C_4 plants found to contain large chloroplasts arranged centripetally. However, they lack grana and hence described as agranal chloroplasts, but some times rudimentary grana may be present and frequently accumulate starch grains (fig. 8.10). The other interesting feature is that these chloroplasts lack Photosystem—II. The mesophyll cells of C_4 plaint leaves, on the other hand, have typical chloroplasts as found in C_3 plants with characteristic large grana along with Photosystem — II. Interestingly enough the chloroplasts of mesophyll cells lack ribulose biphosphate carboxylase activity, which is present in the mesophyll cells of C_3 plants. Nevertheless, the bundle sheath chloroplasts of C_4 plants show Ribulose-biphosphate carboxylase activity and are capable of synthesizing phosphoglyceric acid from CO_2 and RUBP.

The enzymes of Hatch-Slack pathway occur in mesophyll cells while those of Calvin cycle are present in bundle sheath cells. This chloroplast dimorphism is the characteristic feature of Kranz anatomy found in C_4 plants. Bermuda grass (a C_4 plant) however shows a well developed grana in the chloroplasts of bundle sheath cells. If one examines above features of C_4 plants, it is very clear that they show a division of function. Here, mesophyll chloroplastis act as traps of CO_2, which is then transferred to bundle sheath plastids where CO2 is reduced to hexose, sugars.

8.7.3.3. Mechanism of C_4 pathway : The initial reactions of C_4 pathway take place in mesophyll cells. In the first step, CO_2 (actually HCO_3^-) reacts with (the primary acceptor of it) phosphoenol purvate (PEP). In actual terms, CO_2 from atmosphere enters into mesophyll cells combine with H_2O to form bicarbonate ions (HCO_3) and is fixed by primary carboxylation reaction. The carboxylation reaction involving CO_2 and PEP is catalysed by phosphoenol pyruvic carboxylase (PEP Carboxylase) enzyme forming the first product of the C_4 cycle i.e. oxalo-acetic acid (OAA) which is a 4-carbon compound and is a C_4 dicarboxylic acid. Hence it is also described as dicarboxylic acid pathway.

PEP Carboxylase
Phosphoenol Pyruvic acid + CO_2 -------------------------------> Oxaloacetic acid (4C) + Pi

It has been suggested that this carboxylation reaction infact takes place in cytoplasm instead of chloroplast as found in carboxylation reaction of C_3. It is also found that when compared to RUBISCO enzyme, the PEP carboxylase brings about carboxylation at a faster rate. This may be due to more affinity of CO_2 to PEP Case than to RUBISCO. This results in the higher rate of CO_2 assimilation in C_4 pathway (fig.11).

In the second reaction OAA is then rapidly reduced to malic acid (MAL) in the presence of $NADPH_2$ or aminated to form asparatic acid (ASP)

Malate
OAA + $NADPH_2$ -------------------------------> Malic acid + NADP
(4C) dehydrogenase (4C)

Asparate
OAA <-------------------------------> Asparatic acid (4C)
amino transferase

The malate or asparate thus formed in the mresophyll cells are transferred to the neighbouring bundle sheath cells via special channels (probably plasmodesmata) between these cells where further reactions of CO_2 fixation take place.

In the bundle sheath cells (BSC) the malate is now decarboxylated producing pyruvate or pyruic acid and CO_2 by the action of NADP specific malic enzyme. Decarboxylation of malate is special and characteristic reaction of C_4 cycle.

Malic
Malic acid + NADP ---------------------> Pyruvic acid + CO_2 + NADPH + H+
enzyme

8.7.3.4 Regeneration of PEP : Pyruvic acid resulting from the above decarboxylation process is transported back to mesophyll cells where it is converted to phosphoenol pyruvate to keep the cycle going on or to make it ready to under go another cycle of carboxylation. In the presence of phosphate dikinase, ATP and inorganic phosphorus (Pi), pyruvate is converted to PEP.

Pyruvate
Pyruvate + ATP + Pi -------------------------------> PEP + ADP + Pi.
(3c) Phosphate dikinase

The pyrophosphate released in the above reaction is hydrolysed by pyrophosphatase to yield to 2 molecules of inorganic phosphate. Thus formed Pi provides energy for the formation of PEP from Pyruvic acid.

The CO_2 released by decarboxylation in BSC is assimilated photosynthetically through carboxylation with Ribulose biphosphate in the presence of RUBISCO to form phosphoglyceric acid (the second carboxylation reaction). Similar to Calvin cycle (fig 8.9), the subsequent reactions are also identical to C_3 cycle, finally forming one molecule of $C_6H_{12}O_6$ and coinciding with the regeneration of Ribulose Biphosphate.

8.7.3.5 Summary of C_4 pathway : The C_4 pathway involves two phases, 1) Carboxylation reaction resulting in the formation of PEP using PEP carboxylase enzyme in the cytosol of mesophyll cells and 2) decarboxylation of C_4 compound (malate or asparate) by using a decarboxylation enzyme system (Malate enzyme). The CO_2 released is incorporated (II carboxylation) into sugars in the chloroplasts of bundle sheath cells following Calvin Cycle. This means that Hatch-Slack cycle is the feature of mesophyll cells and Calvin cycle (as in C_3 plants) goes on in bundle sheath cells. A wide variety of enzymes take part in these reactions which are shown in Table 8.4.

8.6.4 Bioenergetics of C_3 and C_4 Pathways :

If one examines various reactions of C_3 and C_4 pathways, the actual reduction steps of photosynthesis are same in both C_3 and C_4 cycles. The energy requirements, however differ. Though the requirement of NADPH are the same, but more ATPs are consumed in C_4 than in C_3 cycle.

The overall energy requirements of C_3 can be shown in the following manner.

$$6CO_2 + 18\ ATP + 12\ NADPH + 12\ H_2O \text{ -------> } C_6H_{12}O_6 + 18ADP + 18Pi + 12\ NADP + 6H+$$

Table 8.4 Distribution of Enzymes in mesophyll and bundle sheath cells of C_4 plants

Mesophyll Cells	Bundle Sheath Cells
Phosphoenol Pyvuvate Carboxylase (I carboxylation)	RUBP Carboxylase (II Carboxylation)
Pyruvate Kinase	Malate Enzyme (decarboxylation)
NADP Malate dehydrogenase	Ribose P-isomerase
Pyro — Phosphatase	Ribuloase P Kinase
Phosphoglycerate Kinase	Phosphoglycerate Kinase
NADP-triose Phosphate dehydrogenase	NADP- triose Phosphate dehydrogenase.

The C_4 plants, on the other hand, use additional ATPs for the conversion of Pyruvic acid to Phosphoenol Pyruvacid and for also for its transport. In some plants, one additional ATP is also consumed during the decarboxylation of Oxaloacetic acid to Phosphoenol pyruvic acid by the enzyme phosphoenol pyruvate kinase. The balance equation of C_4 plants would be

$$6CO_2 + 30\ ATP + 12\ NADPH + 12\ H_2O \text{ --> } C_6H_{12}O_6 + 30\ ADP + 30\ Pi + 12\ NADP^+ + 18H^+$$

However, the C_4 plants appear to have greater rates of CO_2 assimilation than C_3 plants though they consume more energy. The C_4 plants are better adopted to environmental stresses than C_3 plants (see Table 8.5). A perusal of the differences between them reveals that the C4 plants are more efficient than the C3 plants in number of anatomical, biochemical and eco-physiological parameters (fig.8.10b). Further, they possess high potential for productivity area, water use and nitrogen assimilation efficiency.

As described above, the C4 plants produce a 4-carbon compound in the form of a carboxylic acid (malic or asparatic acid) which gets transported to bundle sheath, where it undergoes decarboxylation utilizing one of the three known enzymes. Depending upon

Table 8.5 : A list of distinctions between C3 and C4 plants.

C_3 plants	*C_4 plants*
I *General distinctions :*	
Most of the green plants are C_3 plants including rice and wheat (about 90%).	Several tropical grasses and few dicotyledons are C_4 plants (about 10%)
II *Anatomical distinctions* :	
Leaves do not show Kranz anatomy	Leaves show characteristic Kranz anatomy
The bundle sheath cells do not contain chloroplasts	The bundle sheath cells are distinct from the cells of mesophyll. They contain large number of chloroplasts.
Chloroplast dimorphism is not present	Chloroplast dimorphism is present and the chloroplasts of mesophyll cells are smaller than those of bundle sheath cells. The chloroplasts of mesophyll cells have grana but those of bundle sheath are devoid of them or may have rudimentary grana.
III. *Biochemical differences*	
The primary acceptor of CO_2 is RUBP and do not have PEP carboxylase activity (RUBISCO is the carboxylating enzyme)	The CO_2 acceptor in the mesophyll cells is PEP and in the bundle sheath cells it is the RUBP. They show PEP carboxylase activity.
The first stable product of photosynthesis is a 3-carbon compound i.e. PGA	The first product is a 4C compound i.e. OAA
Calvin cycle occurs in all the green cells of the leaf.	Hatch-Slack cycle occurs in the mesophyll cells while Calvin cycle in bundle sheath cells.
The chlorophyll a to b ratio is 2:8	Chlorophyll a to b ratio is 3:9
IV. *Ecophysiological differences*	
C_3 species are better adopted to cool and moist to hot and moist conditions	Adopted to hot, dry or moist conditions.
Lower photosynthetic rates specially at higher light intensities and saturated at 1/4 to 1/3 of full sun light	High photosynthetic rate and increase with increase in light intensity tending to — saturate only at full sun-lights.
Less growth rate at optimum conditions compared to C_4 species.	Optimum growth rate is about 1.5 times higher than C_3 species.
O_2 present in air (=21% O_2) markedly inhibits the photosynthetic process as compared to an external atmosphere containing no oxygen.	The process of photosynthesis is not inhibited in air as compared to an external atmosphere containing no oxygen.
1.5 — 35 mg of CO_2 is fixed for dm^2 leaf area in one hour in full sunlight.	4-80 mg of CO_2 is fixed for dm^2 leaf area/hour in full sun light.

C_3 plants	*C_4 plants*
CO_2 compensation point is 5.0 to 15.0 PPM.	It is 0 — 10 PPM of CO_2.
Rate of transpiration is higher (450 to 950 H_2O g-1 of dry weight)	Less transpiration rates (250 to 350 ºH$_2$O g-1 dry wt).
To fix one molecule of CO_2 3 ATP are consumed.	To fix one molecule of CO_2, 5 ATP are consumed.

enzymes involved in decarboxylation, C_4 plants are categorized into

1. *NADP — ME type* : Using NADP — Malate enzyme

   eg. : Sugarcane and maize.

2. *NAD — ME type :* Using NAD — Malate enzyme

   eg. : *Portulaca, Cynodon* and *Eleusine*

3. *PEP — CK type :* Using PEP carboxy — Kinase enzyme

   eg. : *Panicum* and *Sporolobus*

8.7.5 The Crassulacean acid metabolic path way (Campathway)

Many succulent plants such as *Kalanchoe* and *Bryophyllum* (family Crassulaceae) show formation of an organic acid at night and its disappearance during day light. Further, their stomata are opened during night (Scotoactive) and closed during day time. Hence, most of the gaseous exchange in these plants is at night rather than during light period.

The term Crassulacean Acid Metabolism (CAM) is derived from large succulent family Crassulaceae (A family showing xerophytic adaptations) and all members of this family have CAM pathway. However, this metabolism is not restricted to Crassculaceae alone, but it is also found in about 25 to 30 families like Cactaceae, Agavaceae, Bromeliaceae and Orchidaceae. It is estimated that over 10,000 angiosperms have CAM pathway. However, a succulent plant like *Salicornia* does not show Crassulacean Acid metabolism.

The leaves of CAM plants are made up of spongy mesophyll cells with chloroplasts and vacuoles but do not have Kranz anatomy similar to the one found in C_4 plants. The presence of scotoactive stomata in them is an adaptation for reducing the rates of transpiration and to minimize water loss. Many of these plants like cacti are better adapted to hot, dry and arid climate.

The CAM species like C_4 also have Hatch-Slack pathway and also Calvin cycle. In C4 plants these two pathways are spatially separated, it means that the former occurs in mesophyll cells while later in bundle sheath cells, but in CAM plants both pathways occur in mesophyll cells only but they are separated by time. During night C_4 cycle is operative and C_3 during day time or in day light.

8.7.5.1 CAM pathway and biochemical reactions : The CAM pathway is presented in (fig.8.12). The first reaction in the series which takes place during night time when stomata are opened. The first reaction is the synthesis of oxaloacetic acid (OAA) from phosphoenol pyruvic acid and CO_2 (Carboxylation of PEP). It is observed that the primary acceptor of CO_2 i.e. PEP is derived from the reserve carbohydrates (glucose) through their glycolytic degradation.

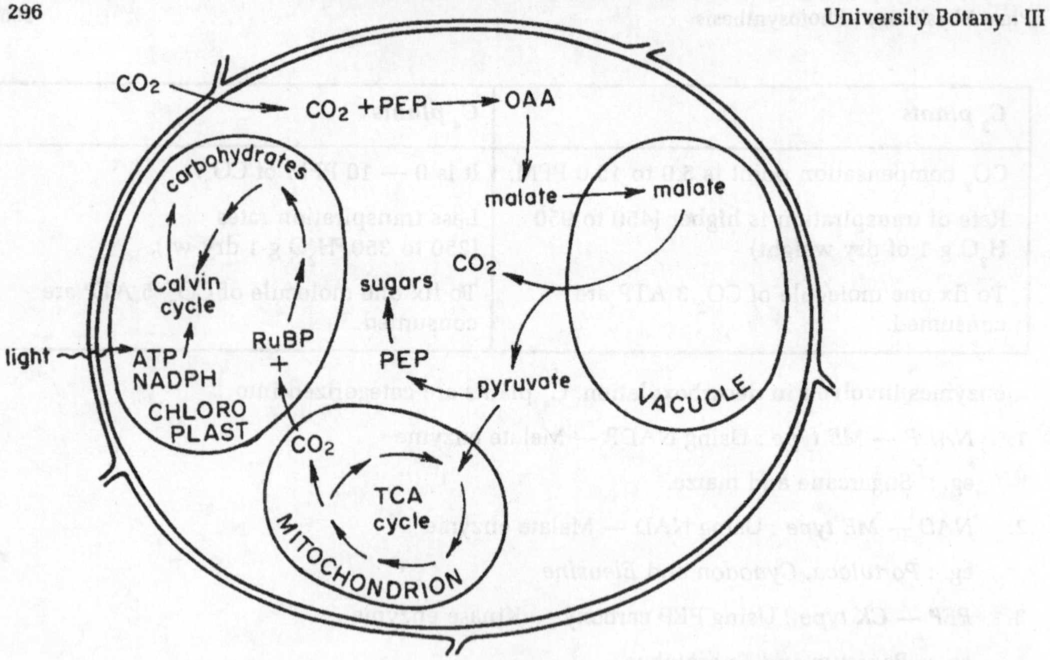

Fig 8.11 The CAM reactions

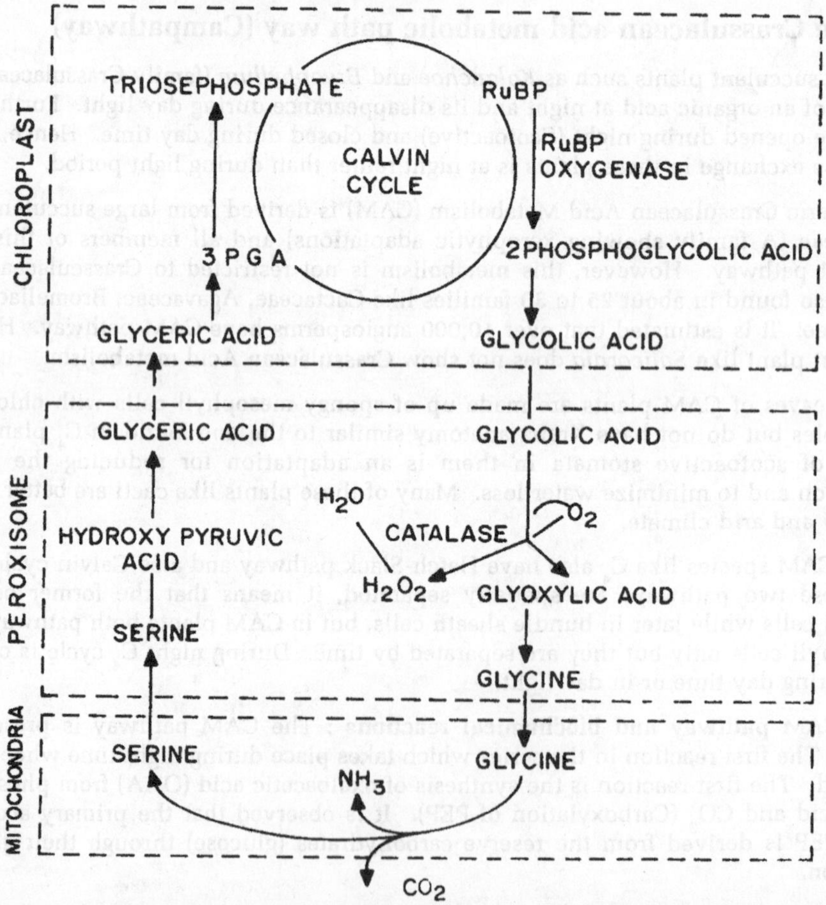

Fig 8.12 C_2 or Photorespiratory Pathway (Observe different biochemical reactions in three cell organelles)

The enzyme PEP carboxylase catalyses the carboxylation of oxaloacetic acid (4C). OAA thus formed is then reduced to malic acid by malate dehydrogenase enzyme.

$$\text{1) PEP} + CO_2 \xrightarrow{\text{PEP carboxylase}} \text{OAA} + \text{Pi}$$

$$\text{2) OAA} + \text{NADPH} \xrightarrow[\text{Malic dehydrogenase}]{} \text{Malic acid} + \text{NADP}$$

Malic acid thus synthesized is transported to vacuoles, where it is stored in the night. The acid is released from the vacuoles during day time is then decarboxylated to pyruvic acid and CO_2. The later is once again is accepted by RUBP in the presence of RUBP carboxylase enzyme. This second carboxylation leads to the formation of phosphoglyceric acid (PGA) and followed by the formation of glucose and regeneration of RUBP following the Calvin cycle reactions.

It is of interest to know that the CAM plants are grouped into two categories based on the decarboxylation of malic acid.

1. **Malic enzyme (ME) CAM Plants :** Here, the pyruvic acid is produced from the malic acid by reductive decarboxylation either in the cytosol or in mitochondria.
2. **Phosphoenol Pyruvate Carbokinase (PEPCK) CAM Plants :** Here, the malic acid is decarboxylated by an enzyme PEPCK. The decarboxylation takes place via the oxidation of malic acid to oxaloacetic acid by malic dehydrogenase enzyme either in cytosol or chloroplast. The OAA is then decarboxylated to PEP.

8.7.5.2 Significance of CAM pathway

1. The special feature of CAM plants is that they can exist for long periods without net uptake of CO_2 in light as they are able to fix CO_2 in dark.
2. They fix CO_2 in the dark rather than in light thus by evolving a mechanism for optimum utilization of CO_2 with minimum loss of water through transpiration. This is considered to be an environmental adaptation as most of the CAM species grow in water scarcity conditions i.e. xeric habitats. This features make them more efficient than C_3 and C_4 plants.

8.8 PHOTORESPIRATION AND GLYCOLATE METABOLISM :

In many plants respiration (oxidation of $C_6H_{12}O_6$) occurs in light as well as in dark. Decker (1959) and Bidwell (1969) have demonstrated that the respiration would be much more in light under photosynthetic conditions than in dark. It was also found that the C_3 species produce three to five times more CO_2 in light than in dark in the presence of CO_2. Decker named this light stimulated release of CO_2 or respiration as Photorespiration. Respiration that occurs in photosynthetic tissue in the presence of light is known as photorespiration. Normal respiration occurs in cytoplasm and mitochondria whereas, photorespiration involve chloroplasts, peroxisomes and mitochondria.

Photorespiration in C_3 plants is much higher than in C_4 plants, where it is either negligible or undetectable. Photorespiration rate results in the direct decrease in CO_2 assimilation and thus leading to decrease in photosynthetic productivity.

Photorespiration is stimulated by various factors like high light intensity, high O_2 level and CO_2 concentration and high temperature. It is observed that high concentration of O_2 in the atmosphere inhibits photosynthesis and accelerates photorespiration.

8.8.1 Biochemical Reactions of Photorespiration : The outline of the reactions of photorespiration are given in (fig. 8.12).

The initial biochemical reaction of Photorespiration takes place in the chloroplast and the substrate for the reaction is Ribulose bi-phosphate (RUBP). It is oxidized to Phosphoglycolate and phosphoglycerate in the presence of RUBP oxygenase (RUBISCO).

$$RUBP + O_2 \xrightarrow{\text{RUBISCO}} 2 \text{ Phospho glycolic acid} + 3 \text{ Phospho glyceric acid}$$

The phosphoglycolate is hydrolysed by a specific chloroplast phosphotase resulting in the formation of glycolic acid.

$$\text{Phosphoglycolate} + H_2O \xrightarrow[\text{Phosphotase}]{} \text{Glycolate} + Pi$$

Glycolate or Glycolic acid is thus considered as the primary substratum for photorespiration. It is synthesised in green leaves in the light but not in dark. When O_2 concentration in the external atmosphere increases, there is an upward increase in glycolic acid concentration and decrease in it when CO_2 concentration is increased. Hence, the high O_2 and low CO_2 favours the formation of glycolic acid and hence photorespiration. Since, glycolic acid is a two —carbon compound, this pathway is also known as C_2 pathway or C_2 oxidative photosynthetic cycle. Further, as both C_2 and C_3 cycles use RUBISCO and regenerate RUBP, these two cycles are now unified under one process, the photosynthetic carbon metabolism.

The glycolate thus formed is then transported to peroxisome, where it is oxidized to glyoxylic acid by the enzyme glycolate oxidase.

$$\text{Glycolic acid} + O_2 \xrightarrow[\text{Glycolate oxidase}]{\text{FMN -}} \text{Glyoxylic acid} + H_2O_2$$

The hydrogen peroxide (H_2O_2) thus produced is broken down by an enzyme catalase into water and O_2. All leaf peroxisomes contain this enzyme.

The glyoxylic acid is then converted to glycine by a transamination reaction with glutamic acid. Glycine (an amino acid) now transported out of peroxisomes into mitochondria, where two molecules of glycine are converted to serine, CO_2 and NH_3.

$$2 \text{ glycine} + H_2O + NAD \xrightarrow[\text{Hydroxy methyl transferase}]{} \text{Serine} + CO_2 + NH_3$$

This liberation of CO_2 in the light is photorespiration. Serine produced in mitochontria migrates back to the peroxisomes, where it is converted to hydroxy pyruvate by an amino transferase and to glyceric acid. The glyceric acid is finally transported out of perioxisomes into chloroplasts, where it is phosphorylated to phosphoglyceric acid which is available for further photosynthetic reactions.

$$\text{Serine} + \text{glyoxylic acid} \xrightarrow[\text{Serine glyoxylate amino transferase}]{} \text{Hydroxy pyruvate} + \text{Glycine}$$

$$\text{Hydroxypryuvate} + NADH \xrightarrow[\text{Hydroxy Pyruvate reductase}]{} \text{Glyceric acid} + NAD$$

$$\text{Glyceric acid} + ATP \longrightarrow 3 \text{ PGA} + ADP + Pi$$

The formation of PGA through glycolic, glyoxylic acids and glycine, serine is known as, glycolate pathway.

8.8.2 Significance of Photorespiration in Green Plants

An examination of biochemical pathway of photorespiration, a large proportion of CO_2 assimilated by photosynthesis is lost through photorespiration. It has been estimated that as much as 50% of photosynthetic potential is lost in photorespiration process resulting in the reduction in dry matter production and photosynthetic yield.

Photorespiration is often measured with the help of CO_2 compensation concentration, which is defined as 'The concentration of CO_2 at which O_2 uptake is balanced by CO_2 release'. In C_3 plants for instance, the CO_2 compensation concentration is as high as 50 ppm while in C_4 plants it is only 10 ppm or less. It is possible to reduce photorespiration in C3 plants by lowering O_2 concentration, since it has been shown that at low O_2 concentration, photorespiration is greatly reduced.

The foregone discussion on photorespiration provides the conclusion that photorespiration always lead to reduction in crop productivity. However, in recent years it has been shown that this process does serve a useful purpose by reducing the impact of 'Photo inhibition', which generally damage photosynthetic apparatus. This makes the C_3 plants well protected from photoinhibition than C_4 plants. There are also some other useful attributes from photorespiration. Some of these are,

1. Amino acids like glycine and serine are synthesized during photorespiratory metabolism which are precursors for many important metabolites such as proteins, chlorophylls and nucleotides.
2. Conversion of glycolate to glyoxylate uses $NADPH_2$ generated in the light phase. Thus photorespiration useful in dissipating excess reducing power.
3. Glycolate formation as the end product also protects plants from the destructive action of aerobic oxidation.

In recent years plant physiologists are making attempts to develop plants with reduced photorespiratory activity with an aim to increase photosynthetic productivity as in plants such as maize and sugarcane (C_4 plants) which have a low photorespiratory activity are found to give high yields. Chemicals which inhibit photorespiration have been identified so that yield reducing photorespiratory activity can be controlled. It has been estimated that as much as 50% yield is reduced due to photorespiration.

8.9 FACTORS AFFECTING PHOTOSYNTHESIS

Like most other processes, photosynthesis is also affected by both external factors (Light, CO_2, temperature water and oxygen) and internal factors. The effect of some of these factors is described here.

8.9.1 The Principle of Limiting Factors : According to Sachs (1860), there are three conditions, i) minimum ii) optimum and iii) maximum for each factor in relation to photosynthesis. Therefore, for a given species, there may be a minimum temperature, below which no photosynthesis takes place, an optimum temparature at which the rate of photosynthesis is maximum and a maximum temperature above which no photosynthesis will take place. This concept of Sachs was replaced by another concept advanced by Blackman (1905). According to this concept or law, 'When a process is conditioned as to its rapidity by a number of separate factors, the rate of the process is limited by pace of the slowest factor'. This law is known as Blackman's law of limiting factor. Blackman found that the rate of photosynthesis varied with the temparature at high light intensities but was unaffected by the temperature at low light. In other words, when photosynthesis was limited by light, it was insensitive to temperature. On the other hand, when light was non-limiting, the process was influenced by the temperature. Similar was the interaction between light and CO_2 concentration when one factor was in abundance, the other factor became limiting for the overall rate of photosynthesis (Fig.8.13).

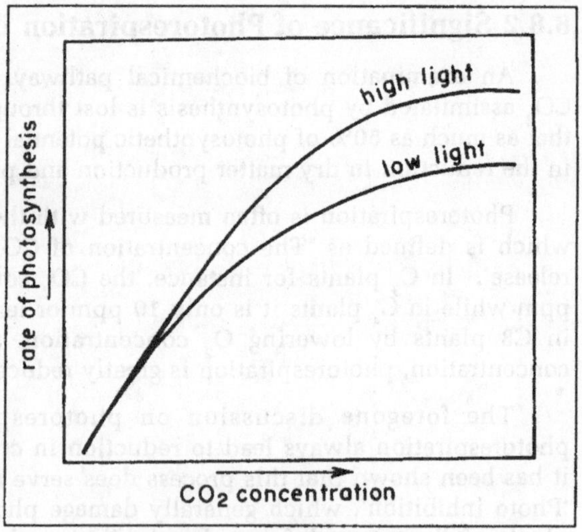

Fig 8.13 Graph showing Blackman's law of limiting factors

The study of effects of various interactive factors (light, CO_2 and temperature) shows that of all the factors, the photosynthetic rate depends on the factor which is present in relatively minimum concentration. This factor is known as limiting factor. All other factors present in optimum or maximum would not be able to increase the rate of photosynthesis until the amount of limiting factor is increased correspondingly.

The principle of limiting factor is applicable to almost all physiological processes. However, it should be realised that different factors are required in different amounts for a particular physiological process to take place. Hence, it is not the absolute quantity or the magnitude of the factors, which is significant, but their relative magnitude in the proportion of the amounts actually required in the process. The quantitatively lowest factor may not necessarily limit the rate of the process, since it may be required only in traces, while the larger amounts of some other factor may be necessary.

8.9.2 Other Environmental Factors Affecting Photosynthesis :

8.9.2.1 Carbondioxide : CO_2 concentration has a very marked influence on the rate of photosynthesis as it is one of the raw materials for photosynthesis. In the atmosphere it is estimated that CO_2 concentration is about 330 PPM far below the requirement for the optimum photosynthesis, hence it becomes a limiting factor. On the other hand, the saturation point of CO_2 is 800 — 1200 PPM, below this saturation point, the rate of photosynthesis increase with the increase in CO_2 concentration. The increase is more rapid upto 1% of Co_2 concentration and beyond this level it slows down due to closure of stomata.

The saturating concentration of CO_2 is higher in C_4 plants than in C_3 ones. If the CO_2 concentration is reduced, the rate of photosynthesis in both C_3 and C_4 plants is slow until it is exactly equal to the rate of photorespiration. At this point, there is no net intake of CO_2, that is, the apparent rate of photosynthesis is zero. In C_3 species, this situation occurs at 40-50 PPM of CO_2. The CO_2 concentration at which CO_2 intake and out put (where photosynthetic fixation just balances respiratory loss) are equal and the rate of photosynthesis is zero in 'the CO_2 compensation point'. In other words, volume of CO_2 consumed in photosynthesis is equal to the volume of CO_2 liberated in photorespiration and conversely there is no net gain or loss and this concentration of CO_2 is known as CO_2 compensation point. Since C_4 plants do not release CO_2 by photorespiration, their CO_2 compensation point is very less in the range of 0 — 10 PPM.

8.9.2.3. Light : Light is also one of the important factors affecting photosynthesis, since the sunlight is the chief source of energy and available plenty, the rate of photosynthesis will increase with increase in light intensity, provided other environmental factors are not limiting. However, when the light intensity decreases, the rate of photosynthesis decreases until a compensation point for light. The effect of this factor on photosynthesis depends on the intensity, quality as well as the duration of light. High light intensity in many plants may inhibit the photosynthesis (solarization) as a result of photoxidation and bleaching. Some vital enzymes are also inactivated at high light intensity.

The quality of light (wave length) also affects the photosynthetic rate. It is observed that maximum rate of photosynthesis is in the range of 600 — 700 nm (orange — red with a peak at about 665 nm) and low rate at 470 — 600 nm range (green light), photosynthesis rate may show an increase when expcsed to longer periods of uninterrupted light.

8.9.2.4 Temperature : The effect of temperature on photosynthesis depends on the species and the environmental conditions under which the plant is growing. Xerophytes can carry on photosynthesis at 50ºC, while mesophytes in the range of 15-35ºC. Interestingly the C_4 plants have higher optimum temperature (30 - 40ºC), while the C_3 plants have higher photosynthetic rate at 15-25ºC and CAM plants at 35ºC.

8.9.2.5 Oxygen concentration : As we know that O_2 is a by product of photosynthesis and its content directly or indirectly affect photosynthesis. The oxygen inhibition of photosynthesis was first observed by warburg (1920) in unicellular alga *Chlorolla*. The inhibiton of photosynthesis by high oxygen concentration is called Warburg effect. The Warbung effect is not only observed in *Chlorella* but also in other C_3 species such as barley, bean, and soybean. But in C_4 species like maize, sugarcane, *Sorghum*, the Warburg effect is not observed since photorespiration is absent in them.

8.9.2.6 Water content : Water is one of the raw materials required for photosynthesis and the amount of water needed by a plant is very small. However, water deficient conditions affect photosynthesis both directly and indirectly. It has been observed C_3 species are more sensitive to water scarcity conditions than the C_4 plants. In general, water deficiency may cause closure of stomata and will result in the decrease of turgor pressure and dehydration of protoplasm. It will also inturn reduce the enzyme activity, Hill reaction, photosynthetic transport and ultimately the NADP reduction.

8.9.2.7 Chemical Pollutants : Due to urbanization and industralization in many countries of the world resulted in the accumulation of toxic chemicals in the environment. This largely affect the plant growth by reducing photosynthetic rates. Air pollutants like SO_2, NO_2, O_3 are responsible for this. Experimental evidences also proved that the accumulation of heavy metals like lead, mercury and cadmium lead to the reduced photosynthesis by causing stomatal closure and also resulting in reduced or limited assimilation of CO_2 by plants. The action of ozone is a cause for concern because of its strong oxidizing actions, damage of selective permeability of membranes of phyosynthesetic cells and cell organelles.

QUESTIONS

A. Answer the following in about 25 lines

1. Primary process in photosynthesis leads to the conversion of light energy into chemical energy. Discuss.
2. What do you understand by 'Z' scheme of light phase. Describe the photochemical reactions.
3. Give a brief account of the C_3 cycle of photosynthesis.
4. Describe CAM pathway
5. Describe sailent features of C_4 species and in what way they differ from C_3 species.
6. Write various biochemical reactions of Hatch-Slack cycle.

B. Answer the following in about 10 lines

1. **Hill reaction**
2. **Emerson enhancement effect**
3. **Cyclic Photophosphrlation**
4. **Light Harvesting complex**
5. **Significance of CAM pathway**
6. **Importance of RUBP carboxylase**
7. **Warburg effect**
8. **Photosystems**
9. **Co_2 compensation**
10. **Thylakoid structure.**

C. Answer the following in 3- 5 lines.

1. **Quantasomes**
2. **Photolysis of water**
3. **Electron carriers**
4. **Fluorescence**
5. **Red drop**
6. **Primary reaction centre**
7. **Peroxisomoes**
8. **Kranz anatomy**
9. **Water oxidizing clock**
10. **Chemical pollutants and Photosynthesis**

9

PLANT PHYSIOLOGY

RESPIRATION

9.1 INTRODUTION

Respiration is an energy releasing process. In respiration the food substances are broken down into simpler elements like CO_2, water and energy contained in the food substances is released in the active form. This energy is used up by the plant in various metabolic activities. Not all the energy released during the respiration, however, is utilised by the plant. Certain amount of energy is released in the form of heat, which is lost into the environment.

In cells during the various metabolic pathways the energy is transformed from one state to another state. Bioenergetics deals with the study of energy transformations from one state to another state during the various physiological processes occurring in the cells.

9.2 THERMODYNAMICS

Every living system requires energy to carry out its various metabolic activities like growth, development etc. The study of energy changes in chemical systems is dealt in "Thermodynamics". All the living systems obey the laws of thermodynamics. The laws of thermodynamics are as follows :

1. **First Law :** It states that energy can neither be created nor destroyed but simply can be converted from one form to other form.
2. **Second Law :** This law states that processes involving energy transformation will not occur spontaneously unless there is degradation of energy from a non-random to a random form. In all energy releasing reactions (Exergonic) certain amount of energy is available for useful work called "Free energy". A certain amount is unavailable for work called "entropy". The entropy increases dis-order in the system.
3. **Third Law :** This law states that all systems, if left to themselves, tend to increase the state of disorganisation and disorder (entropy) unless free energy is provided. The tendency is from orderliness to disorderliness. According to this law the entropy of perfect crystal is 'zero'. The highest state of entropy i.e. disorganisation is death in all living systems.

Energy can be found in different forms viz. Potential energy, Kinetic energy, radiant energy, thermal energy, chemical energy, electrical energy etc. For all these types of energy the ultimate source is sun. The solar energy reaches the earth in the form of electromagnetic waves. In photosynthesis, as studied earlier, the green plants convert the radiant energy from the sun into chemical energy and store it in carbohydrates. These carbohydrates serve as respiratory substrates in respiration. Carbohydrates are broken down into simpler elements releasing the potential energy in the form of Kinetic energy. Hence respiration is described as energy releasing or exoergonic reaction. The photosynthesis is described as endergonic process, because simpler elements like Co_2 and water produce energy rich compunds such as carbohydrates using sunlight, in the chlorophyll containing parts of the plant.

We have to understand "Free energy" (ΔF) enthalpy and entropy (ΔS) and their inter-relationships. The nature of interrelationship among free energy, enthalpy and entropy can be represented by the following equation

$$\Delta F = (\Delta H - T\Delta S)$$

In the above equation

T = Absolute temperature

S = Entropy

We have to understand exergonic and endergonic reactions in the light of entropy.

Photosynthesis :

$$6Co_2 + 12\,H_2O \xrightarrow[\text{Sunlight}]{\text{Chlorophyll}} C_6H_{12}O_6 + 6O_2 + 6H_2O$$

In the above reaction ΔF = 686000 I cal I per mol.

So the energy forms in the above reaction are as follows

ΔF(Free energy) = + 686,000 I cal I per mol.

ΔH (Heat energy) = + 673,000 I cal I per mol.

ΔS (entropy) = - 43.6 cal I per mol.

So in photosynthesis there will be more free energy and less entropy

Respiration :

$$C_6H_{12}O_6 \longrightarrow 6Co_2 + 6\,H_2O$$

Δ DF = - 686,000 cal I per mol.

ΔH = 673,000 I cal I per mol.

ΔS = + 43.6 I cal I per mol.

Thus in respiration free energy decreases and entropy increases.

9.3 PLANT RESPIRATION

All living organisms whether plants or animals require energy for carrying out its various life activities like growth, movement, absorption and accumulation of salts, synthesis of proteins etc. In all these and many other activities, considerable amount of work is done, which involves the expenditure of considerable amount of energy.

Solar energy is the ultimate source of energy available to all living organisms. Green plants trap the solar energy during the photosynthesis, convert it to chemical energy and store it in the carbohydrates synthesised. These carbohydrates contain energy in an inactive form or in potential form. This inactive chemical energy must be transformed to active or kinetic energy before it is being utilised. In respiration the food substances undergo oxidation releasing energy in the active form.

Respiration can be defined as an oxidative process, in which complex organic substances are broken down into simpler elements, releasing the energy. Respiration is a vital process. All living cells must respire. Respiration is not confined to any particular organ or tissue. Hence it is called as "Cellular Respiration". The respiration can be represented by the following equation.

$$C_6H_{12}O_6 + 6O_2 + 6H_2O \text{ -----------------------> } 6\ CO_2 + 12H_2O + 673 \text{ k.cal}$$

This equation shows that one gram molecule of glucose (180 gms) on complete oxidation yields 673,000 calories of energy. Some amount of energy is liberated in the form of heat. In animals this heat energy is utilized for maintaining the body temperature. In plants this heat is lost into the atmosphere in the form of radiation.

Thus, respiration is a complex process, involving the following events.

i. It is an oxidative process, in which the food substances are oxidised either partially or completely to give rise to simpler elements.

ii. It is a catabolic process, wherein the complex substances are broken down into simpler elements like Co_2 and water.

iii. It is an exothermic process, in which the energy is released. One gram molecule of glucose on complete oxidation releases 673 K.cal of energy.

iv. It results in loss of weight.

v. It results in the formation of many intermediate compounds, which are utilised in various metabolic activities like synthesis of amino acids etc.

9.3.1 Types of Respiration : Respiration is of two types i) Aerobic Respiration ii) Anaerobic Respiration

9.3.1.1 Aerobic Respiration : Aerobic respiration takes place in the presence of oxygen. Carbohydrates are completely oxidised to simpler elements like Co_2 and water.

$$C_6H_{12}O_6 + 6O_2 \text{ -----------------------> } 6Co_2 + 6H_2O + 673 \text{ K.cal}$$

Aerobic Respiration occurs commonly in all higher plants and animals.

9.3.1.2 Anaerobic Respiration : It takes place in the absence of oxygen. Food substances are incompletely oxidised to alcohols etc. Anaerobic respiration is rare in higher plants but occur commonly in lower organisms like bacteria and yeasts.

$$C_6H_{12}O_6 \text{ -----------------------> } 2\ C_2H_5OH + 2Co_2 + 28 \text{ K.cal}$$

Anaerobic respiration is injurious to plants, if it is continued for long time. This is due to the accumulation of toxic substances in the cells and also due to the release of low level of energy.

Table 9.1 Differences between aerobic and anaerobic respiration

Aerobic Respiration	Anaerobic Respiration
1. It takes place in the presence of oxygen	1. It takes place in the absence of oxygen
2. It is common in all higher plants	2. It is a rare occurrence in higher plants
3. It goes on throughout the life of the plant	3. It occurs only for short periods in higher plants
4. Carbohydrates are completely oxidised into simpler elements like CO_2 and water	4. It results in the formation of incompletely oxidised substances like alcohol, acids etc.
5. Energy is liberated in larger quantities. One gram molecule of glucose give rise to 673 K.cal of energy	5. Energy is liberated in lesser quantity. One gram molecule of glucose liberates 28 K.cal of energy.
6. Aerobic Respiration is not toxic to the plants	6. It is toxic to the plant, if it is continued for longer times.
7. The end products are CO_2 and water.	7. The end products are alcohols and CO_2.

9.3.2 Respiratory Substrates : Plant cells contain many types of organic substances such as carbohydrates, fats, proteins, organic acids etc, which are oxidised during respiration. These organic substances, which are used in respiration, are called 'respiratory substrates'. Among the respiratory substrates, most abundant and prominent ones are carbohydrates. Generally, carbohydrates such as sugars, starches are the first substrates, which are used up in respiration. If the carbohydrates are totally used up then the fats and lastly proteins are utilised in the respiration process. Fats and proteins are used up in the respiration process only when the cells are facing shortage of carbohydrates.

Carbohydrates : Carbohydrates are the major respiratory substrates in plants. The important carbohydrates, which are oxidised during the respiration, are glucose, fructose, sucrose and starch.

Glucose and fructose are used directly as respiratory substrates.

$$\underset{\text{Glucose \& Fructose.}}{C_6H_{12}O_6} + 6O_2 \text{ ------------------> } 6Co_2 + 6H_2O + \text{Energy}$$

Sucrose is a disaccharide. It is first hydrolysed to glucose and fructose by the action of enzyme 'Invertase'. Glucose and fructose are used up in the respiration.

$$\underset{\text{Sucrose}}{C_{12}H_{22}O_{11}} + H_2O \xrightarrow{\text{Invertase}} \underset{\text{Glucose}}{C_6H_{12}O_6} + \underset{\text{Fructose}}{C_6H_{12}O_6}$$

Starch is a principal reserve food material in plant cells. It is a polysaccharide, composed of glucose. Starch is first hydrolysed to glucose by the action of enzyme "Amylase". Glucose is utilised in the respiration.

$$\underset{\text{Starch}}{(C_6H_{10}O_5)_n} + nH_2O \xrightarrow[\text{Enzyme}]{\text{Amylase}} n\ \underset{\text{Glucose}}{(C_6H_{12}O_6)}$$

Fats : When the carbohydrates are completely exhausted, then fats become respiratory substrates. When the fats act as respiratory substrates, they are first converted to fatty acids and glycerol by the action of enzyme 'Lipase'.

$$\text{Fats} \xrightarrow{\text{Lipase}} \text{Fatty acids} + \text{Glycerol}$$

Fatty acids are broken down to pyruvic acid and the latter enter into Kreb's cycle and is oxidised to Co_2 and water.

Glycerol is phosphorylated to Glycerol 3-phosphate in the presence of an enzyme called ' glycerol kinase'

$$\text{Glycerol} + \text{ATP} \longrightarrow \text{Glycerol — 3 — Phosphate} + \text{ADP}$$

Glycerol — 3-Phosphate is converted to its isomeric compound dihydroxy acetone phosphate in the presence of enzyme " glycerol phosphate dehydrogenate" and the latter enter into glycolysis.

$$\text{Glycerol — 3 — Phosphate} + \text{NAD} \longrightarrow \text{Dihydroxy acetone Phosphate} + \text{NADH} + H^+$$

Organic acids : Many types of organic acids such as malic acid, citric acid, succinic acid etc are found in plant cells in considerable quantity as they are intermediate compounds in Kreb's cycle. They are directly used in the respiratory process. Certain plants like members of Crassulaceae and Cactaceae families synthesize organic acids during the night time and they are being used up during the day time in the respiration releasing Co_2 and water (Crassulean acid metabolism)

$$\underset{\text{Oxalic acid}}{2(COOH)_2} + O_2 \longrightarrow 4\ Co_2 + 2\ H_2O$$

Proteins : Proteins become respiratory substrates when carbohydrates, fats are completely exhausted. Proteins are broken down to amino acids by the action of proteolytic enzymes. These amino acids are converted to organic acids by deamination method. These organic acids enter into Kreb's cycle forming CO_2 and water.

i. $\text{Proteins} \xrightarrow[\text{enzymes}]{\text{Proteolytic}} \text{Amino acids}$

ii. $\text{Amino acids} \xrightarrow{\text{deamination}} \text{Organic acids}$

iii. $\text{Organic acids} \xrightarrow{\text{Kreb's cycle}} Co_2 + H_2O$

9.3.3 Respiratory - Quotient

9.3.3 Respiratory - Quotient : Respiratory quotient (RQ) is defined as "the ratio between the volume of Co_2 given out to the volume of oxygen absorbed by a given mass of tissue in a given period of time at a standard temperature and pressure".

$$\text{R.Q.} = \frac{\text{Volume of } Co_2 \text{ evolved}}{\text{Volume of } O_2 \text{ absorbed}}$$

The value of RQ depends upon the nature of respiratory substrate. RQ value of different substrates is discussed below.

RQ of Carbohydrates : When carbohydrates such as hexoses are respiratory substrates the volume of Co_2 evolved is equal to the volume of O_2 absorbed. Then RQ of the substrate is unity (1) as shown below.

$$C_6H_{12}O_6 + 6O_2 \longrightarrow 6Co_2 + 6H_2O$$

Hexose

$$R.Q = Co_2 / O_2 = 6/6 = 1 \text{ (unity)}$$

RQ of Fats : Fats and proteins contain more hydrogen and less oxygen. The ratio of hydrogen to oxygen atoms is more than 2. So they require more oxygen for complete oxidation and less Co_2 is evolved. Hence the RQ is less than unity. For example

$$C_{18}H_{36}O_2 + 26\,O_2 \longrightarrow 18\,Co_2 + 18H_2\,O$$

Stearic acid

$$RQ = Co_2 / O_2 = 18/26 = 0.7 < 1$$

In some plants like Opuntia, incomplete oxidation of carbohydrates takes place to produce organic acids, which remain unoxidised. In such cases the RQ is always less than one and often zero, as O_2 is absorbed without release of Co_2

$$2C_6H_{12}O_6 + 3\,O_2 \longrightarrow 3C_4\,H_6\,O_5 + 3H_2\,O$$

Glucose

$$RQ = Co_2 / O_2 = 0/3 = \text{zero.}$$

In Bryophyllum leaves the liberated Co_2 is reabsorbed for synthesizing organic acid in the night times, so RQ is less than one. In daytime the Co_2 released during respiration is used up in photosynthesis so Co_2 released is zero, hence RQ is zero.

RQ of organic acids : Organic acids are rich in oxygen so they require less amount of external oxygen for oxidation and more amount of Co_2 is released. So RQ value of organic acids is always more than unity.

$$2C_2H_2O_4 + O_2 \longrightarrow 4\,Co_2 + 2\,H_2\,O$$

oxalic acid

$$RQ = 4/1 = 4$$

The value of RQ deviates from unity due to the following reasons.

i. When other than carbohydrates are used up as respiratory substrates
ii. When incomplete oxidation of respiratory substrates occur
iii. When O_2 absorbed is used in process other than respiration
iv. When liberated Co_2 is again absorbed in some other process
v. When Co_2 is evolved by any other process than respiration.

9.3.4 Mechanism of Respiration : The oxidation of respiratory substrates in respiration takes place in step-wise manner and in a series of reactions, each reaction is catalysed by a specific enzyme. All complex carbohydrates are first converted to hexoses (Glucose or fructose) before actually entering into the respiration process. The oxidation of glucose (respiration) takes place in two phases.

1. Glycolysis
2. Kreb's cycle

First the glucose is converted to pyruvic acid (CH_3 CO COOH) in a series of reactions collectively known as "glycolysis". In the second phase the pyruvate is oxidised to Co_2 and water.

Steps involved in glycolysis are common to both kinds of respiration and proceeds irrespective of the presence or absence of oxygen. The fate of pyruvic acid, however, depends upon the presence or absence of oxygen. In the presence of oxygen complete degradation of pyruvic acid takes place resulting in the formation of Co_2 and water (Kreb's cycle). In the absence of oxygen, incomplete oxidation of pyruvic acid takes place resulting in the formation of ethyl alcohol and Co_2 in anaerobic respiration and lactic acid in the lactic acid fermentation.

Glycolysis, anaerobic respiration and fermentation, occur in the cytoplasm of the cells and Kreb's cycle occur on the oxysomes of mitochondria.

9.3.4 GYLCOLYSIS : The step-wise degradation of glucose into pyruvic acid is called as "Glycolysis". The series of reactions in the glycolysis have been worked out by three scientists namely Embden, Meyerhof and Paranas. Hence,the glycolytic pathway is also known as "EMP Pathway" in honour of three scientists. In glycolysis glucose converts

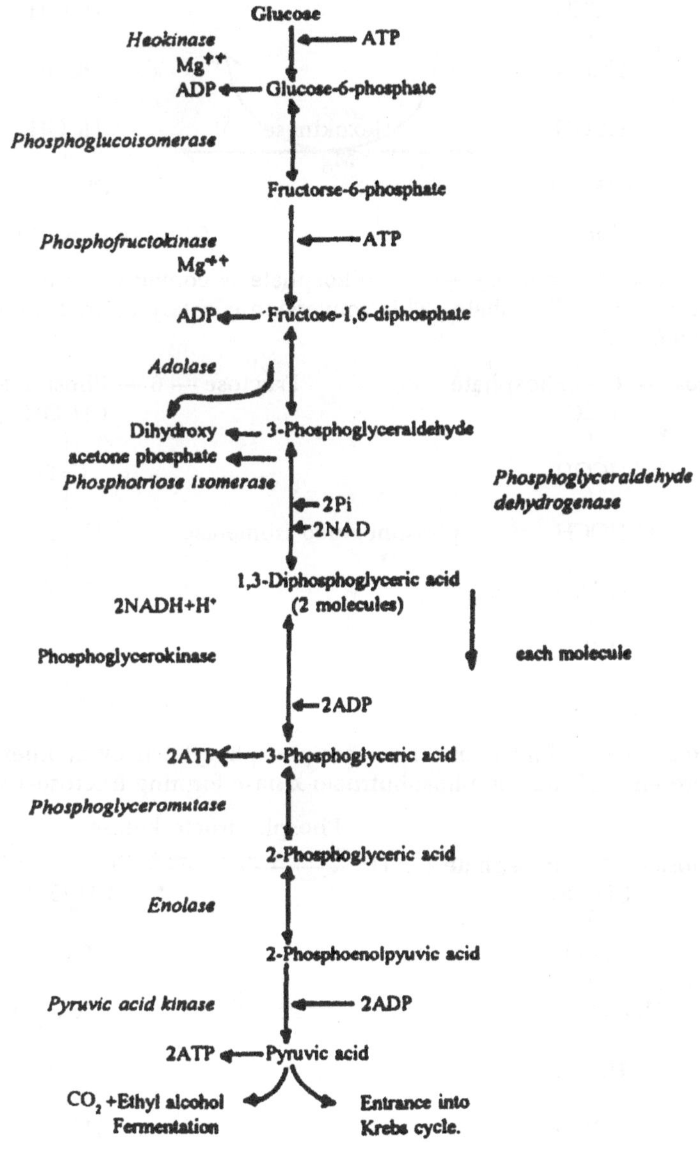

Fig 9.1 Schamatic representation of Glycolysis

into two molecules of pyruvic acid through a series of step-wise reactions as outlined below(fig 9.1).

1. Glucose is first converted to glucose —6- phosphate by reacting with a molecule of ATP in the presence of an enzyme called "Hexokinase", Mg^{++} ions act as co-factors.

hexokinase

Glucose + ATP ---------------------> Glucose-6-Phosphate + ADP

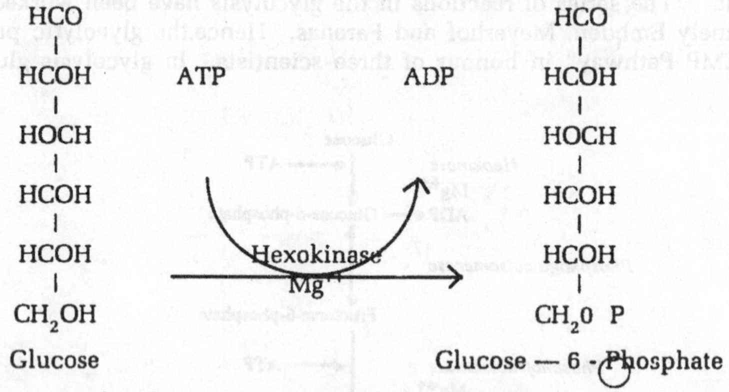

2. In the next step glucose — 6 — Phosphate is converted into its isomeric compound fructose — 6 — Phosphate. This conversion is catalysed by the enzyme 'phosphogluco — isomerase'.

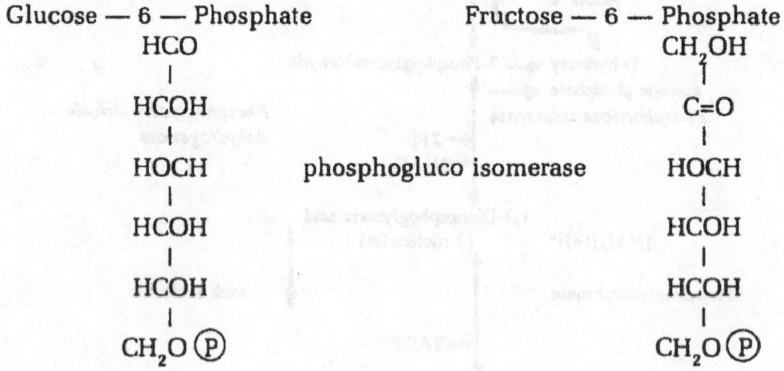

3. Fructose — 6 — Phosphate is further phosphorylated by another molecule of ATP in the presence of enzyme phosphofructo-kinase forming fructose-1-6 — diphosphate

Phospho fructo kinase

Fructose — 6 — Phosphate + ATP ⇌ Fructose-1-6 diphosphate + ADP

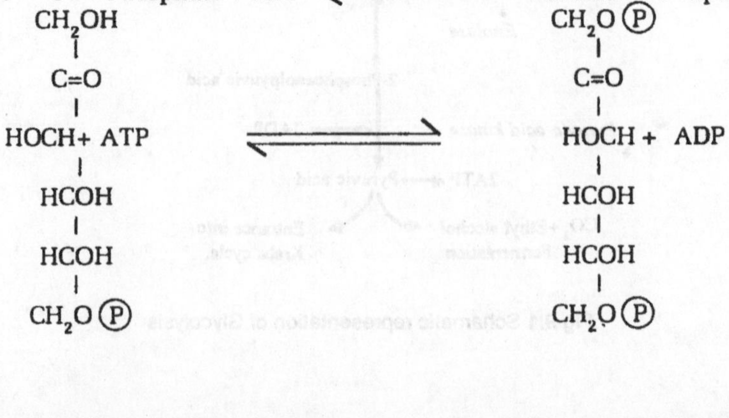

4. Fructose — 1- 6 — diphosphate is broken into two trioses namely one molecule of 3-phospho glyceralalehyde and one molecule of Dihydroxyacetone phosphate (DHAP). These two compunds are isomers and can be inter converted under the influence of enzyme triosephosphate isomerase.

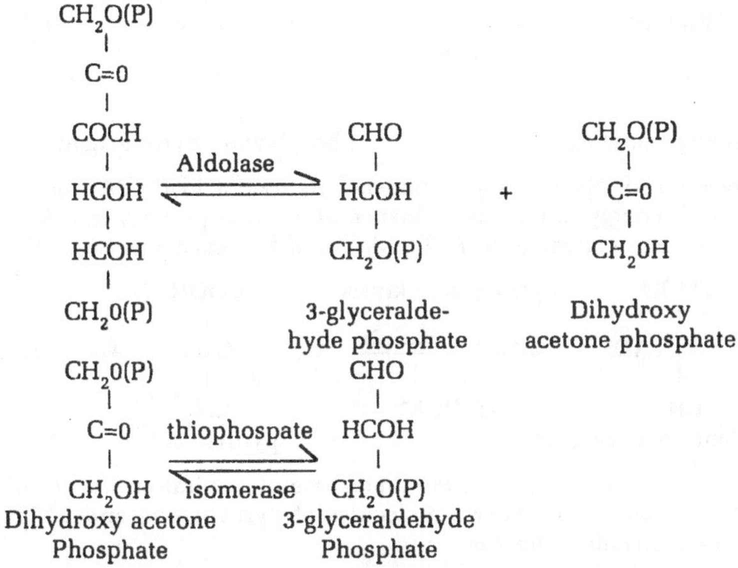

5. Further metabolism is mainly concerned with 3-PGAL, which on reaction with inorganic phosphate is oxidised to 1-3 diphosphoglyceric acid in the presence of enzyme 'Phosphoglyceraldehyde dehydrogenase' and NAD. NAD is reduced to $NADH_2$

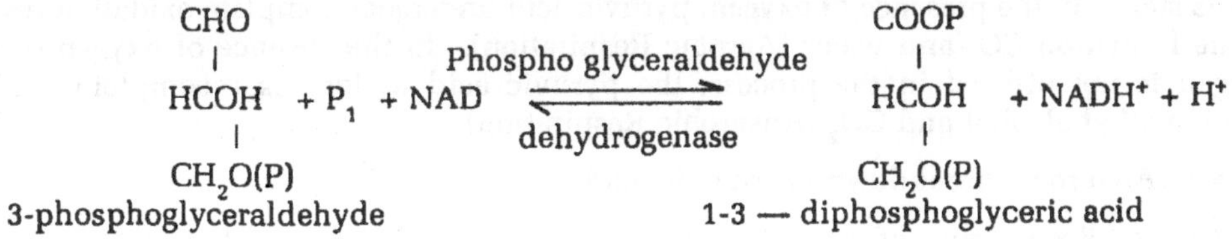

6. In the next reaction 1-3 diphosphoglyceric acid reacts with ADP and looses a phosphate group forming 3-phosphoglyceric acid and ADP is converted to ATP. This reaction is catalysed by the enzyme phosphoglyceric kinase.

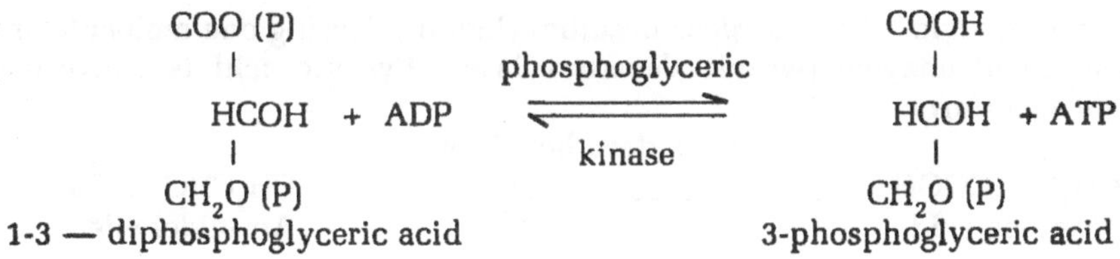

7. 3-phosphoglyceric acid gets transformed to 2-phospho glyceric acid in the presence of enzyme phosphoglycero mutase.

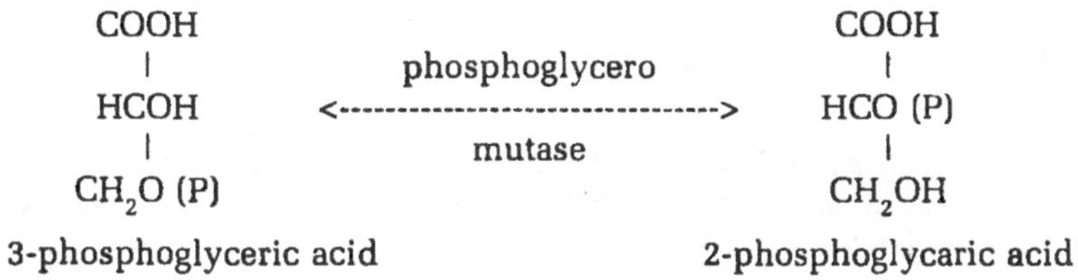

8. In the next step 2-phosphoglyceric acid loses one molecule of water under the influence of enzyme Enolase and is converted to phosphoenol pyruvic acid.

$$\underset{\text{2-phosphoglyceric acid}}{\begin{array}{c}COOH\\ |\\ HCO\,(P)\\ |\\ CH_2OH\end{array}} \;\underset{\text{Enolase}}{\overset{Mg^{++}}{\rightleftharpoons}}\; \underset{\text{phosphoenol pyruvic acid}}{\begin{array}{c}COOH\\ |\\ C\text{-}O\,(P)\\ |\\ CH_2\end{array}} + H_2O$$

9. In the last reaction of glycolysis phosphoenol pyruvic acid (PEPA) looses phosphate group along with energy under the influence of enzyme pyruvic acid Kinase to form pyruvic acid. ADP is converted to ATP. Mg^{++} and K^{+} ions act as Co-factors.

$$\underset{\text{Phosphoenol pyruvic acid}}{\begin{array}{c}COOH\\ |\\ C\text{-}O\,(P) + ADP\\ |\\ CH_2\end{array}} \;\underset{Mg^{++},\,K^{+}}{\overset{\text{Pyruvic acid kinase}}{\rightleftharpoons}}\; \underset{\text{pyruvic acid}}{\begin{array}{c}COOH\\ |\\ C{=}O\\ |\\ CH_3\end{array}} + ATP$$

With the formation of pyruvic acid, the first phase of respiration i.e. glycolysis comes to an end. At the end of glycolysis two molecules of pyruvic acid and 2ATP molecules are formed from one molecule of hexose.

9.3.4.1 Fate of pyruvic acid : Glycolysis ends with the formation of pyruvic acid. Pryuvic acid is a key compund in respiration. Further breakdown of pyruvic acid depends upon the availability of oxygen. In otherwords a bifurcation of the respiratory mechanism occurs at this stage. In the presence of oxygen, pyruvic acid undergoes complete oxidation resulting in the formation CO_2 and water (Aerobic Respiration). In the absence of oxygen or when oxygen is not utilised in the process, the pyruvic acid undergoes incomplete oxidation forming ethyl alcohol and CO_2 (Anaerobic Respiration).

9.3.4.1.1Anaerobic breakdown of pyruvic acid :

9.3.4.1.1.1 Breakdown of pryuvic acid to ethyl alcohol: Anaerobic respiration takes place in the absence of oxygen or when oxygen is not utilised in the process. Under anaerobic conditions pyruvic acid undergoes incomplete oxidation forming ethyl alcohol and Co_2. It involves two steps.

i) In the first step pyruvic acid undergoes decarboxylation releasing one molecule of Co_2 in the presence of enzyme pyruvic decarboxylase. Pyruvic acid is converted to acetaldehyde as follows:

$$\underset{\text{Pyruvic acid}}{CH_3CO\ COOH} \;\overset{\text{Pyruvic decarboxylase}}{\rightleftharpoons}\; \underset{\text{Acetaldehyde}}{CH_3CHO} + Co_2$$

ii) In the second step acetaldehyde is reduced to ethyl alcohol by the enzyme alcohol dehydrogenase and $NADH_2$

$$\underset{\text{Acetaldehyde}}{CH_3CHO} + NADH_2 \;\overset{\text{Alcohol dehydrogenase}}{\rightleftharpoons}\; \underset{\text{Ethylalcohol}}{CH_3CH_2OH} + NAD$$

9.3.4.1.1.2 Break down of pyruvic acid to Lactic acid : Though the break down of pyruvic acid to Lactic acid is not common in higher plants, but it is very common is animal tissues and some bacteria

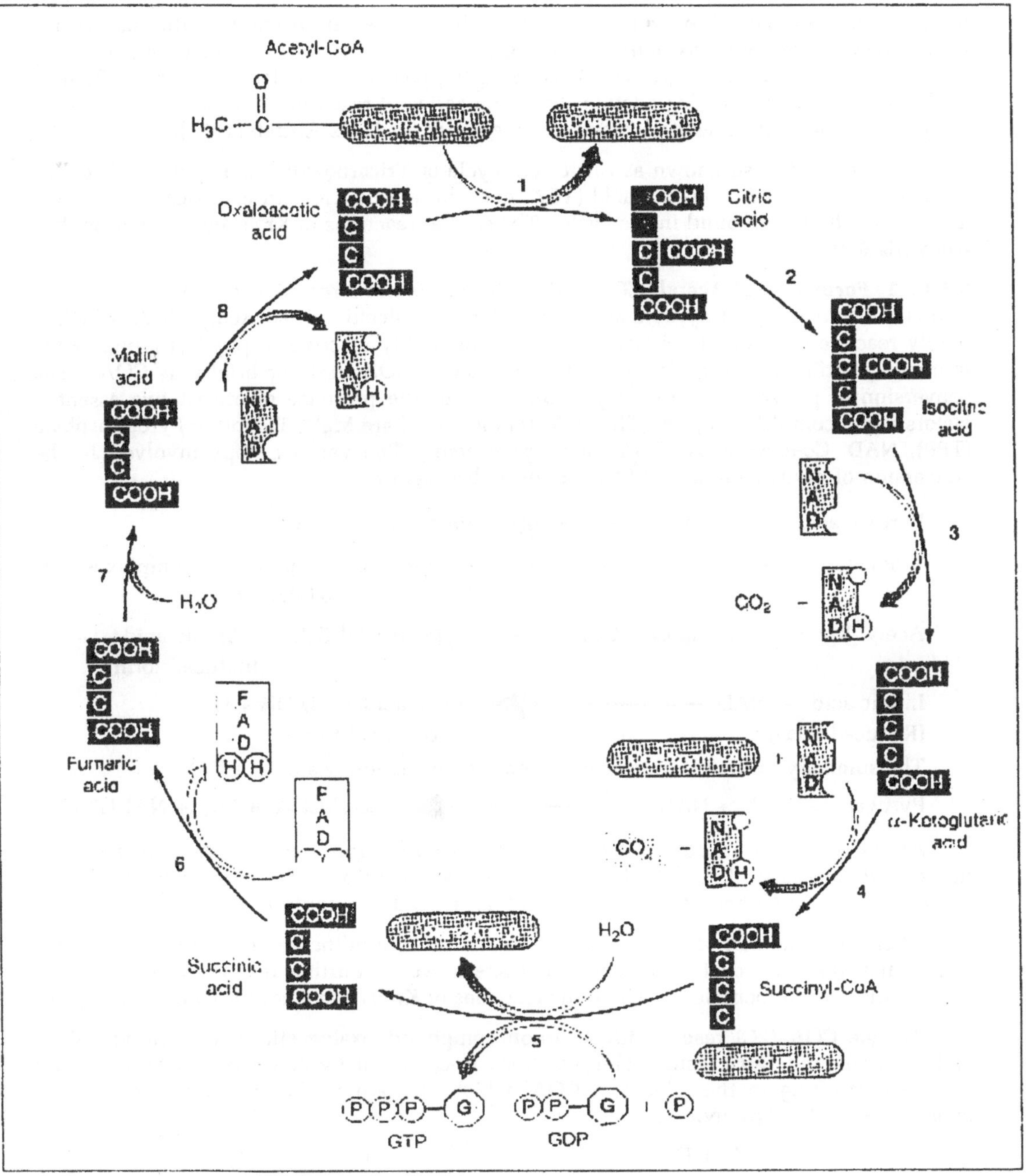

Fig 9.2 Kreb's Cycle

In this case pyruvic acid is converted to Lactic acid in the presence of enzyme lactic dehydrogenase and $NADH_2$.

$$\underset{\text{Pyruvic acid}}{CH_3CO\ COOH} + NADH_2 \xrightleftharpoons{\text{Lactic dehydrogenase}} \underset{\text{Lactic acid}}{CH_3.CHOH.CHOH} + NAD$$

9.3.4.2 KREB'S Cycle (TCA cycle) : In aerobic (the presence of oxygen) conditions pyruvic acid undergoes complete oxidation resulting in the formation of Co_2 and water, releasing considerable amount of energy, which is being trapped in the molecules of ATP. Pyruvic acid undergoes step-wise degradation through a series of reactions. These reactions occur in cyclic manner known as Kreb's cycle, named after Sir Hans A.Kreb (1937).

Krebs cycle is also known as Citric acid cycle or Tricarboxylic acid cycle. It is called citric acid cycle or tricarboxylic acid (TCA) cycle because citric acid is a tricarboxylic acid and it is the first compound in the cycle. The several reactions of Krebs cycle are described below(fig 9.2).

9.3.4.2.1 Formation of Acetyl COA : Pyruvic acid is not a reactive compound. The first step is the conversion of pyruvic acid into activated molecule called Acetyl COA, which is highly reactive compound. Acetyl COA contains highly reactive sulphydryl group at the terminal position of the molecule and is written as COA mSH or briefly as COA. The conversion of pyruvic acid to acetyl COA involves the presence of atleast five essential factors and a complex enzyme. The co-factors involved are Mg^{++}, Thiamine Pyrophosphate (TPP), NAD, Co-enzyme A (COA) and lipoic acid. The various steps involved in the conversion of pryuvate to acetyl COA are described below:

Pyruvic acid + TPP $\longrightarrow$ TPP complex + Co_2

TPP complex + Lipoic acid $\longrightarrow$ Acetyl lipoic acid complex + TPP
(oxidized form)

Acetyl lipoic acid complex + COA $\longrightarrow$ Acetyl COA + Lipoic acid
(reduced form)

Lipoic acid + NAD $\longrightarrow$ Lipoic acid + NADH + H
(Reduced form) (oxidized form)

The summary of the reactions can be represented as follows :

Pyruvate + COA + NAD^+ $\longrightarrow$ Acetyl COA + Co_2 + $NADH^+$ + H_+

Above reaction is called oxidative decarboxylation because in this reaction decarboxylation as well as dehydrogenation (oxidation) of pyruvic acid takes place. The oxidative decarboxylation of the pyruvic acid occurs in the cytoplasm of the cell.

1. After the formation of acetyl COA, it enters into the mitochondria. Acetyl COA is a connecting link between the glycolysis and Krebs cycle. Further reactions take place on the oxysomes of mitochondria, where the enzymes of Krebs cycle are present.

Acetyle COA (2C) reacts with 4-carbon compound oxaloacetic acid forming a 6 — carbon compound Citric acid. This reactions is catalysed by condensing enzyme Citric synthase, resulting in the release of COA, which is again made available to react with another molecule of pyruvic acid.

$$\begin{array}{c} CO\ COOH \\ | \\ CH_2\ COOH \end{array} + CH_3\ CO \sim\sim COA \xrightarrow[\text{synthase}]{\text{citric}} \begin{array}{l} CH_2.\ COOH \\ | \\ C(OH)\ COOH + COA \\ | \\ CH_2\ COOH \end{array}$$

Oxalo acetic acid — Citric acid

2. In the next step citric acid looses one molecule of water to form cis-aconitic acid, which on hydration converted to Isocitric acid. Both reactions are catalysed by the enzyme aconitase.

$$\begin{array}{ccccc} CH_2 . COOH & & CH_2\ COOH & & CH_2 . COOH \\ | & \text{aconitase} & | & \text{aconitase} & | \\ C(OH). COOH & \rightleftharpoons & C.COOH & \rightleftharpoons & CH . COOH \\ | & & | & & | \\ CH_2 . COOH & & CH_2 . COOH & & CHOH . COOH \\ \text{Citric acid} & & \text{Cis . acontic acid} & & \text{Isocitric acid} \end{array}$$

3. Now Isocitric acid undergoes dehydrogenation (oxidation) in the presence of enzyme isocitric dehydrogenase producing oxalosuccinic acid. The hydrogen released is accepted by NADP to produce $NADPH_2$

$$\begin{array}{ccccc} CH_2 . COOH & & & & CH_2 . COOH \\ | & & & \text{isocitric} & | \\ CH. COOH & + NADP & CH.COOH & \rightleftharpoons & +NADPH_2 \\ | & & & \text{dehydrogenase} & | \\ CHOH . COOH & & & Mg^{++} & CO. COOH \\ \text{Isocitric acid} & & & & \text{Oxalo succinic acid} \end{array}$$

4. Oxalosuccinic acid is decarboxylated in the presence of an enzyme decarboxylase, α — Keto—glutaric acid and Co_2 are formed.

$$\begin{array}{ccc} CH_2 . COOH & & CH_2 . COOH \\ | & \text{decarboxylase} & | \\ CH . COOH & \text{------>} & CH_2 . CO . COOH + CO_2 \\ | & & \\ CO . COOH & & \\ \text{Oxalo succinic acid} & & \alpha\text{ - Ketoglutaric acid} \end{array}$$

5. The next step is a complex process. It occurs in two steps.

i) In the first step α - Ketoglutaric acid is converted to succinyl COA and the reaction is catalysed by the enzyme α - Ketoglutaric acid dehydrogenase. It requires TPP, Mg^{++}, NAD, FAD, lipoic acid and COA as co-factors. This step is similar to the oxidation of pyruvic acid to acetyl COA. In this reaction one molecule of Co_2 is released and NAD is reduced to $NADH_2$.

$$\begin{array}{ccccc} CH_2. COOH & & & CH_2 . COOH & \\ | & + COA + NAD & \text{------------------>} & | & + NADH_2 + CO_2 \\ CH_2 . CO . COOH & & & CH_2 . CO\ M\ COA & \\ \alpha\text{ - Ketoglutaric acid} & & & \text{Succinyl COA} & \end{array}$$

ii) In the second step the succinyl COA is hydrolysed to succinic acid by using one molecule of water and COA is regenerated. This reaction takes place in the presence of enzyme succinyl thiokinase. One molecule of GDP (Guanosine diphosphate) is converted to GTP (Guanosine triphosphate). This process is known as substrate phosphorylation. The GTP can react with ADP to form ATP.

$$\begin{array}{l} CH_2 . COOH \\ \quad | \\ CH_2 . CO{\sim}{\sim}COA \end{array} + GDP + P_1 \underset{\text{thiokinase}}{\overset{\text{succinyl}}{\rightleftharpoons}} \begin{array}{l} CH_2 . COOH \\ \quad | \\ CH_2 . COOH \\ \text{Succinic acid} \end{array} + GTP + COA$$

$$GTP + ADP \qquad GDP + ATP$$

6. The succinic acid is oxidized to fumaric acid in the presence of enzyme succinic acid dehydrogenase. FAD is reduced to $FADH_2$

$$\begin{array}{l} CH_2 . COOH \\ \quad | \\ CH_2 . COOH \\ \text{Succinic acid} \end{array} + FAD \underset{\text{dehydrogenase}}{\overset{\text{succinic}}{\rightleftharpoons}} \begin{array}{l} CH . COOH \\ \quad \| \\ CH . COOH \\ \text{Fumaric acid} \end{array} + FADH_2$$

7. Fumaric acid is converted to malic acid with the addition of one molecule of water in the presence of enzyme fumarase.

$$\begin{array}{l} CH . COOH \\ \quad \| \\ CH. COOH \\ \text{Fumaric acid} \end{array} + H_2O \overset{\text{fumarase}}{\rightleftharpoons} \begin{array}{l} CHOH . COOH \\ \quad | \\ CH_2 COOH \\ \text{Malic acid} \end{array}$$

8. In the last step malic acid is converted back to oxaloacetic acid under the influence of enzyme malic acid dehydrogenase.

$$\begin{array}{l} CHOH . COOH \\ \quad | \\ CH_2 . COOH \\ \text{Malic acid} \end{array} + NAD \underset{\text{dehydrogenase}}{\overset{\text{malicacid}}{\rightleftharpoons}} \begin{array}{l} CO . COOH \\ \quad | \\ CH_2 . COOH \\ \text{oxalo acetic acid} \end{array} + NADH_2$$

With the regeneration of oxaloacetic acid the Krebs cycle is completed. The oxaloacetic acid is now made available to react with acetyl COA to form citric acid and the cycle repeats until the entire pyruvic acid is oxidized to Co_2 and water.

Table 9.2. ENZYMES OF RESPIRATION

GLYCOLYSIS

SUBSTRATE	END-PRODUCTS	NAME OF THE ENZYME
Glucose	Glucose — 6- phosphate	Hexokinase
Glucose — 6- phosphate	Fructose — 6- phosphate	Phospho gluco isomerase
Fructose — 6- phosphate	Fructose — 1-6 diphosphate	Phospho fructo kinase
Fructose — 1-6 diphosphate	Glyceraldehyde — 3 phosphate	Aldolase
3-Phospho glyceraldehyde	Dihydroxy acetone Phosphate	Triose phosphate isomerase
Glyceraldehyde 3 — Phosphate	1-3- diphospho glycric acid	Phosphoglyecraldehyde dehydrogenase
1-3- diphospho glyceric acid	3- phospho glyecric acid	Phospho glyceric kinase
3- phospho glyceric acid	2- phospho glyecric acid	Phospho glycerol mutase
2- phospho glyceric acid	phospho enol pyruvic acid	Enolase
Phosphoenol pyruvic acid	pyruvic acid	Pyruvic acid kinase

9.2. KREB'S CYCLE :

SUBSTRATE	END-PRODUCTS	NAME OF THE ENZYME
Oxalo aceticacid + acetyl COA	Citric acid	Citric Synthase
Citric acid	Cis — aconitic acid	Aconitase
Cis — aconitic acid	Isocitric acid	Aconitase
Isocitric acid	Oxalo succinic acid	Isocitric dehydrogenase
Oxalo succinic acid	α - Ketoglutaric acid	Decarboxylase
α - Ketoglutaric acid	Succinyl COA	α - Ketoglutaric acid dehydrogenase
Succinyl COA	Succinic acid	Succinyl thio Kinase
Succinic acid	Fumaric acid	Succinic acid dehydrogenase
Fumaric acid	Malic acid	Fumarase
Malic acid	Oxalo acetic acid	Malic acid dehydrogenase

9.2.B FERMENTATION :

SUBSTRATE	END-PRODUCTS	NAME OF THE ENZYME
Pyruvic acid	Acetaldehyde	Pyruvic decarboxylase
Acetal dehyde	Ethyl alcohol	Alcohol dehydrogenase

9.3.4.3 ELECTRON TRANSPORT SYSTEM : Respiration is a biological oxidative process. The oxidation is carried out by the removal of hydrogen atoms from the oxidizing substrate. The pair of hydrogen atoms ultimately dis-associates into two protons and two electrons.

$$2\,H \text{ ----------------------------> } 2H^{+} + 2e^{-}$$

The protons are released into the medium. Electrons are transferred through a series of electron acceptors. In the last step protons and electrons combine with the oxygen to form water. Thus the molecule of oxygen is used only in the last step, hence it is called "Terminal Oxidation". Simultaneously the energy released during the oxidation is utilized in the formation of ATP. The pathway through which the electrons are transferred or channeled is known as electron transport system (ETS) or Electron transport chain (ETC).

The various electron acceptors of the electron transport system are pyridine nucleotides (NAD and NADP), flavins (FAD), number of cytochromes like cyt b, cyt c, cyt a, cyt a_3.

To explain the mechanism of terminal oxidation, the oxidation of malic acid to oxalo-acetic acid can be taken as an example. In this reaction a pair of hydrogen atoms from malic acid are transferred to NAD, which acts as hydrogen acceptor and is reduced to $NADH_2$ (fig. 9.3). The pair of hydrogen atoms is transferred to flavin adenine dinucleotide (FAD) a co-enzyme of flavo protein enzyme. As a result of which FAD is reduced to $FADH_2$ and $NADH_2$ become oxidized to NAD. In this process the energy released (12.2 K.Cal) is utilized in the formation of ATP molecule by phosphorylation of ADP by an inorganic phosphate (P_1)

$$ADP + Pi \text{ ---------------------------> } ATP$$

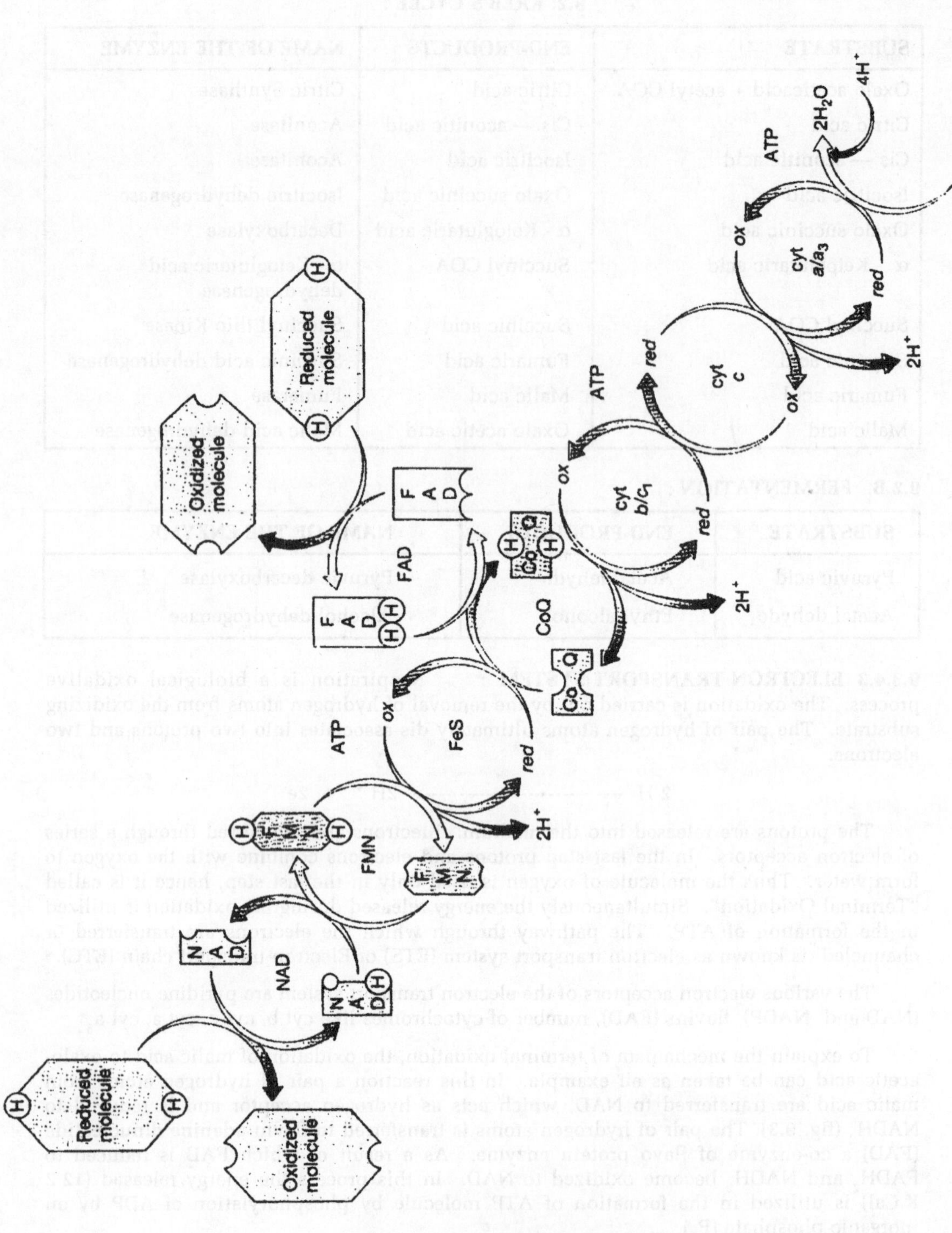

Fig 9.3Electron transport system

The electrons are now transferred to a group of coloured compounds called cytochromes. Cytochromes are conjugated proteins with the prosthetic group containing iron. Atleast four types of cytochromes are involved in the electron transport chain. The sequences of cytochromes are as follows.

Cyt b ------------> Cyt c ----------> Cyt a -------------> Cyt a_3

The oxidized form of cytochrome has ferric (Fe^{+++}) and reduced form has ferrous (Fe^{++}) iron. Ferrous (Fe^{++}) on oxidation by loss of an electron become ferric (Fe^{+++}). Ferric on gain of electron become reduced to Ferrous. Thus Cytochromes can undergo reversible oxidation and reduction. A cytochrome can take only one electron at a time. Each oxidation step requires two electrons, which come from two cytochromes.

During the transfer of a pair of electrons from cytochrome 'b' to cytochrome 'c' certain amount of energy is released (15.2 K.Cal), which is being utilized in the synthesis of another molecule of ATP. The electrons then pass to cyt a and finally cyto a_3, which is commonly called as cytochrome oxidase. In the transfer of electrons from cyt a to cyt a_3, a large amount of energy is released (24.5 K.Cal). Out of this energy 8.K.Cal of energy is utilized for the synthesis of another molecule of ATP, the rest of the energy is released in the form of heat. In the electron transport chain cytochrome oxidase is of immense importance because this alone is capable of oxidation by molecular oxygen. If this cytochrome is inactivated by poisons like cyanides, the cells fail to utilize the oxygen and synthesis of ATP molecules is inhibited.

The electron transport system, as outlined above, is linked with the formation of ATP molecules by phosphorylation process, hence it is called oxidative phosphorylation. Thus when one molecule of $NADH_2$ is oxidized a total of three ATP molecules are synthesized at three different sites in the electron transport chain.

Oxidative phosphorylation occurs in mitochondria, which contain all the essential requirements for electron transport system. In contrast Photophosphorylation occurs in chlorophasts, which proceeds at the expense of light energy trapped by chlorophyll. The light energy is stored in the form of ATP molecules during the photophosphorylation process.

The transfer of electrons in the electorn transport chain occur from higher energy potential level to lower potential energy level. If the potential difference between the two electron carriers is greater, more energy is released. The energy released is equivalent to the difference in the energy levels of the electron donor and electron acceptor. This released energy is utilized in the synthesis of ATP molecules.

9.3 TABLE ATP molecules produced in the oxidation of one molecule of glucose.

Reaction	Reduced co-enzymes produced	ATP Molecules + or -
GLYCOLYSIS		+8
Glucose → Glucose — 6-phosphate	-	-1
Fructose — 6- phosphate → Fructose 1-6- diphosphate	-	-1
3-Phosphoglycer aldehyde → 1-3 — diphospho glyeric acid	2 x $NADH_2$	+6
1-3 diphospho glyceric acid → 3- phospho glyceric acid	-	+2
Phospho enol pyruvic acid → Pyruvic acid	-	+2

Reaction	Reduced co-enzymes produced	ATP Molecules + or -
KREBS CYCLE		+ 30 ATP
Pryuvic acid to Acetyl COA	2 X $NADH_2$	+6
Isocitric acid → oxalo succinic acid	2 X $NADPH_2$	+6
i) α Ketoglutaric acid → Succinyl COA	2 X $NADH_2$	+6
ii) Succinyl COA → Succinic acid	-	+2
Succinic acid → Fumaric acid	2 X $FADH_2$	+4
Malic acid → Oxalo acetic acid	2 X $NADH_2$	+6
Total ATP molecules Produced in the Complete oxidation of one mol. Of glucose		30 + 8 = 38 ATP

9.3.4.4 Fermentation : It is analogus to anaerobic respiration. But fermentation occurs due to the activity of microorganisms. The best known example is the alcoholic fermentation of sugars by yeasts. Alcoholic fermentation was known to ancient man as back of 5000 B.C. The Vedas contain reference about alcoholic drinks (somarasa). Fermentation can be represented by the following equation.

$$C_6H_{12}O_6 \text{ -------------------->} \ 2C_2H_5OH \ + \ 2\ CO_2 \ + \text{Energy}$$

Ethylalcohol

Besides ethyl alcohol, small quantities of other substances such as acids, aldehydes, glycerol are formed during the fermentation.

Alcoholic fermentation is an enzymatic process, carried out by the enzyme complex "Zymase" secreted by the yeast cells. The enzyme complex can be extracted from the yeast cells and can be utilized for bringing out fermentation. Zymase preparations are sold in the market.

Zymase is an iron-containing multi-enzyme complex. Glycolase and carboxylase are the components of the enzyme complex. Several co-enzymes are necessary for the functioning of the enzyme complex.

Lactic acid fermentation is brought out by a number bacteria. Some of them are aerobic and others are anaerobes (eg. Lactose). Both of them are present in Milk. They ferment lactose and form lactic acid. First lactose is hydrolysed to hexose sugars like glucose and galactose. These hexoses are broken down into lactic acid.

$$C_{12}H_{22}0_{11} + H_2O \text{ ---------------->} \ C_6H_{12}O_6 + C_6H_{12}O_6$$

Lactose — glucose — galactose

$$C_6H_{12}O_6 \text{ ------------------>} \ 2C_3H_6O_3$$

Glucose/galactose — Lactic acid

In butyric acid fermentation the end products are butyric acid, CO_2 and hydrogen. This fermentation is brought about by a number of bacteria such as Clostridium butyricum,

C. Pasteurianum etc. All these bacteria are obligate anaerobes. The process can be represented by the following equation.

$$\underset{\text{Glucose}}{C_6H_{12}0_6} \text{ -------------------------> } \underset{\text{Butyric acid}}{C_4H_8O_2} + 2\,H_2 + Co_2$$

Many other fermentative processes are also known. In some process the proteins and other nitrogenous compounds are decomposed by anaerobic bacteria producing H_2 S, ammonia and other products having unpleasant smell. This type of decomposition is called "Putrefaction". It is very common in the decay of dead remains of animals. Putrefaction is beneficial to man as it removes dead bodies of animals and plants on the surface of the earth. Putrefaction process converts dead remains of plants and animals to simpler elements and return back them to soil thereby enriching the fertility of soil.

9.3.4.5 Factors affecting the rate of respiration: Many factors affects the rate of respiration. These factors can be classified into two categories.

1. External factors
2. Internal factors

9.3.4.5.1 External Factors:

a) **Temperature:-** Temperature is one of the most important external factors which influence the rate of respiration. The respiratory rate increases with the temperature Co-efficient (Q10° C) value. A rise of temperature from 0° C to 40°C is accompanied by corresponding increase in the rate of respiration. For every 10° C rise in temperature there is corresponding increase in the rate of respiration by 2 to 2.5 times. The optimum temperature for respiration is around 30⁰ C. Above the optimum temperature, the rate of respiration shows an initial rise followed by a decline. This may be due to the denature of the respiratory enzymes. Lowering the temperature below O⁰ C also causes decline in the rate of respiration. The minimum temperature varies with plants and kind of tissues. Normally the respiration can continue over at —20° C but with slow rate. That is the reason why fruits and vegetables kept in "cold storage" retain their viability and utility.

b) **Oxygen:-** The type of respiration and kind of end products depend upon the availability of oxygen. In the presence of oxygen as described earlier, complete oxidation of carbohydrate take place resulting in the formation of Co_2 and water. In the absence of oxygen, incomplete oxidation of carbohydrates takes place resulting in the formation of alcohols, acids etc.

c) **Carbon dioxide:-** Increase of Co_2 concentration in the atmosphere reduces respiration and inhibition of seed germination, increases their dormancy period. High Co_2 concentration and low oxygen concentration would lower the rate of respiration, thereby increasing their longevity. This character is made use of in storing fruits and vegetables at low temperature and higher Co_2 concentration.

d) **Light:-** Light is not indispensable for respiration, Respiration goes on in day and night irrespective of the presence or absence of the light. However, light has indirect effect on respiration as it controls the photosynthesis production of sugar, which is used in respiration as respiratory substrates. Light also raises temperature, thereby causing the stomata to open, which facilitate the gaseous exchange. In some plants light induces photorespiration. In some plants like sugar cane, maize etc, photorespiration is completely absent.

e) **Water:-** The respiratory rate increases with the increase in water content of the respiring material. It is evident that in dry seeds the respiratory rate is very low but increases rapidly when the seeds absorb water. So the water content of seeds is the important factor in storage of seeds. Little water content of the seeds keep the respiratory rate at minimum thereby, their viability increases.

The effect of water on respiration is due to the following facts:

(i) It converts reserve food materials into soluble sugars, which are used up in respiration.

(ii) It improves the efficiency of respiratory enzymes.

(iii) It helps in maintaining the turgor of the respiring cells.

9.3.4.5.2 Internal Factors:-

a) **Protoplasmic factors:-** The rate of respiration depends upon the quantity and quality of protoplasm in the cells. In younger and dividing cells the respiratory rate is more vigorous than in older cells. In younger cells the protoplasm is thick but in older cells the protoplasm is in thin layer around the large, central vaclole. In general, the rate of respiration decreases with the age of the plant or plant part.

b) **Concentration of respiratory substrates:-** The rate of respiration also depends upon the availability of the respiratory material. When other factors are not limiting, the rate of respiration increases with the increase in respiratory substrates such as sugar etc.

QUESTIONS

A. **Answer the following question in 25 lines.**

1. Give a brief account of mechanism of respiration.
2. Bring out differences between the Aerobic respiration and Anaerobic respiration.
3. Describe briefly the process of Aerobic respiration.
4. Give an account of Embden, Meyerhof, Paranas Pathway.
5. Define Respiratory Quotient. Mention the reasons for deviation of RQ from unity with the suitable examples.
6. Explain the Chemical reaction involved in Kreb's cycle.
7. Enumerate various substrates involved in respiration process. Describe their mode of utilization.
8. Explain the various steps in electron transport system.
9. How do ATP molecules are synthesized?
10. Write an essay on the effects of environmental factors on the rate of respiration.
11. What is fermentation? How does it differ from anaerobic respiration?
12. What is the fate of pyruvic acid if oxygen is not available to it. Describe the process in detail.
13. Define oxidative phosphorylation. How it differs from photophosphorylation.
14. What is entropy? Briefly mention the laws of thermodynamics.
15. Describe the role of enzymes in respiration.
16. Describe the structure and function of Mitochondria in relation to oxidative phosphorylation.
17. Respiration is an oxidative process. Explain?
18. Why are mitochondria are called "Power house of the cell"

B. Short Notes

1. Respiratory Quotient
2. Respiratory substrates
3. Fermentation
4. Anaerobic respiration
5. Lactic acid fermentation
6. Cytochromes.
7. Significance of respiration
8. Terminal oxidation
9. Oxysomes
10. Entropy
11. Pyridine nucleotides
12. Formation of acetyl COA
13. Flavoproteins
14. ATP

C. Define the following in one or two sentences.

1. ETS
2. ATP
3. Metabolic Water
4. Terminal oxidation
5. Entropy
6. Enthalpy
7. Exoergonic reaction
8. NAD
9. Terminal oxidase.

10 NITROGEN METABOLISM

PLANT PHYSIOLOGY

10.1 INTRODUTION

Nitrogen is an important and essential element for the growth of the plant. It comes next to carbon, Hydrogen and Oxygen. Nitrogen is one of the main structural components of living organisms. The element is found in the composition of proteins, Aminoacids, vitamins, Nucleic acids (DNA and RNA), chlorophyll and growth hormones. Leaves contain 1-15% nitrogen of their dry weight.

Though Nitrogen is included in the list of essential mineral nutrient elements, it is not a mineral element in the true sense of the word. The air comprises 79% nitrogen in gaseous form. Plants cannot absorb nitrogen in the gaseous form from the atmosphere. Plants absorb nitrogenous compounds present in the soil. However, some lower organisms like bacteria, cyanobacteria can utilise the gaseous nitrogen directly from the atmosphere.

10.2 SOURCES OF NITROGEN FOR PLANTS

As already explained above, the higher plants cannot utilise the atmospheric nitrogen directly. Therefore, the only source of nitrogen for the plants is soil. Nitrogen can exist in the soil in the form of nitrates (NO^-_3) nitrites (NO_2), ammonia (NH^+_4) or urea or some other nitrogenous compounds. Plants can absorb these nitrogenous compounds directly from the soil by their roots or other absorbing organs.

The most usable source of nitrogen is nitrate ion (NO^-_3). However, nitrite ($N\hat{O}_2$) can also be absorbed by the plants, but nitrites cannot serve as good source of nitrogen for many plants. Ammonia is also directly absorbed by the plants. For plants like rice, wheat, Oats etc., ammonium is a good source when it is supplied in the form of ammonium sulphate ($NH_4)_2$ SO_4) or ammonium nitrate (NH_4 NO_3).

The organic form of nitrogen occurs in the soil in the form of proteins, amino acids, urea and other organic nitrogenous compounds. These organic nitrogenous compounds in the soil are formed by death and decay of plants and animals including man. Aminoacids, which are formed by the decomposition of proteins, are directly absorbed by the plants or it is converted into ammonia by an enzyme urease, produced by soil microrgamisms, before it is being absorbed.

$$NH_2-\overset{\overset{O}{\|}}{C}-NH_2 \xrightarrow{\text{Urease}} 2\,NH_3 + Co_2$$

Urea → Ammonia + carbon-di-oxide

10.3 NITROGEN FIXATION

Molecular nitrogen present in the atmosphere cannot be utilised by the higher plants. Higher plants can utilise the nitorogen when it is present in the form of nitrogenous compounds such as nitrates, nitrites and ammonia etc. But certain lower organisms like bacteria, cyanobacteria can convert free nitrogen into nitrogenous compounds. It is a wellknown fact that by growing leguminous crops, the fertility of the soil can be enriched. The reason for this increase of fertility is due to the presence of bacteria in root nodules of leguminous crops. The bacteria are capable of fixing atmospheric nitrogen.

Nitrogen fixation can be defined as a phenomenon of conversion of free molecular nitrogen into nitrogenous salts to make it available for absorption by the plants. The fixation of nitrogen can be both by physico-chemical process and biological process. Thus nitrogen fixation can be of two types.

1. Physical Nitrogen Fixation
2. Biological Nitrogen Fixation

10.3.1 Physical Nitrogen Fixation : About 10% of total nitrogen fixation occurs due to physical processes such as lightening and thunder storms. During the lightening and thunderstorms heavy electrical discharges takes place. Under the effect of heavy electrical discharges the gaseous nitrogen and oxygen present in the atmosphere combine to form nitric oxide as follows

$$N_2 + O_2 \xrightarrow[\text{discharges}]{\text{Electrical}} 2\text{ NO (Nitric Oxide)}$$

The nitric oxide is again oxidized with O_2 to form nitrogen peroxide (NO_2)

$$2\text{ NO} + \text{O2} \xrightarrow{\text{oxidation}} 2\text{ NO}_2\text{ (Nitrogen Peroxide)}$$

During the rain the nitrogen peroxide combines with rain water to form nitrous acid (HNo_2) and nitric acids (HNo_3). The acids fall on the soil along with the rain water and react with alkaline radicals to form nitrate and nitrite salts, which are soluble in water and can be absorbed by the plants.

10.3.2. Biological Nitrogen Fixation : The conversion of atmospheric nitrogen into nitrogenous compounds through the agency of the living organisms is called "Biological Nitrogen Fixation". The biological nitrogen fixation can be carried out by either free-living micro-organisms or the organisms which live in symbiotic association with the other plants. Thus there are two types of biological nitrogen fixation.

i. Asymbiotic or Non-symbiotic N_2 fixation
ii. Symbiotic N_2 fixation

10.3.2.1. Asymbiotic or Non-symbiotic N_2 fixation : Soil inhabits many free living nitrogen fixing organisms. These include a number of aerobic and anaerobic bacteria and also blue green algae. Nearly 40 species of blue-green algae are known to fix atmospheric nitrogen. These algal members contain 'heterocysts' which are the sites for nitrogen fixation. In addition to above several other micro-organisms are known to fix atmospheric nitrogen. The list of nitrogen fixing organisms is given in table 10.1.

Table : 10.1 Nitrogen fixing biological systems.

1. Free living (Non-symbiotic) systems :	
a) Bacteria	
i. Aerobic bacteria	Several species of Azotobacter, Chroococcum
ii. Anaerobic bactoria	Several species of clostridium and also a few other genera
iii. Photosynthetic bacteria	Several species of Rhodospirillum
iv. Chaemosynthetic	Thiobacillus spp.
b) Blue green algae :	Nostoc,Anabaena, Cylindrospernum Aulosira, Tolypothrix etc.
II. Symbiotic Systems :	
A. Angiosperms	
i. Leguminous Plants rootnodules	Several species of Rhizobium
ii. Non-leguminous plants	Several unidentified species of actinomycetes.
a) Alxus, Casuarina Sp Myrica Sp. Etc	Klebsiella Sp
b) Psychotria Sp	Blue green algae Eg. Nostoc Sp.
B. Gymnosperms	
Cycads	Blue green algae Eg. Anabaena Sp.
C. Ferns	
Azolla Sp	Blue green algae Eg. Nostoc Sp.
D. Non-vascular platns	
Lichens	

10.3.2.2. Symbiotic N_2 fixation : Many bacteria, which live in association with higher plants are capable of N_2 fixation. The best example is root nodules of Leguminosae plants, which harbour N_2 fixing bacteria "*Rhizobium*". The symbiotic N2 fixing bacteria are also known to occur in the roots of plants like *Casuarina*, *Cycas* and *Podocarpus* etc. The leaves of plants like Pavetta, Dioscorea etc also harbour N_2 fixing bacteria. Lichens growing on the surfaces of trunks and branches of trees also known to fix atmospheric nitrogen.

Leguminosae plants.

Leguminosae plants like pea-nut, pea, bean, soya-bean etc produce small, knob like protuberances on the roots called "Root nodules". The root nodules vary considerably in their size and shape. They may be spherical, elongated, flat or finger-like projections. The size of nodules may vary from pin-head to 1 cm in diametre.

The nodule bacteria include many species of *Rhizobium*. Various species of Rhizobium infect and inhabit different species of Leguminous plants. For example Rhizobium leguminosarium infect pea plants; R.Japnica infect soyabean.

The nitrogen fixation by symbiotic organisms contribute to the fertility of the soil for more than the asymbiotic systems. It is estimated that nitrogen fixing root nodule can fix 100-200 times more nitrogen than free-living organisms. This is due to the fact that symbiotic nitrogen fixing bacteria continue to fix nitrogen and the fixed nitrogen is translocated and used up in the synthesis of amino acids and proteins. On the other hand, asymbiotic organisms, generally fix nitrogen to the extent of their own requirements only. The fixed nitrogen is available to the plants only after the death and decay of the free-living organisms.

Our knowledge of symbiotic nitrogen fixation by leguminosae plants goes back to ancient times. The benefits of growing leguminosae plants in enriching the soil fertility was known to ancient people. The Greek philosophers wrote that leguminosae plants can be grown not much for their yield but for their benefits to the crop that follow them.

It was in the early decades of 19th century, the French agronomist **Boussingault** established that the residues of leguminous plants enrich the soil fertility. He suspected that the Leguminosae plants might have received the nitrogen from the atmosphere, but he could not prove the same. In 1888 the two German Scientists Hellriegel and Wilforth demonstrated conclusively that the Leguminosae plants assimilate atmospheric nitrogen because of the presence of root nodules, which harbour certain micro-organisms. In the same year **Beijerinck** isolated bacterium from the root nodules of Legumiuosae plants and identified the bacterium as Rhizobium.

Rhizobium is a soil bacteria, which grow in the vicinity of roots of higher plants. The leguminosae roots secrete some chemical substances into the soil, which stimulate the multiplication of Rhizobium. Rhizobial cells secrete α-Indole acetic acid. The roots are infected by Rhizobial cells. Infection of root-hairs by Rhizobial cells result in curling of root-hairs of plants (fig 10.1).

The mucilaginous threads containing rhizoidal cells grow from root-hairs to cortical region of the root. The Rhizobium penetrates to the cortex through the infection threads. At the same time, it stimulates the cortical cells to divide and redivide to produce swollen area, which later develop into nodule.

The bacterial cells in the nodule multiply rapidly to form spherical structures called bacteroids(fig10.2). The bacteroids contain reddish pigment called Leghaemoglobin (similar to haemoglobin of RBC). The nitrogenase enzyme,which is responsible for the fixation of N_2 is present in bacteroids. Leghaemoglobin acts as oxygen scavenger. It readily combines with oxygen and removes it, which is essential for creating oxygen free atmosphere for nitrogenase enzyme activity. Leghaemoglobin also supplies oxygen, when it is required to bacteroids and stimulates production of ATP by bacteroids. Thus Leghaemoglobin acts as oxygen carrier and reservoir.

10.4 THE BIO-CHEMISTRY OF NITROGEN FIXATION : Nitrogen fixation occurs in the bacteroids. The process requires both a source of electrons and ATP molecules. The host plant provides carbohydrates to bacteroids for the supply of required energy. Carbohydrates formed in leaves during the photosynthesis are transported to the root nodules. Some of the electrons and ATP molecules synthesised during the oxidation of carbohydrates are used to reduce nitrogen to ammonia.

Respiration of carbohydrates in bacteroids leads to reduction of NAD to $NADH_2$ or NADP to $NADPH_2$. In some nitrogen fixing organisms oxidation of pyruvate during respiration causes reduction of protein called flavodoxin. Flavodoxin, $NADH^+_2$ or $NADPH^+_2$

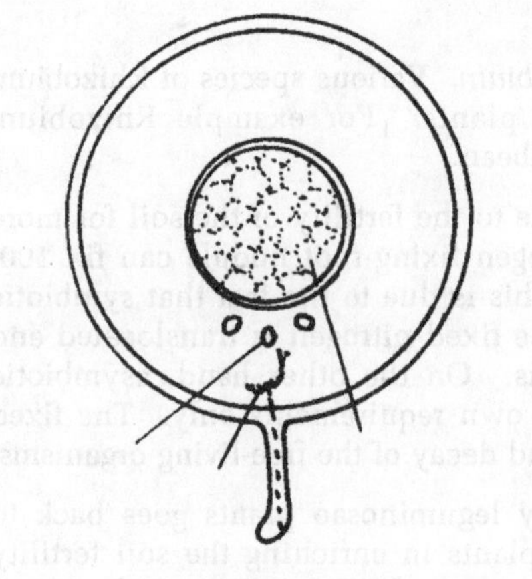

Fig 10.1 Stages in the formation of root module

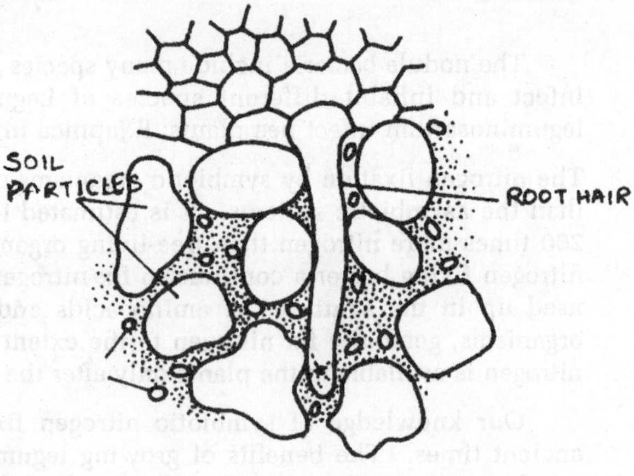

Fig 10.1A Bacterial in the soil

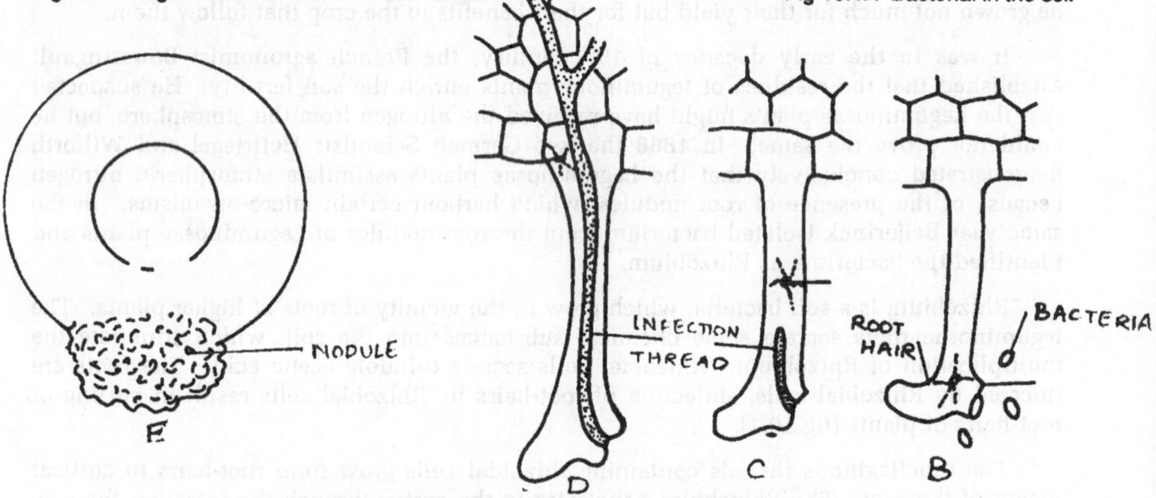

Fig 10.1 B, C, D, E - VArious stages in the formation of module

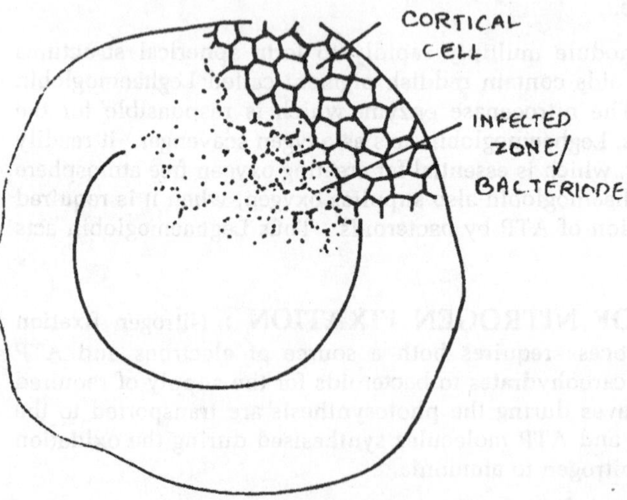

Fig 10.2A T.C. of root showing the infected zone

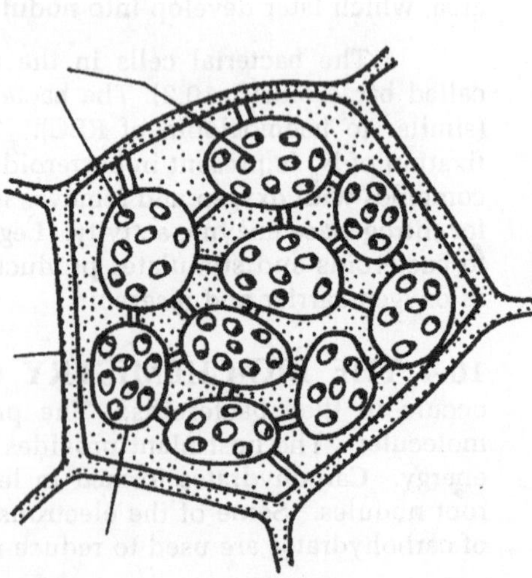

Fig 10.2B Bacteriods in the host cell bounded by sacs

reduces ferredoxin. Nitrogenase enzyme accepts electrons from flavodoxin or ferredoxin or other reducing agents. Nitrogenase enzyme contains two distinct proteins called Fe protein and Fe-Mo protein. Both Molybdenum and Fe of enzyme become reduced upon accepting electrons from ferredoxin and then oxidised by transferring the electrons to nitrogen to form NH3. The electron transport process and utilization of ATP molecules during the reduction of N2 tc NH_3 is given below.

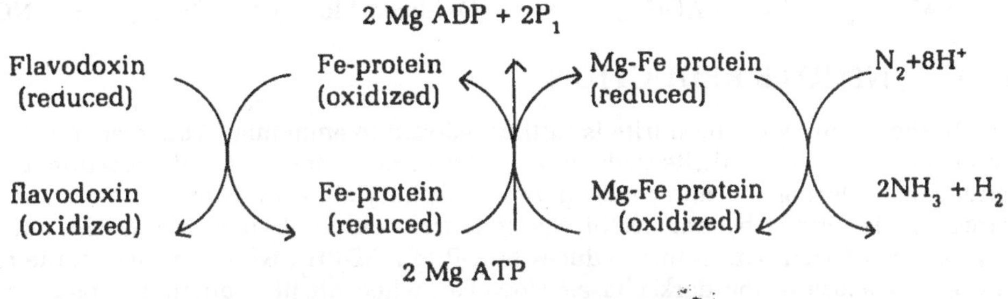

Fig: Electron transport process from flavodoxin to reduction of N_2 to NH_3 coupled with utilization of ATP molecules

10.5. NITRATE REDUCTION

Nitrate is first reduced to ammonia before it is used for the synthesis of amino acids. In nitrates the nitrogen is in a highly oxidised state and in the amino acids the nitrogen is in highly reduced form. So nitrates (NO3) are first reduced and converted to ammonia (NH3) which is later used in the synthesis of amino acids. The overall reaction of nitrate reduction may be represented by the following equation.

$$HNO_3 + 8H^+ \text{ ------------------> } NH_3 + 3H_2O \text{ (Ammonia)}$$

The nitrate reduction mostly occurs in leaves. It can also occur in non-chlorophyllous cells of roots and shoots. Reduction requires both H-donors and ATP molecules. ATP molecules are provided by respiration process. In green leaves ATP molecules synthesized in photophosphorylation are directly used for the nitrate reduction (Burstrom, 1945). The reduction of nitrate to ammonia takes place in two steps.

1. In the first step nitrate is converted to nitrite. This reaction is catalysed by an enzyme called nitrate reductase. This enzyme has been extracted from several plants. In this reduction process NADH +H or NADPH + H acts as electron donor FAD as prosthetic group and Molybdenum (MO) as an activator of enzyme.

The pathway of electron transfer in the reduction of nitrate to nitrite is as follows.

First $NADH_2$ transfers electrons to FAD, which is reduced to $FADH_2$.

$$NADH_2 + FAD \text{ ------------------> } FADH_2 + NAD$$

In the next step the electrons are transferred from $FADH_2$ to Molybdenum.

$$FADH_2 + Mo \text{ ------------------> } Mo\ 2H + FAD$$

In the final step the electrons are transferred from Molybdenum to nitrate. As a result of which the nitrate becomes reduced to nitrite and water is formed.

$$\underset{\text{Nitrate}}{NO_3} + Mo\ 2H \text{ ------------------> } \underset{\text{Nitrite}}{NO_2} + Mo + H_2O$$

The electron transport chain in the reduction of nitrate to nitrite can be shown as follows :

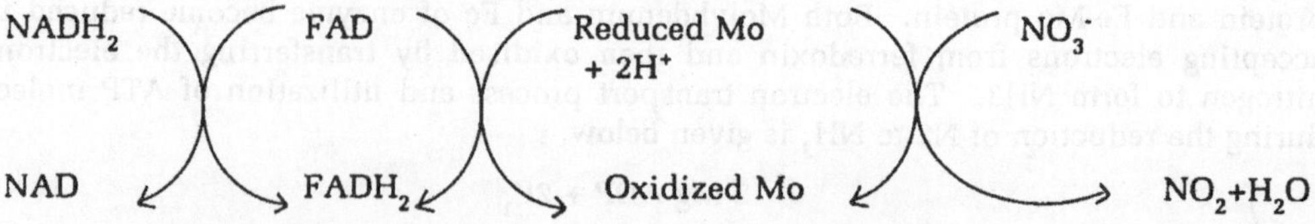

10.6 NITRITE REDUCTION

In the second step the nitrite is further reduced to ammonia. This reaction is catalyzed by an enzyme called nitrite reductase. The electon donor in this reaction is reduced ferredoxin. During the non-cyclic photo-phosphorylation of photosynthesis the reduced ferredoxin is formed by transfer of electrons from chlorophyll to ferredoxin. In the next step reduced ferredoxin in turn reduces NADP to $NADPH_2$, which is used in the reduction of Co_2 to glucose in the dark phase. However, when nitrite reduction is taking place the above sequence is interrupted and reduced ferredoxin is used for reduction of nitrite to ammonia. The above changes can be represented as follows.

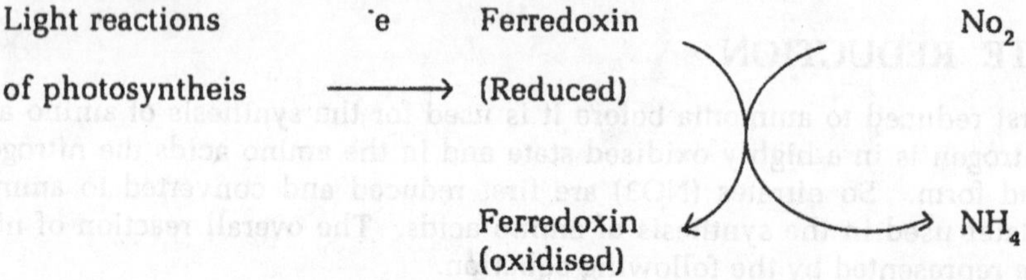

The nature of electron donor in non-green tissues of stem and roots is not known. In roots the reducing power for the reaction comes from respiration. Instead of reduced ferredoxin some other electron donor participate in the reduction process.

The overall reaction of nitrate reduction can be represented as follows.

$$No_2 + 6\ddot{e} + 8H^+ \text{ ------------------> } NH^+_4 + 2H_2O$$

10.7. AMINO ACIDS

Amino acids are derivatives of Organic acids in which one hydrogen atom is replaced by an amino ($-NH_2$) group. The general formula of an amino acid is as follows.

```
        R
        |
H2N — C — COOH
        |
        H
```

Where 'R' represents a side chain attached to α - carbon atom. It may be a simple hydrogen atom or a complex organic group. The simplest amino acid is glycine in which 'R' is a simple hydrogen atom.

```
        H
        |
H2N — C — COOH
        |
        H
```

(Glycine)

The glycine is derived from acetic acid by replacing hydrogen atom of the — CH_3 by an amino group.

Acetic acid	Glycine
$H-\overset{\displaystyle H}{\underset{\displaystyle H}{C}}-COOH$	$H_2N-\overset{\displaystyle H}{\underset{\displaystyle H}{C}}-COOH$

There are about (20) different amino acids, which are known to participate in biosynthesis of proteins. All these twenty amino acids differ only in the type of organic compund attached to α - carbon in place of 'R'. The α - carbon is called as asymmetric carbon atom, because four different types of chemical groups are attached to this carbon. Because of this amino acids show optical isomerism. All the amino acids, which are known to participate in protein synthesis, belong to leaveo rotatory of isomers.

Amino acids are amphoteric compounds, since they show the properties of both acids and bases. This property is due to the presence of atleast one acid group (-COOH) and one or more basic amino ($-NH_2$) groups in the molecule. The formulae of various proteinogenic amino acids are given below.

	Name	Formula
1.	Glycine	NH_2-CH_2-COOH
2.	Alanine	$CH_3-\underset{\displaystyle NH_2}{CH}-COOH$
3.	Valine	$(CH_3)(CH_3)>CH-\underset{\displaystyle \text{ß}NH_2}{CH}-COOH$
4.	Leucine	$(CH_3)(CH_3)>CH-CH_2-\underset{\displaystyle NH_2}{CH}-COOH$
5.	Isolencine	$CH_3-CH_2-\underset{\displaystyle CH_3}{CH}-\underset{\displaystyle NH_2}{CH}-COOH$
6.	Serine	$\underset{\displaystyle OH}{CH_2}-\underset{\displaystyle NH_2}{CH}\cdot COOH$
7.	Threonine	$CH_3\text{-}\underset{\displaystyle OH}{CH}\text{-}\underset{\displaystyle NH_2}{CH}\text{-}COOH$
8.	Phenyl alanine	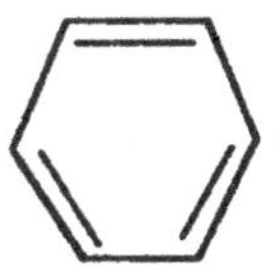 $-CH_2-\underset{\displaystyle NH_2}{CH}-COOH$

and so on

10.7.1 Synthesis of Amino acids :

Synthesis of amino acids takesplace by one of the two methods :

1. Reductive amination
2. Transamination

Mitochondria and chloroplasts are the sites of amino acid synthesis. After formation, the amino acids are moved to other parts of cells. A part of it is utilized in the synthesis of proteins in the ribosomes and part of it is stored in the vacuole. Amino acids are highly mobile. They are transported from one part of the plant to other parts through the phloem.

10.7.1.1 Reductive amination : Keto-acids such as pyruvic acid, oxaloacetic acid and a-ketoglutaric acids play an important role in the synthesis of amino acids by reductive amination method. The above organic acids are formed as intermediates in the Kreb's cycle of respiration.

In this method organic acid directly react with ammonia forming amino acids. The amination is coupled with reduction of organic acid in the presence of reduced co-enzyme $NADH_2$.

Glutamic acid is formed by reductive amination of α-Ketoglutaric acid, which is an intermediate in Kreb's cycle of respiration.

$$\underset{\alpha\text{ - Ketoglutaric acid}}{COOH - (CH_2)_2 - CO - COOH} + NH_3 + NADH_2 \xrightarrow[\text{dehydrogenase}]{\alpha\text{-Ketoglutaric acid}} \underset{\text{Glutamic acid}}{COOH - (CH_2)_2 - HC\text{-}NH_2 - COOH} + H_2O + NAD$$

Similarly pyruvic acid is converted to alanine and oxaloacetic acid is converted to aspartic acid by reductive amination method. The reactions are given below.

$$\underset{\text{Pyruvic acid}}{CH_3 - CO - COOH} + NH_3 + NADH_2 \xrightarrow[\text{(Pyruvic acid dehydrogenase)}]{} \underset{\text{Alanine}}{CH_3 - HC\text{-}NH_2 - COOH} + H_2O + NAD$$

$$\underset{\text{Oxalo acetic acid}}{COOH - CO - CH_2 - COOH} + NH_3 + NADH_2 \longrightarrow \underset{\text{Aspartic acid}}{COOH - HC\text{-}NH_2 - CH_2 - COOH} + H_2O + NAD$$

10.7.1.2 Transamination : Once the above three types of amino acids are formed, they can serve as the base for the synthesis of other types of amino acids by a process known as "Transamination". The reaction is catalysed by an enzyme called "Transaminase".

In this process the amino group (NH_2) of amino acid is transferred to organic acid resulting in the formation of a new amino acids and the original amino acid is converted to organic acid. For example glutamic acid is converted to aspartic acid (AA) by transfer of amino group from glutamic acid to oxaloacetic acid (OA) which is another intermediate compound in Kreb's cycle. The original aminoacid (glutamic) acid is converted to α-ketoglutaric acid.

$$\text{i. Glutamic acid (AA)} + \text{Oxaloacetic acid (OA)} \xrightleftharpoons{\text{Transaminase}} \alpha\text{-ketoglutaric acid (OA)} + \text{Aspartic acid (AA)}$$

$$\text{ii. Glutamic acid} + \text{Pyruvic acid} \xrightleftharpoons{\text{Transaminase}} \text{Alanine (AA)} + \alpha\text{-ketoglutaric acid (OA)}$$

In this way most of the amino acids can be synthesized in the plant cells. They remain dissolved in the cytoplasm of the cell, till they are being picked up and utilized in the synthesis of proteins.

10.7.2 Amides formation : An amide is formed by the substitution of amino (NH_2) group in place of (OH) group of carboxyl group (COOH). This type of reaction is called "amidation". The reverse reaction is called 'deamidation' (hydrolysis). Amides on hydrolysis produce aminoacids and ammonia. For example aspargine and glutamine (amides) are formed from aspartic acid and glutamic acids (amino acids) respectively.

$$\underset{\text{Aspartic acid}}{\begin{array}{c} CH_2.\,COOH \\ | \\ CH\,(NH_2)\,COOH \end{array}} + NH_3 + ATP \rightleftharpoons \underset{\text{Aspargine}}{\begin{array}{c} CH_2.\,CONH_2 \\ | \\ CH(NH_2)\,COOH \end{array}} + H_2O + ADP + P_1$$

$$\underset{\text{Glutamic acid}}{\begin{array}{c} COOH \\ | \\ (CH_2)_2 \\ | \\ CH\,(NH_2)\,COOH \end{array}} + NH_3 + ATP \rightleftharpoons \underset{\text{Glutamine}}{\begin{array}{c} CONH_2 \\ | \\ (CH_2)_2 \\ | \\ CH\,(NH_2)\,COOH \end{array}} + H_2O + ADP + P_1$$

The amides contain more nitrogen than amino acids. Whenever the roots absorb more nitrogen than required it leads to the formation of amides, because if ammonia is accumulated in large quantities, it becomes toxic to the plant. In such conditions the additional ammonia is being utilized in the formation of amides. Thus amides not only protect the protoplasm from the injurious effects of ammonia, but also stores ammonia for future use. Amides serve as supplier of NH_2-group in reductive amination or transamination process. Amides are soluble, so they can be translocated from one place to another place in the plant.

10.8. PROTEIN SYNTHESIS

Bio-synthesis of proteins is a complicated and genetically controlled process. Proteins are polymers of amino acids. About 20 amino acids are known to participate in protein synthesis. **Emil Fischer** has conclusively shown that proteins are formed by condensing together of many amino acids. The amino acids are linked by peptide bonds. The carboxyl group of one amino acid react with the amino group of another amino acid with the elimination of water, resulting in the formation of CO-NH bond, which is called peptide bond. The product is dipeptide

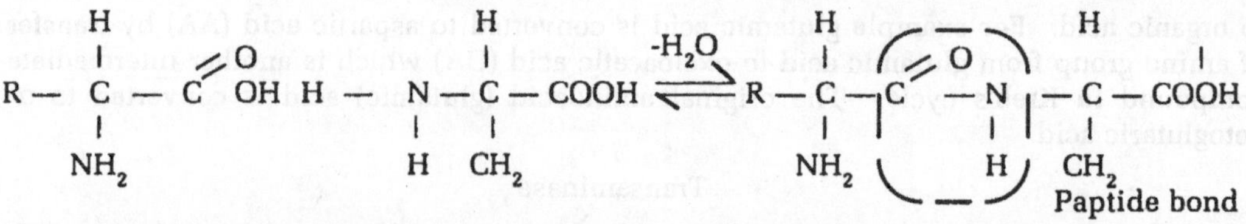

The dipeptide so formed contains an amino group at one end, carboxyl group at other end. It can link with another amino acid forming tripeptide. In this way many amino acids are linked together to form polypeptide chain or protein. Some proteins contain a single polypeptide chain and others may contain more than one polypeptide chains.

Each protein differs from another protein in having specific number and sequence of amino acids. There are thousands of different kinds of proteins, differing from one another in the sequence and relative content of various amino acids. As 26 alphabets in English language can make millions of words, similarly 20 amino acids make millions of types of proteins.

Now the question is as to how the sequence of aminoacids in a protein is determined. Several experiments have revealed that DNA present in the chromosomes of the cell contains "information" or "message" with regard to sequence of aminoacids in a protein. On the basis of "message" contained in a DNA a specific type of RNA is formed, which is designated as "messenger RNA" or m-RNA. This m-RNA contains the information, which is being copied from DNA. According to the 'message' contained in m-RNA specific amino acids are linked together in a particular sequence to form a specific protein. The process of protein synthesis can be summarized as follows.

$$\text{DNA} \xrightarrow[\text{Transcription}]{} \text{m — RNA} \xrightarrow[\text{Translation}]{} \text{Protein}$$

A segment of DNA, which carries the necessary information for the synthesis of a single protein, may be designated as 'gene'. A 'gene' is a section of DNA molecule, which controls the synthesis of a specific protein.

The information in DNA molecule is coded in the form of sequence of nucleotide bases along the strands of DNA molecule. There are four kinds of nucleotides in the DNA molecule. Sugar and phosphate are common to all types of nucleotides. Now it has been established that a sequence of three bases called 'triplet' codes for one specific amino acid. Hence the triplet of three bases of DNA molecule is called "Codon". There will be about 64 different codes, of three bases are taken at a time as 'triplet code'. But only (20) amino acids are known to participate in the synthesis of proteins. So there will be more than one code for each of the (20) amino acids.

10.8.1 Messenger RNA (m-RNA) : The synthesis of proteins takes place at the site of cytoplasmic ribosomes. But the genetic information is contained in the DNA of the nucleus. It cannot move into the cytoplasm. Messenger RNA copies the information contained in DNA. DNA acts as template for the synthesis of messenger RNA. The production of m-RNA from DNA molecule is as follows.

First two strands of DNA molecule separate temporarily at one part of its length. Ribonucleotides are then arranged in a complementary order to the DNA strand. Thus single stranded RNA composing the base sequence complimentary to the base sequence of DNA strand is synthesized. If the DNA strand cantains CATAGTCGA base sequence, it produces RNA with the base sequence of GUA UGA GCU. In RNA thymine is replaced by Uracil. After the formation of messenger RNA the two strands of DNA re-unite. The information contained in DNA in the form of base sequence is copied by messenger RNA.

This m-RNA is separated from DNA template and carried to the ribosomes of the cytoplasm, where the protein synthesis takesplace.

10.8.2 Transfer RNA (t-RNA) : Amino acids are the building blocks of proteins. So the protein synthesis involves end to end union of amino acids by peptide bonds. Polypeptide chains fold and cross link to form proteins. The variation in proteins is due to the variations in the content and sequence of amino acids in the chain.

Proteins contain about (20) different amino acids. Amino acids are present in the cytoplasm of the cell as free units. These amino acids are to be carried to ribosomes, the site of protein synthesis, and are to be linked in the proper sequence as per the inromation contained in the m-RNA.

It has been found that another kind of RNA called Transfer RNA (t-RNA) or soluble RNA (s-RNA) are involved in selecting specific amino acids and carrying them to the site of protein synthesis. The t-RNA functions both as a carrier of amino acid and as well as fixing it in the correct position on m-RNA: Since (20) different amino acids are known to participate in protein synthesis, there are at least (20) types of t-RNA. Each amino acid has its own type of t-RNA.

The t-RNA or s-RNA constitute 15% of total RNA. The t-RNA is smallest consisting of only 74-90 bases. It exhibits a 'clover leaf' model (Phillips, 1969) (fig 10.3).

It has four distinct arms as shown in figure. The bases at the '3 end in all types of t-RNA are in the sequence of CCA and '5 end of the molecule contains guanine (G) base. The amino acids are attached to CCA end. The function of 'G' end is not known. The

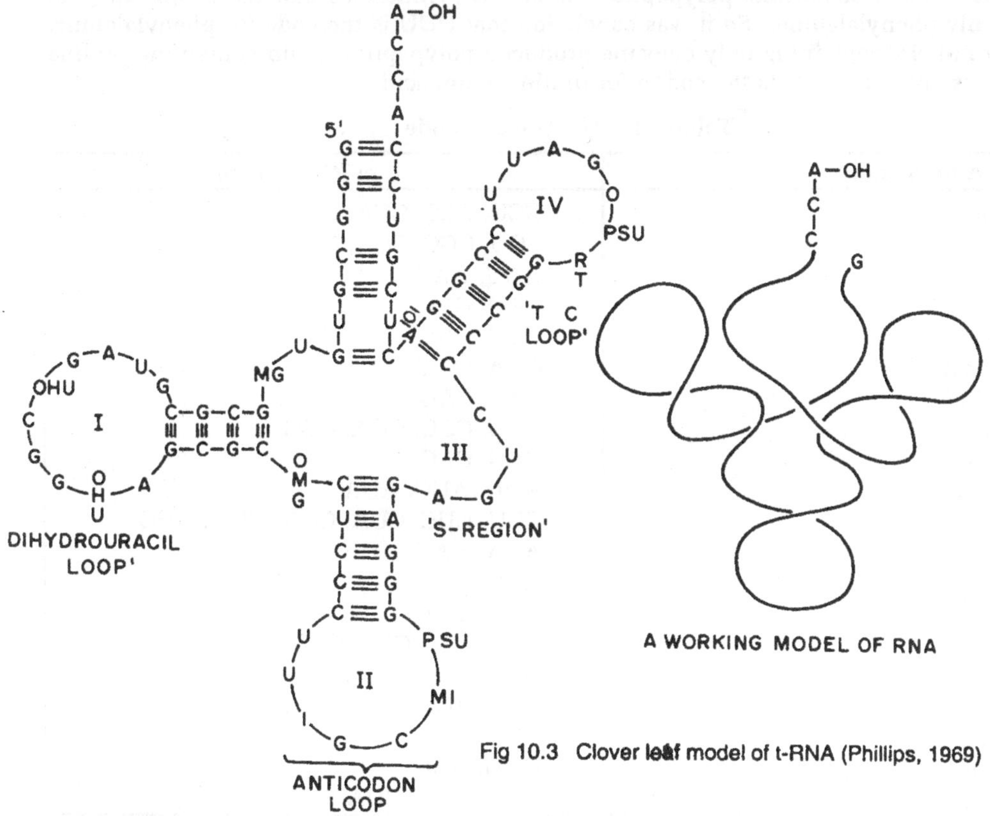

Fig 10.3 Clover leaf model of t-RNA (Phillips, 1969)

loop II generally known as 'Anticodon loop' contains three unpaired bases known as 'Anticodon triplet'. The anticodon forms hydrogen bonding with the specific codon on the m-RNA. The loop — I is known as "Dihydro uracil loop". It is the binding site for the enzyme aminoacyl synthetase. The loop — IV is "Tc loop", which is the site for ribosomal binding. The loop-III is known as "S-region". The main function of t-RNA is to select proper amino acid and transfer to its correct position on m-RNA.

10.8.3. Ribosomal RNA (r-RNA) : Ribosomal RNA is mainly present in ribosomes. It is a constant component of ribosomes and forms integral part of its structure. Ribosomes are nucleo proteins, made up of RNA and proteins in about equal proportion. In prokaryotic cells RNA content is about 60-65% and in eukaryotic cells RNA constitutes about 45%. The molecular weight of RNA runs into millions. Ribosomes of eukaryotic cells contain two un-equal sub-units of 60s and 40s, while that of prokaryotic cells contains sub-units of 50s and 30s. At the time of protein synthesis more than one ribosomes are held together by messenger RNA strand to form polyribosome or polysome. Ribosomes are the 'work benches' for the synthesis of proteins.

10.8.4 Genetic Code : F.H.C. Crick (1962, 1963, 1966) proposed the theory that sequence of three bases codes for single amino acid. Such a triplet of bases occurring on m-RNA are called as 'codons'. **M.W. Nirenberg and J.H. Mathaei** (1961, 62) conducted several experiments to find out the nature of 'genetic code'. These workers succeded in synthesizing a synthetic m-RNA, which contained only one base (uracil). As it contained only one uracil, it is known as 'poly-uracil' m-RNA. To test the effectiveness of the synthetic m-RNA, it was placed in a system, which could synthesize proteins. Such a system is extracted from E.coli. To this system various types of amino acids, necessary ATP molecules were added. After some time polypeptide chain was synthesized and the chain found to contain only phenylalanine. So it was concluded that UUU is the code for phenylalanine. Similarly m-RNA containing only cytosine produced polypeptide chain containing proline amino acids only. So CCC is the codon for proline aminoacid.

Table 10.2 The Genetic code

Amino acid	m-RNA codons
Alanine	GCU, GCC, GCA, GCG
Arginine	CGU, CGC, CGA, CGG, AGA, AGG
Asparagine	AAU, AAC
Aspartic acid	GAU, GAC
Cysteine	UGU, UGC
Glutamic acid	GAA, GAG
Glutamine	CAA, CAG
Glycine	GGU, GGC, GGA, GGG
Histidine	CAU, CAC
Isoleucine	AUU, AUC, AUA
Leucine	CUU, CUC, CUA, CUG, UUA, UUG
Lysine	AAA, AAG
Methionine	AUG
Phenyl- alanine	UUU, UUC
Proline	CCU, CCC, CCA, CCG
Serine	UCU, UCC, UCA, UCG, AGU, AGG
Threonine	ACU, ACC, ACA, ACG
Tryptophan	UGG
Tyrosine	UAU, UAC
20. Valine	GUU, GUC, GUA, GUG

In this way codons for 20 amino acids were determined. There are about 64 codons. All the 64 codons, however, do not code for amino acids. Some codons for example UAA, UAG donot code for any amino acid. There are called 'non-sense' codons. These non-sense codons, infact, are the terminators of the polypeptide chain. They are now called 'ochre' and 'amber' respectively. Like wise AUG is a codon for methionine. It is the initiator of the polypeptide chain.

10.8.5 Synthesis of Polypeptide chain : The first step in the synthesis of polypeptide chain is activation of each amino acid by a molecule of ATP in the presence of an enzyme (E) amino acyl synthetase, resulting in the formation aminoacyl AMP (Aa-AMP) complex and release of pyrophosphate (PP). The enzyme remains attached to the complex forming amino acyl — AMP enzyme complex (Aa — AMP) E. This reaction requires Mg^{++} ions. Each amino acid has its own specific synthetase enzyme.

In the second step the amino acyl-AMP enzyme complex reacts with its specific t-RNA at '3 end. All the t-RNAs have the sequence of same three nucleotides i.e. CCA at the '3 end. The COOH group of Aa-AMP-E complex reacts with OH group of the terminal Adenine residue and bound to the t-RNA by a high energy bond resulting in the formatio of amino Acyl-t-RNA complex. AMP and Enzymes are released.

$$\text{T — RNA + (Aa} \sim \text{AMP)E} \rightleftharpoons \text{Amino acyl — t — RNA + E + AMP}$$

In this complex the amino acid is in the activated state. The released enzyme is used again for activating the similar amino acid. The enzyme, therefore, specifie for a particular amino acid and its specific t-RNA

The messenger — RNA migrates from the nucleus into the cytoplasm, through a pore in a nuclear membrance. First the m-RNA attaches to 30s unit of ribosome by its 5' end. This attachment is carried in the presence of protein factor (F_3).

$$\text{30s + m — RNA} \xrightarrow{F_3} \text{30s — m-RNA}$$

First t-RNA carrying amino acid methionine called t-met t-RNA f ^{met} attaches to 30s m-RNA complex. It is also carried out in the presence of a protein factor (F_1).

$$\text{30s — m — RNA + f-met — t — RNA } f^{met} \xrightarrow{F_1} \text{30s — mRNA — f-met — t RNA } f^{met}$$

At this stage 50s sub-unit of the ribosome associate with its smaller unit and forms 70s ribosome. This combining process occurs in the presence of a protein factor (F_2).

$$\text{30s — m RNA + } f^{met} \text{ — t RNA } f^{met} \text{ + 50s} \xrightarrow{F_1} \text{70s — mRNA — f-met — t RNA } f^{met}.$$

Once this complex is formed the elongation of polypeptide chain depends upon the movement of ribosome along the length of m-RNA, from its 5 end to 3' end. In this way it becomes associated with different parts of m-RNA at different times. Each segment of m-RNA carries a codon for a specific amino acid. Only that specific t-RNA which has complimentary base triplet to m-RNA with its specific amino acid attaches to ribosome at A-site. Elongation of polypeptide chain by joining of various amino acids according to the base pairs (codons) of m-RNA involves mainly three steps.

i) Binding of AA-t RNA (amino acyl complex) at the A-site of the ribosomes.

ii) Formation of peptide bond.

iii) Transfer of peptidyl t-RNA from 'A' to 'P' site.

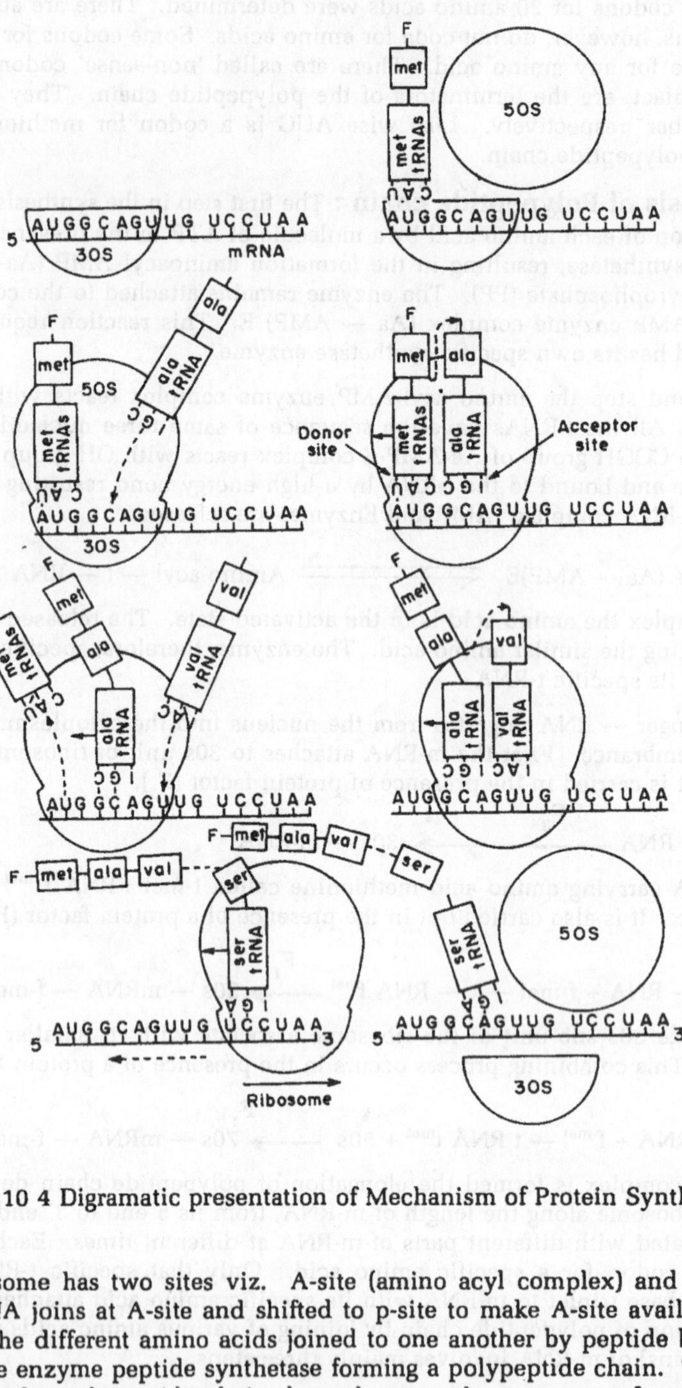

Fig 10 4 Digramatic presentation of Mechanism of Protein Synthesis

Each ribosome has two sites viz. A-site (amino acyl complex) and p-site (peptidyl site). AA-t RNA joins at A-site and shifted to p-site to make A-site available for another amino acid. The different amino acids joined to one another by peptide bonds under the influence of the enzyme peptide synthetase forming a polypeptide chain. The sequence of amino acids in the polypeptide chain depends upon the sequence of codons of m-RNA. Each m-RNA contains the information pertaining to synthesis of a protein. The whole process of protein synthesis occurs strictly in accordance with the information contained in DNA. Once the polypeptide chain formation is completed the two sub-units of ribosome disassociate.

QUESTIONS

A. Answer the following questions in 25 lines.

1. Describe the process of bio-synthesis of proteins.
2. Give a brief account of nitrogen metabolism in plants.
3. Outline the various stages from the absorption of nitrates from the soil, till the formation of proteins.
4. Explain the mechanism of nitrogen fixation by micro-organisms
5. Describe the structure and methods of formation of Amino acids.
6. Write an account of fixation of Nitrogen by plants.
7. Explain the mechanism of Nitrate reduction.
8. What is central dogma? Explain in detail with the suitable illustrations.
9. Describe Amino acid metabolism in plants.
10. What is Genetic Code? Explain in detail.
11. Describe the role played by nucleic acids in the synthesis of proteins.
12. Explain the mechanism of protein synthesis in plants.

B. Short notes

1. Translation
2. Transcription
3. Triplet code
4. Transfer RNA
5. Genetic code
6. Amides
7. Nitrate reduction
8. Rhizobium
9. Root nodules
10. Polysomes
11. Ribosomes
12. Transamination
13. Reductive amination
14. Nitrogenase
15. Symbiotic Nitrogen fixation.

C. Define the following in one or two sentences.

1. Codon
2. Anticodon
3. Peptide bond
4. Symbiosis
5. Leghaemoglobin
6. Nonsense codons
7. Bacteroids
8. Amphoteric Compounds

11 PLANT PHYSIOLOGY
GROWTH AND DEVELOPMENT

11.1 INTRODUCTION

Growth is an essential characteristic of life and living organisms. It is very difficult to define growth in simple terms, because it involves innumerable and unrelated events. Growth refers to permanent increase in size and dry weight of the plant or plant part. It is an irreversible increase in the size. Similarly wood by inbibing water also increase in size. But these cannot be taken as examples of growth. Since in both these cases, they come to normal size upon drying. In certain cases the growth is accompanied by decrease in the dry weight. For example in germination of seeds in the dark, there is an increase in the size of the seedlings, but loss in dry weight results due to respiration and other catabolic processes taking place in the seedlings during the germination. Similarly, Potato tubers, when sprouted in darkness, also result in loss in dry weight. So the growth, in general, may be defined as "Permanent change in any plant or plant part in size, form, weight and volume".

Development, on the other hand, Development refers to qualitative changes in the plant body in the formation of various tissues and organs of plant body such as formation of leaves, flowers and fruits etc.

Growth and development, in fact, cannot be separated into two distinct phenomena. On the other hand, one follows the other in quick succession. Therefore, both growth and development are integrated and interrelated processes.

Morphogenesis means origin of form. The formation of various plant organs such as leaves, flowers, fruits etc. It is the study of development from morphlogical point of view.

11.2 GROWTH REGIONS

In higher plants the typical growth regions are located at the apices of root and shoot. These are known as apical meristems or primary meristems. The apical meristem at the root apex is covered by conical root-cap. The cells of the root cap are originated from calyptrogen meristem. The meristem at the shoot apex, on the other hand, is not covered by any covering from the outside.

Besides the primary meristems, there occur secondary meristems in many plants. Intercalary meristems and lateral meristems are such types of secondary meristems. In bamboo and maize the increase in length of shoot occur due to intercalary meristem. Intercalary meristems are considered as portions of apical meristems separated by permanent tissue. Lateral meristems are situated in the lateral position as longitudinal strips in deeper layers and are knows as vascular cambium and cork – cambium. Lateral meristems are responsible for the secondary growth of the stem and roots.

11.3 PHASES OF GROWTH:

Growth and development include the events of cell division, cell elongation and cell differentiation.

Cell division is an important phase of growth. Every living Organism starts its life cycle with a single cell called zygote. The zygote undergoes repeated cell divisions to form an adult plant. The dividing cells or meristematic cells are characterized by thin walled, isodiametric cells having dense protoplasm with the large nucleus and without or with small vacuoles. The intercellular spaces are absent. Cell divisions are mitotic.

Cell elongation is the second phase of growth. In this phase the newly formed cells begin to enlarge. However, the enlargement is not symmetric in all planes. Enlargement is predominantly in the longitudinal direction. Hence, this phase of growth is referred as "cell elongation". Elongation of the cells is due to the entry of water into the cell due to osmotic effect on account of presence of the large quantities of solutes inside the growing cells. As a result of entry of water into the cell, the turgor pressure of the cell increases resulting in stretching of the cell wall, which is an elastic one. Due to the elongation of the cell wall, the cellulose molecules are deposited to the existing cell wall material. The deposition of new cellulose material takes place in two ways. If the new molecules of cellulose are deposited between the original molecules of cell wall, the phenomenon is called **intussusception**. The other method is **apposition** in which the new material is deposited by the protoplasm as a thin layer to the cell wall already existing. Thus the elongation of cell is due to the stretching of the cell wall due to increase in turgidity of the cell and later deposition of the new cellulose material upon the existing cell wall.

Before cell elongation occurs, several metabolic changes take place. Nucleic acids, proteins, lipids, carbohydrates and other constituents of protoplasm are synthesized. Mitochondria, ribosomes, plastids etc are formed. However, the formation of protoplasm is commensurate with the increase in the volume of the cell. As a result the protoplasm forms a thin layer lining the cell wall with a large central vacuole.

Cell differentiation is the last phase of the growth. In this phase the cells, after enlargement, mature and get differentiated into different kinds of permanent tissue. The cell walls become thick due to the deposition of lignin and suberin.

11.4 GROWTH CURVE:

When the growth of the cell or a plant organ such as leaf, fruit, root or the whole plant is measured at periodical intervals throughout its life and plotted against the time, usually an "s" shaped or sigmoid curve results. It is called as growth curve. The shape of the curve remains the same regardless of whether the growth is measured in terms of length, size, area, volume or weight. The growth rate is very slow in the initial stages and then increases very fast and at the end growth becomes slow till it stops. Thus the whole growth sequence is divided into four phases from the beginning to the end.

11.4.1 Lag Phase of Growth : In this phase the growth rate is very slow. The growth system prepares itself for the grand period of growth to be followed. Hence, more time is taken for little amount of growth.

11.4.2 Log Phase of Growth: In this period the growth rate is exponential. The growth rate follows compound interest law. Hence this phase of growth is also known as "grand period of growth".

11.4.3 Decline Phase of Growth : During this phase the growth rate gradually decreases and the system loses the capacity of growing.

11.4.4 Senescent Phase of Growth : This phase of growth is marked by attainment of maturity of the system. This phase is also called as "Steady Phase".

Thus every plant or plant organ slow in their growth activity in the initial stages and after obtaining the pre-requisite potentialities, the growth rate becomes 'exponential' and at the end the growth rate becomes slow until it stops.

11.5 MEASUREMENT OF GROWTH

Growth in plants can be measured by a number of methods. One or the other method is useful for measuring the growth of the plant depending upon the requirement. The increase in length of the stem can be measured by the following methods.

11.5.1 Increase in Length : Growth is usually measured in terms of increase in length of stem or root. But measurement of length as an index of growth sometimes misleading. Because the seedlings grown in dark are taller compared to the seedlings grown in light, which are short and stout.

11.5.1.1 Direct Method : The length of the plant or plant part can be measured by means of scale at periodical intervals.

11.5.1.2 Horizontal Microscope : A horizontal microscope (Fig 11.1) is fitted close to the growing plant and focussed on a particular point marked at the tip of the plant. After some time the marked point again, focussed by moving microscope tube upwards. The distance covered by the microscope indicates the actual increase in length of the plant.

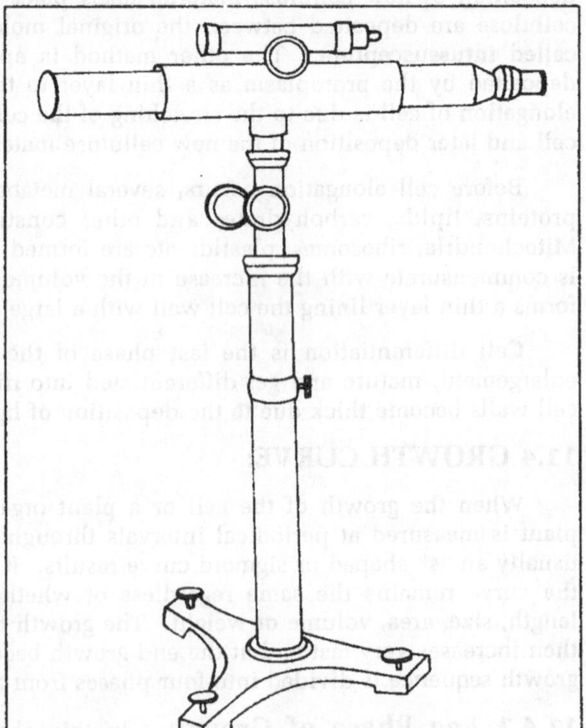

Fig 11.1 Horizontal Microscope

11.5.1.3 Auxano Meters : Several kinds of auxanometers are in use to measure increase in length of the plant. Two of them are described below:

a) Arc auxanometer or **Arc Indicator:** A simple arc auxanometer or Arc indicator is shown in Fig (11.2). A silk thread is taken and one end of the silk thread is tied to the stem tip of the potted plant and to the other end a small weight is tied. The silk thread is hanged on the pulley of auxanometer. The pulley is fixed with a long pointer, which slides over a graduated arc. As the stem grows in

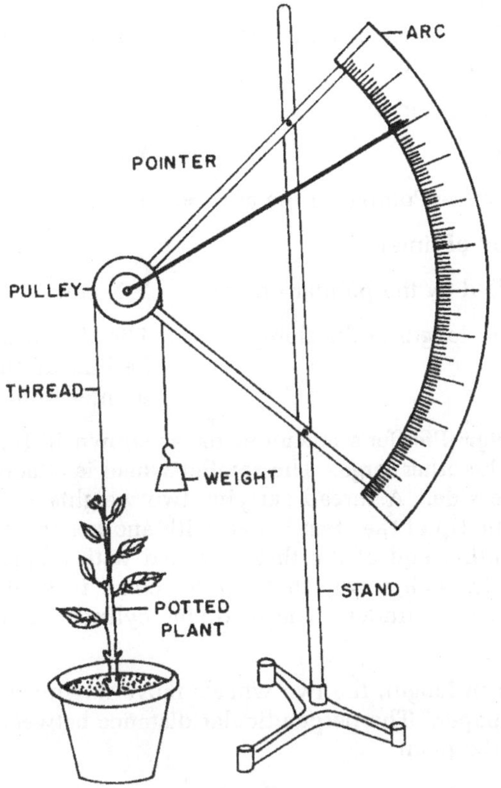

Fig 11.2A Auxano meter or Arc indicator

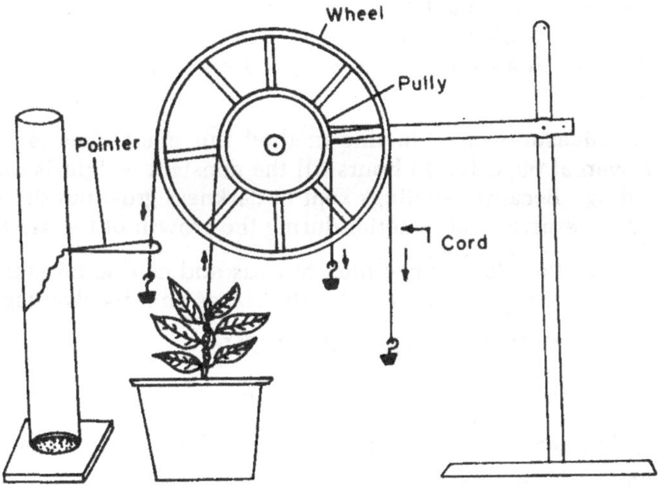

Fig 11.2B Pfeiffer's auxanometer

length the weight pulls the thread down. As a result of which the pointer moves downwards on the graduated arc. The distance can be read on the scale. The actual growth in length of the stem is calculated as follows:

Suppose the initial position of		
Pointer on the arc is	=	X cm
The final position of the Pointer on the arc	=	Y cm
(After Certain period of time)		
The distance travelled by the pointer on arc	=	Y-X cm
The actual growth in length of the stem	=	The distance travelled by pointer X Radius of the pulley length of the pointer from the centre of the pully

b) Pfeiffer's auxanometer: Pfeiffer's auxanometer as shown in (fig 11.2B) consists of two wheels; one small and the other large. The smaller wheel is attached to the central axis of the larger wheel on one side. A thread carrying two weights at two ends is placed over the large wheel. Now the tip of the stem is tied with another thread, which is hanged over the small wheel. The other end of the thread is tied with a small weight. A pointer is attached to this thread in such a way that pointer come in contact with smoked paper, wrapped over the revolving cylinder. The revolving cylinder is rotated at the rate of one rotation per an hour.

As the plant grows in length, the two wheels move downwardly. The pointer moves upwards on the smoke paper. The perpendicular distance between two adjoining marking indicates the growth of the plant.

11.5.2 Fresh Weight Measurement : The weight of the freshly harvested plant is called 'Fresh weight' or 'green weight'. Fresh weight of the plant or plant parts can be taken as one of the criterion for measuring the growth of the plant. This method can be employed in case of irregular bodies such as roots, fruits, tubers, etc. But, this method has certain disadvantages. The fresh weight is often misleading. Fresh Weight of the plant or plant parts depends upon the water content of the plant body. 90% of the fresh weight is due to the water alone. Water content of the plant body depends upon the rate of absorption and transpiration. So the fresh weight of the plant body differ at different times of day. Further the fresh weight diminishes while weighing also because the loss of water takes place constantly.

11.5.3.1 Dry Weight Measurement : In this method the plant material is harvested and dried in hot — air oven at 60°C for 48 hours till the constant weight is obtained. But this method also misleading. Because seedlings kept in darkness grow but dry weight decreases due to consumption of reserve food material during the growth of the seedlings.

Therefore, the growth of the plant is complex process and can be measured by more than one method. Because each method has got its advantages and disadvantages.

11.6 FACTORS AFFECTING THE GROWTH:

The plant growth is influenced by variety of external and internal factors.

11.6.1 External Factors : Amongst the external or environmental factors temperature, light and water supply are the most important factors.

11.6.1.1 Temperature : Temperature affects the plant growth in many ways. Firstly the temperature affects the metabolic activities of the plant. Each bio-chemical process has its own optimum temperature. When the temperature increases beyond the optimum temperature the rate of process decreases and finally the process ceases when the temperature is 60°C and above. The optimum temperature varies from plant species to species. It is higher in tropical plants than in temperature regions. The range of temperature for tropical plants is 10°C to 45°C and for temperate plants it is 5°C to 35°C.

Although temperature affects number of processes, photosynthesis and respiration processes are affected mostly. Rise in temperature increase the rate of photosynthesis as well as rate of respiration. As a result higher day temperature and lower night temperature are conducive for the growth of the plant. Higher temperature during the daytime enhances the rate of photosynthesis and lower temperature at night times reduces the respiratory loss.

11.6.1.2 Light : Light affects most of the processes in plants. For example seed germination, the growth of seedlings, differentiation of tissues, development of the various organs of the plant, flowering etc are affected by the quality and quantity of light.

The light reaches the earth in the form of electromagnetic waves. The wavelength ranges from 10^{-14} to 10^{-6}cms. The radiations with the lowest wavelength are the cosmic rays and the waves with the highest wavelengths are the radio waves. However, it is the wavelengths in the visible spectrum from 400 — 700 nm affects the plant growth. The radiation of lower waves lengths, such as X-rays, cosmic rays and ultra violet rays are not useful for the growth of the plant. On the other hand, they are injurious to the plant and mutagenic. Similarly the plants do no absorb radio waves. The chlorophyll absorbs light mostly from red and blue parts of the visible spectrum. Experiments have shown that plants can grow better when exposed to full spectrum of visible light instead of individual light.

The duration of light in a day affects the plant growth. Such a response of plant to duration of light is known as 'photoperiodism'. The duration of lights in many plants decides the time of flowering after the vegetative growth of the plant.

11.6.1.3 Water Supply : Water plays an important role in the growth of the plant. Water is absolutely necessary for various metabolic activities of the plant and for maintaining turgidity of cells. Water deficit is deterimental to the growth of the plant.

11.6.1.4 Minerals : Plants require many essential elements for the normal growth. The details are discussed elsewhere in the book. When the soil is deficient in any one of the essential elements, the metabolism of the plant is deranged and growth is affected.

11.6.2 Internal Factors

11.6.2.1 Plant Hormones : This is one of the most important internal factors affecting the growth of the plant. The details are discussed in the next pages.

11.7 PLANT GROWTH REGULATORS

All living cells require besides food, water and minerals, certain organic substances in small quantities for the growth of the plant. These are called "accessory organic substances" and can be divided into two categories viz. vitamins and hormones. 'Harmone' term was first used by **Bayliss** and **Starling** (1904). Hormone is a Greek word which means "to urge on" or "to stimulate". **Thimann** (1948) suggested the term 'phytohormone' for hormones occurring in plants.

Hormones are produced in any part of the plant body and are translocated to other parts, where they are used in very small quantities to control or regulate metabolism of plant growth.

Vitamins are also organic compounds, which are used, in small quantities. There is no clear-cut demarcation between vitamins and hormones. Vitamins are synthesized and used in the same part. In contract the hormones are synthesized in one part of the plant body and used up in other part of the plant body i.e. the site of synthesis and action are different in hormones. Vitamins act in association with the enzymes as co-factors. But none of the hormones is known to act as co-factors of enzymes.

Plants produce five kinds of hormones. They are auxins, gibberellins, cytokinins, abscissic acid and ethylene. The hormones are also known as 'plant growth regulators' or "plant growth substances". In addition to these hormones which are universal in occurrence, plants are also known to produce two other hormones; florigen and vernalin. But these have not been extracted and characterized.

11.7.1 Auxins : **The idea of existence of auxins in plants was for the first time conceived by Charles Darwin in 1881. While working on canary grass (phalaris canariensis), he showed that the coleptile of canary grass could bent towards light, when it was unilaterally illuminated. However, the coleoptile failed to bend when the coleoptile tip was covered with an opaque cap. From this he concluded that some substance in the coleoptile tip was transported downwards and caused bending. Similar kind of observations were made by Boysen — Jonsen (1913), while working on oat coleoptile. The oat coleoptile encloses the primary leaf and stem tip (fig 11.3).**

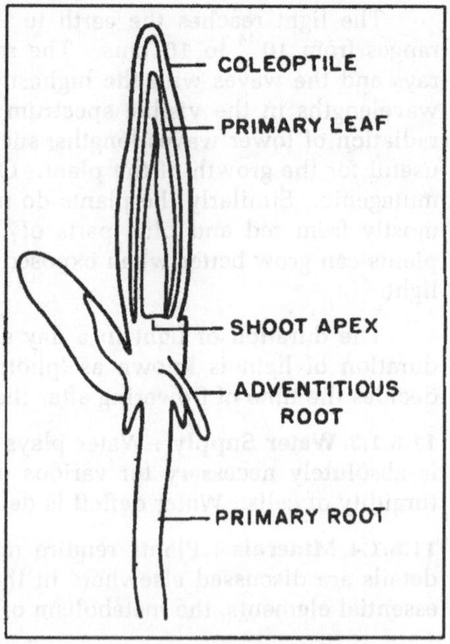

Fig 11.3 Structure of oat coleoptile

Paal (1919) removed the coleoptile tip and replaced on the coleoptile stump towards one side. The side having coleoptile tip grew more than the other side. As a result the coleoptile bend towards the side which was opposite to the side on which the tip was placed. This demonstrates that some substance passed from the tip and this substance caused the growth of the coleoptile.

F.W. Went (1928) did some important work. He was able to collect the substances produced by the oat coleoptile in agar blocks. He removed the several coleoptiles of oat and placed them with their bottom end on a block of agar jelley for some time in the dark. The stimulatory substance present in the coleoptile diffused into the agar blocks, which was later cut into smaller cubes. Agar cubes containing the stimulatory substance were placed on the coleoptile cut stumps on one side. The coleoptile stump showed the curvature towards the side opposite to the side upon which the cube was placed. The angle of curvature produced was found directly proportional to the concentration of hormone in the "Avena Curvaturetest". Now a days this test is used for the quantitative estimation of the ansins.

11.7.1.1 Chemical Nature of Auxins : Kogl and Haagen-Sunit (1931) isolated auxin-a from human urine. It was identified chemically as auxentriolic acid ($C_{18}H_{32}O_3$). Kogl etal (1934) isolated similar substance from corn germ oil and named it auxin-b (Auxanolonic acid; $C_{18}H_{30}O_4$). Later they isolated another auxin from human urine called heteroauxin, which was identified as Indole-3-acetic acid (IAA). The molecular structure of auxin-a, auxin-b, and heteroauxin are given below.(Fig 11.4)

Auxin-a and auxin-b are not known to occur in plants, but heteroauxin (IAA) is known to occur universally in all the plants, including Algae, Fungi, Liverworts, gymnosperms and Angiosperms. Besides IAA, Indole-3 acetaldehyde, Indole-3-pyruvic acid, Indole acetonitrile and Indole-3-ethanol are some other naturally occurring auxins in plants.

In addition to naturally occurring auxins certain other substances are known to produce similar affects on growth. These substances do not occur in plants. They are called synthetic auxins. The important synthetic auxins are Indole-butyric acid (IBA), Naphthelene acetic acid (NAA), 2-4 dichloro phenoxy acetic acid (2-4D) and Para Chlorophenoxy acetic acid (PCPA). These synthetic auxins are inactivated in plants but they are more stable.

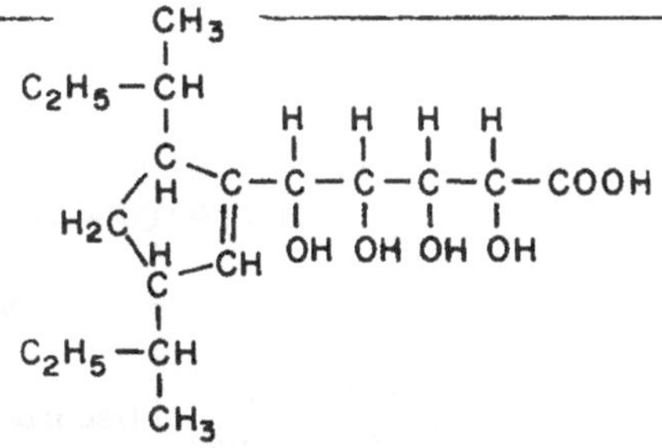

Auxin – a

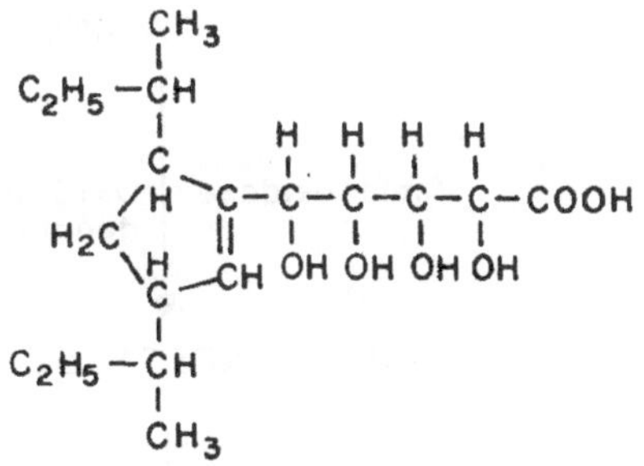

Auxin – b

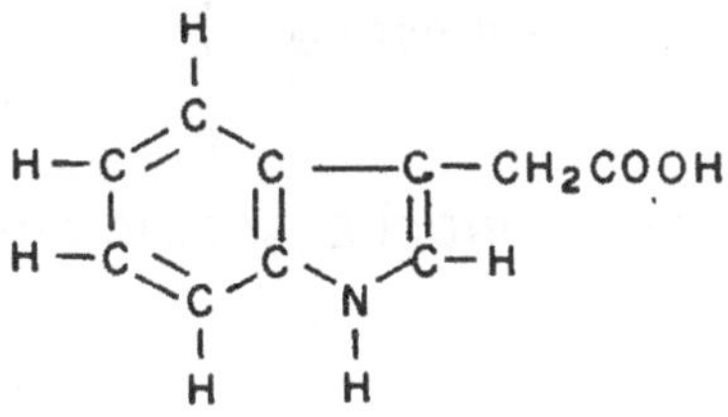

Hetero auxin
(Indole–3–Acetic acid)

Fig 11.4 Structure of Auxins and Hetero auxin

11.7.1.2 Bio-Synthesis of Auxins :

a) Site : The auxins are synthesised in the tip of the coleoptile and stem. Developing leaves, flower buds, fruits, germinating seeds, pollen grains and root nodules of legumes also synthesise auxins.

b) Pathways : The bio-synthetic pathways of auxins are not fully understood but experimental evidence indicate that the starting substance for the synthesis of auxins is an amino acid, trytophan. There are two different pathways. In one pathway trytophan is first converted into indole pyruvic acid, while in another pathway trytophan is converted to tryptamine. In both the path ways indole acetaldehyde is formed. It is finally oxidised to Indole acetic acid (fig. 11.5).

c) Auxin Transport : The auxins in plants move only in one direction, from the tip of the stem to the base. The direction of movement is independent of the orientation of the plant. It means to say that auxin moves downwards if the coleoptile is erect but moves upwards if the coleoptile is inverted. This type of movement is called basipetal polar movement. It is a characteristic feature of auxins. In roots the movement of auxins is from the base of the root to the tip. This type of movement is called "Acropetal polar movement".

Auxins are transported through the living tissues such as parenchyma and phloem. The movement of auxin in phloem is independent of the movement of other organic substances in the phloem sap. Auxin transport is an active process involving the expenditure of energy. The rate of auxin movement is 10-15 mm/hour.

d) Bio-Assay of Auxins : The auxins in plants occur in too low concentrations to be detected by physical or chemical methods. Therefore, they are estimated by using living organisms. Such a technique of measuring the activity of biologically active substances by

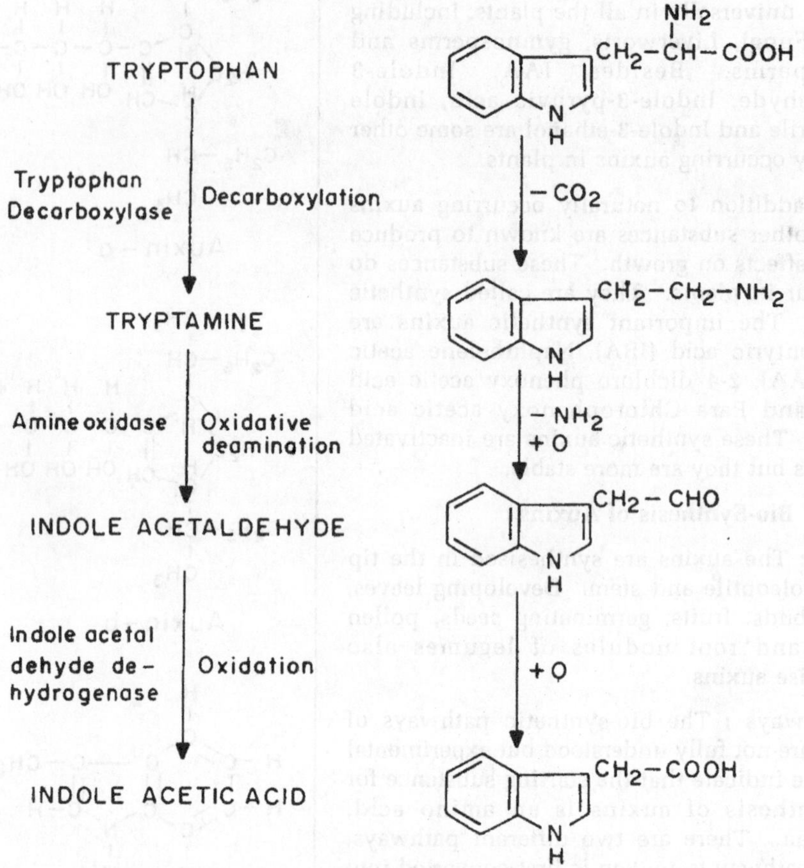

Fig 11.5 Pathway of Tryptopham to IAA

employing living organisms is called 'Bioassay'. The common bio-assay methods for estimation of auxins are (1) Avena curvature test (2) Avena coleoptile straight growth test.

The plant part from which auxin is to be extracted is ground in organic solvent like acetone. The extract is purified and chromatographed. The substance separated on the chromatogram is extracted and is fed to the test plant. The response obtained by the unknown plant extract is compared with the response to a known quantity of IAA. In this way the quantity of auxin present in unknown plant extract is estimated.

11.7.1.3 Physiological Roles of Auxin_The physiological roles of hormones in plants are studied by estimating the endogenous hormonal level of the responding plant part. The correlation between the hormone level of the plant with the response observed in the plant part is taken as positive evidence of the involvement of the hormones in the growth and development process. However, this method is not free from errors. Methods of extraction, purification and estimation of hormones are not fool proof. The losses in hormone levels during these methods are considerable. Therefore, the quantity is estimated is less than the quantity present originally in the plant part. Further, the plant tissue contains more than one type of hormones, and hormones may interact with one another. In view of these the estimation and response of single hormone is always not possible. These points should be taken into consideration while discussing the various roles of hormones.

The various physiological roles of auxin are described below.

I) CELL ELONGATION: The most important characteristic action of auxin is to promote cell elongation. Infact the very discovery of auxin was based on its ability to cause cell elongation in Avena. It is now known that not only in Avena, but in wheat, maize and other cereals the application of auxin caused the elongation of the stem axis. So the auxins are defined as "organic substances which at low concentrations (<0.001) promote the grwoth of the stem axis in length but inhibits the growth of roots. Thus the promotion of shoot growth and inhibition of root-elongation are two important characteristic features of auxins.

The cell elongation is due to loosening of cell wall. The rigid cell wall becomes soften and stretches. The stretching of cell wall is due to:

I. dissolution of the cell wall material

II. Breaking of chemical bonds between cellulose and other cell wall materials.

III. Synthesis of new cell wall material and incorporation of the same in the older cell wall.

Due to loosening of cell wall material the wall pressure decreases, with that the turgor pressure decreases, leading to greater uptake of water by the cell. As a result of all these changes the cell elongates.

II) CELL DIVISION: Auxin causes cell division in meristematic tissues. Effect of auxin on cambium activity was observed by **Snow** (1935) in seedlings and **Soding** (1936) in woody plants. In herbaceous sterms wound healing is promoted by auxins. When the stem is cut new cells are formed around the cut portion of the stem. If young leaves and buds are present above the cut surface of stem, rapid healing of wound takes place because they serve as source of auxins. Tissue culture experiments have shown undoubtedly that auxin promotes the growth of the 'callus' tissue in the culture medium.

III) ROOT FORMATION: It is a common practice that the stem cuttings of rose, jasmine etc readily form adventitious roots when planted in the moist soil. The formation of roots form the basal portion of the stem is promoted when the stem cutting is having young leaves and buds. This clearly indicates that hormones produced by young leaves and buds travel downward and cause root formation from the cut end of the stem. Synthetic auxins like IAA, IBA, 2-4D are more effective than the natural auxins. That is why the synthetic auxins are extensively employed to induce roots on stem cuttings for vegetative propagation of the fruit-yielding and ornamental plants. However, auxins failed to induce root formation in plants like mango, guava and eucalyptus etc. Though auxins induce root formation, it inhibits the root growth.

IV) APICAL DOMINANCE: In many plants lateral and auxillary buds do not grow until the apical bud remains active. This phenomenon is called 'Apical dominance'. If the apical bud is cut off the lateral auxiliary buds start growing into branches.

The apical dominance is due to the auxin production in the shoot-tip. The auxin is transported downwards, reaches the auxiliary buds, the accumulation of auxin in the auxiliary buds inhibits the growth of lateral buds. When the stem-tip is cut off or terminated by inflorescence production the supply of auxin to the lateral buds is stopped, as a consequence of which lateral buds emerge and grow into branches. However, auxin alone is not responsible for apical dominance. Other hormones like cytokinin and gibberellins also play a part in the apical dominance.

There are two theories to explain the apical dominance. **Thimann** (1937) proposed "Direct theory of auxin". According to him lateral buds themselves produce sufficient amount of auxin for their growth. Additional supply of auxin coming from the stem tip make the concentration of auxin in lateral buds supra-optimal, hence the growth of lateral buds is inhibited. If the stem tip is removed, the additional supply of auxin to the lateral buds is cut off, then they grow into lateral branches. However, experimental evidence do not support this theory when the auxin content of the lateral buds is estimated it is found to be minimum when the stem tip is present and increased markedly when the stem tip is removed. Therefore, the validity of the theory is doubted.

Another theory "Nutrient diversion theory" was proposed to explain the mechanism of apical dominace. According to this view there is a nutritive competition between apical and lateral buds. So long as apical bud remains active it receives major part of nutrients and metabolites while the lateral buds are depleted of them. As a result of which the main stem grows and lateral buds growth is arrested. However, the mechanism by which auxin draws the major portion of nutrients and metabolites is not known. It is suggested that the auxin moving down wards from the stem tip inhibits the development of vascular connection between stele and lateral buds. So the lateral buds do not receive the supply of nutrients from the stele, there by the growth of lateral buds is arrested.

V) ABSCISSION OF LEAVES AND FRUITS: Leaves, flowers and fruits fall off after formation of abscission layer. When the leaf is young, the green lamina of young leaf produces auxin. When the leaf becomes old, the synthesis of auxin is decreased, consequently abscission layer is formed and the leaf falls off. If the auxin is applied to damaged or old leaf the formation of abscission layer can be prevented. Similarly in some plants like in apple and citrus, the fruits fall off from the trees, even before they are mature. Application of auxin prevents the falling off flowers and fruits. This character is commercially exploited by fruit growers in preventing premature fall of fruits in apples, pears, citrus, oranges by application of IAA, 2-4D auxins.

VI) FLOWERING: Auxins generally inhibit flowering. But in pine-apple (Ananas Cosmosus) auxins have been found to promote flowering. This plant can be made bloom promptly with the application of 2-4D and NAA. However, the effect of auxin on flowering is indirect, mediated through the formation of ethylene.

In addition to flowering, application of auxin may alter the sex ratio of flowers. In cucurbits auxins increase the number of female flowers and decrease the male flowers. Here also the effect of auxin is indirect and is mediated through the ethylene formation.

VII) PARTHENOCARPY: It is a general observation that after pollination and fertilization the flower abscisses and falls off. However, in several plants such as tomato, brinjal etc the application of auxin causes the formation of fruit without polllination and fertilization, which is called "parthenocarpy" and fruits are called 'parthenocarpic' fruits. Parthenocarpic fruits are seedless. Grapes and Bananas are naturally occuring parthenocarpic fruits. In such plants the ovary itself produces sufficient auxin to form fruit without pollination and fertilizartion. Normally pollen grains and pollen tubes supply the auxins for the formation of fruit.

VIII) WEED KILLING (HERBICIDAL): The unwanted plants or weeds can be eradicated by spraying certain specific auxins. Auxins are selective weed killers and are employed in the eradication of weeds in field crops, lawns and fruits. For example 2-4D is toxic to dicotyledonous broad-leaved plants. So 2-4D can be employed to eradicate weeds in cereals such as wheat, rice etc. It kills the weeds but not harmful to cereals. Many types of herbicides are now known and available in the market and all of them contain some or other synthetic auxins. They are used in very low concentrations, so that they are not toxic to human beings and animals.

IX) GERMINATION: Auxins at very low concentrations promote germination but higher concentrations retard the germination. Potatoes can be prevented from sprouting by application of certain auxins.

11.7.2 Gibberellins

11.7.2 Gibberellins : In Japan there occurred a disease on rice plants known as 'Bakanae disease (Foolish seedling disease)', in which the infected plants were taller, thinner and paler than normal plants and they did not produce any seeds there by the crop yield is considerably reduced. It was later shown that the disease was due to infection of the rice plants by a fungus called "*Gibberella fuzikuroi*" (conidial state known as *Fusarium moniliformae*). The fungus was isolated and grown in culture on synthetic medium. The culture medium was filtered to remove the fungus and filtrate was applied to rice plants. The filtrate produced the same effects as that of the fungus. This led to an important conclusion by **Kurosowa** (1926) that the fungus secretes some chemical substances into the culture filtrate, which was responsible for the unusual elongation of the stem. **Yabuta** and **Sumili** (1938) later isolated the substance and named it "Gibberellin". The chemical was purified, crystallised and active fraction was named as "Gibberellic acid" (**curtic & cross**, 1954).

Although gibberellic acid was originally isolated from the fungus, now it is known to occur universally in Algae, Fungi, Liverworts, and higher plants. Now, as many as 70 different types of gibberellins have been discovered. They are designated as GA_1, GA_2 and so on (fig 11.6). The common gibberellic acid is GA_3. In addition to 70 different gibberellins, plants are known to produce substances, which are structurally similar to gibberellic acid, produce the same effects as gibberellins. Such substances are called gibberellin like substances (Phinney, 1956). Such GA-like substances have been reported from monocot, dicots and gymnosperms.

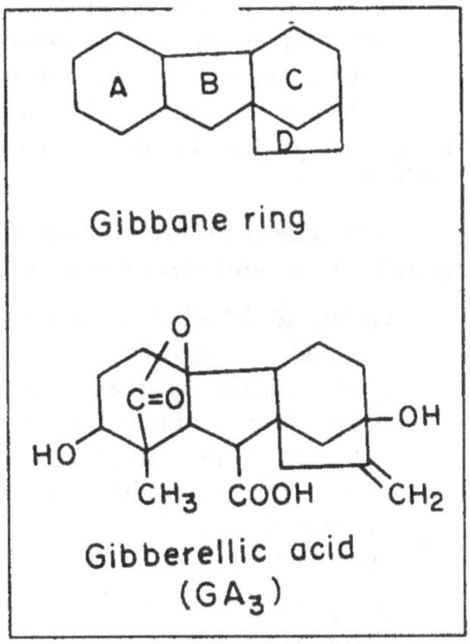

Fig 11.6 Sturcture of GA

11.7.2.1 Chemistry of Gibberellins : Chemically gibberellins are tetracyclic diterpenes. Each molecule consists of gibbane carbon skeleton, which consists of (4) rings, which are designated as A, B, C and D. The basic skeleton of gibberellin molecule is called gibbane ring. All gibberellins including the common gibberellic acid possess the gibbane ring. Some gibberellins like GA_1 to GA_{11} and GA_{19} possess 19 carbon atoms, while others like GA_{12} to GA_{15} have 20 carbon atoms.

Gibberellins are acidic in nature and possess one more carboxylic (COOH) groups. The different gibberellins differ in the presence or absence of hydroxyl (OH) group and the degree of unsaturation of ring 'A' of gibbane skeleton.

11.7.2.2 Bio-Synthesis of Gibberellins : Gibberellins are synthesised in shoot-tip, root-tip, and developing seeds. In the shoot-tip the leaf and bud primordia are responsible for the synthesis of gibberellins.

Acetic acid is the starting material for the bio-synthesis of gibberellins. Three molecule of acetic acid are linked together to form a molecule of mevalonic acid. Kaurene is an intermediate compound. The first gibberellin to be formed is either GA or GA, which give rises to other gibberellins by inter conversions.

The transport of gibberellins, unlike auxins, is non-polar and moves in all directions. It moves both in xylem and phloem tissues.

11.7.2.3 Bio-Assay of Gibberellins : The dwarf mutant varieties of several plants such as Corn, Pea, Cucumber, Beans etc are used for the bio-assay of gibberellins. For example Dward meteor pea variety is sensitive to one-billionth gram of GA_3, so this variety is used for the estimation of gibberellins. Rice seedlings are also used for bio-assay of gibberellins. Small rice seedlings.are placed in a solution containing gibberellin and effect of gibberellin on the growth of the seedlings is measured. Concentration of gibberellins as low as 10^{-6} m can cause detectable response in the seedlings. The different plants differ their response to various kinds of gibberellins.

11.7.2.4 Physiological Roles of Gibberellins

1. STEM GROWTH: The most important characteristic of gibberellins is that it causes stem elongation. Infact, the very discovery of gibberellins was based on the observation of this character on rice platns.

The stem is known to contain several gibberellins. However, the amount of gibberellins in the stem is much less than in leaves, buds, seeds of the same plant. When the GA is applied the stem elongation takes place markedly. The enhanced stem growth is not due to the increase in the number of nodes and internodes, but results from the rapid elongation of internodes. Therefore, the number of nodes and internodes remain the same. The stem elongation is due to both cell division and cell elongation in the nodal and internodal regions.

The effects of GA on stem elongation is most marked in two types of plants namely genetic dwarf and rosette plants.

a) GENETIC DWARFS: In maize four dwarf mutants namely d_1, d_2, d_3 and d_5 were isolated. These dwarf mutants do not grow beyond 1-2 feet. This is because the plants cannot synthesise the gibberellins. The bio-synthetic pathway of gibberellin production is blocked by mutation. In the absence of gibberellin the stem elongation is retarded and the plants remain dwarf. However, when GA is applied the inter nodes elongate rapidly and the plants attain the normal heights. The stem elongation is proportional to the concentration of GA applied, hence this method is used for the estimation of gibberellins.

b) ROSETTE PLANTS: In Carrot, Radish and Cabbage the stem remains short and condensed and the leaves are crowded as the internodes fail to elongate. This gives the plant a "rosette" appearance. The rosette appearance is due to the absence of gibberellins in the stem.

When the GA is applied to the stem of rosette plants the stem elongate. Rapid elongation of the stem of rosette plants is called 'Bolting'. Bolting also occurs in nature at the time of flowering in rosette plants.

2) ROOT FORMATION: Unlike auxins, gibberellins inhibit the formation of adventitious roots on the stem cuttings. Ipomoea and Bryophyllum are execptions. In these plants the application of GA has resulted in the formation of adventitious roots.

3) FLOWERING: Application of gibberllin causes flowering in rosette plants. Rosette plants such as Hyoscyamus and Sasrolus and other plants remain vegetative under short days but flower when exposed to long days. However, application of gibberellin makes the plants to flower even under short days.

Similarly Carrot plants need 'cold treatment' (vernalization) for flowering. However, application of gibberellin makes the plant to flower even without exposing to lower temperatures. Thus gibberellin replaces long day and low temperature requirement for flowering.

4) PARTHENOCARPY: In some plants like Tomato, Apples and pears gibberellin induces parthenocarpy i.e., the formation of fruit without pollination and fertilization. However, the mechanism of the action is not known.

5) SEED DORMANCY: Seeds of some species do not germinate even under favourable conditions. Such seeds are said to be dormant. Application of gibberellin makes the seeds to germinate breaking its dormancy.

Similarly the seeds of apple do not germinate till they are exposed to low temperature (17°C) for few days. When gibberellin is applied to the seeds, the seeds can germinate without being exposed to lower temperature. Thus gibberellin replaces the low temperature requirement for seed germination.

Seeds of Lettuce var Grand rapids do not germinate in dark but readily germinate when they are exposed to sunlight. However, the application of gibberellin causes the seeds to germinate even in the dark. Thus the gibberellin replaces the light requirement for seed germination.

11.7.3 Cytokinis : The process of cell division is also controlled by hormones. Auxins and Gibberellins mainly cause cell elongation. But 'cytokinins' cause cell division. The presence of cell division causing substances suspected in the developing fruits and seeds. The liquid endosperm of coconut, commonly called as coconut milk, is rich in cell division causing substances. Coconut milk causes cell division not only in coconut but also in excised and artificially cultured embryos of Datura and other plants (**Van Overbeek**, 1941). Coconut milk is a complex substance. It contains variety of substances such as auxins, gibberellins, cytokinin, amino acids, nucleic acids etc. The cell division effect may be due to the combined effect of all the substances.

Folk spoog and his associates in the University of Wisconsin attempted to grow pith tissue of tobacco plant invitro by supplying them with nutrients and vitamins. When the

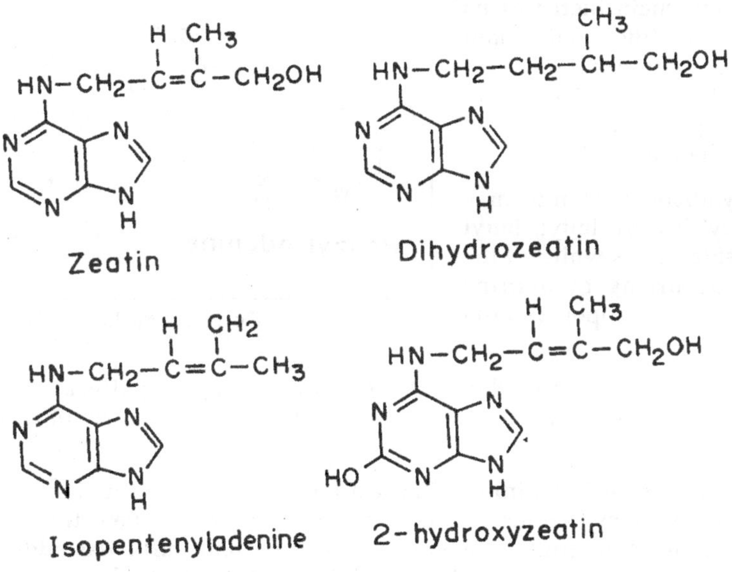

Fig 11.7 Structure of some of the cytokinins

auxin was added to the medium the cell elongation took place but the cells failed to divide. When the pith tissue is cultured along with the pieces of phloem tissue, active cell division started. Therefore, it was thought the phloem tissue could supply the cell division causing factor.

Miller etal (1955) isolated and crystallised the cell division causing substance from autoclaved DNA of sperm of herring fishes. This substance named as "kinetin". It induced cell division in tobacco pith tissue at extremely minute concentration of 1 PPM with the auxin in the medium. Therefore, it became clear that cells require two factors namely Auxin and Kinetin for their division. These two factors interact and bring out cell division. The cell division inducing substances of plants were named as "cytokinins".

Miller (1962) found cell division causing substance in corn germ oil. **Letham** (1963) isolated, crystallised the substance and named it as "Zeatin". Later "Zeatin" was known to occur in many other plants. Kinetin and Zeatin were found to be similar in chemical structure. In addition to them plants were known to contain '4' more cytokinins (Fig 11.7). Cytokinins occur in plants in minute quantities, so it is difficult to extract them in sufficient quantity. Scientists have synthesised two synthetic cytokinins namely kinetin and Benylodenine (BA). Our knowledge on cytokinins is based on the studies of these two syntehtic cytokinins.

Cytokinins occur freely in plants. In higher plants cytokinins occur in apical meristems, developing seeds and fruits, tumors and bacteria nodules. Almost all regions which involve in cell division contain cytokinins.

11.7.3.1 Chemistry of Cytokinins : All cytokinins are derivatives of adenine (fig.11.8). Which is a purine base of nucleic acids (RNA & DNA). All the cytokinins differ from one another in the nature of side chain. The adenine part of the molecule is responsible for its activity. Side chain increases its activity. Even without side chain the adenine part exhibits mild activity. The side chain may contain a ring as in benzyl adenine or straight chain as in Zeatin.

The synthetic cytokinins may have furfuryl benzyl, butyl, butyl or phenyl side chains while all the naturally occurring cytokinins except one have isopentyl side chain.

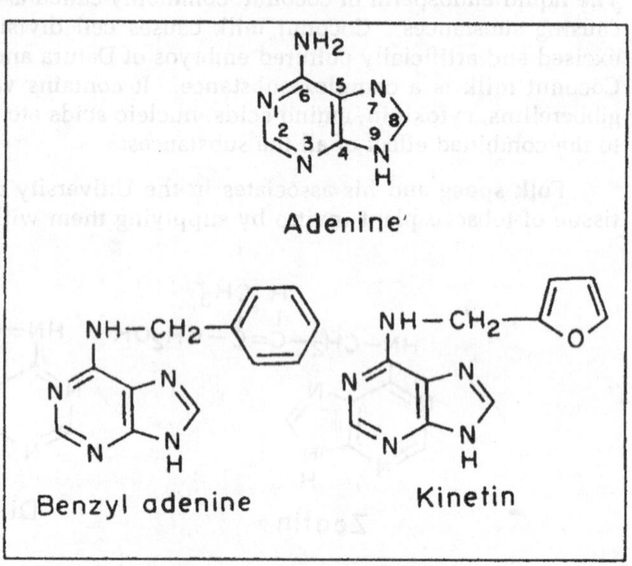

Fig 11.8 Structure of Kinetin

There are evidences to indicate that cytokinins play a significant role in nucleic acid metabolism and protein synthesis. However, the exact mechanism of cytokinin activity is not known.

11.7.3.2 Bioassays for Cytokinins : Steward (1964) developed the first bioassay method for the detection of the cytokinins. He used carrot tissue for the detection of cytokinins. The tissue was cultured on artificial medium. When coconut milk was added to the medium, it caused tremendous increase in the cell division. **Letham** used this test and found that it is sensitive to amounts as low as 0.1 microgram of zeatin per litre of medium. **Miller**

developed another bio-assay method for cytokinins, which is more sensitive than the earlier method. He obtained callus tissue from soybean cotyledon's to which as low as 0.01 microgram of zeatin was added per litre of the medium. It had a measurable growth response.

K. Mothes of Germany found that kinetin and synthetic cytokinins delay the break down of chlorophyll, proteins etc in detached leaves. Taking advantage of this character several bioassay methods for detecting and estimation of cytokinins were developed. Leaf disc method often used for detection and estimation of cytokinins. Leaf discs were cut from the mature leaves of xanthium. Some leaf discs were used for intraction and estimation of chlorophyll. To the remaining leaf discs solution containing cytokinins was added and after few days the chlorophyll content of the remaining leaf discs determined. Leaf discs treated with water taken as controls. The chlorophyll content of the leaf disc was treated cytokinins was found to be more than the chlorophyll content of leaf disc kept as controls. The inhibition of chlorophyll decay is proportional to the concentration of cytokinins used in the solution.

11.7.3.3 Physiological Roles of Cytoknins

CELL DIVISION: The most important characteristic of cytokinins is it stimulated the cell division. It has been demonstrated by tissue culture experiments that sufficient amount of auxins is also required in addition to cytokinins to promote cell division in the callus tissue. Therefore, the presence of auxins and cytokinins are necessary for cell division. The exact mechanism of interaction of two hormones is not fully understood. However, cytokinins are known to increase the synthesis of DNA and m-RNA and auxin promote the synthesis of ribosomal RNA at the time of cell division.

CELL ENLARGEMENT: Cytokinins cause enlargement of leaves and cotyledons. The root is the active side for synthesis of cytokinins. If the root-tip is cutoff, the cotyledons and leaves fail to expand. Application of cytokinins promoted the leaf expansion. Thus the cytokinins replace the requirement of root-tip for leaf expansion. Cytokinins also substitute for the light requirement for leaf expansion.

MORPHOGENESIS: Cytokinins play a vital role in morphogenesis in plants. **Skoog** and **Miller** (1957) observed that the ratio of auxin and cytokinins is a determining factor in initiation of root and shoot buds. The cells of tobacco pith do not grow when auxin or cytokinin alone is added to medium. However, when the auxin and kinetin were added to the medium in the 10:1 ratio pith cells began to grow and developed into callus tissue. But if the ratio of auxin to cytokinin is increased either by decreasing the concentration of cytokinin or increasing the concentration of auxin, the roots differentiated from the callus tissue. Where as high concentration of kinetin and low concentration of auxin promoted the formation of shoot buds, which developed into the tobacco plants. Therefore, it is clear that the interaction of auxin and cytokinin controls the formation of root and shoot buds.

APICAL DOMINANCE: The application of cytokinins reduces the apical dominance, promotes the growth of lateral buds. Thus the effect of cytokinins is antagonistic to the effect of auxins in apical dominance.

Corynebacterium fascians causes a disease called "witches broom" in many plants. The disease is characterised by the loss of apical dominance in the plant and development of lateral branches, which give the appearance of broom. This effect is due to the production of cytokinin by the pathogen on the host plant.

DELAY OF SENESCENCE: Senescence is the phenomenon, in which the leaves loose their chlorophyll, turn yellow and ultimately shed from the plant. During senescence the protiens, nucleic acids, chlorophyll broken down into simpler elements such as sugars, amino acids etc. These substances are transported to other parts of the plants. Application of cytokinins delays the process of senescence by preventing the degradation of chlorophyll, proteins and necleic acids. Therfore, it is noticed that cytokinins level in the leaf lamina is important in regulation of the process of senescence.

OTHER EFFECTS: Cytokinin can break the dormancy of seeds and promote seed germination. Cytokinin also provide resistance to plants under the pathogenesis. Cytokinins also stimulate many enzymes especially those concerned with photosynthesis.

11.7.4 Abscissic Acid (ABA) : Auxins, gibberellins, cytokinins promote growth. These substances are called growth promoters. At the same time there should be some mechanism in plants to check the growth with the space and time for a balanced and co-ordinated growth and development of plant. Scientists were successful in discovering such growth inhibiting substance and named it as Abscissic cid (ABA). It was originally discovered from leaves of Acer, psuedoplatanus and fruits of cotton plant. Now it is known to occur universally in all higher plants.

11.7.4.1 Chemistry of ABA : Chemically ABA is made up of basic 5-carbon isoprene units. Three such isoprence units are present in ABA, hence it is called sesquiterpene, consisting of 15 carbon atoms. It is acidic in nature and has a single carboxylic (COOH) group. The first carbon atom of ABA is asymmetric. As a result of which ABA occurs as optical isomers. Of these dextro isomer (+ABA) occurs in plants. The leavo (-ABA) isomer is synthetic and does not occur in plants.

The starting material for biosynthesis of ABA is mevalonate. Gibberellins are also synthsised from this substance. ABA is synthesised in all parts of the plant body and transported freely in all directions.

11.7.4.2 Physiological Roles of ABA:

GROWTH AND DORMANCY: Application of ABA inhibits the growth of stem, root, leaf and coleoptile. The effect increases with the increase in concentration of ABA. However, this effect is not permanent and plant resumes growth after certain period of time. The inhibitory effect of ABA can be reversed by the application of gibberellic acid. Therefore, ABA and GA are having opposite effects on plants.

Application of ABA makes apical bud dormant in temperate plants like Betula, morus, and fig. When GA is applied dormancy is terminated and growth is resumed.

Application of ABA also inhibits seed germination of several species. Similarly seeds which are dormant are shown to contain large quantities of ABA. Such seeds can be made to germinate with the application of GA. When the seeds start germinating ABA content decreased and GA content increased.

SENESCENCE: The discovery of ABA was due to its effects on the abscission of cotton balls. The ABA content of cotton balls increases at the time of maturity. Application of ABA also causes abscission of debladed petioles of Cotton and other plants. However, application of ABA does not cause abscission of intact leaves attached to the plant.

ABA also promotes senescence of excised leaf discs of many plants. But ABA is ineffective when sprayed on intact leaves in inducing senescence. In temperate woody plants in the autumn, leaf senescence is accompanied by increase in ABA content in leaves.

EFFECT OF DROUGHT ON ABA CONTENT: **Wright and Hiron (1969)** reported that when plants are subjected to drought or water deficit, showed large accumulation of ABA. It is necessary because accumulation of ABA causes the stomata to close. This prevents further loss of water and plants can withstand the effects of drought conditions.

11.8 PHYTOCHROME AND FLOWERING

Flowering is an important phase in the life of the plant. Many plants flower after attaining a minimum vegetative growth. Flowering is controlled by two important environmental factors namely the duration of light (photoperiod) and low temperature (vernalization).

Julien Tournois (1912) was the first to recognise the importance of light in the control of flowering of plants. **Garner & Allard** (1920) showed that flowering can be controlled by artificially increasing or decreasing the duration of sunlight. On the basis of their work, the plants were classified into three categories depending upon their response to day length (photoperiod). These are short-day plants (SDP), Long day plants (LDP) and day neutral plants (DNP), later two more types namely short-Long day plants (SDLDP) and Long-short day plants (LDSDP) were added.

Short day plants (SDP) are those that flower below a certain length of the photoperiod. These plants will flower up to an upper limit of the length of the photoperiod. Beyond that limit these plants do not flower.

The Long-day plants (LDP) are opposite to short-day plants. These plants have a limit of the photoperiod, called critical photoperiod. This is Lower limit for flowering.

A long day plant can flower when it is exposed to a photoperiod more than the limit and never less than the limit. Thus the short-day plants flower below the critical and the long-day plants flower above the critical photoperiod.

However, majority of plants flower under the all photo periods. The length of photoperiod had no effect on flowering. Thus plants are called day neutral plants (DNO).

Some plants do not flower when exposed to either short day or long days. Such plants to be exposed to both short and long days. For example Cestrum nocturnum flower only when long days accompanied by short days. These plants are called Long-short day plants (LDSDP). Some plants like Coreopsis grandiflora should be exposed to short days followed by Long days. These plants are called short-Long day plants (SDLDP).

Phytochrome is the pigment, which absorbs the red and far-red radiations. **Borthwick** and **Hendricks** (1959) descovered this pigment and named as 'phytochrome'. Phytochrome pigment exists in two forms. They are pr and pfr. Pr form absorbs maximum light in the red region of the spectrum at wavelength of 660 mu. Pfr form has the absorption peak in the far-red region at wavelength of 730 mu. These two types of pigments are inter convertible. There are two methods of inter conversion of phyto chromes. They are photo conversion and thermal conversion.

In photo conversion when pr form of phyto chrome absorbs red light of 660 mu, it is converted into pfr form. Similarly when pfr form of phyto chrome absorbs far-red 730mu light of 730mu wave length, it is converted to pr form. Photoconversion is very rapid and is completed within 1-4 minutes.

In addition to this pfr form is converted to pr form in the dark. This method of conversion is very slow and is completed within few hours. This mode of conversion is dependent upon temperature hence it is called 'thermal conversion'

11.8.1 Role of Phyto Chrome in Short-Day Plants

In short day plants pfr form inhibits the process of flowering. However, when plants is exposed to 8 1/2 hours, the conversion of pfr form to pr form is not complete. So the plant fails to flower. When the dark period is interrupted by a brief flash of white light or red light in the middle, whatever pr form is present is converted back to pfr form and the flowering is inhibited. Hence the far-red light inhibits the flowering.

If the red light is accompanied immediately by far-red light, the pfr form is converted to pr form so the plant flowers.

11.8.2 Role of Phytochrome in Long Day Plants :

In long day plants, unlike in short day plants, pfr form causes flowering. In these plants the pfr form accumulates at the end of the photoperiod and causes flowering.

QUESTIONS

A. Answer the following question in 25 lines.

1. What are auxins? Give an account of practical applications of auxins in agriculture and horticulture.
2. Give an account of plant hormones and their role in the life of plant.
3. Describe the physiological role of gibberellins
4. What are cytokinins? Write a brief account on their physiological roles in plants.
5. Describe the growth curve. Comment on facotrs affecting the growth
6. Describe the phases of growth
7. Describe the chemical nature of auxins and bioassay of auxins.
8. Describe the role of phytochrome in flowering
9. Give an account of plant growth regulators
10. Bring about the differences between auxins, gibberellins and cytokinnis with particular reference to their effects on plants.
11. What are growth inhibbitors? Explain the chemistry and physiological roles of growth inhibitors studied by you.
12. How would you differentiate between auxins and gibberellins
13. Discuss the plant hormones and their role in the metabolism of the plant.

B. Short Notes

1. Sigmoid curve
2. Growth regions
3. Auxanometre
4. Arc Indicator
5. Photoperiodism
6. Ethylene
7. Abscissic acid
8. Parthenocarpy
9. Bakanae disease
10. Seed dormancy
11. Apical dominance
12. Avena Curvature test
13. Rosette plants
14. Short day plants
15. Long day plants
16. Zeatin

C. Define the following in one or two sentences

1. Morphogenesis
2. Log phase
3. Senescent phase of growth
4. Bioassay
5. Coleoptile
6. Herbicide
7. Phytochrome
8. Dormancy
9. Bolting
10. Vernalization

12

PLANT PHYSIOLOGY

FRUIT PHYSIOLOGY

12.1 INTRODUCTION:

After pollination and fertilization the ovary of the flower develops into fruit and ovules into seeds. The function of the fruit is to ensure proper dispersal of the seeds that it contains. For this purpose the fruits are fleshy, juicy and attractively coloured. By and large, most of the physiological studies have been confined to fleshy fruits, because of their economic importance.

In most of the fleshy fruits (Eg. tomato, apple, pear etc), the growth follows a single sigmoid curve. However, in some fruits (Eg. Grapes, plum, fig etc) the growth follows a double sigmoid curve (fig.12.1). In such cases there are two phases of rapid growth, interrupted by a phase of slow growth.

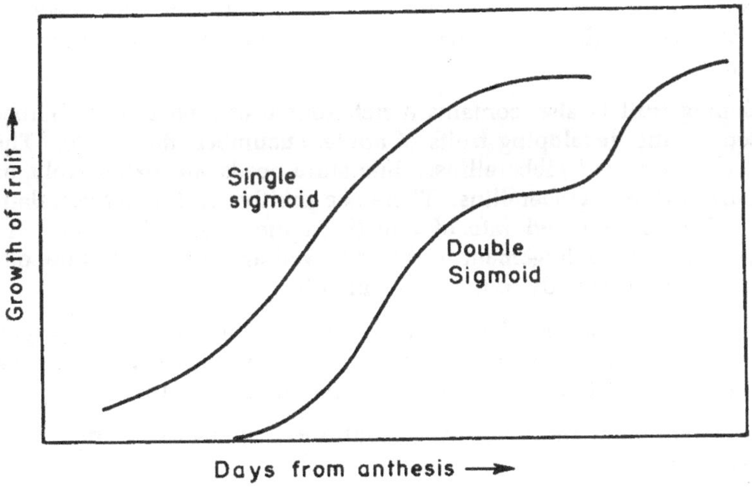

Fig 12.1 Growth of fruit, which follows double signoid curve

Developing seeds contain a rich source of hormones. After pollination and fertilization the ovary of the flower develops into fruit and ovules into seeds. In the absence of pollination the ovary abscices and falls off. Therefore, the pollination provides the necessary stimulus for the production of auxin in the tissues of style and stigma.

Generally the size of the fruit is proportional to the number of seeds it contains. The developing seeds supply the necessary stimulus in the form of auxin for the development of the fruit. The auxin content of the developing fruits increase up to 12 days after pollination, subsequently the auxin content declines gradually (fig. 12.2).

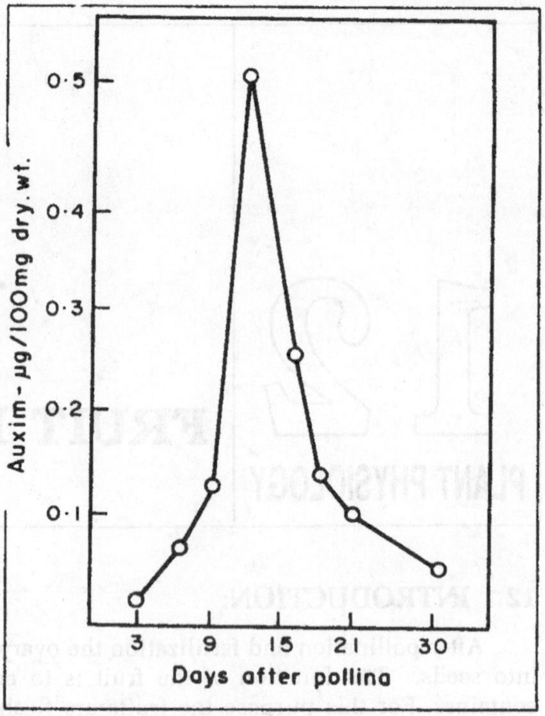

Fig 12.2 Graph showing auxin content in fruit after pollination

In some plants like Banana and Orange the fruit develops without pollination and fertilization. Such fruits are called parthenocarpic fruits. Parthenocarpic fruits are devoid of seeds. So the ovaries of parthenocarpic plants produce more auxin than the normal plants. Therefore, the development of ovary into the fruit is controlled by auxin. The auxins cause the cell enlargement of the ovary wall into the fruit. This cell enlargement is isodiametric. However, the cell enlargement alone is not sufficient to cause fruit development. It also requires nutrients and organic substances. Developing fruit is a metabolic sink. The nutrients flow to the developing fruit. The flow of nutrients to the developing fruit takes place at the cost of the other parts of the plant body. As a result of which vegetative growth of the plant is inhibited. Removal of the developing fruits from the plant revives the vegetative growth of the plant. It is presumed that auxins attract these metabolites by some mechanism, which is not known at present, there by causing the accumulation of the metabolites in the developing fruit.

The developing fruit is also contains a rich source of gibberellins. Gibberellins are reported to occur in the developing fruits of apple, cucumber, melon etc. The immature seeds are the rich source of gibberellins. Immature seeds are extensively used for the isolation and purification of gibberellins. There is a positive and direct correlation between the endogenous GA content and rate of growth in the developing fruit. Like auxins, gibberellins also cause cell enlargement in the fruit wall and attract the flow of metabolites from other parts of the plant body to the developing fruit.

Cytokinins are also reported to occur in large quantities in the developing fruit. In Triticum the cytokinin content is increased upto 3 weeks, subsequently the cytokinin content is declined. Therefore, cytokinins also play an important role in the development of fruit.

Thus the development of the fruit is controlled by many hormones like auxins, gibberellins, and cytokinins.

12.2 FRUIT RIPENING:

The ripening of the fruit is a complex process, involving changes in colour, flavour, aroma, bio-chemical composition etc. Several volatile compounds and pigments are synthesized, which impart colour and flavour to the fruit.

Fruit ripening, infact, is a type of ageing and many people prefer to call fruit ripening as 'fruit ageing'. The fruit ripening process begins after picking of the fruit from the plant. Some of the important changes occurring in fruit ripening are described below:

12.2.1 Cell Wall Softening : The fruit softens with the progress of ripening. The softening of the fruit is due to the enzymatic hydrolysis of cell wall polysaccharides. The cell wall is made up of cellulose, hemicellulose, calcium pectate, polyuronides and glycoproteins. During ripening of the fruit the middle lamella is disorganised. The production of the enzymes like PME (Pectin Methyl Esterase) and PG (Polygalactouronase) increase with the progress of the ripening. Similarly the cellulase (B 1-4 gluconase) enzyme, which was not detected in the unripe fruit, appears during the ripening. The appearance of these enzymes during the ripening process is due to the gene activation.

12.2.2 Changes in Nucleic Acids and Proteins : In many fruits such as apples, melons etc the protein synthesis is presumed on the basis of presence of polysomes in ripening tissues. Similarly the RNA synthesis also increased during the ripening. Therefore, it can be concluded that during the early stages of ripening there is a active bio-synthesis of RNA and proteins. All types of RNA viz t-RNA, m-RNA and r-RNA are reported to appear during the ripening of the fruit.

12.2.3 Changes in Pigment : Fruit ripening is associated with changes in the colour of the fruit. The change in the color of the fruit is due to changes in pigment composition. During ripening the chlorophyll content of the fruit disappears fast, red, yellow and carotenoid pigments get accumulated in chloroplasts. Large quantities of carotenoids are formed a fresh as the fruit matures, as a result of which chloroplasts turn into chromoplasts. The anthocyanin pigments also increase at the time of maturity.

12.2.4 Changes in Flavour : Unripe fruits contain large amounts of starch. During ripening the starch hydrolyses forming sugars. For example in Banana the starch content is about 21% in unripe fruits. But at maturity it contains only 1%. It is accompanied by accumulation of sugars to the extent of 20% by weight. Therefore, it is concluded that during the ripening the enzymes responsible for the conversion of starch to sugars are activated.

The taste of the fruit depends upon the absolute level of sugars and acid contents. The acids are mainly malic and citric acids. Because of presence of these acids, the pH of the fruits is in the acidic range. It can be as low as pH^3 as in lemon. These acids are stored in the vacuoles of fruit cells.

In addition to the sugars and acids, the fruit also produces a large amount of volatile compounds during ripening, which add to the flavour. The volatile compounds include organic acids, esters, alcohols and hydrocarbons. In banana, as many as 200 volatile compounds have been recorded.

12.2.5 Changes in Respiratory Rate : The fruits can be classified into two categories on the basis of their respiratory pattern during ripening. They are climacteric and non-climacteric fruits. In fruits like apples the respiratory rate is low at the time of harvest. When the fruit is stored for ripening, there is a sudden upsurge in the rate of respiration. Such fruits are called "climacteric fruits". This phenomenon has been reported by **Kidd**

and West (1930) in apples. The climacteric rise in respiration indicate the beginning of the degradative changes characteristic of the ripening process. When climacteric rise in respiration commences the process of ripening continues and it cannot be halted. The reason for such rise in respiration is believed to be due to the hydrolysis of starch into sugars and increased availability of sugars for respiration. The climacteric rise in respiration is accompanied by burst of ethylene gas evolution.

A large number of fruits such as grapes and pineapples do not exhibit any climacteric rise in respiratory rate and burst of ethylene gas evolution. These fruits are called as "non-climactric" fruits. In these fruits the ripening is a gradual process. Further such fruits start ripening on the parent plant. After the harvest the ripening declines.

12.3 ROLE OF HORMONES:

Different kinds of plant hormones are known to play an important role in the regulation of fruit ripening.

a) ETHYLENE: Fruits, both climacteric and non-climacteric fruits, produce ethylene gas at its various stages of development. In climacteric fruits the ethylene content is very low during the development of fruits, but suddenly rises after the harvest. However, in non-climacteric fruits such a burst of ethylene production is not noticed after the harvest.

If the ethylene production is controlled in climacteric fruits, ripening is delayed. By this technique bananas could be kept green for many days. If the bananas are stored in carbon dioxide atmosphere, the ripening can be delayed for longer times, since the CO_2 is competitive inhibitor of ethylene.

In non-climacteric fruits, the rise in respiration and production of ethylene is a gradual process. When green oranges are harvested and treated with ethylene, the rise in respiratory rate is slow and gradual, accompanied by loss of green colour and accumulation of reducing sugars. Continuous presence of ethylene is necessary for the maintenance of increased respiration. Removal of ethylene leads to reversal of respiratory rate to its original level. The rise of respiratory rate is proportional to concentration of ethylene.

b) AUXIN: Application of auxins speed up the ripening process in many climacteric fruits such as apples and pears. Auxin stimulates the ethylene production. When bananas dipped in 2-4D solution, it enhanced the ripening process accompanied by enhanced ethylene production. However, the application of auxin has no effect in non-climacteric fruits on ripening process.

c) GIBBERELLINS: The role of gibberellins in fruit ripening is not fully understood. In some fruits like apples and oranges the application of gibberellins enhanced the ethylene production. While in others such as avocado the ethylene production is diminished.

The application of gibberellins, however, cause changes in the color of the fruit. In orange the application of GA, inhibits the loss of chlorophyll, consequently such fruits remain green for longer times. Similar observations were made in tomato and banana. In tomato, the application of GA, delays the appearance of lycopene pigment, which is responsible for the red color of the fruit.

d) ABSCISSIC ACID: The application of ABA hastens the ripening process in banana. Similarly in tomato and grapes also the application of ABA promoted the ripening process. Application of ABA stimulated the production of ethylene in apple. In some fruits the level of ABA rises during the ripening, which is accompanied by rise in ethylene production. The rise in ABA is due to fresh synthesis of ABA. However, the role of ABA in fruit ripening is not fully understood.

12.4 ARTIFICIAL FRUIT RIPENING:

Ethylene is currently used to induce artificial fruit ripening in bananas, tomatoes, mangoes and citrus fruits. Some times carbon monoxide and acetylene are also used for artificial ripening of bananas and mangoes. The production of orange and yellow pigments. The ripening is independent of the maturity of the fruit. To have better quality and taste, it is necessary that optimum matured fruits only be used for artificial ripening.

QUESTIONS

A. Answer the following questions in 25 lines.

1. Describe the process of fruit development after pollination and fertilization.
2. Describe the physiology of fruit ripening.
3. Describe the role of hormones in the regulation of fruit ripening.
4. Bring out various changes during the course of fruit ripening.

B. Short Notes.

1. Double sigmoid curve.
2. Climacteric fruits.
3. Ethylene.
4. Artificial fruit ripening.

13 PLANT PHYSIOLOGY

SEED DORMANCY AND GERMINATION

13.1 STRUCTURE OF SEED

Seed is a reproductive structure characteristic of all seed-bearing (spermatophyta) plants. After pollination followed by fertilization, the ovary of the flower develops into fruit and ovules into seeds.

The dicotyledonous seed (eg. gram seed) is surrounded on the outer side by a thick brown covering called 'testa', inside the testa is a thin white membranous covering called '**tegmen**'. The testa and tegmen constitute the seed coats, developed from the integument of ovule. Towards the pointed end of the gram seed, there is a circular mark called 'Hilum', which represents the point of attachment of the ovule to its stalk called 'funicle'. There is a small pore at one end of the hilum called 'micropyle'. The 'raphe', which represents the line of fusion of the funicle with the integuments, extends from hilum to broader end of the seed. In the middle of raphe is a triangular or rhombiodal area called 'chalaza', which represents the chalazal end of the ovule.

When the seed coats are removed, the embryo is exposed. It consists of two fleshy, rounded structures called **cotyledons**, which contain reserve food materials. These reserve food materials nourish the seedlings till they become autotrophic. The cotyledons are attached to each other by an axis called 'hypocotyl'. At the lower end there is radicle which develop into root. At the other end of the hypocotyl is a bud-like structure called 'plumule', which develops into shoot. Plumule lies in between two cotyledons and radicle slightly projects out of the embryo.

In some seeds the food materials are stored in special tissue called **'Endosperm'**. The endosperm is formed by the triple fusion i.e union of two polar nuclei of the embryo sac with the male gamete. The endospermic seeds are castor, maize, wheat, rice etc.

In cereals the ovary wall is fused with the seed coat forming the 'pericarp'. Such seeds, infact, should be called technically as fruits. Therefore, they are called as grains as wheat grain, rice grain etc.

Thus the angiospormic seed has a miniature plant inside the seed consisting of young root (radicle) and young shoot (plumule), protected by seed coats and provided with reserve food materials in cotyledons.

13.2 SEED DORMANCY

Temporary suspension of growth is called 'dormancy'. Seeds of many plants remain dormant for varying periods ranging from a few months to few years. The dormant phase is to tide over the unfavourable conditions like high temperature, water deficit in the soil. The dormancy is of two types 1. Imposed dormancy 2. Innate dormancy.

13.2.1 Imposed Dormancy : Imposed dormancy is due to the unfavourable conditions of growth such as high temperature, drought etc. This type of dormancy is imposed and induced by external factors. The plant or seed resumes growth after onset of favourable conditions.

13.2.2 Innate Dormancy : Seeds of many plants do not germinate immediately after harvest even though the favourable conditions for growth exists. This type of dormancy is called 'Innate dormancy' or true dormancy. Seed dormancy may be due to various reasons. The chief reasons for seed dormancy are as follows:

13.2.2.1 Hard Seed Coat : Seeds of many plants possess hard seed coat. Hard seed coat is one of the most common factors associated with seed dormancy. Hard seed coat prevents entry of water into the seed. Seed coat is impervious to water because of the presence of thick suberised walls covered with thick, waxy cuticle. In some species the seed coat may be permeable to water but impermeable to gases. Inadequate oxygen supply inhibits the germination. In some plants the seeds may be permeable to both water and oxygen, yet they remain dormant. For example the seeds of pigweed (*Amaranthus retroflexus*) are permeable to water and oxygen but seed coat offer mechanical resistance for the emergency of embryo. In addition, the hard seed coat prevents leaching out of inhibitors from the seed.

13.2.2.2 Immature Embryo : In some plants the seeds are shed from the plant before they are mature. Such seeds do not germinate immediately but remain dormant for some period. During dormancy the development of embryo continues. The time taken for the completion of the development of embryo ranges from few days to few months. After the completion of the development of embryo, the seed starts germination.

13.2.2.3 Ripening Time in Dry Storage : In many cereals such as wheat, rice and barley, the seeds do not germinate immediately after harvest. These need to be stored at ordinary temperature under dry conditions over a period ranging from a few weeks to several months to overcome dormancy. It is ripening time in dry storage. After the dry storage time when the seeds are provided with favourable conditions they germinate promptly.

13.2.2.4 Light Sensitivity : Seeds of some species like *Amarnthus, Betula, Lettuce, Digitalis* remains dormant in dark. Such seeds germinate when they are exposed to light. These are called 'positive photoblastic' seeds. However, there are some species like *Phacelia tanaieliifoliia*, whose seeds fail to germinate in light but germinate in dark. These are called 'Negative photoblastic' seeds. Many cultiivated plants such as tomato, cucumbers are not sensiitive to light and germnate in eiither light or darkness such seeds are called as nonphotoblastic.

Borthwick and his colleagues (1952) worked on the effects of red and far red light on germination. The red light (660nm) is most effective in promoting seed germination. The effect of red light is reversed and seeds do not germinate if it is followed by far red light (730nm). If the last instalment of light is red, the seeds start germinate.

Treatment	Percentage of germination
R	70
R-F	6
R-F-R	74
R-F-R-F	6
R-F-R-F-R	76
R-F-R-F-R-F	7
R-F-R-F-R-F-R	81

R = Red Light F = Far red light

Based on these evidences **Borthwick and Hendricks (1954)** discovered the presence of a new pigment in plants. This pigment was named as 'phytochrome', which is responsible for absorbing red and far red radiations.

Phytochrome exists in two forms namely pr and pfr, the latter is physiologically active. When the lettuce seeds are exposed to red light, the phytochrome 'pr' is converted to 'pfr', which is being active, initiate series of biochemical reactions leading to germination.

13.2.2.5 Chilling Requirement : Seeds of temperate plants like apple, walnut, pinus etc do not germinate immediately after the harvest. Such seeds need to be exposed to chilling temperature (1-5°c) for few weeks. This requirement met in nature when the seeds lie buried in the soil through winter and germinate in the following spring season. Such seeds can be made to germinate soon after harvest by exposing them to low temperatures for few weeks. This phenomenon is called "stratification". Dry seeds cannot be stratified. Only seeds having a minimum amount of moisture can respond to low temperature treatment. Duration of stratification varies from one week to few months depending upon the species. During stratification, ABA content decreases and GA content increases.

13.3 MECHANISM OF DORMANCY

Seeds of many species are known to contain many inhibitory substances. These inhibitory substances are present either in embryo or in the seed coats. The seeds of watermelon, tomato and other fleshy fruits though exposed to sufficient amount of moisture they do not germinate *in situ* when the seed coat is removed the seed starts germination. Several phenolic compounds found in the fruits and seeds of many species cause dormancy.

It is also established that ABA causes dormancy in seeds. The seed coat and pericarp of dormant seeds are shown to contain large quantities of ABA. When the embryos or seed coats are washed with water, the inhibitors are leached out, as a result of which dormancy period is broken with the decrease of ABA, the level of growth promoting substances like gibberellin increase. In hazel seeds (*Coryllus avellana)*, when seed dormancy is terminated by chilling, gibberellin content is increased. The chilling requirement can be replaced by the application of GA. Application of cytokinins also breaks seed dormancy in lettuce and other species.

In Acer seeds (*Acer saccharum*), when the dormancy is terminated by stratification, there has been sharp decline in ABA content and simultaneous increase in the contents of gibberellins and cytokinins. Application of ethylene also causes break of dormancy in pea nut **(Toole eta 1964)**.

13.4 METHODS OF BREAKING SEED DORMANCY

The dormancy of seeds can be broken and dormant seeds can be induced to germinate by one or combination of different methods described below.

13.4.1 Scarification : When the germination of seed is inhibited by hard seed coats or impermeability of seed coat to water or oxygen, dormancy can be broken by scarification. Scarification refers to methods by which the seed coat is softened and it is achieved by abrasion, cutting or chemical treatment. Thus scarification is roughly divided into mechanical scarification and chemical scarification.

In mechanical scarification method, the seed coat is ruptured or punctured mechanically before sowing. The cracks or scratches resulting from such treatment promote germination by decreasing the resistance to entry of water or oxygen and embryo expansion.

Chemical scarification is by using chemicals in the removal or breaking of dormancy. This is the effective way of breaking dormancy. Dipping seeds in strong acids such as sulphuric acid or inorganic solvents like acetone or alcohol and then rinsing the seeds with water can break the dormancy. As in mechanical scarification, chemical scarification also breaks dormancy by weakening the seed coat and dissolving the waxy material present on the seed coat. Treatment with boiling water would also help in some cases in softening the seed coat.

13.4.2 Stratification : Seeds of many temperate plants do not germinate immediately after the harvest, such seeds germinate only after exposing them to chilling temperature. This process is called 'stratification'. Stratification results in the disappearance of inhibitors of germination and increase in the synthesis of germination promoting substances such as gibberellins and cytokinins (fig.13.1). Duration of stratification ranges from one week to few months depending upon the species.

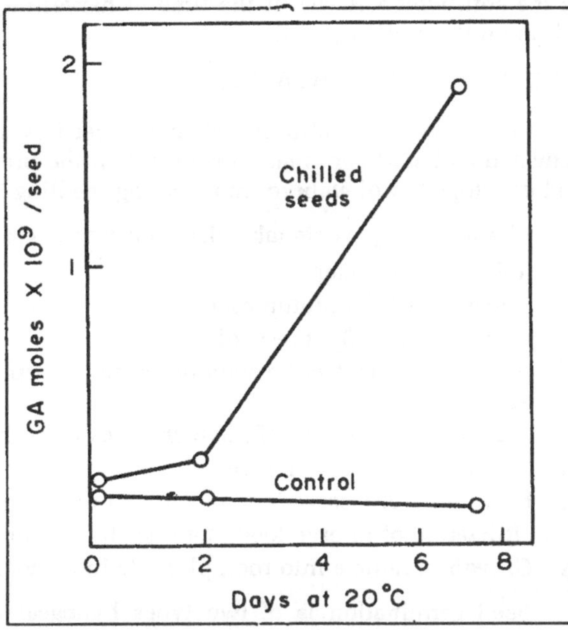

Fig 13.1 Graph showing increase of gibberellin buring stratfication

13.4.3 Light Requirement for Germination : Some seeds germinate upon exposure to light. Such seeds are called 'positive photoblastic'. In certain other seeds light inhibits germination. They require complete darkness for germination. Such seeds are known as "Negative photoblastic".For example lettuce is positive photoblastic. It starts germination after single exposure to light. Lettuce seeds do not germinate in darkness. The length of exposure varies from few seconds to several minutes. In still other seeds germination requires an alternation of light and dark periods.

13.5 BIOLOGICAL SIGNIFICANCE OF DORMANCY

Seed dormancy has great biological significance. After sowing, the seedling can establish itself in the soil and grow into an adult plant only when the environmental conditions are favourable for active growth of the seedling. In the absence of such favourable conditions the survival of seedlings become doubtful. For example the seeds of apple would remain dormant during the winter and start germination in the spring season. This measure would ensure the survival of the seedlings. The seeds of Striga, a root parasite, do not germinate till the seeds come in contact with the host plant. This would ensure the establishment of parasite on the roots of host plant. Seeds of some desert plants remain dormant by accumulating inhibitors. Such seeds germinate only after the inhibitors are leached away by heavy rains. This will ensure the availability of water to the developing seedlings. Therefore, dormancy is a measure to delay the germination till the onset of favourable conditions to ensure survival of seedlings.

Another advantage of dormancy is that the dormancy is not uniform in all the seeds. As a result of which all the seeds do not germinate simultaneously at any given time but continue to germinate over a long period of time. This happens in *Chenopodium* and other weed species. This measure would ensure the greater survival chances for the seedlings. At least some seedlings would survive in one season or other.

Another advantage of seed dormancy is, it avoids germination on mother plant. For example in potato, in the absence of dormancy, the tubers, which are under ground, begin to sprout in the soil, which is wet at the time of harvesting. Dormancy avoids the germination of tubers on the mother plant.

13.6 SEED GERMINATION :

In favourable conditions when the seed is sown in the soil, the embryo becomes activated and start germination resulting in the production of young seedling. The growth and development of embryo into a young seedling is called seed germination.

The following events take place during the germination of the seedlings.

I. Imbibition of water
II. Hydration of the protoplasm
III. Activation and Synthesis of enzymes
IV. Activation of metabolic activities such as respiration, synthesis of proteins and nucleic acids
V. Synthesis and release of Phytochromes from the embryo
VI. Increase in cell division and cell elongation
VII. Hydrolysis of reserve food materials present in the cotyledons and endosperm
VIII. Utilization of reserve food materials by the developing embryo
IX. Growth of radicle into root, plumule into shoot.

Seed germination is of two types hypogeal and epigeal germination. In hypogeal germination the epicotyl region elongates, there by the seed remain under ground. In epigeal germination the hypocotyl region elongates, there by the cytoledons come above the ground.

13.6.1 Factors Affecting Seed Germination

13.6.1.1 Water : The seeds are in highly dehydrated condition. In such condition the embryo remains inactive and metabolic activities are at its lowest ebb. So the absorption of water by germinating seed is the primary requisite for germination. The dry seeds can absorb water by a process called 'Imbibition'. With the absorption of water, the protoplasm becomes hydrated and metabolic activities accelerate leading to germination.

13.6.1.2 Temperature : A suitable temperature is required for germination of seed. the minimum temperature at which the germination starts, varies from one species to another. The optimum temperature for seed germination ranges from 30-35°C.

13.6.1.3 Oxygen : Seeds also require oxygen or air for germination. The oxygen requirement varies from species to species. Some seeds like paddy can germinate under water. In some other species germination rate is increased if the oxygen concentration is enhanced.

13.6.1.4 Light: Seeds of most plants can germinate both in dark and light. Therefore, the light is not necessary for germination. In some species the light breaks dormancy and cause germination, while in others light inhibits germination.

QUESTIONS

A. Answer the following questions in 25 lines.

1. Define seed dormancy. Describe different types of seed dormancy.
2. Enumerate the various causes for seed dormancy.
3. Explain the mechanism of seed dormancy. Add a note on the methods for breaking seed dormancy.
4. Give an account of Biological significance of seed dormancy.
5. Write an essay on factors affecting seed germination.
6. Give an account of various methods of breaking seed dormancy.
7. Explain the causes of dormancy in plants and its significance to.
8. Distinguish between the seed dormancy and quiescence.
9. Describe the changes during the seed germination.

B. Short Notes.

1. Innate dormancy
2. Imposed dormancy
3. Chilling requirement
4. Stratification
5. Scarification
5. Quiescence

C. Define the following in one or two sentences :

1. Epigeal
2. Hypogeal
3. Positive Photoblastic
4. Negative Photoblastic
5. Viability
6. Longevity
7. Quiescence.

14

PLANT PHYSIOLOGY

STRESS PHYSIOLOGY

14.1 INTRODUTION

Plants are subjected to a number of adverse environmental conditions in their life-time. These adverse environmental conditions cause damage to the plant. Stress refers to an environmental factor, which cause potential injury to the plant. Water deficit in the soil; excess of soil solution; presence of high concentration of salts in the soil water; too high or too low temperatures are some of the environmental stresses, often experienced by the plants.

14.2 WATER DEFICIT

Normally the plant cell contains 90 to 95% water by weight. Water deficit or water stress occurs when the rate of water absorption is less or rate of transpiration is more or some times both the factors contribute to the water deficit. Water deficit affects the physiological processes of the plant, which, in turn, adversely affect the crop yield.

Water deficit causes leaf wilting. Leaf wilting may be temporary as in mid-day wilting, which occurs around noon. However, the plant recovers and become normal during cool hours. On the otherhand, when the water stress is prolonged and continuous, the plant shows the signs of permanent wilting. Permanent wilting may be due to lack of adequate soil water. It affects the physiological processes of the plant and ultimately the growth of the plant.

14.3 EFFECTS OF WATER STRESS ON GROWTH AND DEVELOPMENT

Water stress or water deficit is detrimental to the growth and development of the plant. Water deficit inhibits both cell division and cell enlargement. Cell elongation is more sensitive to water stress than cell division. Cell elongation is directly proportional to turgor of cells. Water defiecit causes loss in turgor in cells.

Water stress causes reduction in the size of leaves or pre-mature shedding of leaves. The appearance of new leaves on the plant is also affected. These changes are detrimental to the growth of the plant as it reduces the assimilating area of the plant. The abscission of leaves is due to increase in ABA content of leaves under water stress conditions. Simultaneously the supply of cytokinin from the roots is decreased. The deficiency of cytokinin causes degradation of chlorophyll, proteins and nucleic acids leading to senescence of leaves. Thus increase in ABA content of leaves and decrease in the supply of cytokinin accelerates the falling of leaves.

Water stress also causes inhibition of flowering and fruiting. Plants are extremely sensitive to water stress at the time of flowering. The magnitude of flowering is considerably reduced. Flower buds may abort. Water stress causes heavy shedding of flowers. Consequently the fruit formation is also affected. Seeds become shriveled and seed production is drastically reduced.

14.3.1 Physiological Effects of Water Stress

Water stress affects a number of physiological processes of the plant such as photosynthesis, respiration and hormonal metabolism.

14.3.1.2 Photosynthesis : Water stress reduces the rate of photosynthesis. As already described the water stress causes reduction in the size of the leaves and pre-mature shedding of leaves. As a result of which the photosynthetic yield is considerably reduced. This effect is long lasting and is not reversed even after the relieving of the stress conditions. Water stress also causes increased stomatal closure, there by affecting the diffusion of carbon dioxide into the plant.

In addition to above, the water stress directly affects the process of photosynthesis, possible, by inhibiting enzymes involved in the process. Water stress is reported to inhibit Hills's reaction (Fry, 1970). Both electron transport and photophosphorylation process are also inhibited. Further, water stress damages the photosynthetic apparatus. All these changes result in decreased photosynthetic rate.

14.3.1.3 Respiration : In number of plants the water stress causes the decreased rate of dark respiration. The decrease in rate of respiration is proportional to increase of water stress. Photorespiration is also reduced. However, the photosynthesis is more sensitive to water stress than respiration.

14.3.1.4 Translocation of Solutes : The translocation of photosynthates from leaves to growing regions of the plant is also affected by water stress conditions. The inhibition of translocation flux is due to two reasons :

I. Water stress reduces the rate of photosynesis thereby reducing the amount of photosynthates available for translocation.

II. Water stress inhibits flowering and fruiting, thereby reducing the sink capacity. The transport of photosynthates is from source to sink. Reduction of sink capacity also reduces the rate of translocation of photosynthate flux.

14.3.1.5 Harmonal Metabolism : Water stress inhibits the production of hormones like auxins, gibberellins, cytokinins etc. On the otherhand, the production of ABA and ethylene is enhanced.

Water stress causes wilting of leaves. The wilting causes enhancement of production of ABA in the leaves. The amount of ABA accumulation depends upon the duration of wilting, temperature, and water deficit (fig. 14.1). This phenomenon is common in all angiospermic plants. It is not only observed in intact plants but also in excised leaves as well. The accumulation of ABA in leaves is a protective device to overcome stress. Increase of ABA content in leaves causes the closure of stomata, there by reducing the loss of water from the plant through the transpiration. Under water stress conditions the ethylene production also enhances. Increased ethylene production causes shedding of leaves, inhibition of formation of flowers and fruits.

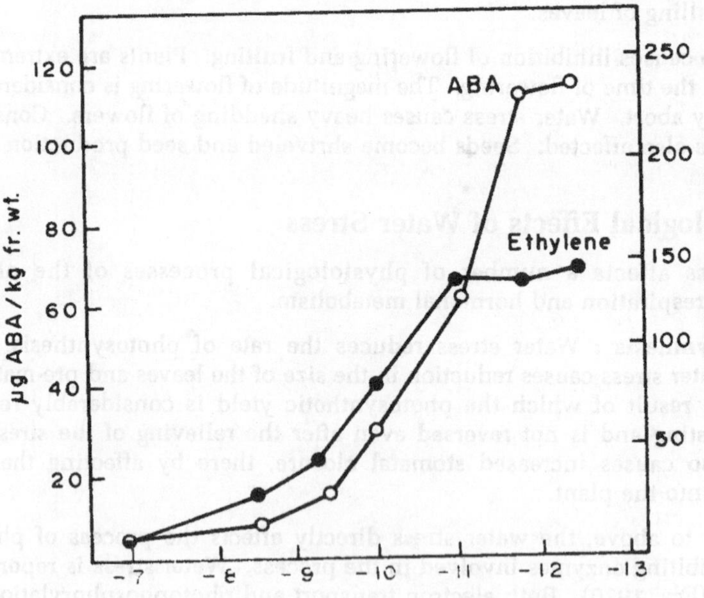

Fig 14.1 Graph showing enhancement of ABA during water stress

14.4 DROUGHT

Drought is considered as deficiency of soil moisture available to plants, resulting in water deficit in plants. As already described water deficit is detrimental to growth and development of plant. So drought can be defined as "deficiency or dearth of water severe enough to check the growth of the plant". Although drought refers to dearth of soil water, the atmospheric factors such as high temperature, low humidity, fast winds increase the severity of drought.

The drought can be classiified into two categories. I. Soil drought II. Atmospheric drought. Soil drought leads to atmospheric drought. If both combine it would become disastrous for the plant than the atmospheric drought alone. Atmospheric drought is due to low atmospheric humidity, high temperature and high wind velacity.

14.4.1 Drought Resistance in Plants : **The drought resistance can be defined as the capacity of the plants to withstand drought conditions. Plants can withstand the drought conditions by three ways.**

14.4.1.1 Drought Avoidance : Many of the desert annual plants adopt drought avoid method. These plants are short-lived (Ephemerals). They complete their life-cycle before the onset of summer. They are, thus, "drought escaping" plants. Such plants do not have any morphological or physiological adaptations to face drought because they avoid dry periods.

14.4.1.2 Drought Resistance by Conserving Tissue Water : Many plants adopt to drought conditions by maintaining high tissue water level either by increasing absorption or by decreasing transpiration. In such plants water is conserved in the tissues by morphological or anatomical adaptations.

One of the morphological adaptations to withstand drought is the well developed extensive root system. Extensive root system enables the plant to absorb the water from larger and deeper areas of the soil. In addition to this plants adopt several methods for

reducing the water loss from the aerial parts of the plant, by regulating shedding of older leaves, developing thick cuticle, sunken stomata, presence of hairs are some of the adaptations to reduce the water loss from the plants.

14.4.2.1 Drought Tolerance : The capacity of the protoplasm to endure dehydration is termed as "drought tolerance" or "drought hardiness". The drought tolerance varies widely. Lichens and moss plants can tolerate complete dry conditions and can be revived by re-watering. These plants can survive under extreme arid conditions for months together and resume their metabolic activities upon rehydration.

More often drought is accompanied by high temperatures. High temperatures cause synthesis of several new proteins, which protect the cells against the thermal damage. Such proteins are called "Heat shock proteins (HSP)". Heat shock proteins are synthesized by plants and animals against the heat stress.

In some plants when HSP are synthesized, normal protein synthesis is suppressed. This is compensatory mechanism to facilitate the synthesis of HSP. When synthesis of HSP is declined, the normal protein synthesis is increased.

14.5 SALT STRESS

All types of soil waters contain salts. Even pure water also contains salts. But excess salinity in soil solution presents a stress condition. Extreme salinity kills the plant due to ion-toxicity and decreased water potential. Moderate salt concentration also causes growth inhibition. There is a great deal of variation among species and even among cultivars in the species in the degree of response to salinity.

The soluble salts in soil solution mainly composed of chlorides and sulphates of sodium, magnesium and calcium. The salinity of the soil is measured in terms of electrical conductivity expressed as decisiemes per metre (ds/m). The soil having greater than 2 ds/m is called 'saline soil'.

The land plants are predominately sensitive to the salinity of the soil water. The salt sensitive plants are called "Glycophytes". The excess salts in the soil water are detrimental to the growth of the plant. The crop yield is reduced considerably.

14.5.1 Effects of Salt Stress : Salt stress causes both morphological and physiological distrubances in the plant.

Salt sensitivity is detrimental to the growth of the plant. Salinity inhibits the seed germination process, the higher levels reduce the percentage of germination. The degree of inhibition increases with the increase in salinity.

The physiological effects include changes in the rate of photosynthesis, respiration, synthesis of nucleic acids etc.

There are several reports regarding the reduction of photosynthetic rate due to salinity. The reduction in photosynthetic rate can be due to I) reduced photosynthetic area II) structural damage to chloroplasts III) Inhibition of light and dark reactions and IV) reduced transport of photosynthetic products.

Salinity effects the rate of respiration. However, in some instances increase in respiratory rate has been observed due to salinity. Such a stimulation of respiration induced by salinity is due to activation of ion-transport system particularly Na+, K+ and ATP ase of the plasmalemma.

Salinity also affects nucleic acid metabolism. The contents of both DNA and RNA decreases. This is due to reduced synthesis and enhanced breakdown of photosynthetic products. Salinity also affects synthesis of proteins.

14.5.2 Mechanism of Salt Resistance : Plants differ widely in their salt resistance. The resistance is maximum in beet roots and minimum in carrots. There is also great variation with regard to resistance in varieties of the species, and also the resistance is not the same in different types of soils.

In the absence of salt resistance in plants, the salts enter the plants and accumulate in different parts of the plant. Accumulation of salts may result in the death of the plant due to ion-toxicity. The plants having salt sensitiveness would limit the accumulation of salts by some or other mechanisms. The following are some of the mechanisms adopted by the plants to prevent accumulation of salts in the plants.

I. Limit the rate of transpiration in order to minimize the entry of ions into the plant. But this would also effect the rate of photosynthesis and also the growth of the plant.

II. Regulate the concentration of required ions (Eg. K+, No_3-) and 'unwanted ions' (Eg . Na+, Cl-) by utilizing the wanted ions in the growth of the plant and discarding the unwanted ions from the plant by shifting them to leaves followed by abscission of leaves.

III. Plants utilize the excess and unwanted ions for the osmoregulation of the cell sap. Although appreciable information is gathered regarding the functioning of the plant under salt stress conditions, however, much work has to be carried out for comprehensive understanding of the mechanism of salt resistance in plants.

14.6 TEMPERATURE STRESS

Both high and low temperatures are harmful to plants. Plants die when they are exposed to temperatures of 50°c to 60°c. Low temperatures are also equally harmful to the growth of the plants. Such harmful effects are called "Cold injury", plants develop various means to resist cold injury.

14.6.1 Effects of Cold Temperatures : Low temperatures cause ice-formation in the cells. First microscopic crystals are formed which coalesce to form bigger crystals. The formations of ice crystals in the cells rupture the protoplasmic membranes and cause damage to various cell organelles. As a matter of fact the ice formed outside the cell is not much injurious as inside the cell.

The freezing temperatures affect the absorption of water by roots. Consequently the deficiency of water causes dehydration of the cells, as a result of which the growth of the plant is inhibited. The enzymes and proteins are denatured.

14.6.2 Prevention of Cold Injury : Cabbage plants exposed to low temperatures (0°c, 5°c and 19°c) for five days developed frost hardiness (Harvey 1930). Some plants change the pattern of growth. The growth is completely ceased or considerably reduced during this period.

The cells develop some physiological adaptations to overcome the cold injury. Cell membranes become permeable, due to which the water transportation become easy and intra-cellular ice-formation is prevented. As a result of which enzymes, proteins are prevented from denaturation.

QUESTIONS

I. Essay type Questions.

1. Describe the effects of water stress on growth and development in plants.
2. What is drought? Explain drought resistance in plants.
3. What are the consequences of salt stress in plants.
4. Explain the cold resistarce in plants.

II. Short notes.

1. Drought tolerance
2. Drought avoidance
3. Cold injury

III. Define the following in one or two sentences.

1. Stress
2. Wilting
3. Cold injury